Mediterranean Marine Avifauna

Population Studies and Conservation

NATO ASI Series

Advanced Science Institutes Series

A series presenting the results of activities sponsored by the NATO Science Committee, which aims at the dissemination of advanced scientific and technological knowledge, with a view to strengthening links between scientific communities.

The Series is published by an international board of publishers in conjunction with the NATO Scientific Affairs Division

A Life Sciences B Physics	Plenum Publishing Corporation London and New York
C Mathematical and Physical Sciences	D. Reidel Publishing Company Dordrecht, Boston, Lancaster and Tokyo
D Behavioural and Social Sciences E Applied Sciences	Martinus Nijhoff Publishers Boston, The Hague, Dordrecht and Lancaster
F Computer and Systems Sciences G Ecological Sciences H Cell Biology	Springer-Verlag Berlin Heidelberg New York London Paris Tokyo

Series G: Ecological Sciences Vol. 12

Mediterranean Marine Avifauna

Population Studies and Conservation

Edited by

MEDMARAVIS

20, Rue Saint Martin, F-75004 Paris, France
and
96, Via S. Satta, I-07041 Alghero, Sardinia

MED MAR AVIS

Associazione Mediterranea per l'Avifauna Marina
Asociación Mediterránea para la Avifauna marina
Mediterranean Marine Bird Association
ASSOCIATION MEDITERRANEENNE POUR L'AVIFAUNE MARINE
جمعية البحر الأبيض المتوسط للطيور البحرية

MEDMARAVIS is a non-profit scientific Association dealing with research and conservation of the Mediterranean Marine Avifauna

and

Xaver Monbailliu

20, Rue Saint Martin, F-75004 Paris, France

Springer-Verlag
Berlin Heidelberg New York London Paris Tokyo
Published in cooperation with NATO Scientific Affairs Divison

Proceedings of the NATO Advanced Workshop on Population Dynamics and Conservation of the Mediterranean Marine Avifauna held at Alghero, Sardinia, March 26–30, 1986

ISBN-13: 978-3-642-70897-8 e-ISBN-13: 978-3-642-70895-4
DOI: 10.1007/978-3-642-70895-4

Library of Congress Cataloging in Publication Data. NATO Advanced Research Workshop on "Population Dynamics and Conservation of the Mediterranean Marine Avifauna" (1986 : Alghero, Sardinia) Mediterranean marine avifauna. (NATO ASI series. Series G, Ecological sciences ; vol. 12) "Proceedings of the NATO Advanced Research Workshop on Population Dynamics and Conservation of the Mediterranean Marine Avifauna held at Alghero, Sardinia, March 26–30, 1986"—T.p. verso. "Published in cooperation with NATO Scientific Affairs Division." 1. Bird populations—Mediterranean Region—Congresses. 2. Birds, Protection of—Mediterranean Region—Congresses. 3. Birds—Mediterranean Region—Congresses. I. MEDMARAVIS (Association) II. Monbailliu, Xaver, 1948-. III. North Atlantic Treaty Organization. Scientific Affairs Division. IV. Title. V. Series: NATO ASI series. Series G, Ecological sciences ; no. 12. QL690.M43N38 1986 598.29'24'091822 86-28025

Softcover reprint of the hardcover 1st edition 1986t

CONTENTS

PREFACE

May I first congratulate, inconditionally, the Association MEDMARAVIS for organizing this symposium and for inviting me representing the Sardinian Ministry of the Environment. I would like to assert that such initiatives help a great deal and in a concrete way, Heads of government departments in charge of land use and natural resources.

At this first symposium of the Mediterranean Marine Avifauna I must congratulate the scientific faculty for proposing such an ample spectrum of high quality papers dealing with diverse parts of the Mediterranean and beyond. I am sure that their contributions will further scientific research, will help to protect our environment and will lead, more especially, to the management and conservation of seabirds.

To know in order to manage has often been said at this symposium. In fact, it is not possible to manage thoughtfully and rationally the natural resources without knowing more profoundly how they function. This reasoning is also valid on a more general scale: regional planning of human activities should not disregard the evaluation of the ecosystem.

We are all witnesses of community struggles defended and won justly because a particular environmental issue was at stake by some imposing development projects though these would have brought employment, wages and sometimes more efficient public services. Today, the concept of ecology as opposed to development has given way to that one of ecology for development. Accordingly the more advanced societies are getting increasingly equipped with ever more complete protection systems. In fact, it is our opinion that civil protection not only means the capacity of facing the big natural disasters such as floods, forest fires or calamities, but implies also, and above all, the capacity of active prevention of negative impact created by local problems of limited dimensions.

- local zoning plans related to the installation of ports, tourist zones, industrial and farming areas, fishing zones, etc.

Thus, the Sardinian administration controls a whole series of activities designed to revaluate and to protect the coastal zones including:

1) hydrological development, soil improvement and afforestation;
2) surveying of the quality of coastal waters adequate for bathing, fishing and shellfish-farming;
3) coastal studies of land use possibilities;
4) safeguarding and reappraisal of brackish lakes linked to the development of fishing activities and of habitat creation for the avifauna;
5) the provision of water treatment plants for coastal effluents which forms part of the regional water purification plan;
6) environmental monitoring of the sea and coasts with regard to human development and carried out by various services including the recently established regional corps for territorial surveillance;
7) the creation of a series of protected coastal areas of scientific interest for its flora, fauna and geomorphology;
8) delimitation of stretches of sea particularly important for the local economy (e.g. coral fishing) or for its ecology (marine parks).

I would like to complete this list of activities by indicating the actions carried out along the seacoast by the Directorate for Public Education in the context of the Landscape Law n° 431 applicable to some coastal sectors. I should not forget the specific programme of our regional office for game research based at Sassari which is also operational in ornithological census work.

May I also indicate the current construction of an ornithological station at the lake of Santa Gilla near Cagliari destinated to become a field laboratory of international importance. Finally I would like to mention that, only very recently, I have presented to the Sardinian Board of Ministers a project for a law which will establish a chain of regional parks and nature reserves. Besides the creation of natural parks, 50 out of the 60 projected nature reserves concern coastal areas or islands. I believe that in this way a great effort will be made for securing better environmental conditions for seabirds.

As far as the lack of finance for research is concerned, deplored by some us us, I believe to be able to assure you that, notwithstanding the restrictive budget requirements, each useful proposal will be examined and, if possible, earmarked for regional funding.

This first symposium, which is a remarkable success, should serve as a reference point to complete our knowledge of the relatively unexplored marine avifauna and to design a relevant conservation strategy. We are determined to show a profound respect for our marine resources, not only because of governmental duties and legal obligations stemming from laws and international conventions, but also because they witness of the particular characteristics of Sardinia and of the whole Mediterranean.

Giorgio CARTA

Sardinian Minister of the Environment

CHAPTER 1 : NATIONAL CENSUS SURVEYS

Chairman: Prof. Bruno MASSA. Dipartimento di Zoologia.
Università di Palermo. Sicilia.

LARIDAE AND STERNIDAE BREEDING IN ITALY: REPORT ON THE 1982-1984 CENSUS PROJECT

Mauro Fasola, Dipartimento Biologia Animale
Pz. Botta 9, 27100 Pavia, Italy

The populations of colonial waterbirds (gulls, terns and other seabirds, and herons) have been accurately censused only in restricted regions of the earth. The continental or world-wide review (Cramp & Simmons, 1977 and 1983; Croxall et al., 1984; Cramp, 1985) include sound data on populations of colonial waterbirds, which are limited to Northern and Central Europe, the coasts of the United States of America and of Canada, and a few other countries, e.g. South Africa. Elsewhere, the data are either incomplete (e.g. the Mediterranean countries) or merely qualitative (most tropical and southern regions).

This lack of data has rapidly been filled in the Mediterranean region: various contributions to this volume (Ferrer, Isenmann, Baha el Din & Meininger, Goutner, and many others) show that waterbirds are now being accurately surveyed in many countries, especially Spain, France, Greece and Egypt.

For Italy, only vague estimates on the populations of colonial waterbirds had been available until the beginning of the eighties. Since then however, all the breeding species have been completely surveyed, except for the Procellariiformes (even then, Massa & Catalisano, this Symposium, give a partial coverage of the latter's distribution): the first national census of the Italian heronries was performed in 1981 (Fasola et al., 1981), and a census of the gulls and terns breeding in Italy was completed between 1982 and 1984. These censuses have interesting results, particularly for certain species of conservational concern, whose Italian populations were previously underestimated: Night Herons Nycticorax nycticorax (17,500 nests in Italy), Little Egrets Egretta garzetta (6,500) and Little Terns Sterna albifrons (6,000) are more abundant here than in any other country of the Western Palearctic. The populations of Larus audouinii (550 nests) and the Slender-billed Gull L. genei (950) constitute a relevant portion of the world populations.

NATO ASI Series, Vol. G 12
Mediterranean Marine Avifauna
Edited by MEDMARAVIS and X. Monbailliu

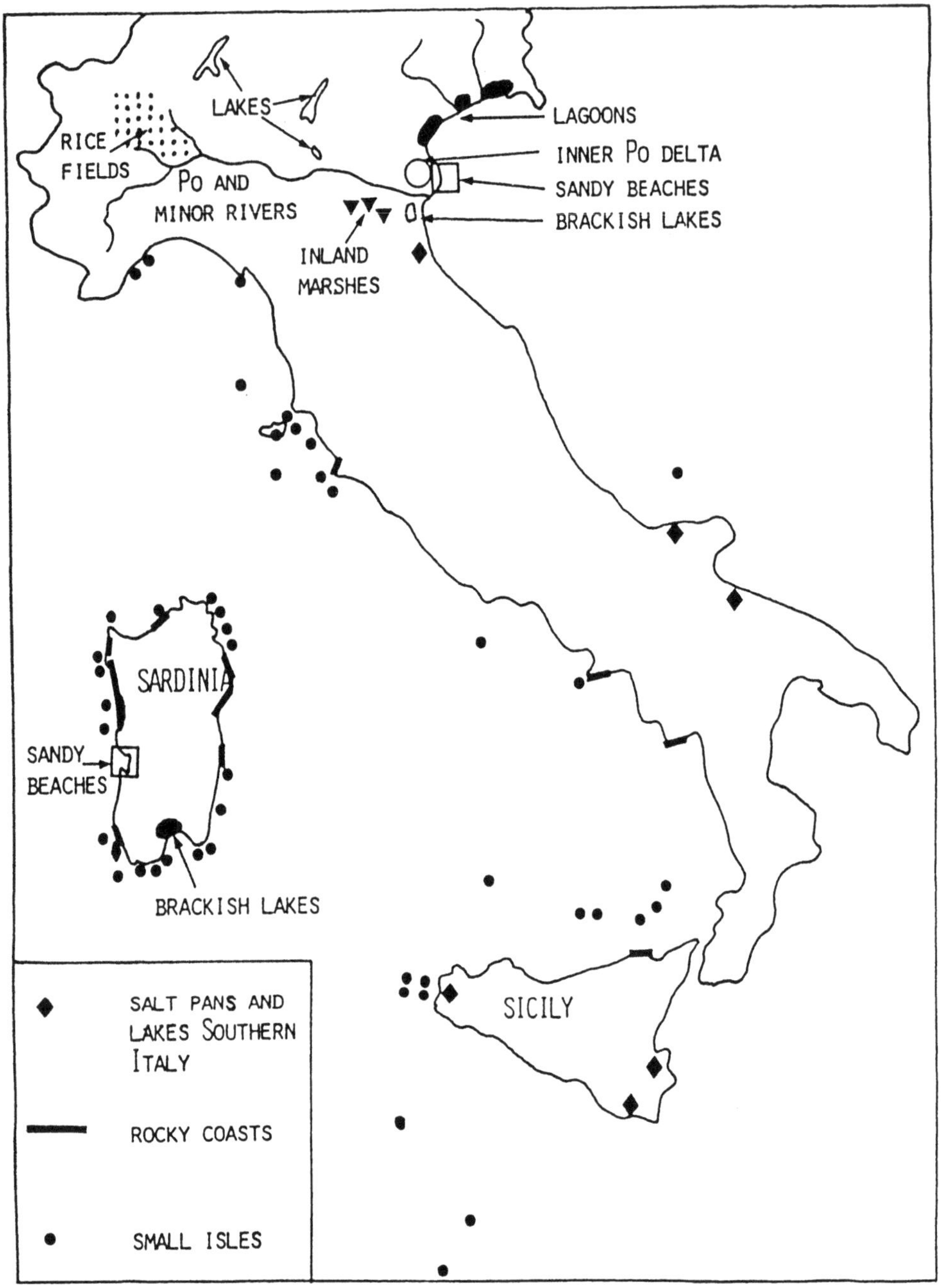

Fig. 1: Main breeding zones of Gulls and Terns in Italy. The zones, defined by ecogeographic criteria, are contiguous areas covered by the same type of aquatic biotope. The map shows only the zones where breeding actually occurred; therefore many coastal and river tracts are not indicated. The same names of the zones are used in tab. 2.

In this paper I summarize some results of the Italian survey of gulls and terns (the "Project Laridae", organized by my Department, and supported by the Istituto Nazionale Biologia Selvaggina). The complete report of the project (Fasola, ed., in press., including monographs by various contributors), describes in detail the distribution, population, habitat, structure of the colonies and nesting habits of each breeding species.

Organization, Aims and Methods of the Survey

Ninety field researchers contributed to the "Project Laridae". The organization of the project was intermediate between the type usually employed in large scale surveys, such as ornithological Atlases (whose research methods and types of data must be simple enough to be suitable for the many contributors), and those employed in individual research (when a specialist may standardize various techniques and record different data). Besides basic data on localization and size of the colonies, the contributors to the "Project Laridae" were asked to study the structure of the colonies and of the nests, and the occurrence of predation and disturbance. We sought to standardize the study techniques by providing each contributor with a list of codified instructions, and four types of data forms to be filled directly in the field.

- Census of the populations

The main aim of the project, the complete census of the breeding populations, was successfully fulfilled. In 1982, preliminary censuses were carried out in Northern Italy; in 1983 the censuses covered the entire country and all the breeding species, except, respectively, Southern Sardinia and Yellow-legged and Audouin's Gulls; in 1984 the census was completed, and was repeated over most of the country. Any group of nests, whose distance from the nearest neighbour nest of other gulls or terns was less than 200 m, was considered to be a colony; nests further than 200 m were assigned to different colonies. Each colony was censused by the technique which best suited the local situation (direct counts of nests, estimates from a distance for inaccessible colonies, counts of the alarm-calling adults for dispersed colonies with mimetic nests). The censuses were usually undertaken once in each colony, at the peak of its occupation. Despite the efforts to standardize the counts, it is acknowledged that these censuses record only "a minimum estimate of the active nests at the peak of breeding". The total population is certainly much more numerous than is indicated by the figures of actual nests censused, because 1) during single counts some nests are overlooked (17% of the nests in gulls, in a test by Ferns & Mudge, 1981); 2) gull and tern breeding extends considerably over spring and summer, so that early and late clutches are missed by peak-breeding counts (Massey & Atwood, 1981; Wanless & Harris, 1984); 3) in long-living species of birds, a high and variable portion of the adults do not breed every year (Coulson, 1984).

	Lm	Lr	Lg	La	Lc	Sn	Ss	Sh	Sa	Ch	Cn
Sweden, Finland, Norway (F,G,O)		366000			156000		1150	37700	585A		145
Iceland, United Kingdom, Ireland (E,F,G)		113000			333626A		15000	10500	1900		
Denmark (F,G)		200000			50000	5	4000	750A	400		200
Belgium, Holland (G)		210000A			53000		7450	10220	700		2000A
East & West Germany (E,F,G,O)	7	134000			28500	51	7950	7400	650		1350
Poland (E,F,O,Q,R)	4	30000			250		300	2200	1080		537
Tchecoslovakia (E,F)		75000						B			B
Switzerland,Austria (E,F,D)		11800			5			387			
Hungary (E,F)	1	20000						200	10	B	B
France (E,F,G,P,U)	7	20000A	12	77	73750A	200	6000	4890	1100	1000	10000
Italy (D)	260	780	948	549	24300	201	149	4608	5980	450	145
Yugoslavia, Albania (E,F,L,W)		1500A			10000			B	B	2000	B
Spain, Portugal (E,F,I, J,Z)		450A	300	1800	56000	900	350	11000	3200	100000	B
Roumania, Bulgary (E,F,S)	B	B	B		B	200	B	B	B	4000	B
Soviet Union (E,F,N)	109000	128000A	21250C		20000A	1600A	53000A	46000A	1565A	6000	2000A
Greece (E,F,L,AA)	3500		40	70	B	400	50	1100	400	300	B
Turkey, Cyprus (E,F,L,T)	15	1000	3600	60	B	400		B	400	1000A	100
Morocco, Chafarinas (K,V)			10	2040	3500				B		
Azores, Canary, Madeira, C.Verde I. (E,M)			6		2000			2000			
Tunisia, Algeria, Lybia, Malta (E,F,L,T)			300	515	3300	25		500	700A		B
Egypt, Israel, Irak (E,F,Y)			200A		20			390	2000		
Mauritania, Senegal (E,F,T)			4200								
TOTAL MEDITERRANEAN	3784	14230	5410	5111	72120	2126	922	20438	10620	4750	10245
TOTAL W. PALEARCTIC	112794	1041530	30866	5111	658236	3982	94250	139145	20025	114750	16332
% ITALY ON MEDITERRANEAN	6.87	5.48	17.52	10.74	33.69	9.45	16.16	22.54	56.13	9.47	1.41
% ITALY ON W. PALEARCTIC	0.23	0.07	3.07	10.74	3.70	5.05	1.58	3.31	29.86	0.39	0.88

Tab. 1: Populations (number of nests) of Laridae and Sternidae regularly breeding in Italy, compared with the populations in the other countries of the Western Palearctic. The species are identified by the initials of their latin names.

For France and Spain, only the populations breeding in the Mediterranean sector of the countries were included in the "TOTAL MEDITERRANEAN". Under Larus cachinnans are included the populations of both the Yellow-legged Gull and the Herring Gull (Larus argentatus. Notes: A incomplete census, B breeding, not quantifiable, C fluctuating up to 100,000 nests. Data sources are: D this project, 1984 breeding season, E. Cramp & Simmons 1983, F. Cramp 1985, G. Evans 1984, H. Barrett & Vader 1984, I. Barcena et al. 1984, J. de Juana 1984, K. deJuana et al. 1984, L. James 1984, M. LeGrand et al. 1984, N. Golovkin 1984, O. Isenmann 1976, P. Muselet 1983, Q. Pinowski & Wesolowski 1981, R. Wesolowski et al. 1984, S. Moller 1975, T. Glutz & Bauer 1982, U. Guyot 1985, V. Berton & Berton 1984, W. Perco, F. pers. com., Y. Baha el Din & Meininger, this symposium, Z. Ferrer, this symposium, AA Goutner, this symposium. When different estimates were available for one country, I retained the most updated one; when a possible range of population size was given, I retained the lower figure.

the "Valli di Comacchio" may not reflect the general trend of the Italian populations. For instance, Common and Little Terns fluctuated widely from 1982 to 1984 in the "Valli di Comacchio" (Brichetti & Foschi, this volume); however, these fluctuations were compensated for in other breeding zones, and the data of the Italian census indicate short-term stable populations in the Common Tern (4,820 nests in 1983 and 4,610 in 1984) and in the Little Tern (6,090 in 1983 and 5,980 in 1984); the Mediterranean Gull in the "Valli di Comacchio" has decreased since 1982, but new colonies have settled in the neighbouring "Valle Bertuzzi", so that the total Italian population has continued to increase (199 nests in 1982, 233 in 1983, 258 in 1984). In summary, during the seventies and the eighties, three new species have colonized Italy (Mediterranean Gull, Slender-billed Gull and Sandwich Tern); Audouin's Gull, the Gull-billed Tern, and the Whiskered Tern (Boldgreghini et al., 1981) have probably remained stable, while the Black Tern has certainly decreased (Brichetti and Martignoni, 1981), and the Black-headed Gull and Yellow-legged Gull have probably increased.

- Wintering populations

A partial winter census was performed in January 1985, over sample zones throughout Italy, except Sardinia (therefore no data were collected on Audouin's Gull which winters in Sardinia and Northern Thyrrenia). The census results for representative zones throughout Italy were extrapolated in order to estimate the total wintering populations (Fasola, 1984). In some species, the wintering populations largely exceeded the breeders (Mediterranean Gull over 10,000 birds in winter, Black-headed Gull around 700,000), owing to the immigration of birds from the North. In the Yellow-legged Gull the winter population (around 100,000 birds) corresponded roughly to the number of the breeders (24,300 pairs) plus, presumably, fledglings. Wintering, non breeding species included the Little Gull *Larus minutus* (around 2,000 birds), the Lesser Black-backed Gull *Larus fuscus* (2,000 birds), and the Common Gull *Larus canus* (3,000); other gulls were scarce.

Distribution

The breeding grounds of gulls and terns in Italy are outlined in fig. 1. The distribution of the breeding species was uneven (tab. 2). This table lists the zones along an approximate gradient going from the continental, freshwater biotopes, through brackish biotopes, to marine biotopes. The value of the "baricentre" of each species (tab. 2) indicates its mean preference along the gradient: Black and Whiskered Terns are strictly continental, Audouin's and Yellow-legged Gulls are markedly marine, while all the other species have similar baricentres (from 7.9 to 9.2) which indicates a distribution centered on brackish environments.

The "habitat breadth" of each species (tab. 2) is positively correlated with its abundance: the Yellow-legged Gull is the

	Lm	Lr	Lg	La	Lc	Sn	Ss	Sh	Sa	Ch	Cn	RICHNESS	DIVERSITY
1 rice fields		35									140	2	0.722
2 small rivers								191	15			2	0.376
3 river Po		5						290	334			3	1.167
4 marshes NW Italy										400	2	2	0.045
5 lakes N Italy					10			12			5	3	1.500
6 "valli", inner Po delta		10			302			804				3	0.912
7 salt pans and lakes Southern Italy & Sicily					11	10			390			3	0.342
8 brackish lakes NW Italy	226	523	10		645	92	123	1943	1505			8	2.219
9 brackish lakes Sardinia		87	851		158	93		200	264			6	2.064
10 lagoons NE Italy		74			2328			1063	656			4	1.495
11 sandy beaches Po delta & Sardinia								18	2737			2	0.057
12 rocky coasts Italy, Sardinia & Sicily					227			2				2	0.072
13 small isles Thyrrenian & Adriatic				91	7183							2	0.097
14 small isles Sicily & Sardinia				458	11946							2	0,228
baricentre	8	7.9	9	13.8	12.9	8.4	8.0	7.6	9.3	4.0	1.2		
habitat breadth	.07	.13	.07	.10	.38	.16	.07	.26	.24	.07	0.8		

Tab. 2: Breeding populations within the 14 ecogeographic zones of Italy depicted in fig. 1. The zones are listed along an approximate gradient from continental to marine. The number of nests in each zone is the 1982-1984 average, whenever data are available. The parameters of the community (whose use is discussed by Blondel, 1979) were calculated as follows. Richness = total no. of species. Diversity $= -\Sigma pi \log_2 pi$. Baricentre $= X1 + 2X2 + \ldots nXn/\Sigma X$. Habitat breadth $= 1/R\Sigma\, p^2 i$. "Habitat" here indicates a general type of environment, at regional scale.

most abundant and has the widest habitat breadth; Little and Common Terns have intermediate abundances and habitat breadths, the other species are low in both parametres. A relation between local abundance and breadth of habitat use has been often found in birds.

The two zones of large brackish lakes (no. 8 and 9 in tab. 2) demonstrate richness of species and diversity of the communities, which are notably higher than in the other zones. The complex "Valli di Comacchio" - Po delta (fig. 2) and its community of breeding gulls and terns have striking similarities with the Rhone delta in Southern France (Brichetti & Isenmann, 1981).

The availability of varied feeding biotopes within a restricted range is the factor which promotes species richness, both in the Rhone delta (Blondel & Isenmann, 1981) and in the Po delta (pers. obs.). During the past three years I have studied the diet and the distribution of foraging gulls and terns, within the whole area shown in fig. 2 (papers in preparation). In the "Valli di Comacchio", 8 species of gulls and terns breed in mixed colonies on small isles of the large brackish lake (fig. 2). During foraging, these birds exploit various biotopes, including the sea (all the gulls, and Sandwich and Common Terns), the fishing boats (gulls and Common Terns), brackish and freshwater ponds, canals and rivers (Mediterranean, Black-headed and Yellow-legged Gulls, Gull-billed, Common and Little Terns), salt pans (Black-headed and Slender-billed Gulls, Common and Little Terns), cultivated fields (Mediterranean and Black-headed Gulls, Gull-billed Terns) and fishing ponds (Yellow-legged Gulls). All these foraging biotopes are within 15 km of the centre of the lake, where most colonies are settled. A different distributional pattern is to be found in the northern part of the Po delta (fig. 2), where the colonies of Common and Little Terns are sharply segregated, the Common Tern breeding only in the small "valli" of the inner delta, and the Little Tern on the outer distribution of the biotopes preferred for foraging: the Common Terns, which breed in the "valli", never feed there and mainly exploit the sea. On the other hand, the Little Terns from the colonies on the beaches go inland to feed on the lagoons, the canals and the small "valli". The distribution of the colonies in terns is probably determined by the interwoven effects of the abundance of suitable prey in the biotopes used for foraging, and of the availability of the preferred nesting habitats. Along the river Po, Common and Little Terns breed in mixed colonies; the distribution and size of the colonies is determined by the structural and hydrological characteristics of the river bed, which affect the availability of suitable feeding zones and of safe nesting sites (Fasola & Bogliani, 1984).

Colonial Strategies

Coloniality in birds is moulded by many pressures, and particularly by the need to avoid predation and by the opportunity to obtain information on the location of food sources from

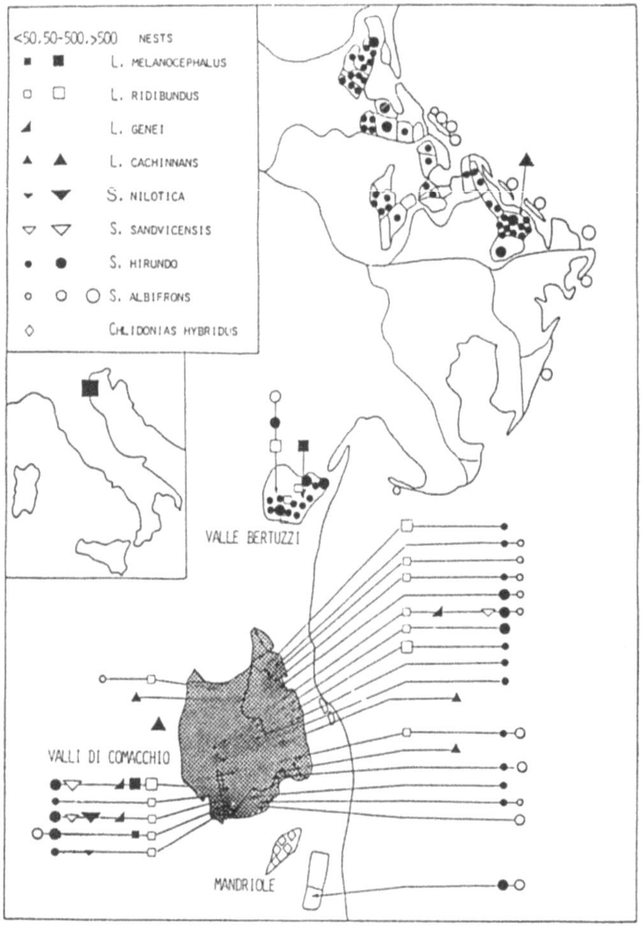

Fig. 2: Distribution of the colonies (species and size of colonies in the inset) on the sandy beaches of the Po delta in 1983 (upper right), the "valli" of the inner delta (upper centre), and in certain neighbouring zones, in 1984. For the "Valli di Comacchio" the precise location of the colonies is not shown; the colonies are settled on small isles and ditches dispersed in the "valli". The "Valle Bertuzzi" and "Valli di Comacchio" are classified in tab. 2 as large brackish lakes, while the freshwater "Mandriole" is one of the inland ponds. "Valle" means a former lagoon whose water circulation is presently man-regulated for fishing and hunting management. Isolated nests, or species with only one nest in a mixed colony, are not shown. In addition to the colonies of Yellow-legged Gulls shown on the map, 200 more pairs bred scatterdly on the inner Po delta, and 300 more pairs on the "Valli di Comacchio" and "Valle Bertuzzi". In 1984, the distribution of the colonies of Little Terns on the Po delta was different: many sites were abandoned (probably owing to increased human disturbance on the beaches), and one very big colony (2,440 nests) settled on a sand bar off the western point of the delta.

colony-mates (Wittenberger & Hunt, 1985). Lind (1963) showed that some species of seabirds (e.g. Sandwich Terns), which clump their nests more than the other terns, rely more on communal defence against predators than on concealment of the nests. Erwin (1977, 1978) found a correlation between distance to foraging areas, gregariousness when feeding, and colony size; the species feeding on patchily-distributed prey and foraging at large distances from the nests, tend to be more colonial. Gotmark (1982) discusses the degree of coloniality in species of gulls in relation to their foraging range and to the risk of predation; larger gulls, which are able to defend their nests individually, form smaller colonies and space their nests at wider distances.

Tab. 3 summarizes some data on the structure of gull and tern colonies in Italy. The largest colonies (large especially in relation to the low total population of the species) are those of the Mediterranean Gull, Slender-billed Gull and Sandwich Tern; on the other hand, Black-headed Gulls, Gull-billed Terns and Black Terns form the smallest colonies. The species which form large colonies also clump their nests as much as is allowed by the peck distance of neighbour breeders; therefore the surface of their colonies is small (tab. 3). The opposite occurs in the species which form small colonies (tab. 3). In accordance with the hypothesis on coloniality discussed above, the "more colonial" species forage at large distances from the colonies (particularly Sandwich Terns, pers. obs.), while some "less colonial" species (e.g. Whiskered Terns, pers. obs.) feed very close to the colonies. The "more colonial" species also breed more frequently in association with other gulls and terns (tab. 3). In addition to the tendency to form large colonies, another factor influencing the association of species within the same colony may be that the small species, like Common and Little Terns, avoid the presence of gulls (tab. 3) which might prey on their nests.

However, colonial strategies are difficult to interpret on a large, geographical scale, as attempted in tab. 3; I hope that the data provided by this survey will stimulate studies of single communities within restricted biotopes, where a comparison of strategies would be more meaningful.

The turnover Index of colony sites (tab. 3) was high in Mediterranean Gulls, Audouin's Gulls, Little Terns and Black Terns. High turnover occurs in species which breed in unstable habitats; these species are said to have a low site tenacity and a high group adherence (McNicholls, 1975). A low site tenacity makes conservation measures more difficult.

Conservation

During the survey, we collected data on predation, human disturbance and other threats to the colonies. Salient among these was the severe disturbance recorded on the colonies of Little Tern in the Po delta (fig. 2), where tourism is now reaching even the most inaccessible beaches of the outer delta.

	Total number of colonies	Total number of nests	Number of nests per colony average	range	Average surface colony	Nearest neighbour distance	Turnover index	% colonies with other gulls	terns
L. melanocephalus	4	226	57.5	2-145	.05	.6	.78	89	100
L. ridibundus	26	734	23.9	2-128	.76	2.7	.37	29	85
L. genei	9	861	92	1-791	.003-.04	.5	.35	66	52
L. audouinii	9	549	51.2	20-160	.19	5.3	high	13	0
L. cachinnans	217	24,300	112.0	1-1000	13.2	6.1	low	13	33
S. nilotica	9	195	18	1-78	.04	1.3	.38	60	40
S. sandvicensis	2	123	46.6	7-117	.002	.5	.42	100	100
S. hirundo	166	4,720	28.3	2-419	.08	4.2	.35	21	13
S. albifrons	90	6,040	87.2	2-2440	1.3	4.2	.52	6	55
C. hybridus	10	400	33	5-210			.46	0	11
C. niger	15	157	7.1	3-21	2.6	14	.83	0	4

Tab. 3: Parameters of coloniality of the gulls and terns breeding in Italy. All the parameters are averages for the years 1982-1984, whenever data were available. The surface of the colonies is in ha, and the nearest neighbour distance of the nests is in m. The Turnover Index (Erwin et al. 1981) is 0.5 (S1/N1 + S2/N2), where S1 and S2 are the number of colony sites occupied only in year 1 or 2 respectively, and N1 and N2 are the total number of colonies. This index (range 0 - 1) expresses the average portion of colonies which changed site between one year and the next.

Presently, no protective measure has been set up in Italy for any colony of gulls or terns. The 1981 census of the heronries in Italy effectively stimulated the protection of colony sites: from 1981 to 1984, 17 heronries (out of the 70 in all the country) were protected by small natural reserves, designed to conserve the breeding habitats. The results of the "Project Laridae", providing a data base for conservation, will hopefully stimulate initiatives of protection of the species of international concern. The conservation of seabirds in Italy is discussed by Bogliani (this volume).

Acknowledgements

I express my gratitude to all the contributors to the Project Laridae, and particularly to the following, who organized local surveys: Francesco Barbieri, Giuseppe Bogliani, Paolo Boldreghini, Pierandrea Brichetti, Marco Lambertini, Ugo Foschi, Bruno Massa, Enrico Meschini, Mario Milone, Fabio Perco, Claudio Pulcher, Helmar Schenk. Peter Evans discussed the data on populations. Mario Spagnesi, Director of the I.N.B.S. supported the project.

References

American Ornithologists' Union 1983. Check-list of North American birds, 6th ed. Allen Press, Lawrence.

Arrigoni degli Oddi, E. (1904). Manuale di ornitologia italiana. Hoepli, Milano.

Barcena, F., Teixeira, A.M. & Bermejo, l. (1984). Breeding seabird populations in the Atlantic sector of the Iberian peninsula. Pp. 335-345, in: Croxall, J.P., Evans, P.G.H., Schreiber, R.W. (eds.) Status and conservation of the world's seabirds. ICBP, Cambridge.

Barrett, R.T. & Vader, W. (1984). Status and conservation of breeding seabirds in Norway. Pp. 323-333, in: Croxall, J.P., Evans, P.G.H., Schreiber, R.W. (eds.) Status and conservation of the world's seabirds. ICBP, Cambridge.

Berthon, D. & Berthon, S. (1984). Compte rendu de l'expédition Balbuzard sur les côtes méditerranéennes du Maroc. Oiseau 54: 201-213.

Blondel, J. (1979). Biogéographie et écologie. Masson, Paris.

Blondel, J. & Isenmann, P. (1981). Guide des oiseaux de Camargue. Delachaux, Neuchatel.

Boldreghini, P., Montanari, F.L. & Spina, F. (1982). Distribuzione e stato del Mignattino piombato (*Chlidonias hybrida* Pallas) in Emilia-Romagna. Atti i Conv. ital. Ornitologia Aulla: 31-36.

Brichetti, P. & Martignoni, C. (1981). Accertata nidificazione di Mignattino (Chlidonias nigra) sul Lago Inferiore di Mantova (Lombardia) ed attuale distribuzione in Italia. Riv. ital. Orn. 51: 113-120.

Brichetti, P. & Isenmann, P. (1981). Studio preliminare sull'evoluzione degli effettivi nidificanti di Laridae e Sternidae nelle valli di Comacchio (Italia) e nella Camargue (Francia). Riv. ital. Orn. 51: 133-161.

Brichetti, P. & Foschi, U. (1986). Prima nidificazione di Sterna bengalensis in Italia. Atti III Conv. ital. Ornitologia (in press).

Clapp, R.B. & Buckley, P.A. (1984). Status and conservation of seabirds in the Southeastern United States. Pp. 135-155 in: Croxall, J.P., Evans, P.G.H., Schreiber, R.W. (eds.) Status and conservation of the world's seabirds. ICBP, Cambridge.

Coulson, J.C. (1984). The population dynamics of the Eider Duck Somateria mollissima and evidence of extensive non-breeding by adult ducks. Ibis 126: 525-543.

Cramp, S. & Simmons, K.E.L. (eds.) (1977 & 1983). The Bird of the Western Palearctic. Vol. I & III. Oxford University Press, Oxford.

Cramp, S. (ed.) (1985). The Birds of the Western Palearctic. Vol. IV. Oxford University Press, Oxford.

Croxall, J.P., Evans,P.G.H. & R.W. Schreiber (eds.) (1984). Status and conservation of the world's seabirds. ICBP, Cambridge.

De Juana, E. (1984). The status and conservation of seabirds in the Spanish Mediterranean. Pp. 347-361 in: Croxall, J.P., Evans, P.G.H. & Schreiber, R.W. (eds.) Status and conservation of the world's seabirds. ICBP, Cambridge.

De Juana, E., Varela, J.M. & Witt, H.H. (1984). The conservation of seabirds at the Chafarinas Islands. Pp. 363-370 in: Croxall, J.P., Evans, P.G.H. & Schreiber, R.W. (eds.) Status and conservation of the world's seabirds. ICBP, Cambridge.

Erwin, R.M. (1977). Foraging and breeding adaptations to different food regimes in three seabirds: the Common Tern Sterna hirundo, Royal Tern Sterna maxima, and Black Skimmer Rynchops niger. Ecology 58: 389-397.

Erwin, R.M. (1978). Coloniality in terns: the role of social feeding. Condor 80: 211-215.

Erwin, R.M., Galli, J. & Burger, J. (1981). Colony site dynamics and habitat use in Atlantic coast seabirds. Auk 98: 550-561.

Evans, P.G.H. (1984). Status and conservation of seabirds in Northwest Europe (excluding Norway and the U.S.S.R.). Pp. 293-321 in: Croxall, J.P., Evans, P.H.G. & Schreiber, R.W. (eds.) Status and conservation of the world's seabirds. ICBP, Cambridge.

Fasola, M., Barbieri, F., Prigioni, C. & Bogliani, G. (1981). Le garzaie in Italia, 1981. Avocetta 5: 107-131.

Fasola, M. (1984). Censimento preliminare dei Laridae svernanti in Italia. Avocetta 8: 57-63.

Fasola, M. (ed.) In press. Distribuzione e popolazione dei Laridae e Sternidae nidificanti in Italia. Ricerche Biologia Selvaggina.

Fasola, M. & Bogliani, G. (1984). Habitat selection and distribution of nesting Common and Little Terns on the Po river (Italy). Colonial Waterbirds 7: 127-133.

Ferns, P.N. & Mudge, G.P. (1981). Accuracy of nest counts at a mixed colony of Herring and Lesser Black-backed Gulls. Bird Study 28: 244-246.

Giglioli, E.H. (1886). Avifauna Italica. Le Monnier, Firenze.

Glutz von Blotzheim, U.N. & Bauer, K.M. (1982). Handbuch der Vögel Mitteleuropas. Vol. VIII. Wiesbaden.

Golovkin, A.N. (1984). Seabirds nesting in the USSR: the status and protection of populations. Pp. 473-486 in: Croxall, J.P., Evans, P.G.H., Schreiber, R.W. (eds.) Status and conservation of the world's seabirds. ICBP, Cambridge.

Gotmark, F. (1982). Coloniality in five Larus gulls: a comparative study. Ornis Svand. 13: 211-224.

Guyot, I. (1985). Quelques données sur la nidification du Goéland d'Audouin Larus audouinii en Corse. In: Oiseaux marins nicheurs du Midi et de la Corse. Annales du C.R.O.P. n° 2.

Isenmann, P. (1976). L'essor démographique et spatial de la Mouette rieuse Larus ridibundus en Europe. L'Oiseau et R.F.O. 46: 25-40.

James, P.C. The status and conservation of seabirds in the Mediterranean Sea. Pp. 371-375 in: Croxall, J.P., Evans, P.G.H. & Schreiber, R.W. (eds.) Status and conservation of the world's seabirds. ICBP, Cambridge.

Le Grand, G., Emmerson, K. & Martin, A. (1984). The status and conservation of seabirds in the Macaronesian Islands. Pp. 377-391 in: Croxall, J.P., Evans, P.G.H. & Schreiber, R.W. (eds.) Status and conservation of the world's seabirds. ICBP, Cambridge.

Lind, H. (1983). Nogle sociale reaktioner hos terner. Dank Orn. Foren. Tidssk. 57: 157-175.

Makatsch, W. (1968). Beobachtungen an einem Brutplatz der Korallmöwe Larus audouinii. J. Orn. 109: 43-56.

Martorelli, G. (1906). Gli Uccelli d'Italia. Rizzoli, Milano.

Massey, B.W. & Atwood, J.L. (1981). Second-wave nesting of the California Least Tern: age composition and reproductive success. Auk 98: 596-605.

McNicholl, M.K. (1975). Larid site tenacity and group adherence in relation to habitat. Auk 92: 98-104.

Moller, A.P. (1975). Sandterns Gelochelidon n. nilotica Gmel. bestandsaendringer i Danmark og analyse af nogle bestandsregulerende faktorer. Dansk orn. Foren. Tidsskr 69: 81-88.

Muselet, T. (1983). Répartition des effectifs de la Sterne pierregarin Sterna hirundo et de la Sterne naine Sterna albifrons nicheuses en France pour l'année 1982. Oiseau 53: 309-322.

Pinowski, J. & Wesolovski, T. (1981). Wplyw regulacji Wisly na avifaune. Pp. 543-567 in: Kajak, Z. (ed.) Ekologiczne podstawy gospodarki srodowiskiem w systemie zlewni Wilsy. Dziekanow Lesny.

Spina, F. (1982). Contribution to the breeding biology of the Whiskered Tern Chlidonias hybrida in Val Campotto (Northern Italy). Avocetta 6: 23-33.

Wanless, S. & Harris, M.P. (1984). Effect of date on counts of nests of Herring and Lesser Black-backed Gulls. Ornis Scand. 15: 89-94.

Wesolowski, T., Glazewska, E., Glazewski, L., Nawrocka, B., Nawrocki, P., Okonska, K. (1984). Rozmieszczenie i liczebnosc ptakow siewkowatych, mew i rybitw gniazdujacych na wyspach Wisly srodkowej. Acta Ornithologica 20: 159-185.

Wittenberger, J.F. & Hunt, G.L. Jr. (1985). The adaptive significance of coloniality in birds. Pp. 1-78 in: Farner, D.S., King, J.R. & Parkes, K.C. Avian Biology. Vol. VIII. Academic Press, London.

SEABIRDS OF THE SOUTHERN SARDINIAN ISLETS

Attilio Mocci Demartis
Istituto di Zoologia ed Anatomia Comparata
viale Poetto 1, 09100 Cagliari

Introduction: historical background

Most of the various papers on the italian insular avifauna published in the second series of the Rivista Italiana di Ornitologia from 1931 to 1983 deal with the tuscan archipelago, Malta and several other smaller archipelagos such as Termiti, Aeolis, Pelagics etc. On the contrary, Sardinia and its small satellite islands have hardly been studied. The only complete work about Sardinian islands is that of Moltoni (1971) dealing with Tavolara, Molara and Molotto. Other previous papers (Bonomi, 1911, 1912) do not give an overall picture of the avifauna of the other small sardinian islands.

Such insular ornithological research was undertaken especially from 1967 to 1975 (fig. 1) thanks to some individual initiatives and to a programme launched in 1970 by the C.N.R., the Italian national research committee, which encouraged and financed some studies carried out by such authors as Moltoni (1968, 1970, 1971). Unfortunately, from 1975 onward the ornithological interest for islands decreased and C.N.R. stopped its stimulating research programme after abandoning its working group dealing with natural ressources conservation. About 1979 ornithological research in islands received a new interest in the Mediterranean. This led to the organization of specialized meetings such as on birds of prey, conservation and bio-geographical distribution of birds on islands. In 1980 a circular bulletin was launched by the scientific staff of the Parc Naturel Régional de Corse dealing with bio-geographical insularity. At present this bulletin does not receive the same interest as when initiated. Following the symposium on mediterranean birds of prey, the interest for mediterranean seabirds has led to the organization of this advanced research workshop. Another programme, the "Progetto Laridae", analyzes the current status of gulls and terns breeding in Italy.

NATO ASI Series, Vol. G 12
Mediterranean Marine Avifauna
Edited by MEDMARAVIS and X. Monbailliu

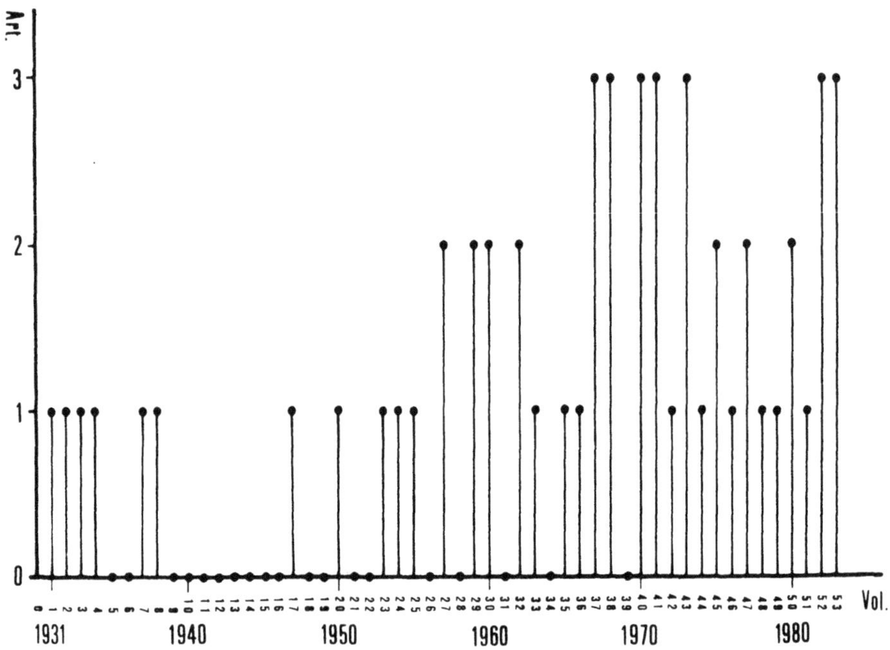

Fig. 1 : Graphic representation of the number of papers on insular avifauna published in the volumes of the second series of the "Rivista Italiana di Ornitologia" (1931-1983).

In Sardinia, since few ornithological studies have been carried out on islets, several island surveys were initiated in 1980. Two continental researchers together with a sardinian ornithologist surveyed the small islands of the Magdalena archipelago within the context of the Laridae Project. Two other biologists studied bird insularity and the breeding ecology of seabirds on the island of Asinara. A team from the zoological institute of Cagliari initiated zoo-geographical research of small western and southern sardinian islands. This paper deals with its first results.

Our research team is composed of Dr. P. Pisano (entomology), Dr. F. Loffredo (arachnology) and myself (ornithology). The Cagliari botanical institute renewed its interest for the flora of some small southern sardinian islands thanks to Dr. L. Massa and several researchers from the same institute. In 1980, a new university research organization, C.U.R.S.I.C. (Centro Universitario Richerche Scientifiche Isola dei Cavoli) was set up by the faculty of mathematics, physical and natural sciences of the university of Cagliari. C.U.R.S.I.C. was created for studying the resources of Isola dei Cavoli and its protection against tourist development.

Physical and climatological conditions of the islands

The five islets studied are situated, going clockwise, at the most south-eastern point of Sardinia (Serpentara and Cavoli), at the south-western point (Rossa, La Vacca) and on the west coast (Maldiventre). The geological strata of Serpentara and of Maldiventre are made up of various Hercynian granites. Cavoli is built up of magmatic and metamorphic rocks of the same period. Isola Rossa is composed of cambrian limestone and dolomites while Isola Vacca consists of andesitic lavas of the inferior Miocene. The surface of these islets ranges from 85.5 ha (Maldiventre) to 9 ha (Vacca). The distances from the Sardinian coast are reported in tab. 1. The bathimetry of the sea stretches does not exceed 100 m in depth. Sea currents around these islets belong in February to the Ausonian cyclonic circuit while in August they form part of the Esperian anticyclonic circuit. The speed of the currents and water temperature are variable from islet to islet and from season to season (see tab. 1). The mean diurnal temperatures vary in the islets considered from 15° to 18°C with a minimum of about 12° - 14° and a maximum of 19° - 21° (tab. 1).

The mean annual atmospheric pressure approaches 1,015 millibar. The predominant winds vary as for direction and intensity from islet to islet. As compared to the weather station of Capo Carbonara, the islands of Cavoli and Serpentara enjoy an annual average of 151 clear days, of 111 cloudy and 103 overcast days. As for Maldiventre, compared to the weather station of Oristano, there are 133 clear days, 112 cloudy and 120 overcast days. Finally at Vacca and Isola Rossa, the degree of nebulosity is derived from the Carloforte weather station: 104 clear days, 172 cloudy and 89 overcast days. In all the islets the relative humidity shifts on the average from 70% to 75%. Annual rainfall varies from 501 to 600 mm. The rainy days vary from islet to islet with peaks never above 70 days. Frost and snow are rare never exceeding 10 days. These conditions indicate a semi-arid subtropical climate at Rossa and Vacca, a sub-tropical climate at Cavoli and Serpentara and a warm-temperate climate at Maldiventre. On all five islets the vegetation consists of garrigue scrubs with Labiatae and accompanied with helio-xerophile species belonging to the Oleo-ceratonion. Physical and climatological data are taken from Serra (1971) and Pinna (1971). Botanical data come from Chiappini (1971).

			Serpentara	Cavoli	Rossa	Vacca	Maldiventre
1	Area	ha	36	42,25	10,50	9	85,56
2	Max. altitude	m	54	40	54	94	18
3	Distance from mainland	Km	3,200	0,700	0,625	2,840	6,850
4	Depth	m	50 - 100	0 - 50	50 - 100	0 - 50	0 - 50
5	Speed of currents in February	mg/h	0,75 - 0,95	0,75 - 0,95	0,3 - 0,55	0,55 - 0,75	0,55 - 0,75
6	Temperature of seawater in February	°C	13,5	13,5	13,5	13,5	13,5
7	Speed of currents in August	mg/h	0,55 - 0,75	0,55 - 0,75	0,1 - 0,3	0,3 - 0,55	0,1 - 0,3
8	Temperature of seawater in August	°C	24 - 24,5	24 - 24,5	24 - 24,5	24 - 24,5	24 - 24,5
9	Average diurnal temperature	°C	17 - 18	17 - 18	15 - 16	15 - 16	16 - 17
10	Minimal diurnal temperature	°C	> 14	> 14	> 14	13 - 14	12 - 13
11	Maximal diurnal temperature	°C	20 - 21	20 - 21	19 - 20	19 - 20	20 - 21
12	Difference in daily temperature	°C	8 - 9	7 - 8	< 6	< 6	8 - 9
13	Difference in annual temperature	°C	14 - 15	14 - 15	<14	<14	14 - 15
14	Atmospheric pressure	mb	1014,9	1014,9	1015,1	1015,2	1015,3
15	Predominant direction of winds	......	N-E ; W	N-E ; W	W ; E	N ; N-W	W ; N-W
16	Clear days	N°	151	151	104	104	133
17	Cloudy days	N°	111	111	172	172	112
18	Covered days	N°	103	103	89	89	120
19	Relative humidity	%	70 - 75	70 - 75	70 - 75	70 - 75	70 - 75
20	Rainfall	mm	501 - 600	501 - 600	501 - 600	501 - 600	501 - 600
21	Rainy days	N°	< 50	< 50	50 - 60	61 - 70	61 - 70
22	Days of frost and snow	N°	< 10	< 10	< 10	< 10	< 10
23	Type of climate		Subtropical	Subtropical	Subtropical semi-arid	Subtropical semi-arid	Temperate warm

Tab. 1: Summary table of the physical and climatic characteristics of the five islets surveyed.

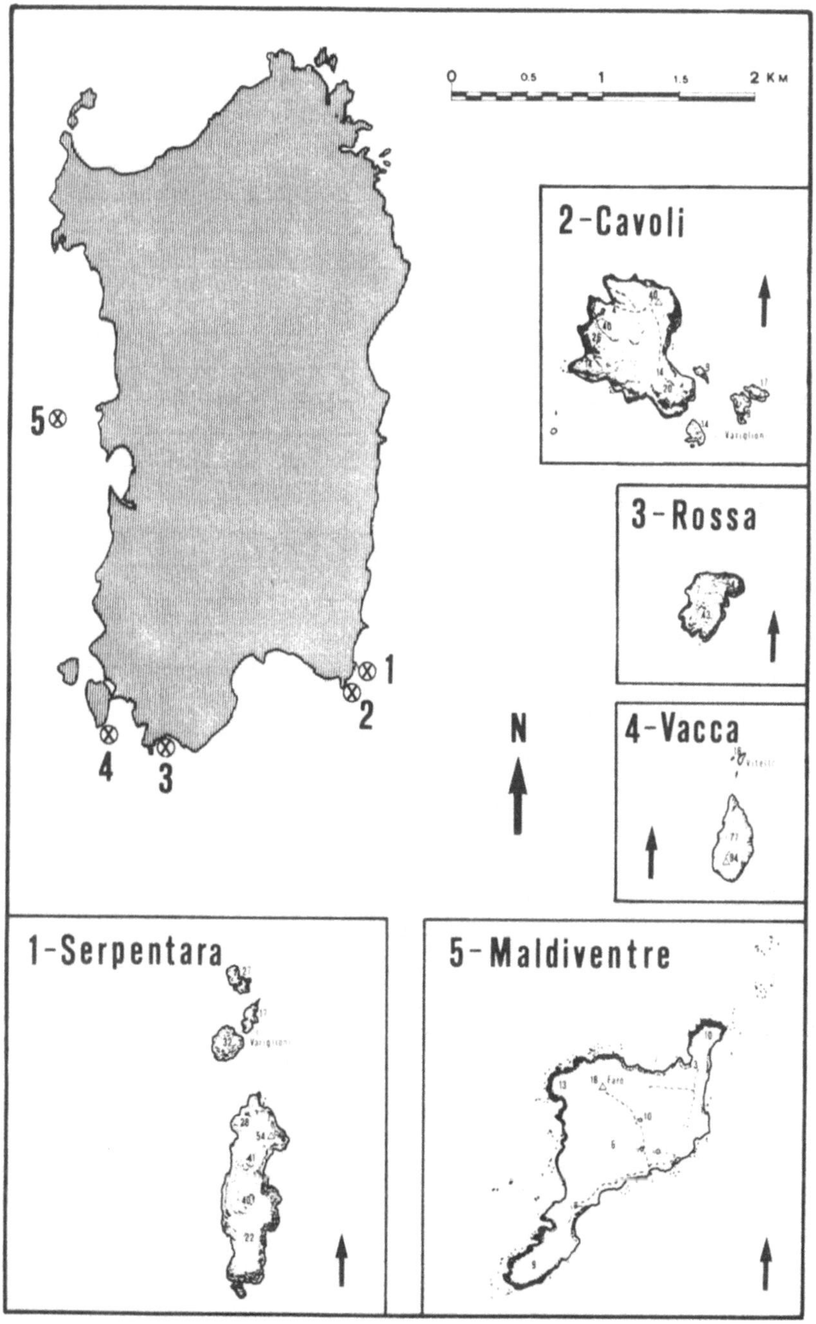

Fig. 2 : Geographical situation of the study area.

	Location	Time	Calonectris diomedea					Puffinus puffinus					Phalacrocorax aristotelis					Larus audouinii					Larus cachinnans				
Date			S	C	R	V	M	S	C	R	V	M	S	C	R	V	M	S	C	R	V	M	S	C	R	V	M
26/5/80	M	6					3															40					250
16/9/81	C	2																						1			
20/5/82	C	4							1										3					46			
27/5/82	C	5							2															110			
9/6/82	S	5						2										2					1200				
13/7/82	M	4															40					21					50
30/9/82	C	2.30																	3					1			
9/12/82	S	2																					110				
7/3/83	S	3						22															1360				
8/3/83	M	2					2										200					4					165
23/3/83	C	3																	4					60			
16/4/83	R	4.30																		25					200		
27/4/83	V	0									28					4					1					5	
10/5/83	C	0		46																							
13/5/83	V	1									3					6										450	
8/6/83	C	2.30																	2					40			
15/6/83	R	3.30																		100					150		
26/7/83	V	2														3										2	
20/4/84	C	4																						136			
14/5/84	M	5					1										15					110					400
22/5/84	R	3																		106					300		
29/5/84	S	3						9					6										2500				
11/6/84	M	3					5										11					67					500
11/12/85	C	3																						70			
Total		73																									

Tab. 2: Number of specimens of the different species studied on the various islets during each visit.

Methods

The zoological research group mentioned above used a rubber motorized dinghy for crossing the sea. Surveys were carried out in 1981 and during the spring and summer of 1982, 1983 and 1984. We went also out at sea during winter when weather conditions made it possible. During the spring of 1985 we did not manage to visit the islands due to bad weather circumstances. All the visits and surveys were carried out in the morning since we had to start going back at 3 pm as to avoid heavy winds before dusk. It was not possible to collect data at the islet of Toro as it is very difficult to land there and because of the unstable wind conditions as Bonomi already observed in 1911.

Taking into account another expedition on 28th May 1980 with other ornithologists (Brichetti et al., 1981), our data are based on 24 visits representing a total of 73 hours of observations in the field. These surveys are summarized in tab. 2 as follows: Isola dei Cavoli: 9 visits or 26 hours; Serpentara: 4 visits or 13 hours; Rossa: 3 visits of 11 hours; Vacca: 3 visits or 3 hours; Maldiventre: 5 visits or 20 hours. If considering the time involved in crossing the sea while surveying marine birds the total amount of time spent during research amounts to 109 hours. Notwithstanding the first results of this study, our research continues covering various zoogeographical aspects.

Results

During our 24 visits to the islets only five seabird species have been recorded: Cory's shearwater (*Calonectris diomedea*), Yelkouan shearwater (*Puffinus puffinus*), shag (*Phalacrocorax aristotelis*), Yellow-legged gull (*Larus cachinnans*) and Audouin's gull (*Larus audouinii*). The data of the number of birds present during each visit are summarized in tab. 2. Our estimates of the number of nests are reported in tab. 3. Population distribution and density differ from islet to islet and underwent variations within the three years of survey.

• *Calonectris diomedea*

This species was seen only four times near the islet of Maldiventre: 3 shearwaters on 26th May 1980; 2 on 8th March 1983 at about 11 am while they were flying and fishing near the islet; 1 specimen on 14th May 1984 on the open sea and 5 birds on 11th June 1984. This species was never seen near the other islets surveyed. In our opinion this is due to the small distance of the islets from the mainland: the larger the distance from the mainland the greater the possibility to meet this species. 46 Cory's shearwaters were observed on 10th May 1983 from a ship liner "Tirrenia" off Isola dei Cavoli at 4 km from the Sardinian coast. Dr. Loffredo observed several Cory's shearwaters on various occasions off the islet of Toro at a distance of 10.3 km off the coast of Sardinia. The same was noticed by Bonomi in 1911.

Number of nests	Calonectris diomedea					Puffinus puffinus					Phalacrocorax aristotelis					Larus audouinii					Larus cachinnans				
	S	C	R	V	M	S	C	R	V	M	S	C	R	V	M	S	C	R	V	M	S	C	R	V	M
26/5/80																				20					≈100
20/5/82							1										3								
27/5/82							2										2					≈11			
9/6/82																					≈600				
7/3/83																					≈650				
8/3/83																									≈75
23/3/83																						≈30			
16/4/83																		≈13					≈80		
13/5/83														1										≈200	
15/6/83																		≈30					≈50		
20/4/84																						≈60			
14/5/84																				51					≈250
22/5/84																		26					≈100		
29/5/84																					≈1000				
11/6/84																				13					

Tab. 3: Number of nests of the different species, counted or estimated during the various years.

• Puffinus puffinus

Even though this species was not found in all the five islets studied, the Yelkouan shearwater is not strictly limited to the west coast of Sardinia as it seems to be the case of the Cory's shearwater. The Yelkouan was also seen in the south-western area at Serpentara and around the islet of Cavoli where it even seems to have bred. In May 1982, narrow tunnels as long as 3 m were found. They are placed among big rocks, covered with bird excrements and strewn with numerous marks of webbed legs showing a middle digit of 5 cm long belonging very likely to the Yelkouan. The only area where this species was not found is the stretch of sea between Isola Rossa and the coastline of Teulada. Maybe this is due to the extreme vicinity of this islet to the Sardinian coast.

• Phalacrocorax aristotelis

Every time we visited Maldiventre we observed shags with a maximum of about 200 specimens on 8th March 1983. Although no nests were found on the islet, the presence of a great number of fledglings implies that the species breeds at Maldiventre. This was also noticed on the islet of Vacca where on 13th May 1983 a fledgling with light plumage was found. We lack data on the presence of this species on the other islets. This may be due to the fact that the islets are too near to the mainland coast and are easily reached by boats disturbing the species. The only exception is Serpentara where 6 shags were observed on 29th May 1984. They probably arrived at the islet coming from the nearby sardinian coast where on 8th June 1983 two shags were seen at Villasimius facing Isola dei Cavoli.

• Larus audouinii

Surprisingly the Audouin's gull was observed on all five islets in fairly large numbers. The largest density is to be found in the western area on Isola di Maldiventre and in the south (Isola Rossa). Some Audouins breed also at Isola dei Cavoli. In May 1980, at Maldiventre,about 20 nests were found (Brichetti et al., 1981), but breeding could not be reconfirmed during the following three years because of timelyness of our visits. However, on 14th May 1984, together with E. Benussi and F. Foschi in the context of the "Progetto Laridae" we counted 51 nests (45 with eggs; 6 empty nests). These nests were located in two subcolonies both in central areas of the islet, far from the coast. The first colony, situated close to the highest point of the islet and south of the lighthouse, contained 45 nests (40 with eggs and 5 empty nests). The second colony, situated in a depression about 200 m from the previous one, contained 6 nests (5 with eggs, 1 empty nest). On a following visit, which was undertaken one month later with other colleagues, there were only 7 nests in the first colony containing unhatched eggs: 6 nests with one sterile egg each and 1 nest with 2 sterile eggs. In the second subcolony we found 6 nests whereof three containing one sterile egg each. Another nest contained a chick and an egg cracking open. A fifth nest had two hatching eggs while the last nest held a dead chick. A large club of adults was also seen in the southern part of Maldiventre.

At Isola Rossa, on 16th April 1983, 25 Audouins were sighted. On a following visit, on 15th June 1983, this group increased to one hundred whereof one tenth were immatures. Since both visits took place before and after the breeding season we did not manage to count the numbers of occupied nests. Yet, the numerical increase of individuals between both visits as well as the percentage of first year birds during our second visit suggest that at least 30 pairs bred there. Such an estimate of 30 nests for 1983 corresponds roughly to the number of nests counted in 1984 on this islet. In fact, on 22nd May 1984, 106 adult Audouins were seen alarming indicating that the colony might have consisted of about 50 to 56 nests whereof only 26 were counted with certainty. In this colony, located on a sheer promontory with a gradient of 70°, the nests placed on the ground and made of plant material, had a mean internal diameter of 17.6 cm and an external one of 25.9 cm. Nine nests contained chicks which were one to three days old. The other 17 nests contained only a variable number of eggs ranging from one to four per nest.

On Isola Vacca we did not notice any evidence of breeding. On 16th April 1983 only one individual was seen. As far as the south-eastern area is concerned, no nests of Audouins were found on Isola Serpentara most likely because of strong competition with the large population of Yellow-legged gulls. At Isola dei Cavoli, on 20th May 1982, two nests with 3 eggs each as well as an empty nest were found. These nests were located on the ground, on small flat surfaces near the sea, at the height of 10 to 15 m above sea level. One nest was built on a rock, the other one on a little bush. However the following year we did not notice any breeding of Audouins owing to the recent increase of Yellow-legged gulls in this islet.

• Larus cachinnans

The Yellow-legged gull is certainly the most numerous seabird species found in all five islets surveyed. The population at Serpentara contained approximately 600 pairs in 1982, 650 pairs in 1983 and 1,000 in 1984 which corresponds respectively to a density of 16.6, 18 and 27.7 breeding pairs per hectare. A similar population increase was noticed on Vacca where in 1983 200 pairs bred corresponding to a density of 22.2 breeding pairs per hectare. Such a high density of Yellow-legged gulls made it impossible for the Audouin's gull to breed at Serpentara and Vacca as shown by what happened on the islet of Cavoli where the population increase of Yellow-legged gulls during the three years of survey was accompanied by the disappearance of the few nests of Audouins after 1982. A similar development occurred on Isola Rossa, where the 50 pairs of Yellow-legged gulls found in 1983 increased their numbers to 100 in 1984. Consequently the number of Audouins decreased from 30 pairs in 1983 to 26 pairs in 1984. Yet, on the island of Maldiventre, in 1984, the relative small number of Yellow-legged gulls (250 pairs), the larger size of the island (85.56 ha) and the resulting low nesting density (2.9 breeding pairs per ha) enabled the Audouin's gull to increase its numbers from 20 pairs in 1980 to 67 pairs in 1984.

The density of Yellow-legged gulls on the various islets is inversely proportional to their distance from the mainland coast. The nearer the coast (Serpentara, Cavoli, Rossa), the easier to find several sources of food in the waste dumps of the towns (Villasimius, Teulada, etc.), in the coastal lakes and lagoons (Notteri, Colostrai, S. Antioco, etc.) and in the cultivated fields near the coast (Castiadas). In colonies situated further away from the mainland this gull is more constrained to find its food in the sea. Tourists do leave behind many food remains on the islets during summer but this does not interfere with the diet of Yellow-legged gulls since they start already breeding in March.

After the breeding season several Yellowlegs tend to leave some of the islets to reach the main coastal lakes of Sardinia. However, on the two eastern islets (Cavoli, Serpentara), gulls do not leave the breeding site since they succeed in exploiting the small amounts of food remains thrown by passengers, against the law, from the liners of the shipping company Tirrenia. These ships pass by the two islets every day on their crossings to and from Cagliari, Naples and Civitavecchia.

Nests of Yellow-legged gulls are built with plant material. They are placed on open grassy areas, in rocky crevices or close to boulders and small isolated bushes. Some nests can also be found in machia vegetation. The nests have an average internal diameter of 23 cm, an external diameter of 33 cm and are situated 2 to 6 cm above the ground. The nests contain 1 to 4 eggs but most of them hold three eggs.

Conclusions

Species composition and numbers of breeding seabirds depend mainly on the size of the islets and their distance to the mainland. The presence of the five species of seabirds considered relies also on the geological composition, the morphology and the nature of the coasts, the geographical position and the extent of isolation from human impact. The interspecific competition among the five seabird species plays also an important role in the demographic structure of the rarer species. Thus the increase of the Yellow-legged gull is responsible for the absence of the Audouin's gull at Serpentara and Vacca and for the diminishing numbers of Audouins at Cavoli. The latter enjoys a population increase where the Yellowlegs are not abundant as is the case on Isola Rossa and Maldiventre.

We should be careful in protecting the environment of these islets by discouraging and putting to an end the speculative development for residential purposes. A good example is Isola dei Cavoli which has become a maritime state property managed by the university of Cagliari as a research centre named C.U.R.S.I.C. This centre is putting up appropriate protective measures for safeguarding the rarer species. A biological population control is required for Yellow-legged gulls by regulating their food sources such as discards of fish from trawlers, domestic waste dumps situated close to the colonies and food thrown by passengers on shipping lines.

References

Bonomi, P. (1911-1912). Notizie di Sardegna. Riv. It. Ornit., A. I (I serie): 90.

Bournonville (de), D. (1964). Observations sur une importante colonie de Goélands d'Audouin (Larus audouinii Payrandeau) au large de la Corse. Le Gerfaut, 54 (4): 439-453.

Brichetti, P.; Foschi, U.F.; Mocci Demartis, A. (1981). Note ornitologiche di una escursione in Sardegna nel maggio 1980. Gli Uccelli d'Italia, 6 (2): 107-114.

Chiappini, M. (1971). Vegetazione. In: Atlante della Sardegna. La Zattera Ed., Cagliari.

Moltoni, E. (1968). Gli uccelli dell'Arcipelago Ponziano (Mar Mediterraneo). Riv. It. Orn., 38: 301-425.

Moltoni, E. (1970). Gli uccelli ad oggi riscontrati nelle isole Linosa, Lampedusa e Lampione (Isole Pelagie, Canale di Sicilia, Mediterraneo). Riv. It. Orn., 40: 77-283.

Moltoni, E. (1971). Gli uccelli ad oggi riscontrati nelle isole di Tavolara, Molara e Molarotto (Sardegna nord-orientale). Riv. It. Orn., 41: 223-372.

Mocci Demartis, A. (in press). Contributo alla conoscenza dell'avifauna delle isole dei Cavoli e di Serpentara (Sardegna sud-orientale). Rendiconti Seminario Facoltà Scienze Università di Cagliari.

Pinna, M. (1971). Tipi di clima. In: Atlante della Sardegna. La Zattera Ed., Cagliari.

Serra, A. (1971). Vento al suolo. In: Atlante della Sardegna. La Zattera Ed., Cagliari.

Steinbacher, J. (1972). Lebensbild eines ungewöhnlichen Vogels. Zur Brutbiologie und Ökologie des Eleonorenfalken (Falco eleonorae). Natur und Museum, 102 (9): 321-330.

THE FIRST BREEDING CASE OF THE STORM PETREL IN GREECE

T. Akriotis - G. Handrinos
"Hellenic Ornithological Society"
P.O. Box 640 52, 157 01 Zographos, Greece

Introduction

Since the first and only unconfirmed record of Leach's Petrel Oceanodroma leucorrhoa for Greece, the Storm Petrel Hydrobates pelagicus remains the only representative of the family Hydrobatidae in Greek waters.

Storm Petrels were first recorded in Greece by Drummond who saw several individuals off Crete on June 19th, 1843 (Drummond 1843). Since then, there are nineteen published and unpublished records of this species, most of them summarized by Bauer et al. (1969). Fig. 1 shows all these records and as it is clearly seen, the majority of them come from the central part of the Aegean Sea and particularly the Cyclades islands. However, it is noteworthy that Storm Petrels have not been recorded there by other ornithologists, even some who studied the avifauna of this island group quite extensively such as Erhard (1858), Reiser (1905) and Wettstein (1938) whilst those records that do exist often refer to quite large numbers. It is evidently a local species.

An interesting point is that most of the observations recorded so far are from birds seen at sea during the summer or late spring. We believe that there is a seasonal bias due to two reasons:

1. Very little seawatching can be carried out from boats during the winter half of the year in Greece due to bad weather conditions, and
2. The northerly winds which blow almost daily from the second half of June to the beginning of September throughout Greece but are especially strong and persistent in the central and southern Aegean. These may at times drive Storm Petrels near shores where they have a higher chance of being seen by an ornithologist.

NATO ASI Series, Vol. G 12
Mediterranean Marine Avifauna
Edited by MEDMARAVIS and X. Monbailliu

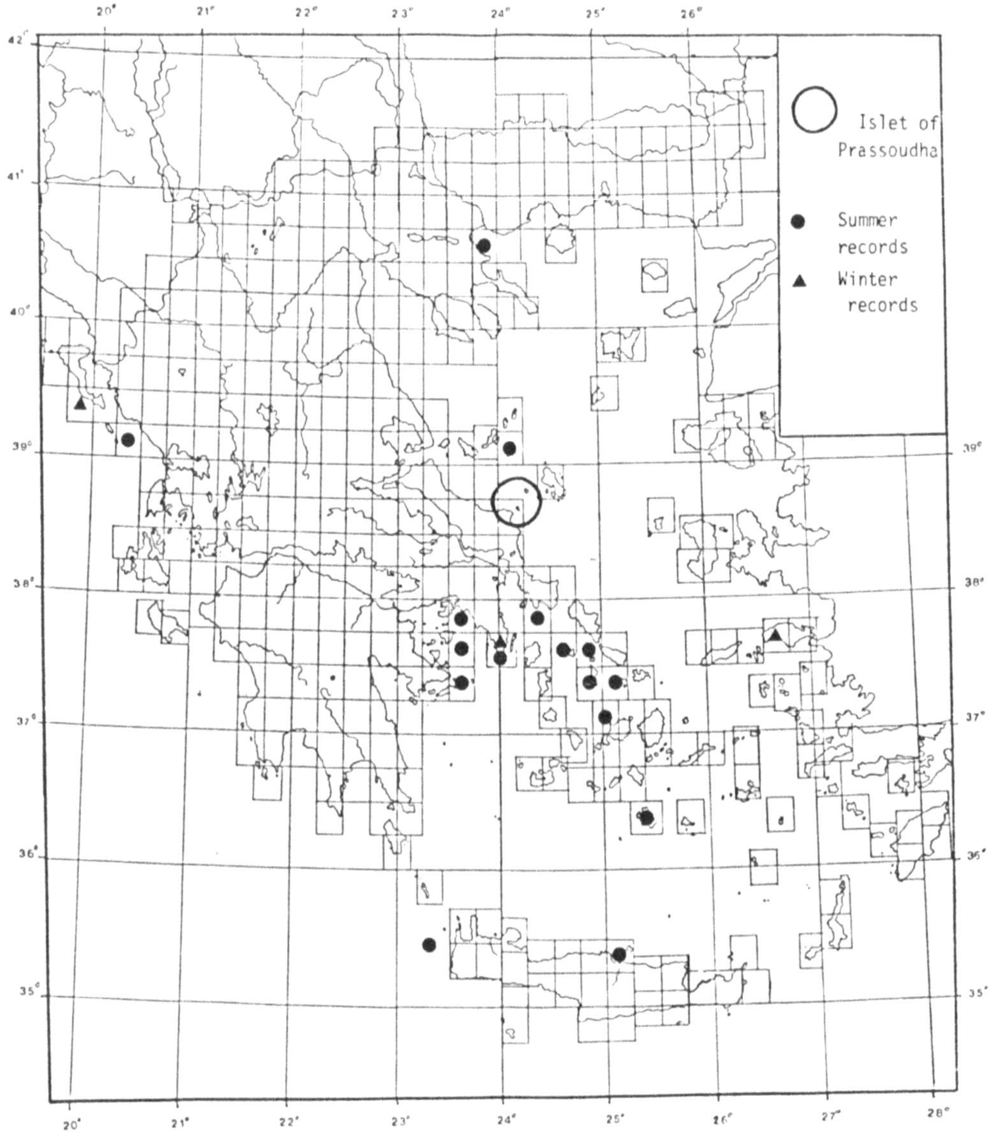

Fig. 1: The records of Storm Petrels (Hydrobates pelagicus) in Greece.

Whatever the case and despite the many records during the breeding season, Storm Petrels have never been actually found nesting in Greek waters, as indicated by Bauer (in Cramp & Simmons, 1977). Stresemann (1943) stated that nesting was probable on the islets off the coast of Crete and Makatsch (1974) that they are rare breeding birds in some islands of the Aegean. None of these however provides any proof of nesting.

The colony on the Islands of Prassoudha (38°40, 24°15 E) (fig.2)

In early August 1982, a small party of hunters and local fishermen visited Prassoudha, a small rocky islet off Kymi on the eastern coast of Euboea, Greece, to shoot rabbits *Oryctolagus cuniculus*.

Mr. E. Yataganas, a licensed taxidermist who accompanied them came back with four birds and two eggs of a species previously unknown to him. On examination of the specimens, G.H. was surprised to see that they were two full-grown birds, two large nestlings still in dark grey down and two eggs of the Storm Petrel. We were unable to obtain any precise information on the islet or the birds themselves but the species seems to be well known to the local fishermen. They said that the birds are always on the islet in the summer and that they call them "vromousses", which literally translated means "the smelly or stinking ones", a fact which shows how familiar they are with this species' characteristic odour.

After two unsuccessful attempts in 1983 due to rough seas, we managed to land on the islet in the afternoon of the 29th of July 1984. The island of Prassoudha lies 4.6 n.m. off the coast and as can be seen from the map there is a large expanse of open sea to the East, potentially suitable habitat for the species. It is approximately 500 m long by 300 m wide and about 50 m above sea level at its highest point near the SE tip of the island. From the highest point at the SE it slopes gently towards the NW all the way to the sea whereas there are cliffs up to 30 m high along the rest of the shore. Large boulders form the shore along the NW side. The bedrock is limestone and most of the soil exists as small amounts of terra rossa in pockets between boulders and slabs of exposed rock. The soil seems to be quite saline and supports some vegetation in the form of sparse garrigue with short *Sarcopoterium spinosum*, *Pistacia lentiscus* and a few isolated stunted and wind-pruned *Olea europea* - the latter almost certainly introduced by man. The most striking feature of the vegetation, however, is a tall member of the genus *Allium* whose dried inflorecences about 1 m high and up to 2 cm thick at the base, still standing upright at the end of July, were very abundant all over the island. The island actually takes its name from this plant, the name coming from the greek word for leek. The above species were the only dominant plants we saw and there was no evidence for the presence of any other species of herbaceous plants, although it would be expected that they would most likely be in the dormant state in mid-summer when we visited the island.

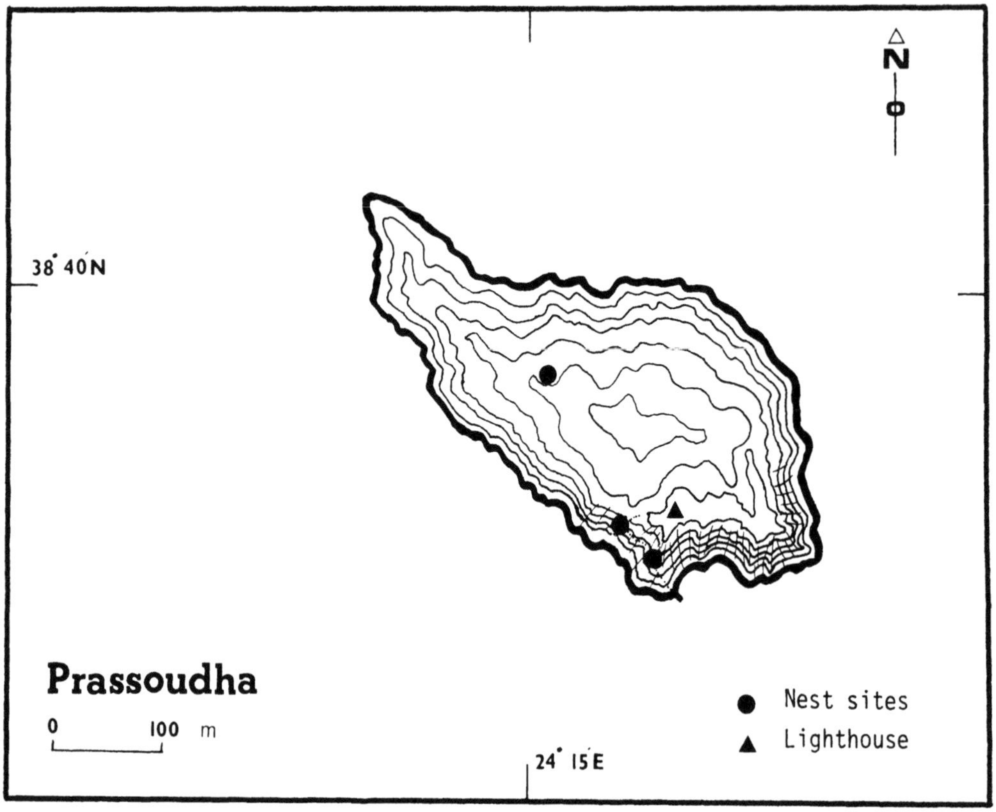

Fig. 2

The soil surface was very often completely bare and the reason for this as well as for the species composition of the vegetation is with little doubt the rabbits Oryctolagus cuniculus which have been introduced in the island by man and appear to be quite numerous. They seem to be the only mammal on the island although the existence of Black Rats Rattus rattus cannot be discounted.

There is a disused lighthouse on the island and a rough track constructed of concrete and dry stone walls leads from the lighthouse down to a small concrete pier. Landing here is extremely difficult due to the exposed position of the island.

Again due to conditions at sea there was little time for a thorough searching and in fact we saw no Storm Petrels at all, either on the islet or at sea during the journey. We located many crevices which seemed to be perfect nest sites. They were mostly horizontal, at ground level and in solid though heavily eroded limestone. They were long and narrow but not very deep, perhaps too small for shearwaters Puffinus or Calonectris but the right size for birds the size of Storm Petrels. Outside one we found the corpse of a long dead adult entangled in a piece of discarded fishing net, plus remains of four other birds. This particular crevice measured about 50 cm x 10 cm at the front and was about 30 cm deep. We also found one unhatched egg and shell fragments of two others. The intact egg was found at the bottom of a vertical slit in the rock, open at the top. It measured 30 cm x 10 cm at the top and was about 20 cm high and the egg simply lied on the soil at the bottom. We found the shell fragments in the dry stone wall at the side of the track that leads from the pier to the lighthouse, about 60 cm up the wall and about 15 m above the sea up the very steep slope. If the breeding of the two years concerned was not extremely different in timing, something which seems unlikely, then there must have been breeding adults and/or nestlings in the nest-holes at the time of our visit and we were expecting to see or hear adults or nestlings. The Storm Petrel, however, is a nocturnal bird at its breeding grounds and in retrospect we think that the success of the hunting party in finding live birds can probably be ascribed to their accidentally causing the birds to call from inside nest-holes frightened by the noise of shooting.

Discussion

Most recent standard works (Voous, 1960, Bauer & Glutz, 1966, Cramp & Simmons, 1977) define the distribution of the species as North Atlantic and Mediterranean, with an offshore, pelagic and suboceanic habitat.

The general biology of the species has been well studied in the Atlantic part of its range, mainly in the British Isles (Cramp et al., 1974, Sharrock, 1976, Cramp & Simmons, 1977). For the Mediterranean populations there is only a small amount of rather fragmentary information published (Sultana et al., 1975, Brichetti, 1981, etc.) which also refers to the western part of

Measurements of Storm Petrels *Hydrobates pelagicus* from Prassoudha, Greece (all in mm.)

Birds, dead, sex unknown (4)

Wing	:	124, 123, 123, 125
Tail	:	68, 67.5, 67, 69
Tarsus	:	29, 30, 30, 31
Bill	:	11.2, 11.0, 11.3, 12.0

Eggs (4)

31.8 x 22.0
31.0 x 21.9
30.1 x 21.9
29.3 x 21.3

Tab. 1

the sea. James (1984) points out that within the Mediterranean basin, the full range and the numbers of the Storm Petrel are very poorly known, while its population trends are not known at all. This is particularly true for the Eastern Mediterranean, where the species has not been studied at all east of 20° E and thus it is difficult to assess its exact status: according to Matvejev and Vasic (1973) the Storm Petrel is sedentary in Yugoslavian waters, while Brichetti (1980) says it is sedentary, dispersive or partially migratory in Italy and Malta. The species is not mentioned at all for Turkey (Vittery et al., 1971) and it is an accidental visitor to Cyprus, with 2 records (Flint & Stewart, 1983).

The migration pattern, if indeed any migration exists, or even the local movements within the Eastern Mediterranean are also not well known (Cramp & Simmons, 1977). Although there are a few winter observations it has not yet been fully documented whether the whole population in this area moves to the Atlantic or merely disperses in various directions.

It is difficult for us to assess exactly the status of Storm Petrels in Greece at the moment as there are too few records. Since, however, most of these records are from late spring and summer and until more data are available we could regard them as summer visitors. For similar reasons, it is at present impossible to say anything about their population size and trends or whether they are vulnerable or threatened. Yataganas who collected the specimens said that there were many birds on Prassoudha during the night but such a statement is too vague. Since the lighthouse does not have a keeper any more and few other people visit the island, the human impact on the colony seems to be negligible. There seem to be very few other possible threats. Only a few Cory's Shearwaters Calonectris diomedea breed on the island, according to the fishermen, and no gulls, although Yellow-legged Gulls Larus cachinnans probably visit the islet frequently. The presence of rats Rattus spp. is unlikely.

So far, the easternmost breeding area in the Mediterranean was the Kvarner archipelago, off the Dalmatian coast in Yugoslavia (Krpan, 1965) and in particular the islet of Vis (43°02 N 16°10 E) (Dif, 1982). As Storm Petrels are faithful to their breeding sites, especially when successful (Cramp & Simmons, 1977), the discovery of the first colony in Greece provides a considerable extension to the breeding range and firmly establishes its presence in the Eastern Mediterranean. Moreover, with Greece having no less than 600 uninhabited islands and islets, it is very likely that many more breeding colonies are yet to be discovered.

Since Prassoudha is not particularly difficult to reach by boat, given the right weather, we plan more visits in order to study the colony and the breeding ecology of the birds.

Acknowledgements

We gratefully acknowledge those who gave us their own unpublished observations and particularly A. Demetropoulos, Ph. Dragoumis, M. Goetlich and T. Martin. Many thanks also to Eust. and Xen. Yataganas, G. Chouliaras and the fisherman Stamatis of Kymi, who took us to the islet.

References

Bauer, K.M. & U.N. Glutz v. Blotzheim (1966). Handbuch der Vögel Mitteleuropas. Band I. Gaviiformes - Phoenicopteriformes.

Bauer, W., O. v. Helversen, M. Hodge & J. Martens (1969). Catalogus Faunae Graeciae. Pars I: Aves. Thessaloniki.

Brichetti, P. (1980). Distribuzione geografice degli uccelli nidificanti in Italia, Corsica et Isole Maltesi. I. Podicipedidae, Procellariidae, Hydrobatidae. Natura Bresciana 16: 82-158.

Brichetti, P. (1981). Deposizione precoce di Uccello delle tempeste Hydrobates pelagicus in Corsica. Avocetta 5: 151-154.

Cramp, S., W.R.P. Bourne & D. Saunders (1974). The seabirds of Britain and Ireland. London.

Cramp, S. & K.E.L. Simmons (eds.) (1977). The birds of the western Palearctic, Vol. 1. Oxford.

Croxall, J.P., P.G.H. Evans & R.W. Schreiber (1984). Status and conservation of the world's seabirds. I.C.B.P. Technical Publication No. 2.

Dif, G. (1982). Les oiseaux de mer d'Europe. Paris.

Drummond, H.M. (1843). List of the birds of the island of Crete ...from 27th April to 18th June, 1843. Ann. Mag. Nat. Hist. 12: 423-427.

Erhard (1858). Katalog der auf den Cykladen einheimischen und überwinternden oder nur durchziehenden Arten von Vögeln. Naumannia 8: 1-26.

Flint, P.R. & P.F. Stewart (1983). The birds of Cyprus. B.O.U. Checklist No. 6.

James, P.C. (1984). The status and conservation of seabirds in the Mediterranean Sea. In Croxall et al. (1984).

Krpan, M. (1965). Ptice otoka visa i ujemu blizich otocica. Larus 16/18: 106-150.

Matvejev, S.D. & V.F. Vasic (1973). Catalogus Faunae Jugoslaviae, Aves. Acad. Scient. Art. Slovenica, Ljubljana.

Makatsch, W. (1974). Die Eier der Vögel Europas. Vol. I. Leipzig.

Reiser, O. (1905). Ornis balcanica. III. Griechenland und die griechischen Inseln (mit Ausnahme von Kreta). Vienna.

Sharrock, J.T.K. (1976). The atlas of breeding birds in Britain and Ireland. Berkhamsted.

Stresemann, E. (1943).Überblick über die Vögel Kretas und den Vogelzug in der Aegäis. J. Orn. 91: 448-514.

Sultana, J., C. Gauci & M. Beaman (1975). A guide to the birds of Malta. Malta Orn. Soc., Valetta.

Vittery, A., R.F. Porter & J.E. Squire (eds.). 1971. Checklist of the birds of Turkey. Orn. Soc. Turkey, Bird Rep. 2: 1-184.

Voous, K.H. (1960). Atlas of European Birds. London.

Wettstein, O.v. (1938). Die Vogelwelt der Aegäis. J. Orn. 86: 9-53.

THE STATUS OF THE SEABIRDS OF THE EXTREME WESTERN MEDITERRANEAN

Eduardo De Juana A., Departamento de Zoologia,
Universidad Autonoma de Madrid, Madrid 34
Andrew M. Paterson, Ed. San Gabriel, 2-4° A
29000 Torremolinos, Spain

Present knowledge of the distribution, relative abundance and movements of most species of seabirds within the Mediterranean is still quite incomplete. This is partly due to the fact that many areas have not yet been adequately surveyed, as there are big differences in the development of ornithology between countries, but also because the existing literature is extremely scattered over many nations and in several languages, thus making difficult or even preventing any attempt of synthesis (see i.e. James, 1984). Regional reviews, such as those presented by Brichetti (1980, 1982) with regard to Italy, Corsica and the Maltese Islands, should be encouraged as a first step to gather the existing information and allow the building up of more comprehensive pictures on the status of Mediterranean seabirds. One of us recently prepared a short review on the status and conservation of the seabirds breeding in the Spanish Mediterranean (De Juana, 1984). In the present paper we deal with all the marine species breeding or visiting the westernmost stretch of the Mediterranean, an area where we have been able to do some fieldwork and to collect most of the relevant existing literature.

The study area (fig. 1) comprises both the Straits of Gibraltar and the so called Alboran Sea. Somewhat arbitrary limits are the towns of Tangier and Tarifa in the West, the Cape of Gata in the North-East and the frontier between Morocco and Algeria in the South-East. This gives a maximal length of some 350 km from East to West, while the width (North to South) varies from 14 km in the Strait to some 190 km in the longitude of Almeria. This paper complements the recent review on the birds of Algeria (Ledant et al., 1981), as well as the one which is presently being made on the birds of Catalonia, Valencia and the Balearic Islands (X. Ferrer, pers. comm.).

NATO ASI Series, Vol. G 12
Mediterranean Marine Avifauna
Edited by MEDMARAVIS and X. Monbailliu

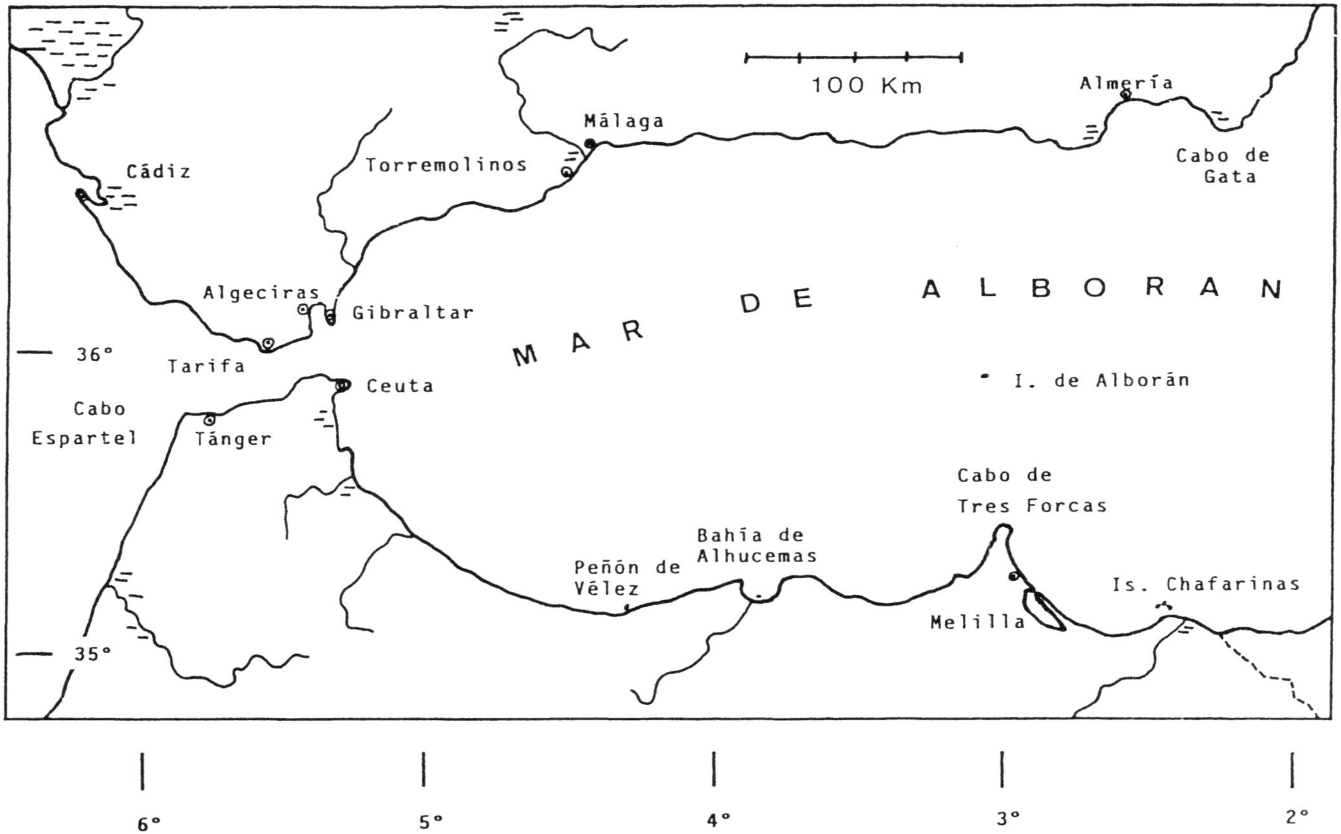

Fig. 1: Geographical situation

Literature is relatively abundant, as this area has since many years aroused interest among ornithologists: being the only existing gate from the Mediterranean to the Atlantic is an obliged way for many seabird species during their migration. It is also a concentration point for storks and raptors passing between Europe and Africa. A substantial part of the information comes from the Rock of Gibraltar, where British ornithologists have been working since long. Irby produced in 1875 his well known "The Ornithology of the Straits of Gibraltar", taking also into account a manuscript on the birds of Tangier prepared earlier by Favier, and more recent reviews include those of Lathbury (1970) and Cortés et al. (1980). Also from Gibraltar, E.F.J. Garcia presented in 1973 a valuable report which solely focused on seabirds ("Seabird activity in the Strait of Gibraltar"). Much information has also been generated at other localities, and special mention should be made of the papers of Pineau & Giraud-Audine (1974-1977) for Tangier, Telleria (1981) for Tarifa, Garrido et al. (1985) for Malaga, and Witt et al. (1984) for the area between Melilla and the Chafarinas Islands.

To the data in literature we add some unpublished information from our own recent work. That of A.M.P. refers mostly to Torremolinos (Malaga province), where he has conducted seawatches on a daily basis from September 1980 to December 1983 (1,025 hours on 895 days). That of E.D.J. usually comes from the Chafarinas Islands (mainly springs of 1984 and 1985, as previous data, years 1976 to 1983, has mostly been included in Witt et al., loc.cit.). There is also some information which has been communicated to us by a number of friends, to all of whom we are most grateful: P.M. Bradley, D. Berral, S. Canadas, H. Castro, J. Doherty, V. Estrada, J.M. Fernandez-Palacios, E.J. Mackrill, J. Manrique, R. Marti, J.M. Miralles and J.M. Varela.

Species Account

For each species we give in order:

a) a general statement on its status within the area;

b) a bibliographic review, comparing this area with the neighbouring ones in the Atlantic and in the Mediterranean;

c) our most relevant new information (after "I").

• Gavia stellata   Red-throated Diver   Colimbo Chico

Accidental or very rare in winter.

The only recent published data appear in Garrido, Alba & Gonzalez (1985) of one bird captured at the mouth of the Rio Guadalhorce in the winter of 1972-73, and in Garrido et al. (1985b) of a single bird observed in the same area on 2.2.1985. In the Strait area, where Irby (1895) considered the species to be common and referred to one captured, there are no recent records.

There are few published observations for the Iberian Peninsula and these refer in their majority to Cataluna and Galicia, where the species appears to be regular. The species has been regular in the Camargue in recent years where it is the least uncommon of the Gaviidae (Blondel & Isenmann, 1981). In Morocco it is known from a single sight record only of one in the Merdja Zerga on 9.1.1974 (Mayaud, 1982). There are no records from Algeria in this century (Ledant et al., 1981).

I - Records from Torremolinos refer to 7 birds on eight dates: 7.3.1981, 4.1.1982 & 5.1.1982 (same bird), 27.11.1982, 29.12.1982, 14.12.1985, 13.1.1986 (immature), 8.2.1986 (adult). (AMP).

● Gavia arctica Black-throated Diver Colimbo Artico

Accidental.

In this century only one record has been published of a flock of 20 off Almeria in January 1960, with 2 of the flock remaining until 20.2.1960 (Cano, 1960). Irby (1895) cited the species as occasional in the Strait but there have been no records since.

As in the case of Gavia stellata, the nearest areas where the species might be considered regular in small numbers are in Galicia, Cataluna and the Camargue. There are very few published records for Iberia. The first Moroccan record was of a single bird on 6.1.1985 at Temara near Rabat (Brit. Birds 78: 638). For Algeria there is only a single record from the last century of a wintering bird captured (Loche in Heim de Balsac & Mayaud, 1962).

I - There is a single record from Torremolinos of one on 27.11. 1982. (AMP).

● Gavia immer Great Northern Diver Colimbo Grande

Accidental.

Irby (1895) recorded Gavia immer as "...occasionally seen in winter" in the Strait, but since then there have been no records for either there or the Sea of Alboran. Nevertheless, the nearby Cadiz Bay would appear to be nearest habitual wintering area, with a maximum 5 birds on 29.1.1983 (Merino & Ruiz, 1984).

This is the least common of the Gaviidae to occur within the Mediterranean (Géroudet, 1959; Cramp & Simmons, 1977), where it appears to be regular only within the Camargue, although the least observed even there (Blondel & Isenmann, 1981). There are a few sightings for Cataluna, the Balearic Islands and Valencia. There appear to be no sightings for North Africa in this century, although Loche (in Heim de Balzac & Mayaud, 1962) considered the species to be "highly accidental" in the last century in Algeria.

I - There are two records from Malaga Bay, an adult being seen 24-25.12.1982 and another single bird 13.1.1986. (AMP).

• Tachybaptus ruficollis   Little Grebe   Zampullin Chico

Common on fresh water, exceptionally on the sea.

Although a common breeding species in both Iberia and North Africa, the occurrence of this species on the sea appears to be exceptional, in opposition to the status shown in the map corresponding to this species in Cramp & Simmons (1977). There appears to be only two bibliographic mentions (Pineau & Giraud-Audine, 1976), who considered the species as occasional on the Tangiers coast. A bird flew into a lighthouse at Gibraltar on 16.8.1952 (Tuke in Cortes et al., 1980). In the oiled seabird counts for 1982 (Bermejo, et al., 1982) a body was found at the mouth of the Rio Guadalhorce, Malaga (E.Alba).

I - The species has never been observed on the sea within the region by the authors.

• Podiceps cristatus   Great Crested Grebe   Somormujo Lavanco

Common on fresh water and semi-saline coastal lakes, but rare on sea.

The species breeds in small numbers on reservoirs on both sides of the Strait (Telleria, 1981) and is recorded as a winter visitor on reservoirs, lakes and estuaries. Greatest numbers occur in Morocco on the Mar Chica (Sebkha Bou Arez), with 120 in January 1963 (Smith, 1965); and 176 on 4.2.1981 (Thevenot et al., 1982). In some areas the species is observed on the sea with frequency. In the Bay of Cadiz, Telleria (1981) observed 200 birds in January 1978. There are large concentrations in the Camargue with up to 1,100 birds recorded, with numbers along the coast (Blondel & Isenmann, 1981). Nevertheless, in the region under consideration there do not appear to exist appropriate marine biotopes and there are few marine observations (in opposition to Cramp & Simmons). There are no records cited for the Strait of Gibraltar and with reference to Malaga, Garrido et al. (1985) indicate that "...isolated birds are seen on the coast...". Brosset (1958) saw one bird at sea from the mouth of the river Moulouyas on 11.11.1956. The species appears to be also exceptional at the sea in Algeria (Ledant et al., 1981).

I - Podiceps cristatus has been recorded on 15 dates from Torremolinos, with 14 between 25.10. and 26.3. (1 on 14.6.). (AMP).

• Podiceps griseigena   Red-necked Grebe   Somormujo Cuellirrojo

Possible accidental, but a lack of satisfactory records.

The old records of Favier for Tangiers (in Irby, 1895), who found the species less common than Podiceps cristatus and "seldom observed on passage", which appears somewhat suspicious, especially when he adds that "...some remain in the country to breed".

The species is accidental within the Iberian region with only four records which correspond to the Mediterranean region. Norman (1971) observed one on the Ebro Delta on 2.5.1969, while a record of one from Barcelona between 16-24.11.1984 is currently with the Spanish Rarities Committee. There are two records for the Balearic Islands, both old, of Jordans (1914) and Munn (1921, in Bernis et al., 1958).

The species is considered to be accidental in the Camargue (Blondel & Isenmann, 1983). There are no records from Morocco and for Algeria, Loche (in Heim de Balzac & Mayaud, 1962) considered the species as "highly accidental" in the 19th century.

• Podiceps auritus   Slavonian Grebe   Zampullin Cuellirrojo

Accidental.

The only record for the Strait area is that of Irby (1895), who recorded that he had seen a specimen collected in the Strait in October 1867. There are few records for the Iberian Peninsula and one for the Spanish Mediterranean of 3 birds in the Delta of the Ebro on 9.12.1973 (Muntaner et al, 1974). The species is classified as "occasional" for the Camargue (Blondel & Isenmann, 1981). There are no records for Morocco or Algeria, with one from the last century for Tunisia (Thomsen & Jacobsen, 1979).

I - A total of 4 birds were seen on three dates off Torremolinos in winter 1980/81 by AMP: 21.12.1980 (1), 10.1.1981 (2) and 25.1.1981 (1). (Records to Spanish Rarities Committee).

• Podiceps nigricollis   Black-necked Grebe   Zampullin Cuellinegro

Regular but scarce in the winter.

Considered by Irby (1895) to be "generally plentiful in Gibraltar Bay" in the winter, the situation appears to have changed considerably according to recent information. Cortes et al. (1980) give only one sighting for Gibraltar in recent years (11.12.1972), although considering that it may be more frequent. Telleria (1981) recorded no more than 3 in the lagoons of Cadiz Bay and Pineau & Giraud-Audine (1976) classify the species as an uncommon and irregular wintering species for Tangiers between the beginning of November and middle of March. For the Malaga coast Garrido et al. (1985) note that it is possible to see small numbers after August along the shore and at the river mouths. Brosset (1958) saw 5 birds on a lagoon at the mouth of the river Moulouyou in October 1956.

I - In total 22-30 birds were observed from Torremolinos (some may have pertained to auritus but were too distant for specific identification). Birds were generally in ones and twos, between 3.9. and 9.4., with 70% between December and February. (AMP).

• Bulweria bulwerii Bulwer's Petrel Petrel de Bulwer

Accidental.

There is only a single record from the area, that of one bird seen from Torremolinos on 7.2.1982 (Paterson, 1984a). There are two more Mediterranean records, one from between Corsica and Genoa in 1898 (Cramp & Simmons, 1977) and the other from Blanes (Barcelona) in April 1984 (Eigenhuis, 1985b). There are six records from Morocco (Brit. Birds 78: 638).

• Calonectris diomedea Cory's Shearwater Pardela Cenicienta

Common summer breeding species, although local; abundant during migration periods with a few wintering individuals.

The only breeding colony in the area is on the Islas Chafarinas, although there are strong indications that in the past there was a colony on Gibraltar (Irby, 1895) and in the Bay of Alhucemas (de Juana et al., 1980). Also, Berthon & Berthon (1984) were told by fishermen of the existence of at least one colony on a coastal islet in the Moroccan coast. In the Chafarinas colony only about 25 nests have been found, although counts of rafts at sunset in different years have shown peaks of 1,800 on 18.7.1979; 2,000 on 31.5.1980 and 3,600 on 6.6.1981, dates which coincide with known occupied nest sites where laying took place in May (de Juana et al., 1984), H.H. Witt estimated 9 - 10,000 birds present at the end of April 1982 (Witt et al., 1984). The other nearest colonies that are known are on the Habibas Islands, Algeria (Jacob & Courbet, 1980) and in the Balearic Islands and Columbretes (de Juana, 1984), while in the Atlantic the species breeds on the Berlengas Islands, Portugal (Barcena et al., 1984).

During the winter months some individuals or groups remain in some areas of the Mediterranean (Brichetti, 1980; Muntaner et al., 1984a), including the Strait of Gibraltar (Garcia, 1973; Comes et al., 1980). It is possible that the species may even be common during this same period in southern Tunisian waters (Deleuil, 1958; Czajkowski in Mayaud, 1982), but negative results are apparently more common for Catalonia (Abello & Castellon, 1984), for Corsica (Brunstein & Thibault, 1980), for Algeria (Burnier, 1979; and Jacob, 1983), such that perhaps the great majority of C. diomedea abandon the Mediterranean during the main winter months of December and January. Considerable numbers of Calonectris diomedea move through the Sea of Alboran

There is some confusion with regard to the incidence and movements of this species in the area, most probably with regard to problems of identification (Bourne et al., in prep.), thus, here the races are discussed separately.

1) P.p. puffinus: There exists no convincing sight record, as the description of Harrison (1960) does not exclude the possibility of P.p. yelkouan. Nevertheless, the species could be occasional in occurrence, as there are recoveries of British ringed birds in the Algarve (Portugal) (Bernis, 1966) and in the Camargue (Blondel & Isenmann, 1981).

2) P.p. mauretanicus: This is the most common race to occur in the area. There are no signs of nesting in the region, suspected by Heim de Balzac & Mayaud (1962), for the Chafarinas but never confirmed by repeated visits to the islands (de Juana et al., 1984), nor for the island of Alboran (de Juana & Varela, 1984), in spite of information given in the distribution map for the species in Cramp & Simmons (1977). Presently breeding is known only from the Balearic Islands where the breeding population does not appear to be significantly high (1,000 - 5,000 pairs, de Juana, 1984).

Breeding takes place early in the year with fledging taking place from the end of June onwards (Araujo et al., 1977). After breeding birds move out to the Atlantic, where they are distributed between southern Morocco (Thevenot et al., 1980) and the North Sea (Cramp & Simmons, 1977), although this race is also seen in different sectors of the Mediterranean, although generally in small numbers, from Provence (Erard, 1962), Corsica and Sardinia (Brichetti, 1980), Algeria (Jacob, 1983) and also Libya (Bundy, 1976). In the colonies on Cabrera (Balearic Islands) the species is quite rare in the summer after breeding is over (Araujo et al., 1977), although it occurs in Menorca throughout the year, with only a possible reduction of numbers during the winter months (Muntaner & Congost, 1979).

Movement through the Strait takes place from May onwards, peaking at the end of June (up to 1,000 birds per hour) (Garcia, 1973). Cortés et al. (1980) place the migration from late August onwards to October, which shows no agreement with either the fledging period in the Balearic Islands or with observations from the Atlantic. Tait (1932) recorded birds off Portuguese coasts from May onwards and Nicholson (1952) in the English Channel after the end of June. Off western French coasts, where there appear to be the greatest concentrations with up to 7 - 8,000 birds (Yésou, pers.comm.), birds are seen from the end of July to the end of September (Mayaud, 1932). Occurrence off British coasts is greatest in September, diminishing rapidly after October.

The inward return to the Mediterranean takes place after December, lasting a considerable period through the Strait (Garcia, 1973), even up to the middle of April (Cortés et al., 1980). Feeny (1960) observed migration off Cape Trafalgar (Cadiz) between 28.3. and 9.4.1959.

I - The race mauretanicus has been observed in all months off the Malaga coast by AMP., although with notable annual differences of timing which present difficulty in explanation, but which are probably more related to meteorological conditions and food supplies, rather than to the normal course of migrations. The only constant is numbers of birds congregating in the bay off Torremolinos in December and up to the middle of January.

The data of the authors strongly support the thesis of an early exit to the Atlantic after May/June with an equally early return. E.J. Mackrill (pers.comm.) has observed birds entering the Mediterranean at Tarifa (Cadiz) in mid-October, although in small numbers, with larger numbers after mid-November, with movement both in and out of the Strait in this period.

Many appear to also remain within the area of the Sea of Alboran between migrations. Thus, in three days of sailing between Ceuta and Cabo de Gata (Almeria) in October 1985, 832 puffinus were counted in the Sea of Alboran (perhaps more than 80% mauretanicus) with the majority in Malaga Bay (198) and off Roquetas (Almeria) with a flock of 430+ birds (EJ).

3) P. puffinus yelkouan

This race breeds abundantly in the eastern Mediterranean, but apparently less so in the western area, where small colonies are to be found as far west as Provence (Vidal, 1985), Corsica (Thibault & Guyot, 1981), Sardinia (Brichetti, 1980) and Algeria (8-10 pairs on El Kala, according to Ledant et al., 1981). Breeding of what might refer to this race has been suspected at times on the Catalan coast (Maluquer Sostres, 1973; Muntaner et al., 1983).

Although post-breeding movements of this race are not well-known (Cramp & Simmons, 1977),it appears that yelkouan disperses widely in a nomadic and haphazard fashion (Brichetti, 1980). In the area of Gibraltar and the Strait it is cited as being abundant after July and up to October (Garcia, 1973; Cortés et al., 1980), being commonest in September (Telleria, 1981). It remains unclear if birds move through in to the Atlantic or not, although Telleria (1981) observed 2,000 leaving the Mediterranean on 29.9.1976 and 1,500-2,000 on 20.9.1977. The waters around Gibraltar and Algeciras appear to be above all a moulting area with large concentrations of up to several thousand birds (rafts of over 2,000), already noted by Bourne & Norris (1966) in September 1964. These birds disappear by the end of October and "only a few stragglers remain during the winter" (Cortés et al., 1980). Jacob (1979) notes movements of small groups along the coasts of central Algeria in October and November, which supposes a movement towards Gibraltar, although there is no evidence of their arrival.

I - Contrary to the region of Algeciras Bay and Gibraltar, Malaga Bay appears to exercise little attraction for P.p. yelkouan. In three years only 85 birds of this race have been observed (as against 10,000 mauretanicus). Fundamentally the dates coincide with those for the Strait, falling between July and October, with 88% in September.

Both sub-species appear to be rare along the North African coast of Alboran. There are no records cited by Heim de Balsac & Mayaud (1962); Brosset (1956-58), Thevenot et al. (1980) and Thevenot et al. (1982). Estrada (1985) saw only 2 birds during a stay on the Penon de Velez between July and December 1984 (single birds 13 and 18.11.1984). For the area between Cabo Tres Forcas and the Islas Chafarinas, Witt et al. (1984) observed isolated individuals or small groups on 3.10.1974, 22.5., 31.5. and 30.6.1980 and 8.6.1981, without distinguishing race. In Algeria the species is common during the winter, especially between October and February, with the majority being yelkouan (Jacob, 1979). Bourne (1957) noted an important concentration east of Oran, Algeria.

I - No birds were recorded in the sea crossing between the Penon de Velez and Melilla in January 1983 (EJ & J. Varela) and none from Melilla in the winter of 1974-75 (EJ). Neither have birds been observed from the Chafarinas between July and December 1983 (J.M. Fernandez-Palacios, pers. comm.). On 6.4.1985 EJ & P.M. Bradley observed two birds from a boat which were on the sea within a raft of C. diomedea near the island of Congreso, Chafarinas. Other recent observations from the Chafarinas area are single birds on 4.7.1983, 3.4.1984, 11.4.1984 and 16.6.1984 (EJ & J.M. Varela).

• Puffinus assimilis   Little Shearwater   Pardela Chica

Accidental.

Generally rare in Europe, the only recent record for Iberia is that of a single bird observed off Torremolinos (Malaga) on 2.11.1981 (Paterson, in de Juana et al., 1985). Within the Mediterranean the only other obserservations known are three from Sicily and mainland Italy (Moltoni & Brichetti, 1978).

• Oceanites oceanicus   Wilson's Storm Petrel   Paino de Wilson

Accidental.

Although the species must appear off western Iberian coasts during the autumn passage (Bernis, 1966; Cramp & Simmons, 1977), there do not appear to be any records more recently published than those of Bourne & Norris (1966), who observed them regularly, in groups of up to 50 birds, in the open ocean to the South-West of Iberia. There are no certain sightings for the Strait, although Garcia (1973) considered it an occasional visitor in late summer and autumn. For the Sea of Alboran there is only the old record of Irby (1895), which refers to a specimen of

Oceanites oceanicus taken by Francisco de los Rios near Malaga. The species has been recorded twice for Sardinia before 1863 and in 1956 (Moltoni & Brichetti, 1978).

I - E.J. Mackrill (pers. comm.) observed 5 during the crossing of the Strait from Algeciras to Ceuta in October 1985.

• Hydrobates pelagicus Storm Petrel Paino Comun

Regular in small numbers, perhaps in all seasons.

There are no nesting records for the region (Irby, 1895, considered it "undoubteded", and Garcia, 1973 "possible"). The nearest Mediterranean colonies are on small islands off Cabo de Palos (Murcia) and in the Balearic Islands (de Juana, 1984). In the Atlantic the nearest breeding area is in Galicia (Barcena et al., 1984). There are no data for North Africa, excepting that of Loche (in Heim de Balsac & Mayaud, 1962), who in the last century found eggs and chicks on the island of La Galite, Tunisia.

There exist little concrete published data on the presence of the species in the Strait. Garcia (1973) indicated that it is seen mainly in late spring and summer and that "...numbers are usually small, but I have seen gatherings of up to 50 birds". Lathbury (1970) cites 50 during a diagonal crossing of the Strait (Gibraltar to Tanger) on 7.8.1966. Cortés et al. (1980) note that most observations have been between May and September, although exhausted birds were picked up on 5.4.1934, autumn 1951 and 19.12.1976. Finlayson (1983) mentions the first records at Gibraltar for the months of March (one on 28.3.1980) and October (another on 24.10.1982). Thevenot et al. (1980 and 1982) cite some birds seen in the Strait on 30.4., 6.9. and 9.9.1979, as well as at least 19 on 29.8.1981. It is of interest that these dates are previous in the year to those given by Cramp & Simmons (1977) for the post breeding migration of North Atlantic birds (September to December). These may well be birds of the Mediterranean race melitensis, which, according to Brichetti (1980) appear to be partial migrants, with movements during the period August-October and February-April. The most important colony by far is on the island of Filfla, Malta, with a population in the order of 10,000 pairs (Sultana et al., 1975).

There is little information available either for the Sea of Alboran. According to Garrido et al. (1985) the species is seen only occasionally and hardly ever near the coast. Vargas & Antunez (1980) note, without any further details, that they observed the species in the Chafarinas in May 1980. Witt et al. (1984) offer the following data: 1 on 6.6.1980 on the Chafarinas; 2 at Cabo Tres Forcas and 1 near Alhucemas on 10.7.1982, 2 on 29.6.1980 near the island of Alboran and observations totalling 10 birds on 29.6.1983 between that island and Cabo Tres Forcas.

For Algeria there exists only a single mention (Ledant et al., 1981) one bird off Reghaia on 21. and 28.1.1978 (Jacob, 1983).

I - The species has been observed only twice off Torremolinos by AMP, both during or immediately after storms, on 8.12. 1981 and 2.1.1982. In the Beached Birds (oiled seabirds) survey of January 1982, the only example collected in the Spanish Mediterranean region was from Almeria Bay (Canadas et al., pers.comm.).

• Oceanodroma leucorhoa   Leach's Storm Petrel   Paino de Leach

Accidental.

This appears on the western Iberian coasts with a certain frequency during periods of gales during the winter months, as can be seen from notes in Ardeola. Much of the records pertain to the provinces of Huelva and Cadiz (Telleria, 1981). Favier (in Irby, 1895) found bodies on the beaches of Tangiers in four winters between 1846 and 1858, as have Pineau & Giraud-Audine (1976) more recently.

Until very recently the only mention for the Sea of Alboran was that of Irby (1895) who cited the species as "recorded from Malaga". During the Spanish Beached Birds Survey of January-February 1982, there were 10 bodies of Oceanodroma leucorhoa recovered on the beaches to the north of Gibraltar (Bermejo et al., 1984a).At Gibraltar itself two birds were seen on 2.1. and 12 corpses picked up on the eastern beach on 10.1. of the same year (Finlayson, 1983). These occurred after heavy storm conditions in the Atlantic. An additional bird was found exhausted at Gibraltar's harbour on 10. November 1982 (Finlayson, pers. comm.).

The species is regarded as an accidental for the Mediterranean from France, Italy, Malta and Israel (Cramp & Simmons, 1977). There are no records from Algeria in this century, although Loche cites a 19th century record in Heim de Balsac & Mayaud (1962). There are two records from the Balearics, one being captured on Menorca and a sight record from Mallorca (Bernis et al., 1958).

• Sula bassana   Gannet   Alcatraz

Common in winter and very common during migration periods which occupy a large part of the year. Occasional birds, usually immatures, remain in the area during the summer.

Only a relatively small percentage of autumn migration birds penetrate into the Mediterranean from the Atlantic through the Strait, where the status has been examined by Thomson (1978), basing the situation on ringing recoveries. Gannets disperse widely throughout the Mediterranean in winter, moving as far as Palestinian coasts, although the numbers decrease rapidly after leaving the Strait area. More than half the recoveries are west of a line from Valencia-Oran, underlining the relative importance of the area, within the Mediterranean for this species in the winter months.

Numbers of Gannets entering the Mediterranean were considered to be in the order of only hundreds of birds by Garcia (1973), although recently Finlayson & Cortés (1984) calculated a migration in the order of 7,500 during the spring migration of 1980, observing from Gibraltar. This is in agreement with the estimates of Bourne (1978) and Nelson (1978), which suppose approximately 4% of the North Atlantic population entering the Mediterranean. A little more than 6% of all recoveries of British ringed birds south of Ouessant are from the Mediterranean according to Thomson (1978). There appears to have been a clear increase in the numbers (Heim de Balsac & Mayaud, 1962; Mayaud, 1965 & 1970), probably due to the increase of colonies in the North East Atlantic (Cramp et al., 1974). There are, for example, no records from the 19th century for Algeria, where the species now appears to be quite common during winter (Ledant et al., 1981), where several hundreds of birds have been recorded in the central coastal section alone (Jacob, 1983). In the 1950's Brosset (1956 & 1958) found the species rare along eastern Moroccan coasts and it was not until the 1960's that regular observations of the Gannet were made off the Camargue (Blondel & Isenmann, 1981). Bernis (1966) considered the species to be rare within the Spanish Mediterranean.

Within the Strait area the autumn migration takes place between the end of August and November (Cortés et al., 1980), with peak movements of juveniles in September and of adults in October (Garcia, 1973; Telleria, 1981; Cortés et al., 1980). Off Algeria, Jacob (1979) noted the first arrival of juveniles around 20.9.

The spring exit to the Atlantic could commence as early as January and be prolonged into June (Finlayson & Cortés, 1984) with peak movements taking place in March and April (Garcia, 1973; Pineau & Giraud-Audine, 1976; Cortés et al., 1980). Peak adult movements take place in March and the beginning of April and the peak immature movements in the latter part of April and May (Finlayson & Cortés, 1984; Jacob, 1983). It is worth noting as an indicator of the rarity of the Gannet in former times that Irby (1895), over several years observation, recorded the earliest autumn date as being 12.10. and the latest spring date as 28.3.

Several authors note the species as being in the area in the winter months. For the Strait, Garcia (1973) notes flocks of more than 100 birds over shoals of fish, although more normal observations from Gibraltar are in the order of 10-15 (Cortés et al., 1980). Thevenot et al. (1980) noted more than 60 during a crossing of the Strait on 24.1.1979.Between the Penon de Velez and Melilla, Witt et al. (1984) counted 83 on 1.2.1983. There is a high number of adult birds; 80% of those which move past Gibraltar in spring (Finlayson & Cortés, 1984) and 76.5% from autumn observations from the Penon de Velez (n = 665) (Estrada, 1985).

Some birds, normally immature, remain in the area during summer (Garcia, 1973; Cortés et al., 1980; Pineau & Giraud-

Audine, 1976; Garrida et al., 1985), as occurs in other parts of the Mediterranean (Spano, 1965; Thomson, 1978; Jacob, 1979; Blondel & Isenmann, 1981; Abello & Castellon, 1985).

I - Observations from Torremolinos are in agreement with the migration patterns noted above, although with a much greater proportion of first year birds. There are relatively few birds normally in January and February, although January 1986 showed large numbers of up to 14 per hour after an abnormally heavy autumn entry. There are very few records, comprising only first summer birds, in July and August. (AMP).

In four days sailing between Ceuta and Cabo de Gata (Almeria) in October 1985, 233 birds were counted (E.J.). The species appears to be abundant in winter off the Chafarinas (J.M. Fernandez-Palacios, pers. comm.), but during spring visits by E.J. and J.M. Varela few have been seen, normally only April, with a maximum of 14 feeding near the islands on 14.4.1984 during gale conditions. There are only three May records, with a group of 10 on 12.5.1976 and another of 25 on 3.5.1979, all referring to immature birds. This last is also the latest recorded date from the Chafarinas. It appears possible that spring passage occurs further out in the open sea (E.J.).

• Sula capensis Cape Gannet Alcatraz del Cabo

Accidental (?)

A record of an adult, which for plumage reasons does not discount the Australian Gannet Sula serrator, was observed by E.J. Mackrill 30 kms east of Tarragona on 28.1.1985. Another bird, also an adult, was observed by E.J. & M.M. Mackrill and A.M.P. off Torremolinos on 8.2.1986. Both records are currently with the Spanish Rarities Committee.

• Sula dactylatra Masked Booby Piquero Blanco

Accidental (?)

A sighting of 10 birds off Rio de Oro in May 1947 is considered to be invalid (Cramp & Simmons, 1977).

An adult was recorded off Sotogrande (Cadiz) on 10.10.1985 by E.J. Mackrill, later being recorded twice off Torremolinos by A.M.P. on 14.12.1985 and 9.1.1986. These three sightings, which must presumably refer to the same bird, are currently with the Spanish Rarities Committee.

• Phalacrocorax carbo Cormorant Cormoran Grande

Scarce wintering species, perhaps rather more common on fresh water. Only the sub-species sinensis has been noted.

In spite of Bernis (1966) considering this species frequent on Iberian coasts and Heim de Balsac & Mayaud (1962) noting it as very common along North African coasts in winter, there are very few observations from this region, which normally refer to 1 or 2 birds only. For the Strait area (where Irby (1895) and other authors considered the species common at the end of the past century) there are few recent sightings from Gibraltar: in April 1957 ("Expedicion Pickvance", Ardeola 4: 187); 16.2.1970 (2 birds), 7.3.1973 (1) and 17.9.1974 (1) (Cortés et al., 1980). There is a sighting of one at Cape Spartel on 30.12.1972 (Pineau & Giraud-Audine, 1976) and four observations from Tarifa in October 1976, which refer to 1-8 birds (Telleria, 1981). There are a few records from nearby Atlantic sites, notably the observation of 18 in Cadiz Bay in January 1978 (Telleria, 1981).

On the Malaga coast Garrido et al. (1985) cite the species as being very scarce, although there are annual sightings of single individuals (Alba & Garrido, 1983, found no birds during four consecutive winters in the area of the mouth of the Rio Guadalhorce (Malaga)). Bernis (1966) noted two recoveries of ringed birds (sub-species sinensis) from the province of Almeria; one from Granada and 3 from Cadiz.

For the North African coasts there are only the observations of Brosset (1956) for the mouth of the River Moulouya, where they used to be seen in small numbers in January and February. There is also a record of 9 from the Mar Chica, Melilla, on 4.2.1981 (Thevenot et al., 1982) and observations of 2-3 from the Chafarinas between 8. and 12.4.1983 (Witt et al., 1984). Further east, in Algeria, it is estimated that some 200 birds winter, spread along the coast and the lagoons and reservoirs of the Tell (Ledant et al., 1981), although Jacob (1983) does not record the species for the central coast area.

All the above observations fall within the extreme dates 17.10. to 12.4. For Algeria the species is seen between early October and early May, with most between October and March, dates which are similar to those for Tunisia (Thomsen & Jacobsen, 1979) and for the Camargue (Blondel & Isenmann, 1981), the Ebro delta (Muntaner et al., 1984b) and Menorca (Muntaner & Congost, 1979), all areas with good wintering populations.

I - Only two observations from the Torremolinos area by A.M.P., one in the area of the mouth of the Rio Guadalhorce (16.-20.10.1982) and one on 17.10.1984. These are the only observations in the region recently outside the area of the Chafarinas. In the Chafarinas, apart from the observations in Witt et al. (1984), more recently in April 1984 a maximum of 4 in 7 days (E.J.). As in previous years the birds were resting in the habitual area on the island of Rey, where in December 1983 up to 14 were counted (J.M. Fernandez-Palacios, pers. comm.), in what could be an habitual wintering area.

• <u>Plalacrocorax aristotelis</u> Shag Cormoran Monudo

Scarce breeding species, with possible small arrivals in winter.

Within the area the species is only known as breeding on Gibraltar, where there are some 5 pairs (Cortés et al., 1980), although Garcia (1973) calculated about 20. Breeding is also possible on the Chafarinas as juveniles have been recorded every year between 1979 and 1983, with maxima of 4 adults and 3 juveniles in July 1982 (de Juana et al., 1984).

In the past the Shag bred in the region of Tangiers (Irby, 1895), but there are no recent observations (Pineau & Giraud-Audine, 1977). It has been seen in Alhucemas Bay (3 on 10.7.1982, Witt et al., 1984), but Berthon & Berthon (1984) do not record the species from Moroccan coasts.

The nearest breeding colonies in the Atlantic are in South-West Portugal according to Teixeira (1984), with some 50 pairs. Within the Mediterranean there are 20-40 pairs in Algeria (Jacob & Courbet, 1980). In the last two years a pair has been observed at the same site on the rocky coast north of Cabo de Gata (Almeria) (Manrique, in. litt.).

Brosset (1956) classified the species as migratory and quite common at the mouth of the River Moulouya between December and February, although Thèvenot et al. (1980), only found one there on 11.11.1979. Groups have been seen in the Mar Chica (12 on 24.3.1981, according to Thevenot et al. (1982) and 55 on 20.1.1985 (Brit. Birds 78: 638)). The possibility exists that these are local birds which may nest on the cliffs to the west of Cabo de Agua where de Juana (1977) observed one adult on 29.12.1975. Neither must the possibility of visiting birds from the colonies in the Balearics be discounted, where there are 645-1,045 pairs (Mayol in de Juana, 1984), or from Sardinia (1,000-1,800 pairs; Schenk & Torre, 1986), or Corsica (approximately 1,000 pairs; Guyot, 1985). There have been ringing recoveries of Corsican birds from North African coasts (Guyot, pers.comm.) and the species is noted as occurring occasionally in areas of the Mediterranean which are distant from the colonies (Blondel & Isenmann, 1981; Brichetti, 1982, Muntaner et al., 1984).

At the mouth of the Rio Guadalhorce (Malaga) the species has been observed in some winters. Alba & Garrido (1983) noted one on 23.12.1978 and also one on 8. and 18.11.1979, while P.J. Kennedy (pers.comm.) noted one on 8.11.1981.

I - Since the observations cited for the Chafarinas in Witt et al. (1984) some more adults and juveniles have been noted in the islands with up to 6 adults, 1 immature and 4 juveniles on 6.7.1983 (AMP & J.M. Varela). Given the irregular nature of the sightings and inability to detect nesting sites, it appears possible that the species does not actually nest on the islands, but move there after breeding from the cliffs of Cabo de Agua, some 4 kms distant (E.J.).

• Stercorarius pomarinus   Pomarine Skua   Pagalo Pomarino

Scarce during migration periods, probably more frequent in autumn. Recent observations show that it might even be regular in the Strait.

For the Strait, Garcia (1973) classified the species as rare in May, September and October, with no more than 2 birds recorded at a time. Cortés et al. (1980) note the species as being regular, although there is only one record from Gibraltar itself, of 4 birds on 21.8.1974. Finlayson (1983) adds 4 new records for Gibraltar: single birds on 2.3., 20.3., 30.4., and 19.9.1980. In the past century Irby (1895) recorded the species from Malaga and with reference to this province, Garrido et al. (1985) indicate that it is seen occasionally from the coast and further out at sea. The only record published for the Sea of Alboran refers to a single bird off Roquetas de Mar (Almeria) on 28.6.1972 (Garcia Rodriguez, 1973).

This species is little recorded for the Iberian peninsula (Bernis, 1966) and until recently was practically unknown in the Mediterranean (Sorci et al., 1971), where it could arrive from either the Atlantic or by overland following riverine systems (Cramp & Simmons, 1983). It is classed as an accidental for Yugoslavia, Greece, Turkey, Lebanon, Israel, Egypt and Malta. The first observations in Cataluna took place only recently (1977-1984) (Martinez-Vilalta et al., 1984; Abello & Castellon, 1985; Eigenhuis, 1985b; Mackrill, 1985), with record dates covering 1 from April, 4 from May, 3 from August and 2 from October. There are only two records from the Camargue in May and June (Blondel & Isenmann, 1981).

For Algeria and Morocco, there is an old mention of one for Tangiers by Favier (in Irby, 1895); an observation on 26.10.1972 to the south of Cape Spartel (Pineau & Giraud-Audine, 1974); one on 29.-30.8.1975 at the mouth of the River Sous (Dubois & Duhautois, 1977) and another on 30.11.1977 in the Bay of Algiers (Jacob, 1983). With reference to the above dates, it is worth noting that this species is most numerous on passage in Europe in May and October (Geroudet, 1982; Cramp & Simmons, 1983).

I - There are five records from Torremolinos, four light phase adults and one immature. Three of these are autumn records: 28.7., 4.8. and 24.9., while the other two are for 6.1. and 22.2. (A.M.P.). Also off Torremolinos, but some distance from the coast, a juvenile bird was seen on 8.10.1975 (E.J.).

Further down towards Gibraltar, at Sotogrande (Cadiz), AMP and E.J. Mackrill observed an adult on 12.10.1975, while E.J. Mackrill (pers.comm.) saw at least one more in November 1985.

• <u>Stercorarius parasiticus</u> Arctic Skua Pagalo Parasito

Common during migration periods, remaining in variable numbers during the winter months, with occasional summer birds.

The Arctic Skua is noted commonly in the Strait area, where it appears to be more frequent during the autumn than in the spring migration (Garcia, 1973). The autumn passage takes place between July and October with maximum intensity in September, while the spring migration takes place in March and April (Cortes et al., 1980). Observers from Irby (1895) onwards note the presence of some wintering birds in the Strait and its western approaches with concrete records of one in Algeciras Bay on 19.1.1978 (Telleria, 1981) and 4 at Cape Spartel and another 4 in Tangiers Bay in January 1963 (Smith, 1965). There are occasionally summer records in the Strait (Garcia, 1973; Cortés et al., 1980).

For the Sea of Alboran, Irby (1895) noted that the species was seen with some frequency off the Malaga coast in March 1882. More recently Alba & Garrido (1983) note the presence of the Arctic Skua off the mouth of the Rio Guadalhorce (Malaga) with a certain frequency, but in low numbers (6 birds in winter 1978-79; 18 in winter 1979-80 and 9 in winter 1980-81), while Garrido et al. (1985) note that it is frequent for the whole of the Malaga coast during both the autumn migration (August to Novemver) and spring (February to April), as well as during the winter months.

From the coasts of Mediterranean Morocco there are records from Tetuan (Heim de Balsac & Mayaud, 1962), Cabo Tres Forcas (Witt et al., 1984), Cabo de Agua (one on 31.8.1963; Smith, 1965) and the mouth of the River Moulouya (regular September to March; Brosset, 1959).

It appears that a quantity of Arctic Skuas regularly enter the Mediterranean where there are records from several countries (Cramp & Simmons, 1983) as far east as Syria, Lebanon and Egypt, leaving open the hypothesis of overland migration. AMP has seen an adult at Neusiedl, Austria, in August, while R.W. Furness (in. litt.) notes a recovery of a British ringed bird in Syria in its first autumn.

In the western Mediterranean it is regularly present, although in small numbers, for Algeria (Jacob, 1983), Tunisia (Thomsen & Jacobsen, 1979), Italy (Sorci et al., 1971) and also, although not noted by Cramp & Simmons (1983) for the south of France (Géroudet, 1982; Blondel & Isenmann, 1982) and also from eastern Sapin (Martinez-Vilalta et al., 1984), with the majority of records falling between August and May.

I - Frequently recorded from the Spanish side of the Sea of Alboran, a total of 306 were recorded between September 1980 and December 1983 alone from Torremolinos (Malaga), normally in ones and twos. Maximum numbers occur in October, being generally scarce after December, with only a very slight increase during March and April (A.M.P.). In October 1985, after four days sailing between Ceuta and Cabo de Gata (Almeria), only 5 birds had been recorded (E.J.). Only one bird has been recorded from the Moroccan coast by the authors, one on 1.2.1983 (also in Witt et al., 1984, but with the date mistakenly given as April). The species has not been seen from the Chafarinas either by the authors or by J.M. Fernandez-Palacios during stays of July to September and November to December 1983 (pers.comm.).

• Stercorarius longicaudus   Long-tailed Skua   Pagalo Rabero

Accidental.

There is no published reference to this species within the Sea of Alboran, while for the Strait there is only the mention of Favier (in Irby, 1895), who noted that the species had been seen twice near Tangiers, in 1846 (an immature) and in October 1858. Garcia (1973) notes that it occurs only rarely in the Strait in autumn, while Cortés et al. (1980) do not even note the species for Gibraltar.

This is the species of skua least observed in western Europe, perhaps as it is also the most pelagic (Cramp & Simmons, 1983). Nevertheless, there are records for Italy, Yugoslavia, Turkey, Israel and Malta (Cramp & Simmons, 1983), which, as in the case of the Arctic Skua, may be due to trans-continental movements following riverine systems (Mayaud, 1983). The species has been recorded on several occasions in the south of France (Blondel & Isenmann, 1981), while observations in Central Europe occur with relative frequency (Cramp & Simmons, 1983; Glutz & Bauer, 1982).

The only published record for the Spanish Mediterranean (Fuset in Maluquer, 1973) has been considered invalid (Martinez-Vilalta et al., 1984). There are only three records for the Atlantic coasts of Morocco in this century (Brit. Birds 78:641).

I - A single immature was observed off Torremolinos on three separate occasions during 23.8.1981 by A.M.P. This record is with the Spanish Rarities Committee.

• Stercorarius skua   Great Skua   Pagalo Grande

Relatively common wintering species, although present throughout the year.

The Great Skua is considered to be the most common of the family in the Strait area, being present throughout the year and especially common during the winter months when westerly gales produce influxes of birds from the Atlantic (up to 40 birds in a crossing between Gibraltar and Tangiers have been noted by Garcia, 1973). Small numbers of migrant birds have been recorded during the normal migration periods, noted as being August to October and March and April (Garcia, 1973; Cortés et al., 1980).

The species is considered as scarce by Garrido et al. (1985) for Malaga province, although observable during all months and more frequently during the migration periods and in winter. Bernis (1966) cites a single bird at Punta Entinas (Almeria) on 11.4.1965.

On the Moroccan side of the Sea of Alboran there is at least one record for Mdiq of a ringed bird recovered (Thevenot et al., 1980); at the Penon de Velez single birds were seen on 12 and 18.11.1983 (Estrada, 1985), and along the coast between the Penon de Velez and Melilla 9 were recorded in five hours sailing on 1.2.1983 (Witt et al., 1984); one at Cabo Tres Forcas on 4.4.1974 (de Juana, 1977) and off the Mar Chica, one on 5.2.1981 (Thevenot et al., 1982); the mouth of the River Moulouya, one on 12.12.1953 (Brosset, 1956) and in general for all the maritime area between Cabo Tres Forcas, the island of Alboran and the Islas Chafarinas (Witt et al., 1984).

The number of Great Skuas that enter the Mediterranean has been estimated by observations from Gibraltar, to be in the order of only a few hundreds (Garcia, 1973). Of 582 recoveries of ringed birds from Scottish colonies only 22 were from the Mediterranean (Furness, 1978). The species is classified as accidental in Yugoslavia, Turkey, Egypt, Lybia and Malta (Cramp & Simmons, 1983). There are observations which appear to indicate a few wintering birds from Algeria (Ledant et al., 1981; Jacob, 1983); Tunisia (Riols, 1969); mainland Italy (Sorci et al., 1971); Corsica (Brunstein & Thibault, 1980) and from the Balearic Islands (Muntaner et al., 1984). Recent off-shore observations from Cataluna indicate that there the species is quite common between October and April (Llorente & Montori, 1982; Martinez-Vilalta et al., 1984; Abello & Castellon, 1985). Actually, the Great Skua tends to avoid the coasts outside the breeding season, unlike the Arctic Skua (Cramp & Simmons, 1983), so that unless observation is undertaken by ship-borne observers it does not appear possible to appreciate the true scale of abundance.

It is also possible, as with the Gannet, that movements into the Mediterranean have been increasing in proportion to the numbers of Great Skuas at the Scottish colonies in this century. In the Shetland Isles alone there were only 37 pairs in 1899, while in 1977 there were 5,700 (Cramp & Simmons, 1983). This would explain in large measure why Favier cited only one record for Tangiers and that Irby considered it uncommon in the Strait (Irby, 1895). Similarly, Heim de Balsac & Mayaud (1962) talk of a Mediterranean entry of only a few birds. Bernis (1966) considered it rare along Spanish Mediterranean coasts, Brosset (1959) supposed it to be an accidental species in eastern Morocco, and Lathbury (1970) gave it as occasional for Gibraltar.

I - In spring journeys to the Chafarinas E.J. & J.M. Varela have observed 35 birds normally singly (and none of S. parasiticus). There are few observations from the Chafarinas themselves, but they are seen with great frequency during sea journeys. Thus, between Melilla and the Chafarinas on 14.4.1976, 5 birds; 4 on 3.5.1979 and 5 on 7.4.1983. Records appear to place the bird as wintering,as nearly 70% of the records fall in April and only one in June (E.J.).

The situation from shore-based observations at Torremolinos is almost the reverse. S. skua is much less common than S. parasiticus, with only 76 records against 306 of the latter in the period September 1980 to December 1983. There is only one winter record of Great Skua, with all other records falling between April and October, with 80% in June and July. This appears to be a late passage of immature and non-breeding birds towards the Atlantic, as all refer to westerly moving individuals or small flocks (A.M.P.). E.J. Mackrill (pers.comm.) considers the species to be frequent at Sotogrande (Cadiz), being much more common than the Arctic Skua.

• Larus melanocephalus   Mediterranean Gull   Gaviota Cabecinegra

Quite common along the Spanish coasts of the Sea of Alboran during the winter months and migration periods, which occupy a large part of the year. Some first summer birds stay during the remaining months. The species appears to be scarce along Moroccan shores.

The species is recorded in small numbers from the Strait during winter months and migration periods. The migration periods are noted as being between the end of June and November and in March and April (Garcia, 1973; Cortés et al., 1980). Comparatively few birds move out into the Atlantic, with the Gull Census of 1984 giving only 177 for Iberian Atlantic coasts (Bermejo et al., in press.) and 395 for Atlantic Morocco (Beaubrun & Thevenot, 1984). Few continue southwards to the Canary Islands and Senegambia (Mayaud, 1983). Other censuses along the Atlantic seaboard of Morocco include 440 in winter 1962-63 (which was notably cold); 100 in winter 1963-64 (Smith, 1965) as well as 106 in December, 1976 (Isenmann, 1978). Telleria (1981) found 221 in Cadiz Bay in January 1978.

The species is known of old from the Spanish coasts of the Sea of Alboran, especially from Malaga province (Saunders, 1871; Lord Lilford in Irby, 1895). In this same region there have been many recoveries of Russian ringed birds (Mayaud, 1954; Bernis, 1967), which comprise some 20% of all those for L. melanocephalus for Iberia. On 26.11.1975, Isenmann (1976) censused 720 birds off Torre del Mar (Malaga) and some 4,000 off Fuengirola at dusk. There are no recent observations which support these numbers, which is surprising considering the known fidelity of this species to traditional wintering areas (Isenmann, 1980). Garrido et al. (1985) give the maximum flock observed personally in Malaga as being 93 birds off the mouth of the Rio Guadalhorce, while the 1984 Gull Census gave only 49 for the whole province. It must be taken into account that this species normally passes the day in the open sea, away from the coasts, and only towards dusk are there concentrations as birds come into roost (Isenmann, 1976).

The Malaga wintering area appears to be quite localized geographically, for excepting those in the area of Gibraltar (9 birds in the 1984 Gull Census), there is a lack of records for other sectors in the region.

Garrido et al. (1985) also note an autumn passage between July and November and a spring passage between February and April, with the presence of immatures in May and June.

For the Mediterranean coasts of Morocco, the little available information appears to indicate that there is no regular wintering area. The Gull Census of 1984 only found one bird between El Jebha and Oued Lao (Beaubrun & Thevenot, 1984). There are observations from the mouth of the River Moulouya (10 birds between 27.10. and 7.11.1954; Brosset, 1956; 1 juvenile on 24.11.1956 and 14 adults on 6.3.1957, Brosset, 1958) and Melilla (one, on 2.2.1983; Witt et al., 1984), as well as recoveries of ringed birds in Tetuan (Heim de Balsac & Mayaud, 1962) and Melilla (Mayaud, 1983). There were no observations of Mediterranean Gulls from Melilla in winter 1974-75 (de Juana, 1977) or from Penon de Velez between July and December 1984 (Estrada, 1985). In Algeria the situation is very similar apparently; very scarce, with only 34 birds censused in winter 1977-78 (Jacob, 1979). Many of the above observations for Morocco could refer more to migrating birds rather than wintering ones. In Algeria the migration periods are given as being between July and November and in March (Schmitt, 1963; Jacob, 1983).

I - At Torremolinos (Malaga) the species is seen all year round in small numbers, with maxima in December and January (up to 226 birds in January 1986), and with minimum numbers in April, May to early June. From mid-April to the middle of July adults are totally absent (A.M.P.). H. Castro (in lit.) refers to a single observation from Cabo de Gata (Almeria), which could be the first for this area (one bird on 7.12.1983).

In the Chafarinas, apart from the mummified body of a first winter bird found in April 1984, the rest of the observations refer to an adult in breeding plumage (possibly the same bird) seen in the breeding colony of <u>L. audouinii</u> in 1980 (27 and 30.5.); 1981 (3. and 4.7.) and 1984 (various dates between 26.4. and 18.5.). Part of these dates have been given in Witt et al. (1984). It is uncertain if the presence of this bird suggests possible attempts at colonization, but in June 1984 an adult Mediterranean Gull was found on a nest with eggs in an Audouin's Gull colony near Mallorca (Balearics) (Capella et al., 1985).

• <u>Larus atricilla</u> Laughing Gull Gaviota Guanaguanare

Accidental.

This American species is known as a vagrant in various European countries and in Morocco (Cramp & Simmons, 1983). For the Iberian peninsula there is a record from Oporto, Portugal (Preiswerk, 1981); a probable from Galicia (SGHN, 1980) and a third from Fuengirola (Malaga), on 7.5.1981 (Barrett, 1983).

• Larus pipixcan Franklin's Gull Gaviota de Franklin

Accidental.

This is another Nearctic species which occurs sporadically in Europe (Cramp & Simmons, 1983). There are two Iberian records, the first from Seville (Bond, 1981) and the other, from within the area under discussion, from Torremolinos, Malaga, of a first winter bird on 29.10.1983 (A.M.P. in de Juana et al., 1985).

• Larus minutus Little Gull Gaviota Enana

An irregular wintering species in small numbers. Post-breeding passage of relatively few birds, but frequent observations in spring, at times with relatively high numbers.

The movements of this species are relatively little known, as are the wintering grounds (Cramp & Simmons, 1983). In the central and western Mediterranean, from peninsular Italy to the coasts of Libya, Tunisia and Algeria, the species is considered to be in one of its principal wintering areas, although observations to support this statement are few. This is perhaps because the birds remain in the open sea during normal meteorological circumstances (Erard, 1960; Mayaud, 1983). Birds were noted in their hundreds off Oran, Algeria, on 5.1.1953 (Bourne, 1957), a record cited with some frequency, but Jacob (1979) counted only 40 in all Algeria between December 1977 and January 1978. Thomsen & Jacobsen (1979) found the species scarce in Tunisia, with considerable variation in numbers between winters. In Italy, the Gull Census of 1984 counted directly only 599 birds for the whole country, estimating 2,000 as a maximum possible (Fasola, 1985).

Recently, the coasts of Mediterranean Spain have been shown to be another possible wintering area of some importance, although again the birds have been observed irregularly. Isenmann (1972, 1976) and the Gull Census of January 1984 (Bermejo et al., in press.) found few or no birds. However, in the Delta of the Llobregat, Barcelona, there have been significant numbers censused in recent years: 1,270, January 1977 (Carrera et al., 1981), 601, 500 and 841 for the respective Januarys of 1979, 1980 and 1981 (Carrera, 1983). Also, in the Ebro Delta 1,000 were counted in February 1971 and another 1,000 in January 1972 (Ferrer, 1977), as well as 200 off the mouth of the Rio Turia, Valencia, in January 1983 (Guillen et al., 1985).

A third wintering area is along the coasts of the European Atlantic, although only of secondary importance according to Erard (1960), but with numbers of 5,000-10,000 birds in autumn in Holland and in December in Brittany, France (Cramp & Simmons, 1983). The autumn passage observed from Cap Gris Nez, France, on the south side of the English Channel, was 3,400 birds in

1970 and 3,300 in 1974 (Oliver, 1977).

There are quite numerous data from the area of the Sea of Alboran on wintering birds. For the Strait, Irby (1895) noted the species "in some numbers, though very irregularly", and Garcia (1973) and Cortes et al. (1980) indicate moderate numbers, with flocks of up to 200 birds during easterly (Levante) gales. Pinaud & Giraud-Audine (1976) note small numbers in winter and Smith (1965) found the species "fairly common" along the Mediterranean coast of Morocco in January 1964, with 35 between Ceuta and Rincon and 300 between Melilla and the mouth of the River Moulouya. In this last area Brosset (1958, 1959) found some corpses and Thevenot et al. (1982) give the sighting of 10 on 5.1.1981 and the collection of a corpse in the February of the same year.

On the Spanish side, Garrido et al. (1985) consider the Little Gull as a scarce wintering species for Malaga. At the mouth of the Rio Guadalhorce, Alba & Garrido (1983) counted only 17 observations of one or two birds in three consecutive winters. Ingram (1960) saw one at Algeciras on 2.2.1958 and Hiraldo et al. (1973) give five winter observations for the province of Almeria,including one of 100 birds on 29.12.1970 at the beach at Villagarcia and another of 60 in the fishing harbour at Almeria on 1.1.1971.

Migration takes place in August-September and March-April (Garcia, 1973; Cortês et al., 1980) and into May (Garrido et al., 1985). These dates are similar to those given for Algeria (Jacob, 1983). Catalonia (Cordero-Tapia, 1983) and the Camargue (Isenmann, 1975). For Gibraltar, Garcia (1973) and Cortês et al. (1980) note few birds during both migration periods, while for Malaga, Garrido et al. (1985) note that while the species is scarce in autumn, it is common in spring, although irregular numerically. This is in agreement with the overall picture for Andalucia (the majority of birds in December, January and April; Hiraldo et al., 1973); Algeria (birds between December and May; Jacob, 1983); Tunisia (Thomsen & Jacobsen, 1979, do not give any autumn records) or the Camargue (daily maxima of 17 in autumn and 1,350 in spring; Isenmann, 1975). This is radically oppposite to the situation on the Atlantic coasts, where the significant passage is in autumn (Cramp & Simmons, 1983).

This situation leads one to think in terms of a looped migration, with birds moving in autumn via the Baltic - North Sea - English Channel, while in spring the movement is through the western Mediterranean. While there is no important passage for Gibraltar, Pineau & Giraud-Audine (1976), who do not note any autumn migration for Tangiers, note a northerly movement in the spring along the Atlantic coast, with flocks of up to 50 birds (between 14.3. and 3.5.).

Literature for the Alboran region shows only two autumn records (one in Algeciras Bay on 8.8.1976; Telleria, 1981; and one at the Mar Chica on 30.8.1963; Smith, 1965). There are several spring records, with a record of 109 on 30.4.1983 at the mouth of the Rio Guadalhorce, Malaga (Garrido et al., 1985). These authors also record flocks of immatures at the same site during the summer months. Summering birds have also been recorded for other areas of Andalucia (Hiraldo et al., 1973) and for the Camargue (Isenmann, 1975) and Algeria (Ledant et al., 1981).

I - From the Torremolinos - Rio Guadalhorce area A.M.P. has recorded only two autumn records (1 immature, 3.11.1980 and 1 adult, 2.10.1983); 15 winter records (12-2), against 1,936 for spring (3-4), with a maximum of 160+ on 8 and 11.4.1982. A total of 29 records only pertain to July and August. There is considerable variaticn in the numbers present from year to year (103 in 1981; 1,820 in 1982 and 48 in 1983). Adults represent only 4.6% of all the annual totals given above (A.M.P.).

At the saltpans of Cabo de Gata (Almeria) there are also considerable spring numbers. H. Castro (pers.comm.) counted 200 on 13.5.1978 and there were 1,000+ over several days in May 1977 (E.J.).

• Larus sabinii   Sabine's Gull   Gaviota de Sabine

Accidental, perhaps occasional.

This species undertakes migrations along the Atlantic coasts of Iberia and Africa, but because of the pelagic nature of the species shore-based observations are rare (Bernis, 1967; Mayaud, 1983; de Juana et al., 1985). The species is recorded as an accidental for several European countries and within the Mediterranean as far as Israel and Egypt (Cramp & Simmons, 1983).

For the area under discussion and adjacent sea areas, there is only a record of an adult at Punta Ceres, Tangiers, on 22.2.1974 (Pineau & Giraud-Audine, 1975). There is also a record of a bird in the Strait, seen during the Gull Census of January 1984 (record with the Spanish Rarities Committee).

• Larus ridibundus   Black-headed Gull   Gaviota Reidora

Very common during migration periods and as a wintering species, rare outside these periods (only 2 months of the year maximum).

Although, in comparative terms, this region is not of outstanding importance for the European wintering population of this species (Isenmann, 1978; Mayaud, 1983, in relation to North Africa), nevertheless, it is the most abundant of the Laridae in the winter months and with the widest distribution.

The Gull Census of January 1984 showed a total of 13,104 Black-headed Gulls for the Spanish side of the Sea of Alboran (Bermejo et al., in press.) and 6,024 birds on the Moroccan side (Beaubrun & Thevenot, 1984). For the Spanish side of Alboran, the figure is broken down by province as follows: 2,200 in Almeria (only censused in the roost at Cabo de Gata); 7,344 in Malaga (only 60 km of coastline censused around Malaga capital); 3,560 in the Mediterranean zone of Cadiz. Taking into account that the coastal areas of Granada were not covered at all (2,100 birds in November 1975, according to Isenmann, 1976), nor large areas of the coastlines of both Almeria and Malaga, one might suppose a figure in excess of 20,000 birds for the whole area.

On the Moroccan side 90% of the birds censused were distributed between the Mar Chica (2,921 birds), Ceuta (1,350), the areas of Restinga-Smir (479 birds), the mouth of the River Martil (388 birds) and Alhucemas Bay (306). As has been noted often in other areas, the species shows a selective preference for areas with a strong human concentration and the greatest concentrations of the Black-headed Gulls occur close to big towns: Malaga (see Isenmann, 1976), Almeria, Algeciras - La Linea - Gibraltar, Ceuta and Melilla.

For purposes of comparison it is of interest that for all the North-West African area a wintering population of the order of 50,000 birds has been estimated (Mayaud, 1983), while the Gull Census of 1984 for the Iberian Peninsula reached the figure of 299,625 birds, of which 143,257 corresponded to the Mediterranean coasts.

The movements of the Black-headed Gull have been described, in greater or lesser detail, by several authors (Irby, 1895; Smith, 1966; Garcia, 1973; Pineau & Giraud-Audine, 1976; Cortés et al., 1980; Telleria, 1981; Thevenot et al., 1982; Garrido et al., 1985) and in general differ little from that given for Iberia by Bernis (1967) or for North-West Africa (Heim de Balsac & Mayaud, 1962).

Birds arrive from the end of July onwards in the region, continuing through to November at least, with arrivals and movements being strongest in October. Greatest numbers are recorded in December and January. First movements away from the area occur in February, being most intense in March. Some birds remain in April and in May only a few are seen, all normally immatures. June is the month in which there is the least probability of finding the species in the region.

There is an overlap between the spring and autumn migratory movements and dispersal within the wintering area. This does not appear to involve very large numbers as, although there are birds which move as far as Equatorial Africa (Cramp & Simmons, 1983), the Atlantic coast of Morocco constitutes the southern limit of the wintering area of major importance (Isenmann, 1978). There the Gull Census of 1984 totalled no more than 10,260 birds. Coastal movements have been recorded from Tarifa (Telleria, 1981) and westerly movements from Gibraltar (Garcia, 1973).

Close to the Sea of Alboran the Black-headed Gull has been found breeding at Fuente de Piedra, Malaga (Studer-Thiersch, 1968), although not in every year (Garrido et al., 1985). Recently the Black-headed Gull has also bred in the saltpans in the Bay of Cadiz (Ruiz Martinez, 1985). There is only one possible breeding recorded for Morocco and North-West Africa, from Oualidia in July 1977 (Dubois, 1979). These southerly expansions are part of the overall extension of range of the Black-headed Gull in Europe (Isenmann, 1976).

I - Numbers of this species may undergo considerable increase in association with periods of very cold weather in more northerly areas. Thus, in January 1985, 22,000 birds were censused in Malaga province alone (J. Crespo, pers.comm. and A.M.P.), perhaps twice the total for more normal winters.

The earliest juvenile recorded from Torremolinos is 12.7., although some adults have been recorded by the end of June. There is a period of 3-4 weeks between the end of May and the end of June when there have been no birds recorded (A.M.P.).

There are very few observations from the Chafarinas, although it is common in the harbour at Cabo de Agua. Extreme dates from Melilla harbour are: 1 immature, 2.5.1981 and 1 adult (with black hood), 30.6.1982 (E.J.).

• Larus genei — Slender-billed Gull — Gaviota Picofina

Occasional or very rare on migration.

There are very few published records of this species for the region. For the Strait, apart from a bird captured in Tangiers by Favier in 1852 (Irby, 1895), there are only the observations of 3 and 9 birds on 2. and 3.10.1973 (Gemra, 1974) and 1 on 8.9.1974 (Garcia-Rua, 1975). On the Spanish side of Alboran there were 2 at the salinas west of Almeria on 26.5.1981 (Witt et al., 1984) and in the not too distant area of Cabo de Palos - Mar Menor, Murcia, 4 on 29.8.1957 (Bruhn & Jeffrey, 1958) and 2 on 28.12.1979 (Gonzalez Barbera, 1981). On the Moroccan side of the region an adult was at Moulouya on 13.5.1956 (Brosset, 1956) and single birds on 4.2. and 14.6.1981 at the Mar Chica (Thevenot et al., 1982).

The ringing results from Russian Slender-billed Gulls show that the important populations from there do not appear to come further west than the coasts of Tunisia (Erard, 1958, 1964; Bernis, 1967; Mayaud, 1983; Cramp & Simmons, 1983). For Algeria, Jacob (1979) found only one bird during the winter 1977-78, although in spring tens to hundreds of birds may be seen to the east of Oran and at Lake Boughzoul, where it may breed (François, 1975; Jacob & Jacob, 1980).

One must therefore look elsewhere for the origin of birds seen within the Alboran region. The answer may lie in the possible movements, as yet little understood, of the small western populations from Sardinia, Camargue, Ebro Delta and Andalucia.

Within the proximity of Alboran the species breeds in small numbers at Fuente de Peidra, Malaga, where there were 11 nests in 1967 (Studer-Thiersch, 1969), although not breeding every year (Garrido et al., 1985). There is also a small population in the Marismas del Guadalquivir, with 18-20 pairs in 1982 (Costa, 1985). Bernis (1967) believes that the species is absent from Andalucia in the winter, while Garrido et al. (1985) say that the Slender-billed Gull is present at Fuente de Piedra in May and June, with small groups or single birds throughout the year.

In the Camargue it is normally present between April and August (Isenmann, 1976c) and in the Ebro Delta from March to July-August (Muntaner et al., 1984). Nevertheless, recently there have been observations from the Ebro Delta, Cordoba and Huelva (some doubtful) from the winter months (Bermejo et al., in press.).

I - From Torremolinos a total of 12 birds on seven dates between 1980 and 1983, with extremes 15.5. and 20.11. (A.M.P.).

• Larus audouinii Audouin's Gull Gaviota de Audouin

Locally common as a nesting species, during migration periods and on the Moroccan coasts, during the winter.

The Audouin's Gull has received considerable attention within the area in recent years (eg: Mayol, 1978; de Juana et al., 1980; Beaubrun, 1983), largely because of its condition as a species under threat (King, 1981). During some years perhaps up to 80% of the world nesting population has been on the Chafarinas, where a large colony was described for the first time in 1966 (Brosset & Olier, 1966) and which in recent years has been above 2,000 pairs (de Juana et al., 1984), and in 1985 there were 2,198 nests (P. Bradley, pers.comm.). The only other colony within the Alboran region was found in 1983 near Al-Hoceima, on the mainland, and consisted of ca. 20 pairs (Berthon & Berthon, 1984). On the island of Alboran, where Lilford found the species breeding in April, 1879 (Irby, 1895), there were no birds seen in June 1982 (de Juana & Varela, 1984).

The nearest nesting populations outside Alboran are in Algeria (ca. 500 pairs, Jacob & Courbet, 1980), Sardinia (300-500 pairs, Schenk & Torre, 1986), Corsica (95-100 pairs, Guyot et al., 1985), Balearic Islands (500 pairs, Mayol, pers.comm.), Columbretes (400 pairs, Carrera, pers.comm.) and the Ebro Delta (1,200 pairs, Ferrer, 1986). Other colonies, all to the east of the above noted, are small. In sum, the population is showing a notable increase over recent years.

The Audouin's Gull can be classified as a partial migrant, with a majority of the birds leaving the Mediterranean in the autumn to winter along the Atlantic coasts of North-West Africa (Smith, 1972; Beaubrun, 1983). Ringing recoveries from the Chafarinas appear to show that a large proportion of birds reach as far as south as Mauretania and Senegal (de Juana et al., in press.). Along Mediterranean coasts of western Europe the species is rare during the winter months (Bermejo et al., in press; de Juana et al., loc.cit.). Along the North African coast significant numbers have been recorded from Algeria (824, Jacob, 1979) and northern Morocco (950-1,000, Beaubrun, 1983), but these figures do not comprise more than a small part of the possible post-breeding population. With what is known about the extent and reproduction indices of the breeding colonies, it is possible that at least 10,000 birds migrate annually through the Alboran region.

The majority of records for the Alboran region correspond to the autumn passage which takes place between the end of June and November, with maximum movement in August. The spring migration may commence by the end of January, showing greatest intensity between mid-February and the beginning of March and finishing in April (Garcia, 1973; Pineau & Giraud-Audine, 1976; Cortés et al., 1980; Telleria, 1981; Beaubrun, 1983).

In the Chafarinas breeding birds arrive from the middle of February onwards and the colony is complete with breeding birds by early April. The post-breeding dispersal from the colony takes place from the beginning of July and lasts until the middle or end of August (de Juana et al., loc.cit.).

In the Gull Census of January 1984, no Audouin's Gulls were recorded on the Spanish coasts of Alboran (although 111 were counted in Murcia and 21 in Huelva), while on the Moroccan side 737 were censused, with 74% of that total between Cabo Tres Forcas and the Algerian frontier, the majority of the remainder occurring between El Jebha and Ceuta (Beaubrun & Thevenot, 1984). In these areas there are also concentrations of non-breeding birds during the spring and summer (Beaubrun, 1983).

I - A clear increase of records has been seen at Torremolinos (A.M.P.): 21 in 1981; 104 in 1982 and 185 in 1983. This increase could have a direct relationship with the important increase in the Ebro Delta colony (36 pairs in 1981 and 1,200 in 1985; Ferrer, 1986). A similar increase has been recorded at the saltpans at Cabo de Gata (H. Castro & J. Manrique, pers.comm.). There have been 18 reports of colour ringed birds from the Chafarinas in the Spanish side of the Strait although 100 km further east, at Torremolinos, there has been only one (E.J. & J.M. Varela).

An interesting feature of the Torremolinos observations is that while 306 birds were observed in the autumn (1981-83), only 2 were recorded in the winter months and 2 more in the spring. It seems most likely that the spring migration bypasses Malaga Bay. The earliest record of a juvenile from Torremolinos is 20.7. (A.M.P.).

● <u>Larus canus</u> Common Gull Gaviota Cana

Occasional or rare in winter; more frequently recorded in the Strait during abnormally cold weather.

The species was not recorded in the Alboran region during the Gull Census of 1984, the region being considerably to the South of the normal wintering area of the Common Gull (Cramp & Simmons, 1983). In Iberia it is normally scarce along northern coasts, although in colder winters it may become relatively common and there are many records of ringed birds (Bernis, 1967). Such recoveries descend quantitatively along the Atlantic coast to the Strait (1 from Huelva, 4 from Cadiz), but for the Mediterranean there are no recoveries further south than Catalonia. There is also a recovery from Kenitra (Morocco). A few observations exist from Tunisia (Thomsen & Jacobsen, 1979), Algeria (Ledant et al., 1981), Morocco (Smith, 1965; Dubois & Duhautois, 1977; Thevenot et al., 1982) and also from Mauretania (Mayaud, 1983). Nearly all observations fall between November and March, which is in agreement with the recovery pattern from Iberia (97% between November and March, 83% between December and February) according to Bernis (1967).

The Common Gull has been noted above all in the Strait region of Alboran. Irby (1895) considered it to be "during some winters common". (In the past century Loche and Taczanowski described it as common in Algeria, which, according to Mayaud, 1983, could be explained by the greater incidence of colder winters in that century). In the present the species is considered to be very irregular and scarce at Gibraltar, with observations between November and March (Cortés et al., 1980). According to Garrido et al. (1985), the species is seldom observed in Malaga province, giving the sighting of a sub-adult between 3 and 7. 7.1978 at the mouth of the Rio Guadalhorce. For the same site Garrido & Alba (1985a) give notice of an adult on 4. and 11.8. 1984. For the Moroccan shores of Alboran there is only one published observation, a juvenile at the mouth of the River Moulouya on 5.1.1981 (Thevenot et al., 1982).

Although Common Gulls are typically wintering birds in the area, some isolated individuals may be seen in summer, such as those noted for the Rio Guadalhorce, or the 2 seen at the mouth of the River Sous on 29.8.1975 (Dubois & Duhautois, 1977). The Camargue, apparently the only Mediterranean area where the species is regular, there are observations throughout the year, although most commonly in the period December-March (Blondel & Isenmann, 1981).

I - Within the Alboran region there is no observation by the authors on the Moroccan side. From the Spanish side, 10 birds on seven dates from Torremolinos, all spring records between 16.2. and 22.5. (1980-83), of which only one refers to an adult bird (A.M.P.). There is also a record of 3 from Cabo de Gata (Almeria) on 15.3.1981 (H. Castro, pers.comm.).

• <u>Larus fuscus</u> Lesser Black-backed Gull Gaviota Sombria

The most commonly seen gull on the Spanish side after L. ridibundus during winter and migration periods, which occupy the greater part of the year. On the Moroccan side of Alboran the species appears to be comparatively scarce, except in the area of the Strait.

Only a small part of the populations of this species which migrate along the Atlantic coasts of the Palearctic (subspecies graellsii and intermedius) winter within the western Mediterranean region. Recoveries of British ringed birds show that on occasions the species reaches Tunisia, Sicily, Sardinia and peninsular Italy (Harris, 1962; Mayaud, 1983; Cramp & Simmons, 1983). Within the Spanish Mediterranean, the species is common within the Alboran region, becoming rarer further north to Barcelona (Isenmann, 1976; Carrera et al., 1981). The species is common in North Africa as far as Algeria (1,755 adults censused in winter 1977-78; Jacob, 1979) but appears to be rare in the north of Tunisia (Thomsen & Jacobsen, 1979).

The January 1984 Gull Census showed a total of 5,844 birds on the Spanish side of Alboran (Bermejo et al., in Press), although taking into account the deficiencies of the census (see: L. ridibundus), the estimate of 8,000-10,000 birds does not seem out of place. On the Moroccan side 952 adults were counted and a total of 1,500 was estimated (Beaubrun & Thevenot, 1984). The species is more common on both sides of the Strait and in near proximity, although nowhere near so numerous as in the Atlantic (27,450 between the Algarve, Portugal, and Cadiz, with 20,000+ on the coasts of Atlantic Morocco). On the northern side of Alboran more than 90% of the net total was censused between the provinces of Cadiz and Malaga, while on the southern side 96% were recorded between Tangiers and El Jebha, with no more than 27 birds between Cabo Tres Forcas and the Algerian border.

In the Strait the post-breeding migration is seen from July through to October, with maximum movements in August and September (Cortés et al., 1980), although other authors note much movement in the Atlantic in October (Smith, 1965; Telleria, 1981) and the greatest numbers do not appear to be reached until December and January in the wintering areas of Cadiz (Telleria, 1981) or Algeria (Jacob, 1983). Harris (1962) showed how, in agreement with a region-time analysis of ringing recoveries, the southerly movements continue into November. Also, in the middle of winter, cold weather associated movements involving thousands of birds may occur, such as those recorded at Casablanca on 28.-29.12.1962 (Smith, 1965); at Gibraltar at the beginning of January 1969 (Garcia, 1973) and at Tangiers on 27.12.1972 (Pineau & Giraud-Audine, 1976).

The spring passage northwards is noted in the Strait especially in March and April (Garcia, 1973), although there are movements from February onwards (Pineau & Giraud-Audine, 1976; Cortés et al., 1980; Thevenot et al., 1979), lasting until May (Heim de Balsac & Mayaud, 1962; Jacob, 1983). June is the month in which there is the least probability of encoutering the species in the region and is also the only month in which there are no ringing recoveries for Iberia (Bernis, 1967). Occasional summering birds, usually immatures, are noted by Heim de Balsac & Mayaud (1962), Smith (1965), Jacob (1983) and Garrido et al. (1985).

An adult was seen in the colony of Yellow-legged Gulls at Gibraltar on 4.6.1977 (Cortés et al., 1980) and in the Chafarinas an adult was also seen in the colony of Yellow-legged Gulls between 4. and 30.6.1980 (Witt et al., 1984).

I - In hard winters the numbers of the Lesser Black-backed Gulls on the Spanish side of Alboran may increase very considerably, as with L. ridibundus. Such happened in January 1985, when it was estimated that the population was between 20,000 and 25,000 L. fuscus - some four or five times the normal for Malaga province (J. Crespo & A.M.P.).

● Larus argentatus Herring Gull Gaviota Argentea

Occasional (?)

With regard to the Herring Gulls, the authors accord specific status to the southern, yellow-legged birds under the name of L. cachinnans, which is therefore treated separately. Pink legged argentatus birds have been seen on rare occasions in these latitudes, although there are ringing recoveries of British, Dutch and North-West French birds from Iberia (Bernis, 1967) and Morocco (Mayaud, 1983; Thevenot et al., 1980, 1982), as there are some observations from Italy (Moltoni & Brichetti, 1978). Harris (1964), in an analysis of the recoveries of British ringed argentatus, casts doubt on these data by supposing confusion between chicks of argentatus and fuscus in mixed colonies on the part of the ringers, L. fuscus being clearly migratory. Mayaud (1983) arrived at a similar conclusion. Bernis (1967) showed 14 recoveries in Iberia, of which 13 were from the Atlantic coast and 1 from the Mediterranean (Valencia) and Telleria (1981) gave 7 recoveries from the Atlantic side of the Strait.

Within the Strait, Cortés et al. (1980) consider the species as occasional in the winter, as it is also noted by Garcia (1973) and Pineau & Giraud-Audine (1976). There are no data for the Sea of Alboran, although for the Mar Menor (Murcia) there is a record for 8.4.1972 (Navarro Medina, 1973) and further north Isenmann (1976) saw one in Castellon on 25.11.1975. On the Atlantic coast of Morocco Thevenot et al. (1982) record an adult at Massa on 12.12.1981.

I - There are no observations by the authors within Alboran.

• <u>Larus cachinnans</u> Yellow-legged Gull Gaviota Patiamarilla

A common resident on the Spanish side of Alboran; very common and widely distributed on the Moroccan side.

The nesting population within Alboran is estimated to be 5,500 pairs minimum, with perhaps 85% on the Moroccan side of the region. This figure compares with recent estimates of 15,000+ pairs for the Spanish Mediterranean, including the Balearic Islands (de Juana, 1984); over 2,500 in Algeria (Jacob & Courbet, 1980); 600-700 in Tunisia (Czajkowski in Mayaud, 1983); 5,000-8,000 in Sardinia (Schenk & Torre, 1986); 2,250 in Corsica and 22,200 ($\pm$ 1,000) in Mediterranean France (Guyot, Launay & Vidal, 1985).

Along the northern coastline of Alboran the known colonies are at Gibraltar (ca. 600 pairs; Cortés et al., 1980), in the area of Punta de la Mona, Almunecar, Granada (J. Garzon, in de Juana, 1984) and between Calahonda and Castell de Ferro, Granada (ca. 20 pairs, according to Gonzalez Cachinero, 1985). There are further colonies nearby in the provinces of Cadiz, Almeria and Murcia.

On the African coast of Alboran the greatest concentration by far is on the Chafarinas, with a nesting population of the order of 3,500 pairs in 1980 (de Juana et al., 1984), but there are small to medium sized colonies widely distributed along the coast as indicated by Valverde (1956). A recent census organized by Beaubrun and Soto (in Thevenot et al., 1982) showed the following: 250 adults on the island of Leila or Peregil in the Strait; ca. 10 pairs on the cliffs of Benzu, below Djebel Musa; 138-148 pairs at Cabo Negro; 15 pairs dispersed between Tetuan and El Jebha; 20-25 pairs at Punta Pescadores, El Jebha; 800 adults at Cala Iris (some 500 pairs in the islet there, Berthon & Berthon, 1984); 30 pairs on the islet of Badis; 10-15 pairs at Al Hoceima and at least 3 pairs on the cliffs of Ras Quebdana (Cabo de Agua). In total ca. 1,200 pairs outside the Chafarinas (Beaubrun & Thevenot, 1984). Finally, on the island of Alboran itself on 23.6.1982, de Juana & Varela counted 280 adults and 30 empty nests (1984).

Concerning the winter distribution, the results of the Gull Census of January 1984 gave the following: 2,501 birds on the Spanish side (Bermejo et al., in press.) and nearly 3,500 on the Moroccan side (Beaubrun & Thevenot, 1984). The greatest concentrations were in the vicinity of the breeding colonies. At the saltpans of Cabo de Gata 450 were counted. 2,005 in the area Algeciras - Gibraltar (1,738 were consused at Gibraltar on 14.1.1979), 663 adults in Ceuta, 423 adults between El Jebha and Al Hoceima and 1,108 between Cabo Tres Forcas and the Algerian frontier. This last figure appears small in comparison with the numbers that breed on the Chafarinas, but it is worth noting that in the extreme west of Algeria, Jacob (1979) censused 5,900 adults in the winter 1977-78, in a relatively small area where Jacob & Courbet (1980) encountered only some 1,000-1,200 pairs breeding. This may indicate a small movement to this region

There have been two great "wrecks" of this sort since 1950. The first was studied by Jouanin (1957) with respect to the situation in France. This produced also many records for Catalonia (Palaus, 1957). The second "wreck" took place in January-February 1984, and was responsible for a large number of deaths (at least 30,000 birds dead or dying on French beaches, Brit. Birds 77: 237) and unusual records from France and Iberia (Ardeola, 31: 144-145 and Ardeola 32: 418; Bermejo et al., in press; Brit. Birds 77: 589 and 78: 642), also from Morocco (Brit. Birds 78: 642) and Italy (Fasola, 1985).

In the Strait, Favier considered the Kittiwake to be common in winter, although Irby thought it to be so in some winters, depending upon the weather conditions pertaining (Irby, 1895). Garcia (1973) and Cortés et al. (1980) talk of influxes after westerly gales. Garrido et al. (1985) note the presence of the Kittiwake after westerly gales as scarce, with the majority of the birds being sick, oiled or dead on the beaches. During three winters Alba & Garrido (1983) counted 5 birds off the mouth of the Rio Guadalhorce, Malaga, all in December 1978 and January 1979 (2 dead, 1 oiled, 1 sick and 1 apparently healthy). The Beached Birds Survey, carried out in February along Spanish Mediterranean coasts found 0 oiled/dead in 1982; 4 in 1983 and 24 in 1984 (de Juana & Varela in prep.).

From Mediterranean Morocco, Brosset (1959) found tens of corpses around the mouth of the River Moulouya in December 1956 and January 1957, dates which coincide with an influx into the Mediterranean, and Beaubrun & Thevenot (1984) record various observations from Ceuta, Cala Iris, Mar Chica and the Chafarinas in February and March 1984, date of the last influx.

Nevertheless, the presence of the Kittiwake in the western Mediterranean could well be regular, and not accidental or scarce, as recent work from ships appears to indicate (see: Llorente & Montori, 1982, for Catalonia; Witt et al., 1984, for Moroccan waters). The species has been recorded practically every winter for the last ten years within the Alboran region and the western Mediterranean in general: 1974-75 (Sargatal & Llinas, 1978), 1975-76 (Isenmann, 1976); 1976-77 (Thomsen & Jacobsen, 1979); 1977-78 (Sargatal & Llinas, 1978; Jacob, 1979, 1983; Isenmann & Czajkowski, 1978); 1978-79 (Carrera et al., 1981; Alba & Garrido, 1983; Jacob, 1983); 1979-80 (Llorente & Montori, 1982); 1980-81 (Llorente & Montori, 1982); 1982-83 (Bermejo, de Juana & Varela, in prep.; Witt et al., 1984) and 1983-84 (Abello & Castellon, 1985; Fasola, 1985; Bermejo et al., in press).

Regarding the length of stay by the Kittiwake in the Mediterranean, Favier (in Irby, 1895), Garcia (1973) and Cortés et al. (1980) are in agreement that birds are in the area between November and March. Records from Algeria also fall within these months (Ledant et al., 1981). For Malaga, Garrido et al. (1985) note adults between November and March, as with preceding authors, but state that immatures occur in all seasons. There are a few observations from October, April and May cited in Brosset (1959), Sargatal & Llinas (1978), Pineau & Giraud-Audine (1976) and Blondel & Isenmann (1981) for the Camargue. Brosset (1956) notes a juvenile at the mouth of the River Moulouya on 18.8. which is remarkably early.

I - There are five autumn observations from Torremolinos, all between 22.-26.9.1980, and four spring records, 1 on 3.3. 1982; 1 on 6.3.1983 and 2 on 10.3.1983.

The record of 64 on 1.3.1983 in Witt et al. (1984) was a result of a sea journey between Penon de Alhucema and Cabo Tres Forcas of 4 hours duration. The record was broken down as follows: 8 single birds, three groups of 2; two groups of 3, one of 4 and another of ca. 40, this last feeding in association with 39 Gannets S. bassana and many Dolphin spp Delphinus or Stenella (E.J. & J.M. Varela). There have been none recorded on the spring journeys to the Chafarinas, where J.M. Fernandez-Palacios (pers.comm.) saw only 4-5 in November and December 1983, but "many" during the influx of February 1984, including some within the harbour taking fishermen's bait.

• Sterna nilotica Gull-billed Tern Pagaza Piconegra

Regular and quite common during migration periods, although there are few observations of the species along the coasts.

The Gull-billed Tern does not appear to breed in northern Morocco (Beaubrun & Thevenot, 1984), although there is a large breeding population in southern Iberia. There is an important breeding colony at Fuente de Piedra, Malaga, at only 55 km inland from Malaga Bay. This colony shows large fluctuations, having reached a total of ca. 3,500 birds according to Vargas et al. (1978), while Garrido et al. (1985) record it reaching 1,000 pairs. Within the Andalucia region there are important colonies in the Marismas del Guadalquivir (Bernis, 1967). It is therefore not unusual that the species is recorded during migration, although normally along river or on lakes and ponds, such as the mouth of the Rio Guadalhorce, Malaga, or the River Moulouya; the lake at Smir, Mar Chica, etc. being seen rarely over the sea. Irby (1895) noted that he had never seen it over the sea and the first records for Gibraltar are from 1980 (Finlayson, 1983).

Migrant Gull-billed Terns are seen in North Africa between the end of July and November and between February and June (Mayaud, 1983). Records from the Alboran area are concentrated in the periods August-September and April-May. Garrido et al. (1985) state that birds arrive at Fuente de Piedra between the end of March and May, with laying commencing in mid-April, while the autumn passage takes place in the period July-September. These are dates similar to those given for the Camargue (Blondel & Isenmann, 1981) and for the Ebro Delta (Muntaner et al., 1984). Brosse & Jacquemard-Brosse (1962) cite this species on early October from San Pedro de Alcantua (Malaga). Cabo & Camacho (1981) record birds in the Mar Chica between March and November and from the same site there is a record of 2 on 5.2.1981 (Thevenot et al., 1982).

There have been rare records of wintering birds along the Atlantic coast of Morocco (Smith, 1965; de Juana, 1974; Beaubrun & Thevenot, 1984), but within the Alboran region there is only the published record of one bird at Smir on 5.12.1973 (de Juana, 1974). There are apparently no published winter records for the Iberian peninsula.

I - There are only 25 observations from the coast at Torremolinos, which in spring fall mainly in May (extreme dates: 6.3.-19.6.) and in the autumn in August and September. There are three records of single birds which fall in the winter months: 2.12.1980, 17.1.1983 (first winter bird); 31.12.1983 (adult) (A.M.P.).

• Sterna caspia Caspian Tern Pagaza Piquirroja

Regular in small numbers during migration periods, with a few wintering birds. It appears to be more frequent in the area of the Strait.

The nearest important wintering areas are in Tunisia and the south of Mauretania and Senegal (Mayaud, 1956; Staav, 1977; Cramp, 1985), but small wintering areas occur in the extreme South-West of Iberia (Alés et al., 1984) and the western coast of Morocco (Blondel, 1964; Smith, 1965, Isenmann, 1978, etc.), areas from which there have been some winter recoveries of birds ringed in the Baltic (Bernis, 1967; Staav, 1977; Mayaud, 1983). In the Gull Census of January 1984, 44 birds were counted between the Algarve (Portugal) and Huelva (Bermejo et al., in press) and 11 on the Atlantic coasts of Morocco (Beaubrun & Thevenot, 1984). A further 2 were found on the Mediterranean coast of Morocco, one at Bou Ahmed (to the West of El Jebha) and one in the Mar Chica. The only other wintering records for the Alboran area are: 3 in the Mar Chica on 4.2.1981 (Thevenot et al., 1982), and 1 at the saltpans of Cerrillos (Almeria) between 4. and 7.1.1984 (Canadas et al., 1985). The species is noted as being regular in winter at Tangiers (Garcia, 1973; Pineau & Giraud-Audine, 1976).

The Caspian Tern appears to be much more frequent during the migration periods in the Atlantic. Bernis (1977) gives details of 7 ringing recoveries between Cadiz and the Algarve while there are none between Malaga and Almeria. There are autumn records for Algeciras (3.8.1976 and 6.10.1977, Telleria, 1981; and 19.9.1981, Paterson, 1985a), Gibraltar (9.9.1976, Cortés et al., 1980; 4.9.1980 and 1.10.1981, Finlayson, 1983), Torremolinos (25.10.1980 and 21.8.1981, Paterson, loc.cit.), the mouth of the Rio Guadalhorce (2.9.1979, Garrido & Alba, 1981 and 3.10.1982, Alba, 1984) and the coast at San Pedro de Alcantua, in Malaga (in 6 days from 3. to 14.10., Brosse & Jacquemard-Brosse, 1962). At the mouth of the Smir, a bird in first winter plumage on 13.7.1954 (Valverde, 1954) and at Melilla, one on 8.11.1974 (de Juana, 1977). These last two observations mark the extreme dates for passage. For the Camargue, Isenmann (1973) records its occurrence between August and November, while for central Algeria, Jacob (1983) notes records between 29.9. and 4.11.). For the Ebro Delta there are observations in every month from March to November, but more often in April and September (Martinez & Muntaner, 1979).

The species is even more scarce in the Alboran region during the spring passage, which appears to be the rule throughout the western Mediterranean. The only published data are: for the mouth of the Rio Guadalhorce (1 on 13.4.1985, Alba & Garrido, 1985); Torremolinos (1 on 18.5.1981, Paterson, loc.cit.) and the mouth of the Moulouya (1 on 28.5.1981, Thevenot et al., 1982). A Torremolinos record of one on 12.6.1981 (Paterson, loc. cit.) could better refer to a summering bird rather than a passage migrant.

• Sterna maxima Royal Tern Charran Real

Occasional.

The post-breeding dispersal of the important population on the Banc d'Arguin (Mauretania) regularly reaches northern Morocco (Heim de Balsac & Mayaud, 1962; Thevenot et al., 1982; Mayaud, 1983) and eventually the Iberian peninsula (Bernis, 1967). The species is regularly recorded as far north as Tangiers (Irby, 1895; Pineau & Giraud-Audine, 1974), but to date there have been few recorded from southern Spain and Mediterranean Morocco.

There are records from Sanlucar de Barrameda (Cadiz) of 2 on 25.4.1970 and 27.4.1971 (Allen et al., 1971 & 1972) and single birds have been seen from Isla Cristina (Huelva) on 4.7.1977 and 24.9.1977 (Cabot, 1978).

Within the Alboran region one at Estepona (Malaga) on 16. 10.1979 (Martin, 1981); at the mouth of the Rio Guadalhorce on 20.9.1982 (Alba, 1982); at Algeciras on 19.9.1981 and from Torremolinos on 14.7.1983 (Paterson, 1984b).

On the Moroccan side of the Sea of Alboran there is a record from Ceuta in Mayaud (1983) and also an observation from Oued Lao of one on 8.2.1981 (Thevenot et al., 1982).

The majority of these records fall between July and October, which coincides with the records from Atlantic Morocco. (The breeding season on the Banc d'Arguin finishes in June-July, Mayaud, 1983).

• Sterna bengalensis Lesser Crested Tern Charran Bengales

Regular and quite common on the Moroccan side of Alboran and in the area of the Strait during migration, but rare or occasional on the Spanish side of the region. Possibly a rare wintering species.

The only important breeding area in the Mediterranean is on the island of Garah in the Gulf of Sirte, Libya, where Moltoni censused some 2,000 birds. There is no recent information on this site (Heim de Balsac & Mayaud, 1962; Mayaud, 1983). In the past century it appears that two pairs nested on an island off Tripoli (Lebanon) (Mayaud, loc.cit.). Recently a pair bred in the Ebro Delta in 1979 (Chokomian, 1981) and perhaps in subsequent years (Muntaner et al., 1984). Ferrer (1986) cites 2 pairs breeding in the region in 1985.

Post-breeding migration of the Mediterranean population is westwards to the Strait and south to the wintering grounds off West Africa in the tropic zone (Cramp, 1985). The birds move through the Alboran region, with an abundance of sightings referring to the autumn passage on the Moroccon side of the Strait. Favier (in Irby, 1895) found the species rare at Tangiers, but frequent at Larache between September and November and Pineau & Giraud-Audine (1974) found the Lesser Crested Tern regular in small numbers (1-6 individuals) between 17.8. and 27.10.1973 (with a flock of 10 on 3.10. and 45 on 4.10.1973 at Smir). At the mouth of the River Moulouya Brosset (1956) considered the species regular between 22.8. and early October, with a flock of 30 on 23.10. September and October appear to be the months of heaviest migration in Algeria (Schmidt, 1963; Jacob, 1983) and on the Atlantic coast of Morocco (Smith, 1965). The few birds which cross to the Spanish side of Alboran and the Strait also generally occur in these months. There is a record for the mouth of the Rio Guadalhorce (one adult, 11-13.10.1979, Alba, 1981a) with the rest of the records occurring in the vicinity of the Strait at Algeciras (2 on 20.10.1972, Allen, 1973; 1 on 7.10. 1977, Telleria, 1981); Tarifa (1 at the end of September 1976, and 1 on 28.8.1977, Telleria, loc.cit., also 1 on 22.8.1980, Callebaut & Sneyers, 1981); Barbate (1 on 26.8.1977, Telleria, loc.cit.), and recently, also at Gibraltar (2 birds each on 26.9.1981, 10.10.1981 and 2.9.1982, Finlayson, 1983).

The spring migration is noted in North Africa in May and June especially (Jacob, 1983; Thomsen & Jacobsen, 1979; Bundy & Morgan, 1969). There are few records from Alboran for the spring period although birds occur from March onwards (ca. 30 on 23.3. at the mouth of the River Moulouya, Brosset, 1956; and 5 on 24.3.1981 within the Mar Chica, Thevenot et al., 1982), while at the other extreme of the spring migration 3 at Tangiers on 22.7.1974 (Pineau & Giraud-Audine, 1975). On the Spanish side of Alboran and the Strait, Irby (1895) relates the capture of 2 males at Tarifa on 20.4.1873 and also at Tarifa one bird was observed on 15.5.1969 (Salmon, 1971). Out in the Atlantic there is a record of one at Sanlucar de Barrameda on 8.4.1964 (Peltzer, 1966).

The record of a bird at the Mar Chica on 6.11.1979 (Thevenot et al., 1980) corresponds to a migrant, as Jacob (1983) gives the extreme autumn date as 4.11. in Algeria. The record of 2 at Melilla on 2.2.1983 (Witt et al., 1984) and the single bird in the Mar Chica on 18.2.1981 (Thevenot et al., 1982) may more probably refer to wintering birds rather than early spring migrants.

I - One record from Torremolinos of an adult on 31.5.1982 (A.M.P.).

• Sterna sandvicensis   Sandwich Tern   Charran Patinegro

Reasonably common in winter; very common during migrations, with many summering birds.

In the Alboran region there is a convergence of Sandwich Terns from three principal sources (Müller, 1959; Bernis, 1967; Cramp, 1985). There are birds of Ukrainian origin, with a nesting population of 20,000+ pairs according to Isenmann (1980) and ca. 40,000 pairs according to Golovkin (in Cramp, 1985), which migrate westwards along European and North African coasts, with the major wintering area being the central-western Mediterranean. A small percentage (6% of ringing recoveries) moves out into the Atlantic and along Moroccan and Iberian coasts, with ringing recoveries as far as the Ivory Coast on rare occasions (Cramp, 1985). On the other hand, the birds from the Camargue and the Ebro Delta (1,500+ pairs according to Isenmann, 1986 and Ferrer, 1986) appear to move out to the Atlantic (Blondel & Isenmann, 1981). Lastly there is the big European Atlantic population from Britain, Holland, Denmark, etc., which largely winters off West Africa to the south of Mauretania, although some remain as far north as the Gironde, France; Iberia and North-West Africa and which, in some numbers, penetrate the western Mediterranean, reaching as far east as Algeria and Tunisia. There are ringing recoveries of birds from the Atlantic Europe region in these areas, as well as in Mediterranean Spain.

The Sandwich Terns which winter in small numbers within the Alboran region may be of both Russian and European origin. Bernis (1967), of five ringing recoveries from Malaga and Almeria, noted that 3 were of Ukrainian origin, 1 Scottish and 1 Danish.

The numbers of Sandwich Terns that winter are small, perhaps no more than 200 birds, and very scattered, as in other areas of the Mediterranean (Isenmann, 1976, 1978, 1980). The Gull Census of January 1984 showed ca. 70 along the Spanish shores of Alboran and the Strait (census incomplete) (Bermejo et al., in press) and 84 along the Moroccan coast of the region (Beaubrun & Thevenot, 1984). Vilagrasa et al. (1982) calculated a population of at least 3,000 Sandwich Terns within the western Mediterranean, taking into account the census of 1,700 from Algeria (Jacob, 1979).

Areas of concentration within Alboran may include the saltpans of Almeria province (pers.obs.); the region around the mouth of the Rio Guadalhorce, Malaga (maximum wintering number 9 on 1.1.1978, Alba & Garrido, 1983); Algeciras Bay (Cortés et al., 1980); the southern side of the Mar Chica where 4 were seen on 3.12.1980 and 51 on 4.2.1981 (Thevenot et al., 1982) and 20+ on 28.12.1975 (de Juana, 1977) and the mouth of the River Moulouya, 9 on 14.1.1962 on 3.2. and 20 on 5.2.1981 (Thevenot et al., 1982).

birds. Pineau & Giraud-Audine (1974, 1975) do not even mention the species for Tangiers in the autumn and note few observations for the spring passage. For Gibraltar, Garcia (1973) talks of a small passage and Cortés et al. (1980) of moderate numbers.

On the Moroccan shores of Alboran records are either scarce or absent from most authors. Brosset (1956, 1959), Estrada (1985) and Thevenot et al. (1980, 1982) do not record its presence at all, while Witt et al. (1984) cite only single birds or small groups in April and May seen in the harbour at Melilla, the Mar Chica and at Cabo de Agua and the mouth of the River Moulouya. Cabo & Camacho (1981) record the species in the Mar Chica between July and September, with a maximum of 100 birds, and is scarcer still in Algeria, where Ledant et al. (1981) recount three observations in recent years, while Jacob (1983) does not hesitate to classify it as accidental.

With regard to the timing of the migrations, the autumn passage is cited as being between July and October at Gibraltar, with greatest intensity in September (Cortés et al., 1980), and at Malaga from the middle of July to the end of October (Garrido et al., 1985). Breeding birds from the Camargue and Ebro Delta start moving out in July, with the main exodus in August with a few remaining into September (Blondel & Isenmann, 1981; Muntaner et al., 1984).

For the Atlantic coast of Morocco, Thevenot et al. (1982) give much later dates, which the authors believe more probably correspond to northern European birds, from the end of September to mid-November (Cramp, 1985).

The spring passage at Malaga takes place between late March and May (Garrido et al., 1985). At Gibraltar, Garcia (1973) gave April as the month, while Cortés et al. (1980) expanded the period until June. These are dates which agree well with spring arrivals at the Ebro Delta and the Camargue (Blondel & Isenmann, 1981; Muntaner et al., 1984) and with the observations of Witt et al. (1984) for Mediterranean Morocco. For Atlantic Morocco, Thevenot et al. (1982) give correspondingly earlier dates, from mid-March to early April.

Wintering birds are not uncommon in Iberia (Erard & Vielliard, 1975; Isidoro, 1970; Melendro & Rodriguez Valverde, 1975; Ferrer, 1977; Ena & Purroy, 1982) and neither is there a lack of ringing recoveries for this season (Muselet, 1982). During the Gull Census of January 1984, 58 were counted along Atlantic coasts and 7 within the Mediterranean,but none from Alboran. The only published winter observation appears to be of 1 at Gibraltar on 25.1.1979 (Cortés et al., 1980).

On the Moroccan side of the Strait, Smith (1965) recorded a flock of 80 birds, probably S. hirundo, on the beach at Cape Spartel on 11.1.1963 and another one was seen at Ceuta on 13.12.1977 (Suarez et al., 1981).

The only nesting record known occurred at the Mar Chica in 1978, when a chick was found in June, according to Cabo & Camamacho (1981). There does not appear to have been any attempt at breeding since (Beaubrun & Thevenot, 1984). The nearest known colonies to the region are in the Marismas del Guadalquivir (ca. 100 pairs; Barcena et al., 1984) while further away there are colonies in Tunisia,Sardinia, Camargue, and within the Spanish Mediterranean at the Ebro Delta and Albufera de Valencia (Thomsen & Jacobsen, 1979; Schenk & Torre, 1986; Blondel & Isenmann, 1981; Muntaner et al., 1984; Estacion Ornitologica de la Albufera, pers.comm.).

I - At Torremolinos the first birds are normally seen in March, although there is a record of one on 7.2.1982. Passage is concentrated between mid-March and mid-April, finishing quite sharply. There are occasional observations for May, June and July. The autumn passage is equally concentrated, between August and mid-September. There are only two records for October, 9.10.1981 and 23.10.1982 (A.M.P.).

• Sterna paradisaea Arctic Tern Charran Artico

Occasional.

There are extremely few references in literature to the Arctic Tern in the region, perhaps because in some cases it goes unobserved amongst the far commoner Common Tern. Irby (1895) mentions the capture of a winter plumage bird in the Strait, and for the same area, Garcia (1973) notes that only three have been identified with certainty, in April, August and September, while for Gibraltar itself, there is only one record: 1 on 13.10.1979 (Cortés et al., 1980). For Malaga, Garrido et al. (1985) claim to have seen the species during autumn passage at the end of August and beginning of September. Heim de Balsac & Mayaud (1962) quote the existence of a juvenile in the collection of the British Museum at Tring, England, ticketted as coming from Tangiers on 21.8.1888.

The scarcity of published data extends to neighbouring Atlantic regions, from where there is a single ringing recovery, noted in Bernis (1967), from the Algarve; 2 birds from Cadiz province, 1 at Bonanza on 14.4.1959 (Feeny, 1960) and 1 at Cadiz on 27.12.1977 (Solis, 1978). There is a single observation from North-West Morocco, between Tangiers and Arcila, of 2 birds on 10.4.1974 (Pineau & Giraud-Audine, 1975).

In spite of the paucity of records the Arctic Tern must pass the western approaches to the Strait in some numbers, as the migration route utilized by the European populations (estimated at 106,000 pairs for Britain and Scandinavia alone) and by a part of the North American and Siberian populations (Cramp, 1985; Bernis, 1967; Heim de Balsac & Mayaud, 1962; Mayaud, 1984) passes the Atlantic Iberian and Moroccan coasts. Without doubt, during normal weather conditions the majority of passage takes place out in the open sea, out of the sight of land (see: observations by Bourne & Norris, 1966, referring to the voyage Britain-Gibraltar), much more so than in the case of the Common Tern. Thus records from Alboran are scarce, with a little more chance in the vicinity of the Strait. The species is noted as an accidental within the Mediterranean for Algeria (no records this century according to Ledant et al., 1981), Italy (Moltoni & Brichetti, 1978), Cyprus and Turkey (Cramp, 1985).

• Sterna fuscata Sooty Tern Charran Sombrio

Accidental.

The Sooty Tern has a wide circum-tropical and sub-tropical distribution and has been recorded as an accidental in several European countries (Bernis, 1967; Cramp, 1985), in Tunisia (Thomsen & Jacobsen, 1979) and recently in Morocco (as "possibles", 7 at Oualidia on 17.4.1981 and 1 at Sidi Kaouki on 11.10.1981; Thevenot et al., 1982).

There are three records for the Iberian peninsula: 1 captured in Leon province around 1941 (Marquez, 1959); 1 observed and photographed at the Marsismas del Guadalquivir on 18.5.1963 (Meeth, 1965) and one within Alboran, an immature observed at the mouth of the Rio Guadalhorce on 27. and 28.8.1979 (Alba, 1981b).

Bernis (1967) cautions the possibility of confusion with the similar S. anaethetus, which breeds in some numbers on the Banc d'Arguin, and for which there are 9 records from the British Isles (Cramp, 1985).

• Sterna albifrons Little Tern Charrancito

Moderately common during migration periods, with some pairs nesting regularly within the area of the Mar Chica, Melilla and occasionally, in the saltpans of Almeria.

The Little Tern breeds locally in both North Africa and in Iberia in small numbers (Heim de Balsac & Mayaud, 1962; Bernis, 1967). Within the Alboran region breeding has been proved at the saltpans of Cerrillos, Almeria (J. Miralles, pers.comm.). Some 50 pairs nested around the Mar Chica in 1981 (Thevenot et al., 1982) and 8 pairs were seen carrying out courtship feeding on 3.5.1982 (Witt et al., 1984). The species could well also nest in other areas of the Moroccan coast, such as the marshes of Restinga-Smir (Beaubrun & Thevenot, 1984). In the past century a few pairs used to nest at the mouth of the Rio Guadiaro (Sotogrande, ENE of Gibraltar) and at other points along the Spanish coast of the Strait area. The nearest important breeding nucleus is between the saltpans of Cadiz and the Marismas del Guadalquivir, which has been estimated recently at 2,000 pairs (Barcena et al., 1984). A few pairs nest in other neighbouring areas, such as the mouth of the River Lucus and other points of Morocco to the south of Tangiers (Valverde, 1956; Pineau & Giraud-Audine, 1977), the coast of Algeria (Jacob & Courbet, 1980) and the saltpans of Murcia and Alicante, Spain (pers.obs.).

Through ringing recoveries the passage of Little Terns through Iberian waters may be appreciated clearly (Bernis, 1967). The important populations from Britain, the North Sea area and the Baltic move alont the Atlantic coast of Iberia. The area of Alboran is served above all by the populations from the western Mediterranean which move to the Strait and to tropical West Africa (Cramp, 1985). The populations of greates importance which move through the Alboran-Strait region are those of the Ebro (650 pairs in 1985, Ferrer, 1986), the Camargue (250-550 pairs in recent years, Blondel & Isenmann, 1981) and Sardinia (350-500 pairs, Schenk & Torre, 1986).

The species is noted as "very scarce but regular" by Cortés et al. (1980) for Gibraltar between July and October (Garcia, 1973). Finlayson (1983) mentions a first early July record for Gibraltar: 8 birds on 5.7.1982. Brosset (1956) found it common at the mouth of the River Moulouya between 8.8. and 20.10. and Garrido et al. (1985) mention frequent observations for Malaga between the end of July and late October. For central Algeria, Jacob (1983) cited 4 birds as late as 2.11. The peak passage period appears to be in September (Telleria, 1981, for Tarifa; Pineau & Giraud-Audine, 1977, for Tangiers and Smith, 1965, for the Atlantic coast of Morocco). September is also the month in which the majority of birds leave the Ebro colony and also the Camargue (Muntaner et al., 1984; Blondel & Isenmann, 1981). The recoveries quoted by Bernis (1967) for Iberia (see above) all refer to the autumn period, falling in August (5), September (12) and October (1).

The spring passage, which takes place over a relatively short period, extends from early April to the end of May (Garrido et al., 1985), or even into early June (Brosset, 1956, gives extreme spring dates for the mouth of the River Moulouya as 4.4. and 4.6.). Peak passage appears to be at the end of April (Pineau & Giraud-Audine, 1977) and the beginning of May (Glanville & Walker, 1963, for Punta Sabinal, Almeria; Jacob, 1983, for central Algeria). On the Ebro Delta and the Camargue breeding birds start to arrive during the second and third weeks of April (Muntaner et al., 1984; Blondel & Isenmann, 1981).

Although the normal wintering area does not appear to begin before the latitude of Senegal (Morel & Roux, 1966) there are records of birds in the middle of winter from both North Africa (Dubois & Duhautois, 1977; Isenmann, 1978; Thomsen & Jacobsen, 1979; Jacob, 1983) and Iberia (Ena & Purroy, 1982; Bermejo et al., in press), but there are no published records for the Alboran area.

I - The earliest observation from Torremolinos is for 11.4. The spring movement, in agreement with other authors, is concentrated between mid-April and mid-May. There are some observations from June and one of 2 birds on 14.7.1982. The autumn passage starts in August and peaks to the end of this month or in early September. Last observation is of 15.10. (A.M.P.).

For the area Chafarinas - Melilla (some data have been published in Witt et al., 1984) there is a small movement in late April and early May, with a maximum of 14 on 27.4.1982 (E.J. & J.M. Varela). Up to 14 birds were seen fishing along the shoreline of Melilla beach in the second week of July 1984, almost undoubtedly from the Mar Chica population (J.M. Varela & P.-C. Beaubrun, pers.comm.).

• Uria aalge Guillemot Arao Comun

Occasional or very scarce in winter.

The normal wintering area of the Guillemot is somewhat to the north of the Strait and Alboran (Cramp, 1985). Mead (1974), in his analysis of British ringing recoveries of the species (without doubt the origin of the great majority of the Guillemots in Iberia, Bernis, 1967) found that only two had occurred further south than the Bay of Biscay, one from Lisbon and the other at Gibraltar. Some reach the Strait area and penetrate the Mediterranean, where the species is cited as being of rare occurrence for Italy and Malta (Moltoni & Brichetti, 1978). It may be that some distant birds are confused with Razorbills (Telleria, 1981), but comparison with that species shows 60 Mediterranean recoveries against 1 Guillemot (Mead, 1974). Similarly, the Beached Birds Survey of 1981, 1982 and 1983 showed a total of 50 Alca torda, 3 Fratercula arctica and no Uria aalge (de Juana & Varela, in press). There do not appear to be any records for Mediterranean North Africa and only two for the Atlantic coasts of Morocco (Brit. Birds 78: 642; Mayaud, 1984).

In the Strait area, where Irby (1895) cited the species as occasional in the winter during periods of westerly gales, neither Garcia (1973) nor Telleria (1981) note any personal records, although Garcia gives the record of Oliver of 7 flying west to the Atlantic on 17.4.1967. Cortés et al. (1980) note its occurrence at times during migration for the Strait, but for Gibraltar there were only two inshore records: single birds on 24.3. 1967 and 9.9.1974. Finlayson (1983) adds the following ones: 1 bird on 1.3.1980, 3 on 7.3.1980, 1 on 10.3.1980 and 12 on 1.4.1980.

For Alboran there is only one published observation, an oiled corpse in Malaga harbour on 17.7.1982 (Garrido et al., 1985).

I - A bird at Torremolinos on 17.-18.1.1981 (A.M.P.).

• Alca torda Razorbill Alca Comun

Common during migration periods, scarce to quite common in winter with variable abundance.

Almost all the Razorbills which reach Iberia and North Africa and which penetrate the Mediterranean must belong to the important British population (estimated at 144,000 pairs by Cramp et al., 1974). Amongst recoveries of ringed birds in Iberia, 54 were of British origin and one, somewhat doubtful, from southern Sweden (Bernis, 1967). The three recoveries from North Africa are also British birds (Heim de Balsac & Mayaud, 1962). It is no coincidence, therefore, that in the analysis by Mead (1974) there is a strong agreement with the other data, although with possible differences in the rates of communications of recoveries between countries. Mead noted that of 59 Mediterranean recoveries of British birds, 31 were from Iberia, 21 from France and only 5 from North Africa, 1 from the Balearic Islands and 1 from Corsica. The recoveries are especially concentrated in a sea area of high oceanographic productivity in the Gulf of Lyons.

The Razorbill is considered to be regular and quite abundant during winter off the Camargue, with concentration of up to 1,000 birds (Blondel & Isenmann, 1981), while off Catalan coasts Abello & Castellon (1984) give the species as very common in winter. There is a record of a concentration of 1,500-2,000 birds off Gava, Barcelona (Lascurain, in Ardeola 20: 315).

Off North African coasts the situation is very different. Heim de Balsac & Mayaud (1962) give three observations for Algeria and Jacob (1983) did not record any during winter 1976-77 and only a small number the following winter. At Tunisia, it is a very rare and irregular species, with no record in recent years (Thomsen & Jacobsen, 1979), excepting for a recovery at Tunis in January 1982 (Mayaud, 1984). For the Balearic Islands it is considered to be regular, although scarce and localized (Muntaner et al., 1984) and a similar status is accorded to Italy (Moltoni & Birchetti, 1978). Accidental records exist for Yugoslavia and Egypt (Cramp, 1985).

In the Alboran region an equal asymmetry exists in the distribution of records between northern and southern coasts. Thus, there are nine ringing recoveries from the Spanish side and only one on the Moroccan (Mead, 1974). Off northern shores it is given as occurring in small numbers at Gibraltar (Garcia, 1973) and as "regular and numerous" by Cortés et al. (1980). For Malaga, Garrido et al. (1985) record it as being frequent and even numerous in some years.

From the Moroccan shores of the Strait area and Alboran there are only the following published observations: Favier (in Irby, 1895) noted the species for Tangiers, as do Pineau & Giraud-Audine (1976) who record a total of 10 between 5.12.1973 and 14.2.1974; for the mouth of the River Moulouya, Brosset (1956) found some corpses; for the Chafarinas, 1 on 24.4.1982 (Witt et al., 1984) and Estrada (1985) recorded the species on fifteen occasions in November and December 1984, from the Penon de Vélez with a maximum 9 birds.

The abundance of the Razorbill in Alboran varies from year to year (Irby, 1895; Garcia, 1973; Garrido et al., 1985), an aspect which is in agreement with the pattern of recoveries given by Mead (1974), who wrote that British birds "consistenly

Harris (1984) notes two recoveries for the Balearic Islands and Bernis et al. (1958) considered it an uncommon winter visitor to the area, while Muntaner & Congost (1979) consider that it is exceptional in Menorca. For Catalonia, after many transects by ship, Abello & Castellon (1985) saw only one bird, on 1.2.1983, while they found the Razorbill very common in the same areas.

The situation for Alboran is very similar, although there are some indications that the Puffin may be common in open sea. There are at least three ringing recoveries: one in January from the Granada coast (Bernis, 1967) and two, also in winter, from Morocco (Harris, 1984). Additionally, various bodies have been collected on the Moroccan beaches. Brosset (1956) collected several at the mouth of the River Moulouya; Pineau & Giraud-Audine (1976) 3 in the region of Tangiers; Alba & Garrido (1983) 8 at the mouth of the Rio Guadalhorce, Malaga; the Beached Birds Survey of February 1982, 3 along the coasts of Cadiz and Malaga (de Juana & Varela, in press). Valverde (1956) also noted a body on the beach near Tetuan. On the other hand, observations of living birds are almost non-existing in the winter months, with only Cortés et al. (1980) noting that the species is "also recorded in the winter months".

At Tarifa, Telleria (1981) noted movements after the beginning of October, becoming daily after the middle of October. According to Cortés et al (1980) the passage extends into December.

For the spring passage, Cortés et al (1980) give a long period between February and the beginning of June with greatest movement in April. Other authors cite birds still present in May (Favier, in Irby, 1895; Van Oordt, in Bernis, 1967; Stead, 1959). There is a record from the Camargue in June (Blondel & Isenmann, 1981) and from Castellon, 5 on 23.7. at Peniscola (Pons Oliveras, 1968). Moltoni & Brichetti (1978) give the species as occasional in summer in Italy.

I - From Torremolinos, one observation (5 birds) in November and 25 birds between 1.3. and 31.5., plus 2 on 11.7.1983 (A.M.P.). On voyages to the Chafarinas in the spring only one bird has been seen, on 30.3.1984, flying west at 40 miles South-East of Malaga (E.J. & J.M. Varela).

References

Abello, P. & Castellon, A. (1984). Some notes on seabirds in the Catalan Sea. Rapp. Comm. int. Mer Médit., XXIX Congrès - Assemblée Plénière, Lucerne.

Alba, E. (1981a). (Note on Sterna bengalensis). Ardeola, 28: 159.

Alba, E. (1981b). (Note on Sterna fuscata). Ardeola, 28: 159.

Alba, E. (1982). (Note on Sterna maxima). Ardeola, 29: 186.

Alba, E. (1984). (Note on Hydroprogne caspia). Ardeola, 30: 120.

Alba, E. & Garrido, M. (1982). (Note on Uria aalge). Ardeola, 29: 187.

Alba, E. & Garrido, M. (1983). Observaciones invernales en la desembocadura del Rio Guadalhorce, anos 1977 a 1981. Alytes, 1: 225-244. Adenex, Mérida.

Alba, E. & Garrido, M. (1985). (Note on Sterna caspia). Ardeola, 32: 418.

Alba, E. & Garrido, J.A. (1980). Aves capturadas y observadas en la desembocadura del Rio Guadalhorce. Jabega, 32: 65-70. Excma. Diputacion Provincial de Malaga.

Alés, E. et al. (1984). (Note on Hydroprogne caspia). Ardeola, 31: 145.

Allen, F.G.H. (1973). Royal and Lesser Crested Terns in Cadiz Province. Seabird Report, 3: 36.

Allen, F.G.H.; Hoogerheide, J. & Loterijman, J.A. (1971). Observacion de Sterna maxima en Cadiz. Ardeola, 15: 144.

Allen, F.G.H.; Hoogerheide, J. & Loterijman, J.A. (1972). Sterna maxima in Cadiz Province. Ardeola, 16: 262-263.

Araujo, J.; Munoz-Cobo, J. & Purroy, F.J. (1977). Las rapaces y aves marinas del archipielago de Cabrera. Naturalia Hispanica, 12 (94 pp.), ICONA, Madrid.

Baker, R.R. (1980). The significance of the Lesser Black-backed Gull to models of bird migration. Bird Study, 27: 41-50.

Barcena, F.; Teixeira, A.M. & Bermejo, A. (1984). Breeding seabird populations in the Atlantic sector of the Iberian Peninsula. ICBP Technical Publication No. 2: 335-345.

Barrett, R.A. (1983). (Note on Larus atricilla). Ardeola, 29: 186.

Beaubrun, P.-Ch. (1983). Le Goéland d'Audouin (Larus audouinii Payr.) sur les côtes du Maroc. L'Oiseau et R.F.O., 53: 209-226.

Beaubrun, P.-Ch. & Thevenot, M. (1984). Anatidas y Laridos de Marruecos del Mediterraneo. Communication to the I Congreso Hispano Africano de las Culturas del Mediterraneo, Melilla.

Bermejo, A.; Carrera, E.; De Juana, E. & Teixeira, A.M. (in press): Primer censo general de gaviotas y charranes (Aves: Laridae) invernantes en la Peninsula Iberica (enero de 1984). Ardeola 33.

Bermejo, A.; De Juana, E. & Varela, J.M. (1984a). (Note on Oceanodroma leucorhoa). Ardeola, 30: 115-116.

Bermejo, A.; De Juana, E. & Varela, J.M. (1984b). (Note on Alle alle). Ardeola, 30: 120.

Bernis, F. (1965). Observacion de Puffinus griseus en el Mediterraneo Espanol. Ardeola, 11: 143.

Bernis, F. (1966). Aves Migradoras Ibericas, fasciculo 1. Sociedad Espanola de Ornitologia, Madrid.

Bernis, F. (1967). Aves Migradoras Ibericas, fasciculo 5. Sociedad Espanola de Ornitologia, Madrid.

Bernis, F. y colaboradores (1958). Guion de la avifauna balear. Ardeola, 4: 25-97.

Berthon, D. & Berthon, S. (1984). Compte rendu de l'expédition Balbuzard sur les côtes méditerranéennes du Maroc. L'Oiseau et R.F.O., 54: 201-213.

Blondel, J. et C. (1964). Remarques sur l'hivernage des limicoles et autres oiseaux aquatiques au Maroc (Janvier 1964). Alauda, 32: 250-279.

Blondel, J. & Isenmann, P. (1981). Guide des oiseaux de Camargue. Delachaux et Niestlé. Neuchâtel.

Bond, T.E. (1981). (Note on Larus pipixcan). Ardeola, 28: 158.

Bosano, V.A. (1981). Gull counts at Gibraltar. Alectoris, 3: 33-38.

Bourne, W.R.P. (1957). Manx Shearwater, Little Gulls and other seabirds wintering off the Algerian coast. Ibis, 99: 117-118.

Bourne, W.R.P. (1978). The seabirds of the eastern North Atlantic. Ibis, 120: 117-119.

Bourne, W.R.P. & Norris, A.Y. (1966): Observaciones durante una traversia de ida y vuelta entre Gran Bretana y Gibraltar, septiembre 1964. Ardeola, 11: 57-63.

Brichetti, P. (1980). Distribuzione geografica degli uccelli nidificanti in Italia, Corsica e Isole Maltesi, I. Parte introduttiva, famiglie Podicipedidae, Procellariidae, Hydrobatidae. Natura Bresciana, Ann. Mus. Civ. Sc. Nat. Brescia, 16: 82-158.

Brichetti, P. (1982). Distribuzione geografica degli uccelli nidificanti in Italia, Corsica e Isole Maltesi, 2. Famiglie Phalacrocoracidae, Ciconiidae, Treskiornithidae. Natura Bresciana, Ann. Mus. Civ. Sc. Nat. Brescia, 19: 97-157.

Brosse, J. & Jacquemard-Brosse, S. (1962). Notes sur l'avifaune de Gibraltar au moment des migrations d'automne. L'Oiseau et R.F.O., 32: 228-239.

Brosset, A. (1956). Les oiseaux du Maroc oriental de la Méditerranée à Berguent. Alauda, 24: 161-205.

Brosset, A. (1958). Les oiseaux de l'embouchure de la Moulouya, Maroc oriental. Alauda, 26: 36-47.

Brosset, A. (1959). Les oiseaux de l'embouchure de la Moulouya (Maroc oriental), les migrateurs. Alauda, 27: 36-60.

Brosset, A. & Olier, A. (1966). Les îles Chaffarines, lieu de reproduction d'une importante colonie de Goélands d'Audouin, *Larus audouinii*. Alauda, 34: 187-190.

Bruhn, J.F.W. & Jeffrey, B. (1958). Notas sobre aves de Murcia y Cadiz, observaciones de fin de verano. Ardeola, 4: 109-117.

Brunstein, D. & Thibault, J.-C. (1980). Oiseaux de mer hivernant en Corse, 1979-80. Association des Amis du Parc Naturel Régional de Corse, Ajaccio.

Bundy, G. (1976). The Birds of Libya. British Ornithologists Union, London.

Bundy, G. & Morgan, J.H. (1969). Notes on Tripolitanian birds. Bull. Br. Orn. Club, 89: 139-144.

Burnier, E. (1979). Notes sur l'ornithologie algérienne. Alauda, 47: 93-102.

Cabo, J.M. & Camacho, I. (1981). Aves acuaticas de la Mar Chica de Melilla, julio 1977, junio 1978 y verano 1979. Mediterranea, 5: 23-33.

Cabot, J. & Aguilera, E. (1974). Observacion de Pagaza Piquirroja (*Hydroprogne tschegrava*) en las marismas de Isla Cristina (Huelva). Ardeola, 20: 361.

Callebaut, E. & Sneyers, A. (1981). (Note on *Sterna bengalensis*). Ardeola, 28: 159.

Cano, A. (1960). El Colimbo Artico en Almeria. Ardeola, 60: 373-374.

Canadas, S.; Castro, H.; Manrique, J. & Miralles, J.M. (1985). (Note on *Sterna caspia*). Ardeola, 32: 418.

Capella, L.; Mayol, J. & Pons, M. (1985). Reproduccion de la Gaviota Cabecinegra en las Islas Baleares. Ardeola, 32: 416.

Carrera, E. (1983). Hivernada de Larids (Aves: *Laridae*) al litoral del Principat i del Pais Valencia (1980 i 1981). Butll. Inst. Cat. Hist. Nat., 49 (Sec. Zool., 5): 143-150.

Carrera, E.; Ferrer, X.; Martinez, A. & Muntaner, J. (1981). Invernada de Laridos en el litoral meidterraneo catalan y levantino. Ardeola, 28: 35-50.

Chokomian, M. (1981). Note on Sterna bengalensis. Ardeola, 28: 159.

Cordero-Tapia, P.J. (1983). Las aves del Maresme. Catalogi, status y fenologia. Universitat de Barcelona.

Cortés, J.E.; Finlayson, J.C.; Mosquera, M.A. & Garcia, E.F.J. (1980). The Birds of Gibraltar, Gibraltar Bookshop, Gibraltar.

Coulsen, J.C. (1966). The movements of the Kittiwake. Bird Study, 13: 107-115.

Cramp, S. (1985). The Birds of the Western Palearctic, vol. IV, Terns to Woodpeckers. Oxford University Press, Oxford.

Cramp, S.; Bourne, W.R.P. & Saunders, D. (1974). The Seabirds of Britain and Ireland. Collins, London.

Cramp, S. & Simmons, K.E.L. (1977). The Birds of the Western Palearctic, vol. I, Ostrich to Ducks. Oxford University Press, Oxford.

Cramp, S. & Simmons, K.E.L. (1983). The Birds of the Western Palearctic, vol. III, Waders to Gulls. Oxford University Press, Oxford.

Cuyas Robinson, J. (1969). Aves marinas y aves migrantes en el mar balear. Ardeola, 13: 205-209.

De Juana, E. (1974). Datos invernales sobre aves de Marruecos (diciembre 1973). Ardeola, 20: 267-286.

De Juana, E. (1977). Nuevos datos de invierno sobre aves de Marruecos. Ardeola, 23: 49-62.

De Juana, E. (1984). The status and conservation of seabirds in the Spanish Mediterranean. ICBP Technical Publication No. 2: 347-361.

de Juana, E.; Bradley, P.M.; Varela, J. & Witt, H.-H. (in press) Sobre los movimientos migratorios de la Gaviota de Audouin (Larus audouinii), Ardeola, 33.

de Juana, E.; Bueno, J.M.; Carbonell, M.; Pérez-Mellado, V. & Varela, J.M. (1980). Aspectos de la alimentacion y biologia de reproduccion de Larus audouinii Payr. en su gran colonia de cria de las Islas Chafarinas (ano 1976). Bol. Estacion Central Ecol. 8: 53-65.

De Juana, E. y el Comité de Rarezas de la S.E.O. (1985). Observaciones homologadas de aves raras en Espana, informe de 1984. Ardeola, 32: 137-143.

De Juana, E. & Varela, J.M. (1984). Una visita a la Isla de Alboran (junio de 1982). La Garcilla, 63: 45-47.

De Juana, E. & Varela, J.M. (1984). Tres anos de inspeccion costera de aves petroleadas en el Mediterraneo espanol. Communication to III Reunion del Grupo Ibérico de Aves Marinas, Perlora (Asturias).

De Juana, E.; Varela, J.M. & Witt, H.-H. (1984). The conservation of seabirds at the Chafarinas Islands. ICBP Technical Publication No. 2: 363-370.

De Juana, E.; Varela, J.M. & Witt, H.-H. (1980). Le Puffin Cendré Calonectris diomedea nicheur aux Iles Chaffarines. Alauda, 48: 27-31.

Deleuil, R. (1958). Sur les oiseaux de mer des côtes tunisiennes. L'Oiseau et R.F.O., 28: 228-232.

Dubois, P. (1979). Précision du statut de quelques espèces observées au Maroc. Alauda, 47: 43-45.

Dubois, P. & Duhautois, L. (1977). Notes sur l'ornithologie marocaine. Alauda, 45: 285-291.

Eigenhuis, K.J. (1985a). Bulwer's Petrels at Costa Brava in April 1984. Dutch Birding, 7: 20-21.

Eigenhuis, K.J. (1985b). (Note on Stercorarius pomarinus). Ardeola, 32: 416.

Ena, V. & Purroy, F.J. (1982). Censos invernales de aves acuaticas en Espana (enero 1978, 79 y 80). ICONA, Madrid.

Erard, Ch. (1958). Sur les zones de reproduction et d'hivernage et les migrations du Goéland Railleur, Larus genei. Alauda, 26: 86-104.

Erard, Ch. (1960). Sur l'aire de reproduction, les zones d'hivernage et les migrations de la Mouette pygmée (Larus minutus). Alauda, 28: 196-228.

Erard, Ch. (1962). Notes sur la migration des oiseaux de mer à Beauduc, Camargue. Alauda, 30: 217-225.

Erard, Ch. (1964). Compléments à l'étude de l'aire de reproduction et des migrations du Goéland railleur, Larus genei. Alauda, 32: 283-296.

Erard, Ch. & Vielliard, J. (1966). Comentarios sobre avifauna invernal en el Oriente Espanol. Ardeola, 11: 95-100.

Estrada, V. (1985). Nueva informacion fenologica y faunistica del Penon de Vélez de la Gomera (Norte de Marruecos). La Garcilla, 65: 56-59.

Fasola, M. (1985). Censimento preliminare dei Laridae svernanti in Italia.

Feeny, P.P. y cols. (1960). Sobre aves del sur de Espana con especial referencia a migracion (primavera 1959). Ardeola, 6: 125-150.

Ferrer, X. (1977). Introduccio ornitologica al Delta de l'Ebre. Treb. Inst. Cat. Hist. Nat., 8: 227-302.

Ferrer, X. (1986). Fluctuations of the Gull and Tern populations in the Ebro Delta, NE Spain (1960-1985). (In this same volume).

Finlayson, J.C. (1983). Recent trends and changes in the status of the birds of Gibraltar. Alectoris, 5: 2-9.

Finlayson, J.C. & Cortés, J.E. (1984). The migration of Gannets Sula bassana past Gibraltar in spring. Seabird, 7: 19-22.

François, J. (1975). L'avifaune annuelle du lac de Boughzoul (Algérie). Alauda, 43: 125-133.

Furness, R. (1978). Movements and mortality rates of Great Skuas ringed in Scotland. Bird Study, 25: 229-238.

Garcia, E.F.J. (1973). Seabird activity in the Strait of Gibraltar: a progress report. Seabird Report, 3: 30-36.

Garcia Rodriguez, L. (1973). Observacion de Stercorarius pomarinus en Almeria. Ardeola, 19: 22.

Garcia Rua, A.E. (1975). Migrantes y migracion visible en la zona de Gibraltar (anos 1972-1974). Ardeola, vol. esp.: 627-655.

Garrido, M. & Alba, E. (1981). (Note on Hydroprogne caspia). Ardeola, 28: 159.

Garrido, M. & Alba, E. (1985a). (Note on Larus canus). Ardeola, 32: 417.

Garrido, M. & Alba, E. (1985b). (Note on Sterna dougallii). Ardeola, 32: 418.

Garrido, M.; Alba, E. & Gonzalez, J.M. (1985). Aves de las zonas humedas malaguenas. Excma. Diputacion Provincial, Malaga.

Garrido, M.; Alba, E.; Léon, S. & Ruiz, R. (1985). (Note on Gavia stellata). Ardeola, 22: 409.

Gemra (1974). Gaviota Picofina (Larus genei) y Gaviota de Audouin (L. audouinii) en el litoral gaditano. Ardeola, 20: 360.

Géroudet, P. (1959). Les Palmipèdes. 2nd. ed. Delachaux et Niestlé, Neuchâtel.

Glanville, D. & Walker, C. (1963). Migracion primaveral en Almeria, ano 1960. Ardeola, 8: 131-141.

Glutz von Blotzheim, U.N. & Bauer, K.M. (1982). Handbuch der Vögel Mitteleuropas. Band 8/I. Akademische Verlagsgesellschaft. Wiesbaden.

Gonzalez Barbera, G. (1981). (Note on Larus genei). Ardeola, 28: 159.

Gonzalez Cachinero, J.M. (1985). (Note on Larus argentatus). Ardeola, 32: 417.

Guillen, A., Gimenez, M. & Navarro, V. (1985). (Note on Larus minutus). Ardeola, 32: 416.

Guyot, I. (1985). La reproduction du Cormoran huppé Phalacrocorax aristotelis en Corse. In: Oiseaux marins nicheurs du Midi et de la Corse, pp. 70-76. Annales du C.R.O.P. No. 2, Aix-en-Provence.

Guyot, I.; Launnay, G. & Vidal, P. (1985). Oiseaux de mer nicheurs du Midi et de la Corse: évolution et importance des effectifs. In: Oiseaux de mer nicheurs du Midi et de la Corse, pp. 31-47. Annales du C.R.O.P. No. 2, Aix-en-Provence.

Harris, M.P. (1962). Migration of the British Lesser Black-backed Gull as shown by ringing data. Bird Study, 9: 174-182.

Harris, M.P. (1964). Recoveries of ringed Herring Gulls. Bird Study, 11: 183-191.

Harris, M.P. (1984). The Puffin. T & A.D. Poyser, Calton (United Kingdom).

Harrison, J.G. (1960). Notas sobre algunas aves en Andalucia (mayo 1959). Ardeola, 6: 340-343.

Heim de Balsac, H. & Mayaud, N. (1962). Les oiseaux du Nord-Ouest de l'Afrique. Paul Lechevalier, Paris.

Hiraldo, F.; Herrera, C.M. & Hidalgo, J. (1973). Sobre Larus minutus en Andalucia. Ardeola, 19: 97-100.

Ingram, C. (1960). Observacion de Gaviota enana en Algeciras (Cadiz). Ardeola, 6: 384.

Irby, L.H. (1895). The Ornithology of the Straits of Gibraltar, 2nd ed. Taylor & Francis, London.

Isenmann, P. (1973). Le passage de la Sterne caspienne Hydroprogne caspia en 1971 et 1972 en Camargue. Alauda, 41: 365-370.

Isenmann, P. (1975). Observations sur la Mouette Pygmée (Larus minutus) en Camargue, de 1971 à 1974. Terre et Vie, 29: 77-88.

Isenmann, P. (1976a). Note sur le stationnement hivernal des Laridés sur la côte méditerranéenne d'Espagne. L'Oiseau et R.F.O., 46: 135-142.

Isenmann, P. (1976b). Contribution à l'étude de la biologie de la reproduction et de l'étho-écologie du Goéland railleur Larus genei. Ardea, 64: 48-61.

Isenmann, P. (1976 & 1977). L'essor démographique et spatial de la Mouette rieuse Larus ridibundus en Europe. L'Oiseau et R.F.O., 46: 337-366 and 47: 25-40.

Isenmann, P. (1978). Note sur les stationnements de Laridés sur la côte atlantique du Maroc en décembre 1976. Bull. Inst. Scientifique Rabat, 2: 77-86.

Isenmann, P. (1980). Résultats d'un premier recensement de Laridés hivernants sur le littoral méditerranéen français (février 1979). L'Oiseau et R.F.O., 50: 161-163.

Isenmann, P. (1986). Thirty years of fluctuations of the Laridae populations in the Camargue (1956-1985) (in this volume).

Isenmann, P. & Czajkowski, M.A. (1978). Note sur un recensement de Laridés entre Nice et Naples en décembre 1977. Riv. ital. Orn. 48: 143-148.

Isidoro, A. (1970). Anilhagem de aves na reserva ornitologica de Mindelo (Vila do Conde) (1964-65-66). Cyanopica, 3: 59-93.

Jacob, J.-P. (1979). Résultats d'un recensement hivernal de Laridés en Algérie. Le Gerfaut, 69: 425-436.

Jacob, J.-P. (1983). Oiseaux de mer de la côte centrale d'Algérie. Alauda, 51: 48-63.

Jacob, J.-P. & Courbet, B. (1980). Oiseaux de mer nicheurs sur la côte algérienne. Le Gerfaut, 70: 385-401.

Jacob, J.-P. & Jacob, A. (1980). Nouvelles données sur l'avifaune du Lac de Boughzoul (Algérie). Alauda, 48: 209-219.

James, P.C. (1984). The status and conservation of seabirds in the Mediterranean Sea. ICBP Technical Publication No. 2: 371-375.

Jouanin, Ch. (1957). L'irruption en France de Mouettes tridactyles en février 1957. L'Oiseau et R.F.O., 27.

King, W.B. (1981). Endangered Birds of the World, The ICBP Red Data Book. Smithsonian Institution Press, Washington D.C.

Lathbury, G. (1970). A review of the birds of Gibraltar and its surrounding waters. Ibis,112: 25-43.

Ledant, J.-P.; Jacob, J.-P.; Jacobs, P.; Mahler, F.; Ochando, B. & Roché, J. (1981): Mise à jour de l'avifaune algérienne. Le Gerfaut, 71: 295-398.

Lévêque,R. (1960). Nota sobre migradores observados durante un viaje por Espana (primavera 1959). Ardeola, 6: 348-351.

Lind, P.B. (1957). Nota sobre algunas aves de Andalucia, con una observacion de Sylvia curruca. Ardeola, 3: 307-310.

Llorente, G. & Montori, A. (1982). Algunas observaciones de Rissa tridactyla (Linn.) (Aves, Laridae) y Stercorarius skua (Brünich) (Aves, Stercorariidae) en el litoral barcelonés. P. Dept. Zool. Barcelona, 7: 81-83.

Mackrill, E.J. (1985). (Note on Stercorarius pomarinus). Ardeola, 32: 416.

Maluquer, S. & Pons, J.R. (1961). La Avifauna de la isla de Buda en primavera-verano de 1961. Ardeola, 7: 79-111.

Maluquer Sostres, J. (1973). Els ocells de les terres catalanes, 2nd ed. Barcino, Barcelona.

Marquez, J. (1959). Captura de Sterna fuscata en la Peninsula Iberica. Ardeola, 5: 213-215.

Martin, J. (1981). (Note on Hydroprogne caspia). Ardeola, 28: 159.

Martinez, A. & Muntaner, J. (1979). Migration de la Sterne caspienne Hydroprogne caspia par le Delta de l'Ebre. Alauda, 47: 29-33.

Martinez-Vilalta, A.; Ferrer, X. & Carboneres, C. (1984). Situacion de los pagalos (Stercorarius sp.) en le litoral catalan (NE de la Peninsula Ibérica). Misc. Zool., 8: 217-223.

Mayaud, N. (1954). Sur les migrations et l'hivernage de Larus melanocephalus Temminck. Alauda, 22: 225-245.

Mayaud, N. (1965). Contribution à l'ornithologie du Nord-Ouest de l'Afrique. Alauda, 33: 34-50.

Mayaud, N. (1970). Additions et contributions à l'avifaune du Nord-Ouest de l'Afrique. Alauda, 38: 27-43.

Mayaud, N. (1982). Les oiseaux du Nord-Ouest de l'Afrique. Notes complémentaires. Alauda, 50: 45-67, 114-145 & 286-309.

Mayaud, N. (1983). Les oiseaux du Nord-Ouest de l'Afrique. Notes complémentaires. Alauda, 51: 271-301.

Mayaud, N. (1984). Les oiseaux du Nord-Ouest de l'Afrique. Notes complémentaires. Alauda, 52: 266-284.

Mayol, J. (1978). Observaciones sobre la Gaviota de Audouin, Larus audouinii Payr., en el Mediterraneo occidental (primavera de 1978). Naturalia Hispanica, 20. ICONA, Madrid.

Mead, C.D. (1974). The results of ringing auks in Britain and Ireland. Bird Study, 21: 45-86.

Meeth, P. (1965). Observacion de Charran Sombrio (Sterna fuscata) en Andalucia. Ardeola, 10: 65-66.

Melendro, J. & Rodriguez Valverde, A. (1977). Censo de Laridos y Pelecanidos invernantes en la costa de Pontevedra (diciembre 1974). Ardeola, 22: 75-83.

Merino, L.A. & Ruiz, F.J. (1984). (Note on Gavia immer). Ardeola, 30: 115.

Moltoni, M. & Brichetti, P. (1978). Elenco degli uccelli italiani. Riv. ital. Orn., 48: 65-142.

Morel, G. & Roux, F. (1966). Les migrateurs palearctiques au Sénégal. 1. Non Passereaux. Terre et Vie, 20: 19-72, 143-176.

Müller, H. (1959). Die Zugverhältnisse der europäischen Brandseeschwalben (Sterna sandvicensis) nach Beringungsergebnissen. Vogelwarte, 20: 91-115.

Muntaner, J. & Congost, J. (1979). Avifauna de Menorca. Treb. Mus. Zool. Barcelona No. 1 (173 pp.).

Muntaner, J.; Escandell, A.; Ramos, E. & Orfila, G. (1984). Adicion y revision faunistica de las especies de la publicacion "Avifauna de Menorca". In: Avifauna de Menorca, 2 Edicion. Treb. Mus. Zool. Barcelona No. 1.

Muntaner, J.; Ferrer, X. & Martinez-Vilalta, A. (1984). Atlas dels ocells nidificants de Catalunya i Andorra. Ketres, Barcelona.

Muntaner, J.; Ferrer, X. & Roda, F. (1974). Censo de aves acuaticas en el Delta del Ebro. Invierno 1973-1974. Ardeola, 20: 179-185.

Muselet, D. (1982). Les quartiers d'hivernage des Sternes pierregarins (Sterna hirundo) européennes. L'Oiseau et R.F.O., 52: 219-235.

Nadal, B. (1961). Lista sistematica de aves recientemente capturadas en la isla de Mallorca. Ardeola, 7: 234-240.

Nadal, B. (1965). Aves poco comunes capturadas en Mallorca de 1961 a 1964. Ardeola, 10: 54-55.

Nadal, B. (1971). Lista de capturas de aves raras o poco comunes en Mallorca, periodo 1966-1969. Ardeola, 15: 95-98.

Navarro Medina, J.-D. (1973). Datos sucintos sobre aves acuaticas del Mar Menor. Ardeola, 19: 57-60.

Nelson, J.B. (1978). The Gannet. T. & A.D. Poyser, Berkhamsted (United Kingdom).

Nisbet, I.C.T. (1960). Sobre algunas aves observadas en Andalucia en abril de 1960. Ardeola, 6: 211-219.

Norman, R. (1971). Resultados de una visita al Delta del Ebro, primavera 1969. Ardeola, 15: 85-90.

Olioso, G. (1980). Observation d'un Mergule nain Alle alle en Méditerranée. Alauda, 48: 257.

Oliver, P.J. (1977). Le passage d'automne de la Mouette pygmée Larus minutus au Cap Gris-Nez. Alauda, 45: 191-196.

Palaus, J. (1957). Datos adicionales sobre Rissa tridactyla en Cataluna. Ardeola, 3: 315-316.

Paterson, A.M. (1984a). (Note on Bulweria bulwerii). Ardeola, 30: 115.

Paterson, A.M. (1984b). (Note on Sterna maxima). Ardeola, 30: 120.

Paterson, A.M. (1985a). (Note on Sterna caspia). Ardeola, 32: 418.

Paterson, A.M. (1985b). (Note on Sterna dougallii). Ardeola, 32: 418.

Peltzer, R. (1966). Noticia sobre una expedicion al Guadalquivir, primavera 1964. Ardeola, 11: 114-116.

Pineau, J. & Giraud-Audine, M. (1974). Notes sur les migrateurs traversant l'extrême nord-ouest du Maroc. Alauda, 42: 159-188.

Pineau, J. & Giraud-Audine, M. (1975). Notes complémentaires sur les migrations dans l'extrême nord-ouest du Maroc. Alauda, 43: 135-141.

Pineau, J. & Giraud-Audine, M. (1976). Notes sur les oiseaux hivernant dans l'extrême nord-ouest du Maroc et sur leurs mouvements. Alauda, 44: 47-75.

Pineau, J. & Giraud-Audine, M. (1977). Notes sur les oiseaux nicheurs dans l'extrême nord-ouest du Maroc: reproduction et mouvements. Alauda, 45: 75-103.

Pons Oliveras, J.R. (1968). Tardio acantonamiento de Frailecillos (Fratercula arctica) en la costa mediterranea espanola. Ardeola, 12: 239.

Preiswerk, G. (1981). Protocol of the observation of a Laughing Gull (Larus atricilla) at Porto (Portugal). Cyanopica, 2: 40-42.

Riols, Ch. (1969). Note sur le Grand Labbe Stercorarius skua en Méditerranée. Alauda, 37: 162-163.

Ruiz Martinez, F.J. (1985). Note on Larus ridibundus. Ardeola, 32: 416-417.

Salmon, H.M. (1971). Sterna bengalensis en Cadiz. Ardeola, 15: 144.

Sargatal, J. & Llinas, R. (1978). Els ocells de l'Emporda. Grafiques Montserrat, Figueres.

Saunders, H. (1871). A List of the Birds of Southern Spain. Ibis (1871): 54-68, 205-225, 348-402.

Schenk, H. & Torre, A. (1986). Breeding distribution, numbers and conservation of seabirds in Sardinia, 1978-85 (in this volume).

Schmitt, B. (1963). Notes d'Alger. Alauda, 31: 218-221.

Smith, K.D. (1965). On the birds of Morocco. Ibis, 107: 493-526.

Smith, K.D. (1972). The winter distribution of Larus audouinii. Bull. Br. Orn. Club, 92: 34-37.

Sociedade Galega de Historia Natural (1980). Laridos no comunes en las costas gallegas. Brana, 1-2: 151-160.

Solis, F. (1978). Note on Sterna paradiseae. Ardeola, 24: 260.

Solis, F. & Lubian, J.M. (1983). Note on Alle alle. Ardeola, 29: 187.

Sorci, G.; Massa, B. & Cangialosi, G. (1971). Il Genere Stercorarius. Brisson è regolare nel Mediterraneo. Riv. ital. Orn., 41: 161-198.

Spano, S. (1965). La Sula (Sula b. bassana) in Italia. Riv. ital. Orn., 35: 1-

Staav, R. (1977). Etude du passage de la Sterne caspienne Hydroprogne caspia en Méditerranée à partir des reprises d'oiseaux bagués en Suède. Alauda, 45: 265-270.

Stead, Ph.J. (1959). Notas ornitologicas de Andalucia. Ardeola, 5: 179-184.

Stewart, N.F. (1965). Nota sobre aves de Andalucia (enero-marzo de 1965). Ardeola, 10: 38-39.

Studer-Thiersch, A. & P. (1968). Sobre aves acuaticas en una laguna del interior de Andalucia (observaciones de marzo à junio de 1967). Ardeola, 14: 167-174.

Suarez, E.; Garcia, A. & Berral, D. (1981). Note on Sterna hirundo. Ardeola, 28: 159.

Sultana, J.; Gauci, C. & Beaman, M. (1975). A guide to the Birds of Malta. Lux Press, La Valeta.

Tait, W. (1924). Birds of Portugal. H.G. & G. Witherby, London.

Telleria, J.L. (1980). Autumn migration of the Cory's Shearwater through the Straits of Gibraltar. Bird Study, 27: 21-26.

Telleria, J.L. (1981). La Migracion de las Aves en el Estrecho de Gibraltar. Vol. II: Aves no Planeadoras. Universidad Complutense, Madrid.

Thévenot, M.; Beaubrun, P.; Baouab, R.E. & Bergier, P. (1982). Compte rendu d'ornithologie marocaine, année 1981. Institut Scientifique, Rabat.

Thévenot, M.; Bergier, P. & Beaubrun, P. (1980). Compte rendu d'ornithologie marocaine, année 1979. Institut Scientifique, Rabat.

Thibault, J.-C. & Guyot, I. (1981). Répartition et effectifs des oiseaux de mer nicheurs en Corse. L'Oiseau et R.F.O., 51: 101-114.

Thomsen, P. & Jacobsen, P. (1979). The Birds of Tunisia. Jelling Bogtrykkeri, Copenhagen.

Thomsen, A.L. (1978). The migration of the Gannet into the Mediterranean. In: A Bundle of Feathers, Delhi.

Valverde, J.A. (1954). Hydroprogne caspia y Plegadis falcinellus en Marruecos. Ardeola, 1: 127-128.

Valverde, J.A. (1955). Aves de Marruecos espanol en julio. Ardeola, 2: 87-114, 213-240.

Vargas, J.M. & Antunez, A. (1980). Inventario faunistico de Chafarinas. Jabega, 32: 60-64. Excma. Diputacion Provincial, Malaga.

Vargas, J.M.; Antunez, A. & Blasco, M. (1978). Comportamiento reproductivo y alimentario de la Pagaza Piconegra (Gelochelidon nilotica) en la Laguna de Fuentedepiedra, Malaga. Ardeola, 24: 227-231.

Vidal, P. (1985). Premières observations sur la biologie de reproduction du Puffin des Anglais yelkouan Puffinus puffinus yelkouan sur les îles d'Hyères. In: Oiseaux marins nicheurs du Midi et de la Corse, pp. 58-62. Annales du C.R.O.P. No. 2, Aix-en-Provence

Vilagrasa, F.X.; Carrera, E. & Pardo, R. (1982). Hivernage de la Sterne caugek Sterna sandvicensis sur les côtes de la Catalogne et du Levant (Espagne). Alauda, 50: 108-113.

Witt, H.-H.; De Juana, E. & Varela, J.M. (1984). Seevögel (Procellariformes, Pelecaniformes, Lariformes) der marokkanischen Nordküste. Vorkommen und Brut. Beitr. Vogelkd., Jena, 30: 81-89.

SEABIRDS ALONG THE MEDITERRANEAN SEA COAST OF EGYPT

Peter L. Meininger
Foundation for Ornithological Research in Egypt,
Grevelingenstraat 127, 4335 XE Middelburg, The Netherlands

Sherif M. Baha el Din
Ornithological Society of Egypt, 4 Ismail el Mazni Street,
flat no. 8, Heliopolis, Cairo, Egypt

Introduction

Until a few years ago there was a paucity of ornithological information from the north coast of Egypt, including the Nile Delta lakes. Several recent expeditions to this region revealed new and important information, particularly on quantitative aspects of the Mediterranean coastal avifauna. Systematic counts of migrating waterbirds along the north coast of Sinai have also been carried out during six autumns between 1973 and 1982 (Paran & Paz 1980, Paran 1980, Y. Paran in lit. , Petersen & Sørensen 1980, 1981a, 1981b, Baha el Din & Salama 1984). Counts of waterbirds have been carried out in the Nile Delta lakes in the winters of 1978/79, 1979/80 and in November 1981 (Meininger & Mullié 1981a, 1981b, Bennett et al. 1982). Also, preliminary surveys of breeding birds were made in the spring of 1983 in the Nile Delta lakes (Meininger et al. 1986) and in the spring of 1985 along the north coast of Sinai (Fry et al. 1985).

In this paper we have used published information and unpublished records submitted to the Birds of Egypt Project (hereafter BEP). One aim of the BEP, originally initiated by PLM and Wim C. Mullié, is an up to date book on Egyptian birds (Goodman & Meininger in prep.). The individuals who submitted material to the BEP have been cited under the individual records.

This paper should be considered a preliminary summary of the present knowledge of the seabirds along the Egyptian Mediterranean Sea coast. It should be emphasized that relatively little is known and that future studies will undoubtedly reveal much information.

NATO ASI Series, Vol. G 12
Mediterranean Marine Avifauna
Edited by MEDMARAVIS and X. Monbailliu

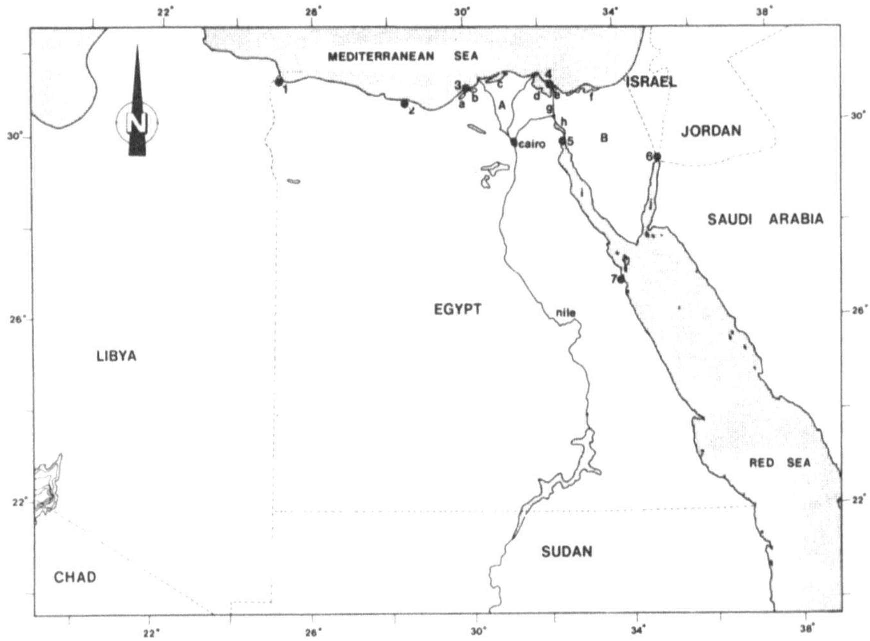

Fig. 1: Localities mentioned in the text.

A. Nile Delta
B. Sinai

a. Lake Maryut
b. Lake Idku
c. Lake Burullus
d. Lake Manzala
e. El Malaha
f. Lake Bardawil
g. Suez Canal
h. Bitter Lakes
i. Gulf of Suez
j. Gulf of Aqaba

1. Salum
2. El Alamein
3. Alexandria
4. Port Said
5. Suez
6. Eilat, Israel
7. Hurghada

Fig. 1 shows the geographical position of the localities mentioned in the text.

Breeding Seabirds

Several seabird species are known to breed along the coast and on islands of the Egyptian portion of the Red Sea. These include Red-billed Tropicbird *Phaethon aethereus*, Brown Booby *Sula leucogaster*, Sooty Gull *Larus hemprichii*, White-eyed Gull *L. Leucophtalmus*, Caspian Tern *Sterna caspia*, Swift Tern *S. bergii*, Lesser Crested Tern *S. bengalensis*, White-cheecked Tern *S. repressa* and Bridled Tern *S. anaethetus* (Jennings et al. 1985, Goodman & Storer in press.). None of these species breed along the Mediterranean Sea coast of Egypt and only White-eyed Gull, Swift Tern and Lesser-crested Tern have been occasionally observed there.

The Egyptian Mediterranean coast virtually lacks islands and steep rocky cliffs and the seabird species breeding in the northern Red Sea, generally the most northern representatives of Indian Ocean species, do not occur there. Consequently the Egyptian Mediterranean breeding seabird community is composed of only a few species. Only one species, the Little Tern *S. albifrons*, has a population of international significance.

The following seabird species are known to breed along the Mediterranean Sea coast of Egypt:

- Slender-billed Gull - *Larus genei*

El Negoumi et al. (1950) mentioned a breeding colony on Gezira Dib near Port Said (probably in Lake Manzala). In 1979, 200-400 nests were found in El Malaha, in north-western Sinai, east of Port Said (Meininger & Mullié 1981a). No other Egyptian breeding sites for this species are known, but considering the continuous presence of adult Slender-billed Gulls throughout the year, including during the breeding season, in the environs of Port Said, it seems likely that it regularly breeds in El Malaha.

- Yellow-legged Gull - *Larus cachinnans*

In February 1980, pairs, apparently on territory, were observed in a lagoon east of El Alamein (P. Isenmann). On 7 May 1985 this lagoon was visited and a colony of 10 to 15 pairs was found on an island. Local inhabitants also reported that they regularly collect eggs in a nearby sand dune area (S.M. Goodman, PLM). This is the only evidence of this species breeding in Egypt. The total Egyptian population is estimated at 20-50 pairs.

- Little Tern - *Sterna albifrons*

The Little Tern is a locally common breeding resident along the Mediterranean coast. The total Egyptian population is tentatively estimated at 1,800 to 2,400 pairs, of which 800-1,000 nest in the Nile Delta (Meininger et al., 1986) and 800-1,200 in northern Sinai (based on Fry et al., 1985; Dunnet et al., 1986). The coastal area between Alexandria and Salum is virtually unexplored, but is known to hold several small breeding colonies. A few pairs breed (irregularly?) along the Bitter Lakes, and inland along Lake Qarun.

Migrant and Wintering Seabirds

In autumn a considerable passage of waterbirds takes place along the north coast of Sinai, and preliminary estimates indicate that 130,000 to 330,000 marine and freshwater birds, of over 80 species pass through the region (Baha el Din & Salama 1984). In tab. 1 we have summarized the species and numbers of seabirds observed during six systematic autumn counts between 1973 and 1982.

In the species accounts below, the following abbreviations have been used to indicate the status of each seabird species along the Egyptian Mediterranean coast:

RB	Resident breeder
MB	Migrant breeder
PV	Passage visitor
WV	Winter visitor
AV	Accidental visitor (up to five known records)

The following abundance categories have been used, which indicate the maximum number of migrating and wintering birds present on one day (only along the Mediterranean coast, including the Nile Delta lakes):

ab	abundant	> 100,000
co	common	10,000 - 100,000
fc	fairly common	1,000 - 10,000
sc	scarce	100 - 1,000
ra	rare	10 - 100
vr	very rare	1 - 10

• Cory's Shearwater - Calonectris diomedea ra PV ra WV

Small numbers are regularly observed at sea throughout the year. In autumn it is frequently observed along the north Sinai coast (tab. 1), with a daily maximum of 38 on 28 September 1980 (Petersen & Sørensen 1981a). There are several records from the northern portion of the Red Sea, mainly in spring.

• Sooty Shearwater - Puffinus griseus vr PV

There are a few records from the north Sinai coast: one observed on 2 September 1974, two on 14 September 1974, two on 12 November 1975 (Paran 1979), and single individuals on 30 August and 12 September 1981 (Petersen & Sørensen 1981b) and on 16 September 1982 (Baha el Din & Salama, 1984).

• Manx Shearwater - Puffinus puffinus ra PV

Small numbers are regularly noted at sea, mainly in autumn (tab. 1).

• Gannet - Sula bassana ra PV ra WV

There are several records of small numbers between late August and late April. The largest concentration appears to be twenty, five miles off the coast, near Port Said in early February 1961 (Jones 1961).

• Shag - Phalacrocorax aristotelis AV

Meinertzhagen (1930) saw a few at Salum and Sidi Omar in February and March and picked up two dead specimens. The only records from the country is one oiled bird near Hurghada, Red Sea, on 20 November 1982 (Baha el Din & Saleh 1984).

• Great Skua - Stercorarius skua AV

In September 1978 one Great Skua was seen off the north coast of Sinai (U. Paz). This species and/or the South Polar Skua S. (skua) mackormickii has been observed on several occasions between March and July near Port Said (Mörzer Bruijns & Voous 1965, Bourne 1967).

• Pomarine Skua - Stercorarius pomarinus ra PV ra WV

• Arctic Skua - Stercorarius parasiticus ra PV ra WV

Both species have been regularly recorded in small numbers along the Mediterranean coast from mid-August to mid-winter (tab. 1). It seems plausible that these birds, almost exclusively adult or immature birds (but no first-year!), move overland across Sinai to the Red Sea. In spring both species occur regularly in the northern end of the Red Sea, with significant concentrations near Eilat, Israel. At this locality mostly adult light-phased birds are seen migrating in a northerly direction, thus leaving the sea and continuing their migration overland (Meininger & Sørensen 1986). Along the Mediterranean coast small numbers have been observed in spring and there are several records of immature Pomarine Skuas in June and July (Meininger & Dielissen 1979, Meininger & Sørensen 1986).

• Long-tailed Skua - Stercorarius longicaudus vr PV

There are at least five autumn records (involving nine individuals) from the north coast of Sinai, and smaller numbers have been regularly recorded in spring in the northern Red Sea (Meininger & Sørensen 1986).

• White-eyed Gull - Larus leucophtalmus AV

In spring and summer this species, which breeds in the Red Sea, occurs regularly in the Suez Canal, sometimes in considerable numbers (hundreds). Some birds follow the Suez Canal to Port Said, thus reaching the Mediterranean Sea. At Port Said one was seen on 2 April 1960 (Bourne & Radford 1961) and three on 28 June 1977 (Meininger & Dielissen 1979).

• Great Black-headed Gull - Larus ichthyaetus vr PV vr WV

It is regularly observed in small numbers along the Mediterranean coast of the Nile Delta from mid-November to mid-March, and occasionally in the summer. The largest reported concentration in Egypt is 230 near Port Said on 26 December 1979, composed mainly of first-year birds (PLM).

• Little Gull - Larus minutus sc WV

Small numbers occur along the coast and on the Nile Delta lakes between mid-September and early May. The largest known concentrations in Egypt are 130 (40% imm.) on Lake Idku on 28 January 1979 (Meininger et al., 1979) and 260 on Lake Burullus on 7 February 1986 (SMB, D. Merrie).

• Sabine's Gull - Larus sabini AV

One was observed in the late 1970's at Lake Bardawil (P. Schlüter) and an adult was noted off northern Sinai on 17 September 1980 (Petersen & Sørensen 1981a).

• Black-headed Gull - Larus ridibundus ab PV ab PV

Abundant along the coast and on coastal lakes from early September to early May and scarce non-breeding summer visitor. There may well be several hundred thousand wintering in the Nile Delta.

• Slender-billed Gull - Larus genei sc RB co PV co WV

See previous section for details on breeding. This species is common along the sea coast and on the lakes from August to May, and probably over 10,000 winter in the Nile Delta. Thousands of non-breeding birds summer in this area. Egypt is an important wintering area for Slender-billed Gulls breeding in the Black Sea (Erard 1958).

• Audouin's Gull - Larus audouinii vr PV

There are at least four recent records in total from the Mediterranean coast in March, September and November.

• Common Gull - Larus canus ra WV

Small numbers have been observed between mid-September and mid-March.

• Lesser Black-backed Gull - Larus fuscus co PV sc WV

This species is common on both autumn and spring passage. In mid-winter hundreds are present on the lakes of the Nile Delta and along the Mediterranean coast. A few remain in the area during the summer months, mostly immatures.

• Yellow-legged Gull - Larus cachinnans ra RB fc WV

See previous section for details on breeding. Thousands winter along the Mediterranean coast and on the Nile Delta lakes, and hundreds remain during the summer months.

• Kittiwake - Rissa tridactyla vr WV

There are a few records from the Mediterranean coast: Boyd (1917) observed a few immatures at Port Said on 15 March 1916; three were seen along the north coast of Sinai in September 1978 (Y. Paran); and a dead immature, in a partly decayed state, was found near Baltim, Lake Burullus, on 10 May 1983 (PLM, U.G. Sørensen). This species may be more common in Egyptian waters than our current information suggests.

• Gull-billed Tern - Sterna nilotica sc PV ra WV

Meinertzhagen (1930) considered this species a common winter visitor to Egypt, on both inland and coastal waters, from early November to March. He stated that it was probably the most common tern in Egypt. These statements are striking. In recent years the Gull-billed Tern is one of the less common terns in Egypt. This apparent decrease is probably related to the recent decline of breeding birds in the USSR, eastern Europe and Turkey (Møller 1975). Presently the Gull-billed Tern is a scarce passage visitor throughout the country from late August to mid-May (see also tab. 1). The only recent winter record is of nine near Port Said on 10 January 1980 (Meininger & Mullié 1981a).

• Caspian Tern - Sterna caspia sc PV sc WV

This species is a regular migrant and winter visitor in small numbers to the area between mid-August and mid-March. Several Swedish ringed birds have been recovered in the Nile Delta and along the Mediterranean coast, and Scandinavian breeding birds winter in these areas (Staav 1977).

• Swift Tern - Sterna bergii AV

This species breeds on islands in the Red Sea. Meinertzhagen (1930) mentioned that a specimen was shot at Port Said on 19 March (year unknown).

• Lesser-crested Tern - Sterna bengalensis AV

There is a small breeding population of S. b. torressii along the Mediterranean coast of Libya (Bundy 1976). One specimen collected in the Gulf of Suez has been assigned to this subspecies (Cramp 1985). S. b. bengalensis breeds on islands in the Red Sea. Records from the Nile Delta and the Suez Canal suggest some movements between the Red Sea and the Mediterranean Sea, but it is not clear which form(s) is involved.

Meinertzhagen (1930) mentioned specimens collected at Port Said in May and at Alexandria in June. A single bird was seen at Lake Manzala on 10 October 1943 (Meiklejohn 1944), one off north Sinai on 23 September 1981 (Petersen & Sørensen 1981b) and six in the Suez Canal north of El Kubri on 9 June 1984 (E. Haase).

• Sandwich Tern - Sterna sandvicensis fc PV fc WV

From mid-August to mid-May Sandwich Terns are regularly noted along the Mediterranean coast and on adjacent saline lakes (see also table 1). In January 1980 about 900 were counted in El Malaha and Lake Manzala (Meininger & Mullié 1981a). Small numbers have been observed in June and July near Port Said and Alexandria (Meininger & Dielissen 1979).

• Common Tern - Sterna hirundo fc PV

This species is a regular migrant from early August to mid-October (tab. 1) and from mid-March to mid-June, both along the coast and inland along the Nile. Small numbers remain throughout the summer, but it is completely absent in winter.

• Little Tern - Sterna albifrons fc MB fc PV

For information on breeding see previous section. This species is a regular migrant from mid-August to late October (tab. 1) and from late March to late May. It is rarely observed inland.

• Whiskered Tern - Chlidonias hybridus co PV co WV

This species is a common passage visitor, along both the sea coast and inland, from late August to early June. Meinertzhagen (1930) considered the Whiskered Tern only a passage visitor. However, in recent years it has become clear that it winters in large numbers in the Nile Delta. Counts include: 17,500 on Lake Burullus from 14 to 16 February 1980 and 7,080 on Lake Manzala from 6 to 8 February 1979 (Meininger & Mullié 1981b). A recent estimate of the European breeding population of this species was 18,000 pairs (Mees 1979). Thus, the Nile Delta is probably one of the most, if not the most, important wintering areas for the Whiskered Tern breeding populations of the Western Paleartic.

• Black Tern - Chlidonias niger sc PV

A small number of migrants are observed along the sea coast and inland, between early August and mid-November (tab.1) and between mid-March and early May. It occasionally winters in the Nile Delta lakes.

• White-winged Black Tern - Chlidonias leucopterus co PV

This species is a common passage visitor, along both the sea coast and inland, between mid-July and early November (tab. 1) and from mid-March to late May.

• Razorbill - Alca torda AV

A Razorbill was found dead on the beach near Alexandria in the winter of 1908/09 (Ticehurst 1912). It is impossible to know if the bird reached Egyptian waters alive.

Threats to seabirds along the mediterranean sea coast of Egypt

The virtual absence of seabird (gulls and terns) breeding colonies in the Nile Delta lakes is striking. It is probably related to the lack of isolated and undisturbed breeding sites. Throughout the Nile Delta the human presence is considerable, e.g. there are a large number of small fishing boats and on even the smallest islands there are settlements (Meininger et al. 1968). The Slender-billed Gull is only known to breed in El Malaha, an isolated area east of the Suez Canal. In 1979 all the gull eggs were collected by fishermen from a colony of 200-400 nests (Meininger & Mullié 1981a), and this practice is probably common. Eggs are regularly collected at the Yellow-legged Gull colony near El Alamein. Little Tern eggs are probably also collected with some frequency for human consumption, although this needs confirmation.

In the Nile Delta lakes, and also elsewhere, large numbers of waterbirds are being trapped or shot. In Lake Manzala, between mid-September and mid-March, 98,000 to 162,000 waterbirds are caught or shot annually, of which 1,100 to 1,600 are gulls and terns. This would include 120 to 180 Slender-billed Gulls, 100 to 170 Whiskered Terns and the balance composed of Black-headed, Lesser Black-backed and Yellow-legged Gulls. (Mullié & Meininger 1983). Although the actual numbers of seabirds killed are relatively low, hunting probably has a negative effect in many areas.

The main threat to important seabird areas in Egypt comes from large scale land reclamation projects, which - if not stopped in time - will destroy entire coastal ecosystems. Up to now at least 25% (700 km2) of the coastal wetlands in the Nile Delta have already been reclaimed, while there are plans for reclamation of other large areas. (Meininger & Mullié 1981a, b). In 1972 the total area of reclaimed saltmarsh in the Nile Delta was already estimated at about 4,000 km2 (George 1972).

Relatively little information is available about the impact of chemical and physical pollution of Egyptian coastal waters. Wastes from large industries, for instance near Alexandria and Port Said, and human sewage are discharged on a large scale and without purification in the Mediterranean Sea. This likely has a negative impact on the environment (e.g. Saad 1973) and consequently on the seabird populations.

Large quantities of organochlorine insecticides and polychlorinated biphenyls (PCBs) are discharged with wastewaters via the

	13 Aug- 29 Sept 1973 *1	15 Aug- 5 Oct 1978 *2	11-20 Oct 1979 *3	7-30 Sept 1980 *4	16 Aug- 24 Sept 1981 *5	2-24 Sept 1982 *6
Calonectris diomedea	..	-	-	85	1	13
Puffinus griseus	-	-	-	-	2	1
Puffinus puffinus	111	249	-	48	153	29
Sula bassana	-	-	1	-	-	-
Stercorarius pomarinus	-	-	3	16	5	3
S. parasiticus	62	66	25	60	116	70
S. longicaudus	-	-	-	-	7	1
Larus melanocephalus	-	-	-	-	-	-
L. minutus	31	2	-	2	4	-
L. sabini	-	-	-	1	-	-
L. ridibundus	435	321	-	98	138	383
L. genei	944	1123	187	1638	.2011	515
L. audounii	-	-	-	1	1	-
L. fuscus	761	.1156	95	539	634	222
L. cachinnans	643	.950	344	606	948	810
Rissa tridactyla	-	3	-	-	-	-
Larus ssp.	1325	1095	-	-	410	182
Sterna nilotica	211	410	-	91	198	54
Sterna caspia	57	85	-	57	48	14
S. bengalensis	-	-	-	-	1	-
S. sandvicensis	-	20	108	55	85	316
S. hirundo	3794	2769	-	3964	.12433	2898
S. albifrons	659	.1810	-	721	1347	872
Chlidonias hybridus	370	896	-	802	.950	96
C. niger	242	124	9	155	127	39
C. leucopterus	7657	8835	-	3222	.18436	3478
Sterna/Chlidonias ssp.	3498	3906	-	-	73	-

after: *1 Paran & Paz 1980, *2 Y. Paran in lit., *3 Petersen & Sørensen 1980, *4 Petersen & Sørensen 1981a, *5 Petersen & Sørensen 1981b, *6 Baha el Din & Salama 1984

Tab. 1: Total numbers of seabirds observed migrating along the north coast of Sinai, Egypt, in six autumns between 1973 and 1982.

Nile River and the coastal lakes in the Mediterranean Sea. Uptake and accumulation of such chlorinated hydrocarbons by sediments, micro-organisms and fish led to the build up of such compounds in the food chain. In fish samples, collected off Port Said, several orchanochlorine insecticides were found, mainly HCB, Lindane and DDT; DDT and its metabolites were the most frequently distributed insecticides in the fish studied and presented the highest residues compared to other insecticides (i.e. 299 µg/kg wet weight muscle tissue total DDT in Trichurus sp.). The residue levels of PCBs were highest in Solea vulgaris. (El Dib & Badawy 1985).

The levels of DDE (metabolite of DDT) found in the eggs of Audouin's Gull, especially in the western Mediterranean, give reason for concern about the reproductive success of this species, which could be implemented by further contamination of this marine environment by DDT (Bijleveld et al. 1979). However, Bourne & Bogan (1980) thought these levels not to be remarkably high for a member of the genus Larus, which they considered to be generally rather resistent to this form of contamination. In seabirds fairly similar levels of the commoner contaminants have been found throughout the Mediterranean Sea, including several mg/kg of PCBs, rather less DDE, and only traces of Dieldrin and other compounds (Bourne & Bogan 1980). No birds or eggs from Egypt have been examined, but the highest DDE levels were found in the Eastern Mediterranean: in a Yellow-legged Gull egg from Cyprus and in a Little Gull collected in winter in Malta (Bourne & Bogan 1980).

DDE concentrations in Cormorant Phalacrocorax carbo eggs from the Delta of the Danube were found to be in the same range as is associated with a 15% reduction in eggshell thickness (Fossi et al. 1984). Cormorants winter in large numbers in the Nile Delta lakes. It is clearly desirable to investigate the situation in Egyptian waters, especially concerning fish-eating breeders as the Little Tern and Slender-billed Gull. We agree with Fossi et al. (1984), that the interaction of different pollutants and the influence of those interactions on biological response should be more thoroughly investigated, especially under natural conditions.

It is clearly desirable to investigate the situation in Egyptian waters, especially concerning a fish-eating breeder as the Little Tern.

Egyptian coastal waters are on an important shipping lane and many ships pass through the Mediterranean Sea on their way to or from the Suez Canal. Spills from oil shipment or accidents are potentially serious threats to the local seabird populations. This threat will increase with the placement of oil platforms in 1985 off the north coast of Sinai (Fry et al. 1985).

In the nineteen seventies convincing evidence was found of extensive surface oil pollution in the Mediterranean Sea: the sea off Egypt belonged to the areas with the highest densities of oil slicks observed (Levy et al. 1981). Oiled birds have been found with some frequency along the Mediterranean coast of Egypt, but nothing is known about the total numbers involved.

Recommendations

It is recommended that in the near future studies will be made of:

- distribution and numbers of breading seabirds along the Mediterranean Sea coast of Egypt,

- factors threatening seabird populations in Egypt: oil pollution, pesticides, reclamation, hunting, egg collecting.

None of the lakes in the Nile Delta and northern Sinai, important for breeding and non-breeding seabirds, have a legal status as a nature reserve. We consider that at least the following areas should be protected as wildlife reserves: Lake Burullus, parts of Lake Manzala and Lake Bardawil and El Malaha.

Acknowledgements

This paper is mainly based on results of various expeditions, which were financially supported by the Prins Bernhard Fonds, International Waterfowl Research Bureau, Netherlands Foundation for International Nature Conservation, Netherlands Foundation for International Bird Protection, Holy Land Conservation Fund, U.S. Fish and Wildlife Service, Egyptian Wildlife Service, British Petroleum Egypt Branch and Ornithological Society of the Middle East.

We are most grateful to Steven M. Goodman (Ann Arbor, Michigan, USA) and to Wim C. Mullié (Wageningen, The Netherlands) for their valuable comments on the manuscript, and to all observers who submitted observations for the Birds of Egypt Project.

References

Baha el Din, S.M. & M.A. Saleh (1983). Report of the ornithological results of the Egyptian Red Sea pollution expedition. Ornithological Society of Egypt, Cairo.

Baha el Din, S.M. & W. Salama (1984). Waterbird migration study at Zaranik, Sinai, 2-24 September 1982. Courser 1: 28-43.

Bennett, C.A., R.G. Bijlsma & R. Stouthamer (1982). Survey of waterbirds on some Egyptian wetlands, autumn 1981. Report, Bennekom.

Bijleveld, M.F.I.J., P. Goeldin & J. Mayol (1979). Persistent pollutants in Audouin's Gull (Larus audouinii) in the Western Mediterranean: a case-study with wide implications? Environm. Cons. 6: 139-142.

Bourne, W.R.P. (1967). Observations of seabirds and a review of literature. Sea Swallow 19: 51-76.

Bourne, W.R.P. & J.A. Bogan (1980). Organochlorines in Mediterranean seabirds. Environm. Cons. 7: 277-278.

Bourne, W.R.P. & M.C. Radford (1961). Notes on seabird reports received during 1960. Sea Swallow 14: 7-27.

Boyd, A.W. (1917). Bird of the Suez Canal Zone and the Sinai Peninsula. Ibis 1917 (10): 539-557.

Bundy, G. (1976). The Birds of Libya. London.

Cramp, S. (ed.) (1985). The Birds of the Western Palearctic, IV. Oxford.

Dunnet, G. (1985). Report Bardawil, October 1985.

El-Dib, M.A. & M.I. Badawy (1985). Organichlorine insecticides and PCBs in water, sediment, and fish from the Mediterranean Sea. Bull. Environ. Cotam. Toxicol. 34: 216-227.

El Negoumi, A.; H.F. Zain el Din; M.A. el Monery & M.K. Fayed (1950). Birds of Egypt (in arabic). Dar el Fekr el Aribi, Cairo.

Erard, C. (1958). Sur les zones de reproduction et d'hivernage et les migrations du Goéland Railleur (Larus genei). Alauda 26: 86-104.

Fossi, C.; S. Focardi & A. Renzoni (1984). Trace-metals and chlorinated hydrocarbons in Brids' Eggs from the Delta of the Danube. Environmental Conservation 11: 345-350.

Fry, C.H.; G.A.M. Atta; S. Baha el Din & C.R. McKay (1985). Bardawil Lagoon baseline environmental studies: 1985 spring ornithological survey. Report, Aberdeen.

George, C.J. (1972). The role of the Aswan High Dam in changing the fisheries of the South-eastern Mediterranean. In: Farvar, M.T. & J.P. Milton (eds.). The Careless technology. The Natural History Press, New York.

Goodman, S.M. & P.L. Meininger (eds.) in prep. The Birds of Egypt.

Goodman, S.M. & R.W. Storer, in press. The seabirds of the Egyptian Red Sea and adjacent waters, with notes on selected Ciconiiformes. Gerfaut, in press.

Jennings, M.C.; P.C. Heathcote; D. Parr & S.M. Baha el Din (1985). Ornithological survey of the Ras Dib area and the islands at the mouth of the Gulf of Suez, Egypt. Report, Marlborough.

Jones, M.E. (1961). Extracts from an interesting voyage: Gannets off Port Said. Sea Swallow 14: 44-45.

Levy, E.M., M. Ehrhardt, D. Kohnke, E. Sobtchenko, T. Suzuoki & A. Tokuhiro (1981). Global Oil Pollution. Intergovernmental Oceanographic Commission, Unesco, Paris.

Mees, G.F. (1979). Distribution and numbers of Whiskered Tern Chlidonias hybridus (Pallas) in Europe and North-Africa (in Dutch; French and German summary). Zool. Bijdr. Rijksmus. Nat. Hist. Leiden, No. 26.

Meiklejohn, M.F.M. (1944). Bird notes from Egypt, autumn 1943. Bull. Zool. Soc. Egypt 6: 16-18.

Meinertzhagen, R. (1930). Nicoll's Birds of Egypt. Hugh Rees, London.

Meininger, P.L. & G.C. Dielissen (1979). Ornithological observations in Egypt in 1977 and 1978 (in Dutch, English summary). Veldornitol. tijdschr. 2: 78-86.

Meininger, P.L. & W.C. Mullié (1981a). The significance of Egyptian wetlands for wintering waterbirds. Holy Land Conservation Fund, New York.

Meininger, P.L. & W.C. Mullié (1981b). Egyptian wetlands as threatened wintering areas for waterbirds. Sandgrouse 3: 62-77.

Meininger, P.L., W.C. Mullié, J. van der Kamp & B. Spaans (1979). Report of the Netherlands Ornithological Expedition to Egypt in January and February 1979. Report, Middelburg.

Meininger, P.L. & U.G. Sørensen (1986). The occurrence of skuas Stercorariidae in the Middle East, with special reference to Egypt and the Northern Red Sea. Vogelwarte, in press.

Meininger, P.L., U.G. Sørensen & G.A.M. Atta (1986). Breeding birds of the lakes in the Nile Delta, Egypt. Sandgrouse, in press.

Møller, A.P. (1975). The breeding population of Gull-billed Terns Sterna nilotica Gmel. in 1972 in Europe, Africa and western Asia, with a review of fluctuations during the present century. Dansk orn. Foren. Tidsskr. 69: 1-8.

Mörzer Bruijns, W.F.J. & K.H. Voous (1965). Great Skuas (Stercorarius skua) in the Northern Indian Ocean. Ardea 53: 80-81.

Mullié, W.C. & P.L. Meininger (1983). Waterbird trapping and hunting in Lake Manzala, Egypt, with and outline of its economic significance. Biol. Conservation 27: 23-43.

Paran, Y. (1979). Sooty Shearwaters (Puffinus griseus) in the Eastern Mediterranean. Bull. Zool. Soc. Middle East 2: 5-6.

Paran, Y. (1980). Some notes on waterbirds observed in Egypt and north Sinai - 1978/79. Bull. Orn. Soc. Middle East 4: 2-5.

Paran, Y. & U. Paz (1980). Autumn migration of waterbirds on the North Coast of Sinai. Acta XVII Congr. Int. Orn., II: 1387-1388.

Petersen, I. & U.G. Sørensen (1980). Ornithological observation from the Bardawil-lagoon, North Sinai, Egypt. October 1979. Report, Copenhagen.

Petersen, I. & U.G. Sørensen (1981a). Migration studies from the eastern part of the lagoon "Sabkhet el Bardawil" on the North-coast of the Sinai-Peninsula, 7-30 September 1980. Report, Copenhagen.

Saad, A.H.M. (1973). Catastrophic effects of pollution on Egyptian waters near Alexandria. Atti 5° Coll. int. oceanogr. med. Messina: 553-572.

Staav, R. (1977). Etude du passage de la Sterne Caspienne Hydroprogne caspia en Méditerrannée à partir des reprises d'oiseaux bagués en Suède. Alauda 45: 265-270.

Ticehurst, C.B. (1912). The birds of Lower Egypt. The Zoologist 1912: 41-59.

IMPORTANCE OF THE MEDITERRANEAN FOR WINTERING CORMORANTS PHALACROCORAX CARBO SINENSIS

M.R. van Eerden and M.J. Munsterman
Rijksdienst voor de IJsselmeerpolders
PO Box 600, 8200 AP Lelystad, The Netherlands

Introduction

In this paper we summarize the present information available for wintering Cormorants Phalacrocorax carbo sinensis in the Mediterranean. During recent years (1980-1985) we followed sinensis birds originating from colonies in North-West Europe on their wintering grounds in the Mediterranean. A more detailed programme is in progress. Here we present some preliminary results of the project on winter feeding ecology, which pinpoint also the major questions about migration strategies of this subspecies.

As a breeding bird Ph. c. sinensis occurs both in the North-West European continent as well as in more central and eastern European localities. The present breeding status of the subspecies along the Mediterranean coasts is little known, especially for the eastern parts. Only two small colonies occur nowadays: one in Sardinia, another one in Aragon, North-East Spain (X. Ferrer, pers.comm.). New settlements have been found recently in the Po Delta (P. Brichetti, pers.comm.). Formerly breeding took place at Chikli, Lac de Tunis, Tunisia (Heim de Balsac & Mayaud, 1962). In Greece and Yugoslavia, birds might breed on islets in the Adriatic Sea, possibly also in the Aegean region. But the majority of the birds in eastern Europe seem to breed inland (Cramp et al., 1977; Hansen, 1984). As a fish eating species the birds nest close to the water, often on islets in rivers and streams, along the shores of large lakes, or in wetlands accompanying such lakes. The subspecies is well adapted to breeding in trees, contrary to Ph. c. carbo from the British Isles and Norway which nests on bare rocky islets and cliffs (Lack, 1945; Røv, 1984). Partly because of the spectacular population increase in recent years after a long period during which Ph. c. sinensis was considered a threa-

NATO ASI Series, Vol. G 12
Mediterranean Marine Avifauna
Edited by MEDMARAVIS and X. Monbailliu

tened subspecies, studies are now underway in colonies in Denmark, the Netherlands, Germany (FRG and GDR), Sweden and Poland. In this part of the breeding range a lot of birds have been ringed and the population is subject to continuous monitoring (Gregersen, 1982; Hald Mortensen, in prep.). In central and eastern Europe the current situation is less clear. Regular census data of most of the colonies are not available and it is unknown whether any overall trend in numbers of this part of the population occurs (Yugoslavia, J. Muzinic, in lit.; Rumania, I. Munteanu, pers.comm.; Greece, A. Crivelli, pers.comm.). A marked increase here does not seem likely however.

In winter, sinensis migrates south further than carbo which is sedentary or wanders along the Atlantic coasts. Sinensis from northern Europe disperses all over the Mediterranean countries including North Africa. At this time of the year some birds forage at sea, other being faithful to freshwater habitats as in summer. We are convinced that data on winter feeding ecology are of prime importance in judging both present and future development of bird numbers. The Mediterranean inshore waters are known to hold a large number of birds in winter. The question arises to what extent costs and benefits during winter might be of influence on mortality rate and reproductive performance of the birds the following summer. So far our data mainly concentrate on the western part of the Mediterranean basin, but hopefully this knowledge will be extended in more detail to the eastern Mediterranean as well.

Status and migration

The present status of the species in the Mediterranean in winter has so far not been studied through specific census programmes. However data from various ringing programmes in northern Europe underline the importance of the area.

In colonies in the Netherlands a lot of birds were ringed already in the first half of this century. Out of c. 4,300 young ringed c. 850 were recovered dead in subsequent years. Fig. 1 depicts all recoveries of dead birds during winter, here defined as the period between 1 November and 1 March (1920-1960). As can be seen the Cormorants disperse over a wide range during winter. The Atlantic coasts of France, the Iberian Peninsula and the Alpine lakes form major halting areas for Dutch birds. In the Mediterranean two regions are highlighted: the Rhône Delta in southern France as well as the coast of Tunisia. There are few recoveries from Morocco, Algeria and from Italy, except for Sardinia. Hardly any recoveries were recorded from the Adriatic coasts of Yugoslavia and Greece.

The data for birds ringed in Danish colonies in the westermost Baltic (Gregersen, 1982) show a more or less similar, very scattered pattern. Nevertheless, obviously fewer birds have been reported from the Atlantic coasts of France, Spain and Portugal and very few come from countries in the western Maghreb region as compared to the Dutch recoveries.

Again the Rhône Delta and Tunisia hold a large proportion but contrary to the Dutch birds also Italy and Yugoslavia provide wintering areas for these birds. The picture for the Swedish colonies in Kalmarsund very clearly suggests a southward movement right across central Europe reaching the Adriatic Sea east of the Alps (fig. 1). Compared to the Dutch data a clear shift to the east is obvious and this distinction also holds true with regard to the Danish colonies (cf. Andersson et al., 1984). Apparently very few birds from the Swedish breeding colonies fly southwest to reach France or the western Mediterranean. The recovery pattern shows, besides the wide range in wintering areas, the trend that birds from different colonies in North-West Europe (Netherlands - Sweden being c. 900 km apart) might segregate in winter as well, although the segregation is not absolute.

The birds from GDR and Poland also follow mainly the route east of the Alps to wintering areas along the Adriatic, in Greece and Tunisia (Bauer & Glutz von Blotzheim, 1966; Gromadzka, pers. comm.).

To indicate the timing of migration, fig. 2 shows as an example a set of Danish recoveries arranged by month for four areas in Europe. According to Gregersen (1982) up to mid-October most birds are still present in northern and central Europe. From October onward southern Europe is reached and especially in January birds are recovered from the southernmost wintering areas in the Mediterranean.

The onset of spring migration is not easily discerned from recovery data alone. A lot of birds are recovered in their first autumn, but already by midwinter the highest mortality seems to be over leading to a steep drop in recovery rate. Therefore the spring migration is somewhat masked. A further reason for a decline in recoveries in this period is the closure of the hunting season in most European countries in February or March. Duck hunting until the end of March is only allowed in Tunisia, Greece and Rumania (Lampio, 1982). Frequent counts in the Camargue point to a northward migration of Cormorants in March, supported by direct observations of migrating birds over the Cévennes hills (E. Carp, pers. comm.).

Most adult birds arrive in the colonies in North-West Europe during the second half of March and the beginning of April, some birds being present already in January/February. The arrival of subadult and second summer birds occurs somewhat later, in late April and May (Gregersen, 1982; van Eerden, unpubl.). As can been seen in fig. 2, in March recoveries have still been reported from the southern part of the Mediterranean. Birds from the Cagliari area on the south coast of Sardinia leave mainly in the last decade of March and the beginning of April (A. Mocci Demartis, H. Schenk, pers.comm.). At Lac de Tunis large numbers are present until the end of March (A. Bal Khaddi, K. Nassr, pers.comm.).

Fig. 1: Ring recoveries of birds reported dead originating from colonies in the Netherlands (A), Denmark (B) and Sweden (C). Each dot represents 1, 5 or 10 recoveries depending on size. Panels A and B refer to the period 1 November - 1 March, C depicts all recoveries. Redrawn after Gregersen (1982) and Andersson et al. (1984).

Numbers, habitat in winter

As stated above no overall census data are available for the species in its winterquarters, although January counts of waders and waterfowl often include Cormorants. From the recovery data two wintering sites turned up to be of major interest, southern France and Tunisia. In southern France we collected information on numbers, foraging behaviour and flight range in December 1982, December 1983, January 1985 and January 1986. The coasts of Tunisia were visited twice intensively, in February 1984 (van Dijk et al., 1986) and in January 1985. For comparison we also visited two areas that seemed poor for wintering Cormorants, the Italian Riviera and Sicily, in January 1986. Additional data are provided by Im and Hafner (1984).

• Southern France, Italian Riviera

In southern France the Camargue forms an area of international importance for waterfowl, thanks to its great variety of aquatic habitats. For a detailed ecological description of the Camargue we refer to Blondel & Isenmann (1981). Fresh and brackish lakes, ditches and pools, saline shallow areas in close proximity to the sea together with the riverine influence of the Rhône (which is never frozen over) make the Camargue a multiple choice fishing area, which means a predictable wintering place for Cormorants. These various habitats provide fish populations of several species and/or age class composition (A. Crivelli, pers.comm.). Throughout the season regular shifts in foraging areas occur which may be the result of differences in availability of prey (fish).

Four large roosts serve the birds as a resting place overnight. The foraging flights to and from these roosts were recorded following individuals and small flocks by car. Especially the early morning and late afternoon passage provided enough data to reconstruct flyways from individual observations. The major foraging flights have been summarized in fig. 3 (direction and distance observed). As can be seen the Cormorants spread out from the roosts both to the sea as well as to inland sites. Each of the four roosts is known to hold between 2,000 and 3,000 birds in winter. The total winter population (1980-1985) has varied between 6,000 and 9,000 birds (Pasquet, 1982; Im & Hafner, 1984). Flight range was recorded up to 40 km one way:Etangs de Perpignan - Bassin de Thau. Regular flights range from 20 to 30 km. These four "primary roosts" are connected to 10 to 15 "secondary roosts" where birds rest at daytime between foraging trips. Sandbanks, islets, poles of fyke nets, wrecked vessels and rocks surrounded by water can serve for this purpose. During periods of severe frost (as in January 1985) a large part of the wetlands can be frozen. In the winter of 1985 even the saline Etang de Vaccarès was completely covered with ice. The roosts are then abandoned and the birds sleep in coastal regions or in hypersaline water-logged saltpans (Salin de Giraud, Golfe de Fos).

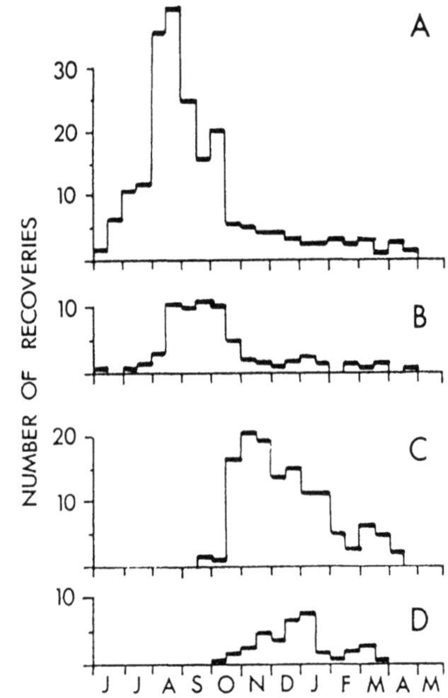

Fig. 2: Distribution of recovered birds from Danish colonies according to time of year, for 4 regions in Europe. Notice major shift to the Mediterranean in the second half of October. A = Breeding area, B = Baltic and North Sea, C = Northern Mediterranean, D = Southern Mediterranean. Redrawn with permission from Gregersen (1982).

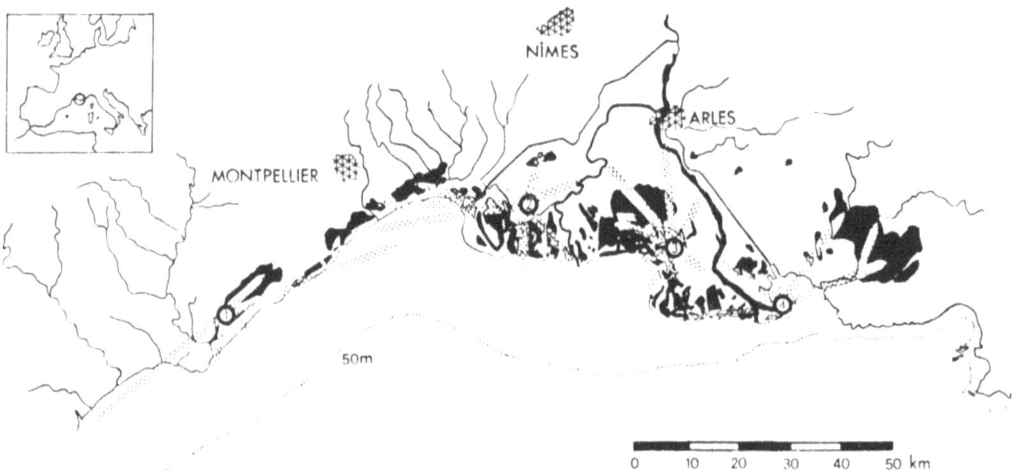

Fig. 3: Wintering area in southern France, Rhône Delta, showing patchy distribution of fishing grounds in the shallow Mediterranean (salt), Etangs and salines (salt, brackish, fresh) and river Rhône (fresh). The main foraging areas exploited from 4 main roosts have been indicated by arrows (1981-1986). 1 = Bassin de Thau, 2 = Etang de Lairan, 3 = Etang de la Dame, 4 = Golfe de Fos.

The situation in southern France described here seems typical for the most frequented coastal areas in the western Mediterranean: Tajo Estuary Portugal (Teixeira, 1984), Ebro Delta Spain (X. Ferrer, pers.comm.). To make a comparison with the delta habitats the steep Mediterranean coasts further east between Marseille and Genoa were visited in January 1986. A maximum number of 100 birds was present there, a number strongly diverging from that of the large scale shallows between Perpignan and Marseille.

• Tunisia, Sicily, Sicilian Channel Islands

Further to the south the data on ring recoveries suggested the Tunisian coast to hold a large proportion of wintering Cormorants. Especially Lac de Tunis has been known for a long time to be an important wintering area (Heim de Balsac & Mayaud, 1962). Recently almost the entire coast was visited during winter twice, with special attention to Cormorants. One Dutch expedition collected data on the Golfe de Gabès which is difficult to get to (van Dijk et al., 1986). The authors concentrated on the rest of the coast in January 1985 and paid special attention to the Iles de Kerkennah, south-east of Sfax, which were already known as a wintering place (Gaultier, 1981). The two census periods revealed a complete picture of the winter distribution of the species.

Similarly to the situation in southern France the birds roost in coastal waters, saltpans and lakes. In the north, Lac de Tunis is by far the most important place for Cormorants. Birds were seen fishing on the lake itself but also went to sea to forage. The roost here is the old Punic fortress on the islet of Chikli. In recent years (1980-1985) this roost may hold up to 3,000 birds. Further north smaller numbers are found, Lac de Bizerte being the most important roost holding 450 birds in January 1985. The most prominent wintering area along the Tunisian coast however is to be found in the Golfe de Gabès, including Kerkennah Islands. Along the entire coasts of this area Cormorants were found, some sleeping scattered on fyke poles and other fishing gear, others concentrating on certain roosting sites e.g. the Salines de Thyna, which can hold over 1,000 birds. The largest roosts however occur in islets east of Ile de Gremdi, Kerkennah. Here six uninhabited islets (several hectares each, low sparse vegetation of halophytes: Salicornia, Halimione, etc.) were investigated with the help of local fishermen. These roosts have been known by the inhabitants of the Kerkennah Isles a long time already. Two complete counts, one in the morning of 24th January 1985 during take off of the Cormorants and one the following afternoon and evening during their return, revealed 11,850 and 10,770 birds respectively. Furthermore east on the edge of an offshore plateau a series of wrecked vessels was known to hold a large number of Cormorants too. This population was conservatively estimated at 8,000 birds in January 1985 (A.B. Abdessatiar, M. Khelif, pers. comm.). These data reveal that the entire population wintering in the Golfe de Gabès, including offshore waters and Kerkennah, probably was about 23,000 birds. Together with data along the

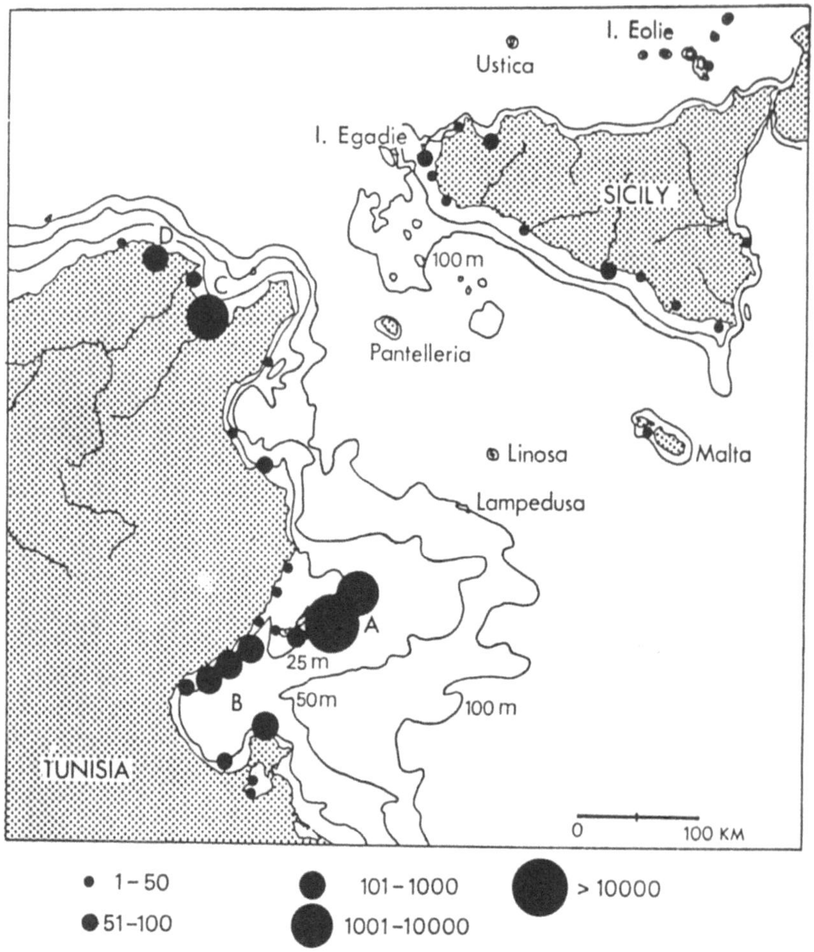

Fig. 4: Wintering area in the Southern Mediterranean, along the Sicilian Channel, showing distribution of bird numbers in relation to the presence of shallow water as indicated by depth contours of 25,50 and 100 m.
A = Iles de Kerkennah, B = Golfe de Gabès, C = Lac de Tunis, D = Lac de Bizerte.

rest of the coast, including the waters around Djerba and Bir-het el Bibane, we estimate for Tunisia a total wintering population of about 25,000 to 30,000 birds in recent years.

For comparison we visited the other side of the Sicilian Channel between 12 and 29 January 1986. The entire coast of Sicily was studied and special attention was paid to areas with small islets, dams and other structures which could serve Cormorants as a resting place. We estimate the total population on Sicily in January 1986 at c. 700 birds, most of them present in inland lakes and on the more shallow coasts in the south and west of the island. The archipelagoes of the Eolic and Egadic Islands do not hold large numbers either (Moltoni and Frugis, 1967; Sorcie et al., 1973). Also the isolated island of Ustica is scarsely visited by Cormorants (Ajpma; 1959) as is the case for the Pelagic Islands: Linosa, Lampeduas and Lampione (B. Massa, pers.comm.; Moltoni, 1970) and for Pantelleria (Moltoni, 1973). The situation of the Maltese islands has been investigated in more detail by Sultana and colleagues, who estimated about 25 birds maximally to winter there (J. Sultana, pers. comm.). The data have been plotted in fig. 4 and as can be seen at a glance there is an obvious difference in wintering numbers of Ph. c. sinensis on either side of the Sicilian Channel. Especially the more shallow seas indicated in fig. 4 by the 25, 50 and 100 m isobaths seem to be an overall prerequisite for any significant bird numbers in winter.

Food and feeding habitats in winter

Cormorants are almost entirely piscivorous. Especially smaller fishes (10 to 100 g) are taken, but sometimes large prey are caught (400 g *Cyprinus carpio*, Osieck, 1981; 500 g *Anguilla anguilla*, van Dobben, 1952).

Data from northern Europe during summer show a wide variety of prey species that occur in the diet of these birds. According to the local abundance of certain fish preys Cormorants seem to be highly opportunistic in their food choice. Both gregarious and solitary living fish can be hunted and the birds themselves may forage alone or in flocks (Madsen and Spärck, 1950; Jonsson, 1979; van Dobben, 1952; Gregersen, 1982; pers.obs.). So far we can not explain all these differences but it seems likely that the birds might show an overall adaptation to the different behavioural patterns of their prey.

In winter the situation does not differ much from that in summer. Social fishing has been observed in the Mediterranean: Camargue, southern France (Im and Hafner, 1984, pers.obs.), Tuscany, Italy (Baccetti, pers.comm.) and Tunisia (Lac de Tunis, Kerkennah).

According to the number of birds present, flocks may consist of tens to hundreds or up to several thousands of birds. Generally this phenomenon of social fishing is met in shallow areas where the sea bottom is flat, which allows the birds to concentrate their prey with little risk of escape (Van Eerden and Voslamber,in prep.). Few data are available about prey choice in Mediterranean waters. In their study on wintering Cormorants in the Camargue;Im and Hafner (1984) collected pellets on different roosts. In these pellets the residue of fish can be detected easily and especially otoliths, pharyngeal bones and jaw structures prove to be diagnostic for species determination. Moreover from the length of otoliths and pharyngeal bones the length (and thus weight) of the fishes involved can be reconstructed (see for methods Doornbos, 1979). In tab. 1 the data from Im and Hafner have been rearranged using only the overall mean of four samples from January/February to indicate the general picture for this area. As can be seen from tab. 1 fish species from salt, brackish and fresh water habitat are the constituents of the Cormorants' diet. Out of 432 pellets (representing one daily ration each) over 4,700 fish preys were identified. Only a few species are of major importance.

On the basis of number of prey captured, Sand Smelt Atherina boyeri and Sun Perch Lepomis gibbosus rank first with 52 and 23% respectively, followed by Mullet Mugil spp.(14%). The latter genus (four species present in Camargue) was however present in about half of the meals studied. No exact data exist about the proportion of the prey species according to weight. Due to identification problems Im and Hafner (1984) could not make this comparison in detail. Because of large differences in weight of the species involved we can make a gross approximation by using a range in weight of prey fish eaten. This range is based on our observations on captures of fishing Cormorants and on the abundance of common prey fish in Camargue waters (Crivelli, 1981; R. Britton, pers.comm.). On the basis of weight Mullet Mugil spp. was estimated to form about 40 to 60% of the birds' diet, followed by Carp Cyprinus carpio (15 to 30%) and Sun Perch Lepomis gibbosus (15 to 20%).

For Tunisia some preliminary information is available from the Kerkennah Islands. Here our analyses of the pellets point also to few species of fish taken of which Diplodus sargus (Sparidae) forms the major part (more than 75% on basis of weight, 1984/85 samples). This fish occurs preferentially in the Posidonia sea weed zone which again confirms the Cormorants' preference for inshore waters. Moreover the diet on Kerkennah seems more uniform than in the Camargue. This might be due to the abundance of certain fish preys in a less complex habitat than we described for southern France, but also to the very prominent habit of social fishing of Cormorants as compared to the Rhône Delta.

Movements of individual birds

In Denmark, the Netherlands, Poland and GDR coloured marking schemes using PVC rings on birds are in progress. Inscriptions can be read in the field by telescope up to 400 m. Because especially Danish birds have been colour ringed intensively, their prominent presence in the field does not necessarily mean

Species	Fish habitat	Approx. weight range (grams)	Frequence in diet (%)		
			Meal	Number of prey	Weight of prey
Atherina boyeri Sand Smelt	Salt	2-4	11	52	1-4
Mugil sp. Mullet	Salt/brackish	75-150	52	14	40-60
Solea solea Sole	Salt/brackish	20-50	1	<1	<1
Anguilla anguilla - Eel	Brackish/fresh	30-100	4	1	1-3
Cyrprinus carpio Carp	Fresh	75-150	17	7	15-30
Rutilus rutilus Roach	Fresh	75-150	3	4	5-10
Lepomis gibbosus Sun Perch	Fresh	20-50	12	23	15-20

Tab. 1: Diet of Cormorants in the Camargue, southern France during winter according to Im & Hafner (1984). Data have been rearranged using overall mean for 432 pellets with more than 4,700 prey fish. (Including Scardinius erythropthalmus.

that the origin of the various flocks is predominantly Danish. Especially with regard to wintering birds the potential information based on sightings is immense and some preliminary findings will be given here in order to indicate the type of questions that we hope to answer in future.

One of the problems to tackle is that of site tenacity in winter. A striking example of a wintering bird showing up every year at the same place is WHITE 715. This bird was born in 1982 in Vorsø, Denmark. It was seen in four subsequent winters at Marais d'Olonne, Vendée, along the French Atlantic coast (P. Yésou, in lit.): 3 January 1983 - 16 March 1983, 9 November 1983 - 13 March 1984, 29 November 1984 - 14 March 1985, 14 November 1985 - 14 March 1986. The similarity in time schedules from year to year is remarkable. Other birds are only noted for shorter times, which indicates passage to other areas. Only regular observations throughout a series of years can test the hypothesis that also certain stopover sites might be used every year by the same birds. A bird suggesting this pattern is RED 510. Born in Denmark, Brandesgardsø 1984, it was observed on the 2nd and

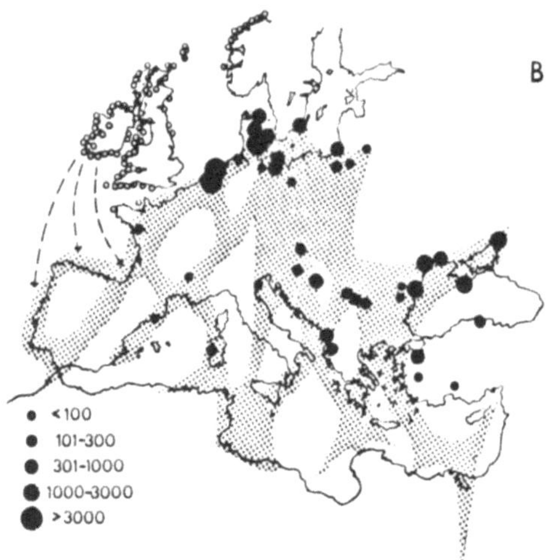

Fig. 5: A. Overview of potential wintering habitat for Cormorants on an European scale. Shallow areas (<50 m depth) are indicated in black as well as the course of the main rivers. The January isotherms of 0, 5 and 10°C have been shown by broken lines.

B. Map of main colonies of sinensis (filled dots) and carbo (open dots) in Europe with suggested flyways to prime wintering areas. The situation in eastern Europe is only little known as to position and size of colonies.

8th November 1984 at Marais d'Olonne, Vendée. Thereafter it only reappeared on 13th January 1985. One year later this bird returned to the same place where it was seen on 21st and 28th November 1985 and was spotted again on January 21st, 1986 (P. Yésou, in lit.). The resemblance of the time-tables of different years suggests the development of annually repeated itineraries to and from more or less known feeding areas. A typical "long range" example of site tenacity to stopover sites is GREEN 179. This bird was born in the colony Toftesø, Jutland Denmark, where it was ringed on June 9th 1984. On August 4th 1984 it was seen flying at Havnø, a site near the colony. On November 20th it was present at Veerse Meer, South-West Netherlands, and on January 19th 1985 in Lac de Tunis. The next year the bird returned to Toftesø where it was seen twice on August 9th and 14th. Again it turned up at Veerse Meer on October 21st 1985, presumably on its way to the Mediterranean. More data will have to be collected in future in order to quantify the occurrence of site tenacity in winter and to unravel other possible migration patterns.

Future prospects

One of the major tasks in future will be to gather more information about wintering areas and stopover sites of Cormorants. Both counts as well as observations on colour ringed Cormorants can give an indication of the origin of the birds at a certain site. For example out of c. 9,200 birds marked with colour rings in Denmark and the Netherlands over 8,700 were from colonies in Denmark (c. 95%). Field estimates of the proportion of Danish birds from the Rhône Delta (1982-1986) and Tunisia (1985) in winter were 41/52 (79%) and 30/36 (83%) respectively, the rest being of Dutch origin. Unless large differences exist in mortality rates and/or rate of ring loss this points to a relative predominance of Cormorants of Dutch origin in these two western Mediterranean sites.

In fig. 5 an attempt has been made to visualize the different options the birds have to choose from for selecting a wintering site on an European scale. As we have shown, the wintering sites of Cormorants are especially found in shallow waters, in lakes, lagoons, estuaries and sometimes at sea. In panel A an overview is given of the occurrence of shallow waters (50 m isobaths) and the course of the most important large rivers. The greater depth of the Mediterranean as compared to the North Sea and Baltic is obvious. However if winter temperatures are taken into account, a large part of the suitable areas in northern Europe are not predictable wintering sites because of frost in the fishing grounds. Generally speaking between the January isotherms of 0° and +5°C extensive ice cover of fresh water may occur in some years, forcing the birds to move south. Suitable wintering areas as far as depth and temperature are concerned can be found in southern France, Tunisia, North-East Spain, around Sardinia, along the West coasts of Italy, along both sides of the Adriatic and parts of the Aegean Sea. Further South the coasts of Libya and Egypt may be suitable wintering areas for Cormorants. Of course depth and temperature are not

Fig. 6: Differences in stage of prenuptial moult as indicated through visual sampling of populations at different latitudes. Notice the advanced stage of populations in the Mediterranean compared to those in central Europe. Type 5 birds with completed moult show the opposite trend being relatively more numerous towards the breeding colony. (Absolute numbers are highest in the south). Data refer to the situation in the first decade of February 1985. A = Tunisia, B = Camargue, C = Valence, D = Rheinau, E = Lac Leman, F = IJsselmeer, G = Colony Oostvaardersplassen.

the only factors to qualify a wintering site. Safe roosting places and predictability of fish abundance are also important.

In panel B of fig. 5 the main colonies of both sinensis and carbo have been depicted. For eastern Europe and USSR this information is only indicative for the distribution and size of the colonies. Following the findings presented above, the birds from different colonies in northern Europe segregate in winter. The major halting places and wintering areas have been connected to suggest the main migration pathways. The birds from eastern Europe are expected to migrate mainly southeast and south to reach the coasts of Greece, Libya and Egypt. Also sinensis that has been seen migrating along the river Nile to Sudan (P.L. Meininger, pers.comm.) might belong to this group. Further counts are needed to verify the patterns suggested above.

Another point of interest for future research will be to study the different migration strategies of the Cormorants, some birds wintering relatively close to the colony, others migrating far south. If site tenacity proves to be a widespread characteristic of individuals, then it might be possible to compare the reproductive performance and mortality rates of birds known to have wintered at different places in order to evaluate costs and benefits of the different migration strategies. For the reproductive output of a pair two items are of prime importance: the timing of start of breeding and their nutritional reserves which are interrelated. Early breeders have in general more offspring than later arriving pairs (Gregersen, 1982). Birds with greater nutritional reserves are believed to perform better than birds with a lesser body condition (Drent and Daan, 1980). Concerning the timing of start of breeding,close to the colony wintering birds might be at advantage over long range migrants. The former birds are able to respond readily to the onset of spring and to start breeding as soon as possible. With respect to nutritional reserves the picture is less clear. On the one hand the long distance migrants might be able to fatten up faster because of better feeding conditions at lower latitudes (winterspawning of fish, greater activity of fish due to higher temperatures). Moreover the birds themselves may benefit from a reduction of their daily energy expenditure (higher temperatures, increased solar radiation). However compared to birds wintering close to the colony, the long range migrants experience higher costs of migration thus diminishing their initial reserves. On the other hand the birds wintering closer to the colony have to cope with lower prey availability and are at risk during cold spells when their waters may be closed off by frost. Besides these two extremes intermediate patterns of timing and distance covered may occur. As yet it is unknown to what extent the hypothesis of Pienkowski et al. (1985) on frog-leap migration and sexual segregation (subdominant individuals being forced to migrate further south to avoid competition with dominant individuals) may hold true for Cormorants.

In order to study the different migration strategies data on colour ringed birds will be of paramount importance. In this context we present some preliminary data which point to the existence of differences amongst individuals wintering at different latitudes. A first example concerns the time of return to the colony.

In winter 1984-1985 we read a number of individuals from Danish colonies at various places in Europe. Because ring reading effort is greatest in Vorsø we selected birds from that place. We compared data of return in Vorsø for birds we spotted in Lac Leman, Switzerland and Tunisia. For 4 Swiss birds that were seen in the colony, out of 7 observed, the average date of arrival at Vorsø was April 5th (median value, range 28.3. - 18.4.). For 4 Tunisian wintering birds, out of 11 read, the date of return in Denmark was April 23rd (range 3.4. - 5.5.). So the long-distance migrating Tunisian birds tended to arrive later that year than the ones from Switzerland which is in agreement with the suggestions presented above.

Other data suggesting the existence of differences between Cormorants from different wintering areas are concerned with the stage of prenuptial moult. Cormorants of the sinensis group develop in spring conspicuous white patches of feathers on their flanks, together with the formation of elongated white feathers on the head and neck. We distinguished 5 types in the field, that have been depicted in fig. 6. During the first ten days of February 1985 we travelled from south to north on our way back from Tunisia and visited 7 sites which held Cormorants. At these sites, Tunisia (Lac de Tunis, Kerkennah), the Camargue, Rhône near Valence, Lac Leman near Villeneuve, Rhine near Rheinau, IJsselmeer near Harderwijk and in the colony Oostvaardersplassen, we made counts of the proportion of the plumage types. At each site at least 100 adult birds were inspected (in the colony only 30, no more birds being present at that time). Fig. 6 shows that in Tunisia more birds have started moulting head and flank feathers than at any other place visited. Further north fewer birds have started moulting towards breeding plumage, especially on the Swiss and German sites.

So in general the long distance migrants seem to have started their prenuptial moult earlier than birds wintering further north. This may be induced by increased daylength in the southern wintering areas (Lofts, 1975; Mewaldt & King, 1978). Surprisingly enough, going north however, an ever increasing proportion of type 5 birds, with completed moult, was recorded. At the IJsselmeer, at 25 km distance from the colony, and in the colony itself 39 respectively 80% belonged to this category. Because of the lack of a similar trend in type 4 birds it seems less likely that the well advanced moulters originated from the population present at those sites. More likely we think that the birds moved north from a more southerly wintering area after or during the last stages of moult completion. During our period of study in the Mediterranean two birds indeed proved to have migrated north in late winter already. GREEN 180 (Toftesø, Denmark, 1984) was seen in Lac de Tunis on January 19th 1985 and already on February 16th 1985 it was seen in IJmuiden, the Netherlands, before reaching Denmark, where it was observed again.

WHITE 387 (Vorsø, Denmark, 1982) was seen at Lac Leman near Villeneuve, Switzerland, on February 5th 1985. Only 12 days later it proved to be back in the Western Baltic, Fehmarn Sund, near Burg (FRG).

To sum up, we hope to have indicated by the examples given that the type of information based on field observations of individual birds is worthwhile. Especially in the years to come we would like to welcome observers throughout the winter range of Ph. c. sinensis to collect data on the following subjects:

- Field counts in winter, especially on roosts in January;
- Readings of colour marked birds, October - April, with notice on plumage-type;
- Characterization of local populations: proportion of juveniles October-April, plumage types (cf. fig. 6) in adults, first decade of February;
- Study of the diet, October-April, through analysis of pellets.

Acknowledgements

During the various stages of the project we experienced great hospitality from staff at Station Biologique de la Tour du Valat. Dr. Bun Hort Im introduced us to several roosts and feeding areas, often difficult to get to. Dr. Alan Johnson helped in getting permission to enter Salin de Giraud and he as well as John Walmsley proved to be a sound base for knowledge about the species and the Camargue area. Mr. A. Bal Khaddi, Office National de Pêche, Tunis, kindly gave us the opportunity to study Cormorants at Chikli, Lac de Tunis which was of great help for ring readings. Drs. Gré Blokzijl, Zoological Laboratory Groningen University arranged the usage of a portable mistral-proof hide. Mr. Rinse Wassenaar supplied data of recoveries of Dutch Cormorants from the Euring system. Dr. Rudolf Drent and Dr. Heinz Hafner were of great help in improving the manuscript. The co-operation with our Danish colleagues concerning the exchange of information about colour marked birds was a great stimulus for us. Jens Gregersen permitted kindly to use data from his wonderful book which were redrawn in fig. 1 and 2.

References

Ajola, G. (1959). Gli Uccelli dell'Isola di Ustica. Riv. It. Ornit. 29: 89-128.

Andersson, G.; J. Karlsson and N. Kjellén (1984). Storskarven Phalacrocorax carbo i Skane. Tidigare förekomst och nutida uppträndande. Anser 23: 109-124.

Bauer, K.M. and Glutz von Blotzheim, U.N. (1966). Handbuch der Vögel Mitteleuropas, Band 1. Akad. Verlagsgesellschaft, Frankfurt.

Blondel, J. and Isenmann, P. (1981). Guide des Oiseaux de Camargue. Delachaux and Niestlé, Neuchâtel-Paris.

Brichetti, P. (1982). Distribuzione geografica degli Uccelli nidificanti in Italia, Corsica e Isola Maltesi. Natura Bresciana 19: 97-157.

Cramp, S. and Simmons, K.E.L. (eds.) (1977). Handbook of the birds of Europe, the Middle East and North Africa, volume 1. Oxford University Press, Oxford.

Crivelli, A. (1981). Les peuplements de poissons de la Camargue. Terre et Vie 35: 617-671.

Dobben, W.H. van (1952). The food of the Cormorant in the Netherlands. Ardea 40: 1-63.

Doornbos, G. (1979). Winter food habits of Smew (Mergus albellus L.) in Lake IJssel, The Netherlands: species and size selection in relation to fish stocks. Ardea 67: 42-48.

Dijk, A.J. van; Dijk, K. van; Dijksen, L.J.; Spanje, T.M. van and Wymenga, E. (1986). Waders and Waterfowl in the Gulf of Gabès, Tunisia, from January to March 1984. Report of the WIWO-Tunisia Project 1984.

Drent, R.H. and Daan, S. (1980). The prudent parent. Energetic adjustments in avian breeding. Ardea 68: 225-252.

Gaultier, T. (1981). L'Avifauna des Iles Kerkennah. Report Université de Tunis, Belvédère.

Gregersen, J. (1982). Skarvens Kyster. Bygd, Esbjerg.

Hansen, K. (1984). The distribution and numbers of the Southern Cormorant Phalacrocorax carbo sinensis in Europe. Dansk. Orn. Foren. Tidsskr. 78: 29-40.

Heim de Balsac, H. and Mayaud, D. (1962). Les Oiseaux du Nord-Ouest de l'Afrique. Paris.

Im, B.H. and Hafner, H. (1984). Impact des Oiseaux piscivores et plus particulièrement du Grand Cormoran (Phalacrocorax carbo sinensis) sur les exploitations piscicoles en Camargue, France. Report C.E.E. ENV-491-F, 83 pp.

Jonsson, B. (1979). Skarvarna och yrkesfisket i Kalmarsund. Calidris 8: 171-200.

Lack, D. (1945). The ecology of closely related species with special reference to Cormorant and Shag. J. Anim. Ecol. 14: 12-16.

Lampio, T. (1982). Hunting rationalization - a tool for management of waterfowl populations. Proc. 2nd Tech. Meeting on Western Palearctic Migratory Bird Management, Paris, 1979: 107-112. IWRB.

Lofts, B. (1975). Environmental control of reproduction. Symp. zool. Soc. Lond. 35: 177-197.

Madsen, F.J. and Spärck, R. (1950). On the feeding habits of the Southern Cormorant in Denmark. Dan. Rev. Game Biol. 1: 45-76.

Mewaldt, L.R. and King, J.R. (1978). Latitudinal variation of post-nuptial molt in Pacific coast White-Crowned Sparrows. Auk 95: 168-174.

Moltoni, E. and Frugis, S. (1967). Gli Uccelli delle Isole Eolie (Messina, Sicilia). Riv. It. Ornit. 37: 91-234.

Moltoni, E. (1970). Gli Uccelli ad oggi riscontrati nelle Isole di Linosa, Lampedusa e Lampione (Pelagie). Riv. It. Ornit. 40: 77-283.

Osieck, E.R. (1981). Verjaging van Aalscholvers en Blauwe Reigers op de Viskwekerij Lelystad. Report SBB, 87 pp.

Pasquet, E. (1982).Statut hivernal actuel des Plongeons, Grèbes et Grand Cormoran. Report C.R.B.P.O., Paris, 37 pp.

Pienkowski, M.W.; Evans, P.R. and Townshend, D.J. (1985). Leapfrog and other migration patterns of waders: a critique of the Alerstam and Högstedt hypothesis, and some alternatives. Ornis Scandinavica 16: 61-70.

Røv, N. (1984). Clutch size and hatching time in colonies of Cormorant _Phalacrocorax carbo carbo_ in Central Norway 1982-1984. Fauna Norv. Ser. C., Cinclus 7: 124-126.

Sorci, G.; Massa, B. and Cangialosi, G. (1973). Avifauna delle Isole Egadi con notizie riguardanti quelle della prov. di Trapani (Sicilia). Riv. It. Ornit. 43: 1-119.

Teixeira, A.M. (1984). Sobre a ocorrência e distribuiçao em Portugal do Corvo-Marinho _Phalacrocorax carbo_. Cyanopica 3: 215-223.

Thomsen, P. and Jacobsen, P. (1979). The Birds of Tunisia. Copenhagen.

STATUS AND CONSERVATION OF THE STORM PETREL HYDROBATES PELAGICUS IN SICILY

Bruno Massa & Adelaide Catalisano
Istituto di Zoologia dell'Università di Palermo

Mediterranean Distribution

According to James (1984) there are few data available about the status and the distribution of the Storm Petrel in the Mediterranean Sea. Bibliographical research shows us how this species is generally decreasing in the whole Mediterranean, and even disappeared in many breeding sites. Fig. 1 gives the actual situation. We report the following comments.

- Aegean Sea

The only known breeding site quoted by Handrinos (1984) has been found in an islet off eastern Euboea. Akriotis & Handrinos (1986) refer that only a few pairs of Storm Petrels breed there.

- Adriatic Sea

Absent in the Italian coasts, the Storm Petrel has been reported as a breeder along dalmatian coasts by Matveiev & Vasic (1973).

- Sicilian Channel

The greatest colony in the Mediterranean is on the Isla Filfla (Malta). According to Sultana & Gauci (1982) the population there is over 10,000 pairs. The Storm Petrel also breeds on the rock of Lampione, off Lampedusa (few pairs). Breeding at Pantelleria is doubtful. In Isola Marettimo (Egadi) we reported a big colony (between 500 and 1,000 pairs: Massa & Catalisano, in press). A few pairs breed in Isola Levanzo (Egadi). According to Thomsen & Jacobsen (1979) the Storm Petrel disappeared in the small islands and along the coasts of Tunisia.

NATO ASI Series, Vol. G 12
Mediterranean Marine Avifauna
Edited by MEDMARAVIS and X. Monbailliu

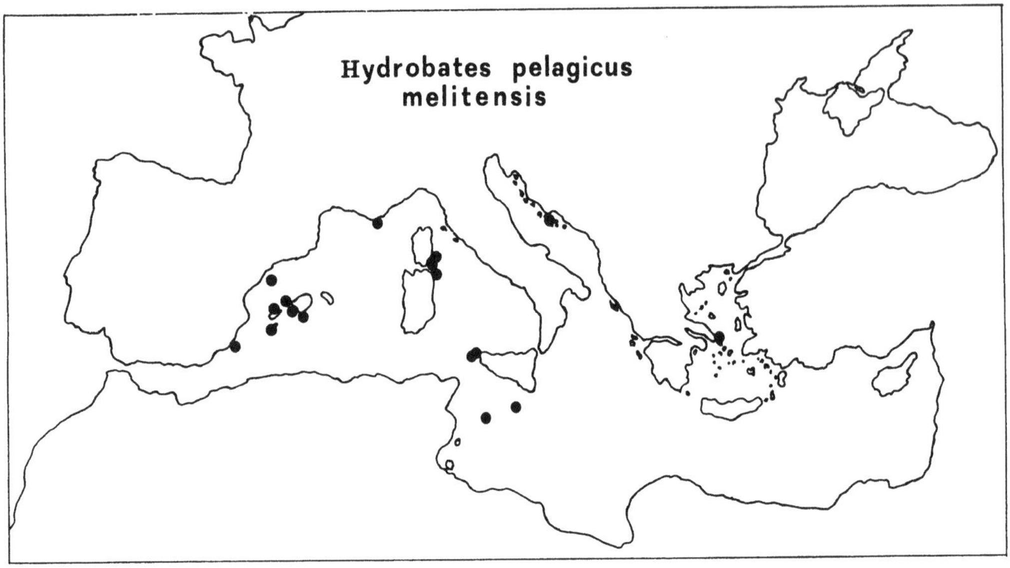

Fig. 1: Mediterranean distribution of the Storm Petrel, according to contemporary authors.

• Tyrrhenian Sea

There is no evidence of breeding in Isola di Ustica and the Aeolian Islands, and even if there were, the Storm Petrel would surely have a very little population. We did not find any evidence of breeding along the islets off the Italian coasts. Brichetti (1985) reports as only possible sites the Tuscan islets and the small islands off Latium and Campania coasts.

• Sardinia and Corsica

Torre (1983) reports the Storm Petrel decreasing, though quite frequent. Besides he states that there are few data about its breeding in Sardinia. Brichetti (1985) reports six probable breeding sites and Schenk & Torre (1986) 150-300 possible pairs in Sardinia. Guyot et al. (1985) report only three breeding sites in Corsica, two of them are constituted by several tens of pairs and one of less than 10 pairs.

• Mediterranean French coasts

Guyot et al. (1985) stated that one of the five known French breeding sites in the Mediterranean still exists. This only colony is constituted by a variable number of pairs (from 13 to 50). See also Walmsley in this volume.

• Mediterranean Spanish coasts

According to de Juana (1984) the only Spanish breeding sites are concentrated in the islets of Medas, Columbretes and Cabrera, and inshore islets of Alicante and Maiorca. They consist of hundreds of pairs (de Juana & Mayol, pers. comm.).

• Coasts of Algeria and Morocco

According to de Juana et al. (1980), Jacob & Courbet (1980), Ledant et al. (1981) and Berthon & Berthon (1984), there is no evidence of breeding of Storm Petrels along these coasts.

In conclusion, it seems that the Mediterranean Storm Petrel now breeds only on few islets. There are probably less than 30 colonies, which consist mainly of very few pairs and rarely of over ten pairs (less than 100). There are only two big colonies, both of them in the Sicilian Channel, the biggest at Filfla and the smallest at Marettimo. We can estimate that outside these colonies, 800-1,000 pairs of Storm Petrels still breed in the Mediterranean.

Historical Situation in Sicily

Doderlein (1869-74) stated that the Storm Petrel was fairly common in the Sicilian sea. Martorelli (1906) considered it the most common Petrel in the Mediterranean Sea, especially throughout the winter months. It is interesting to indicate what Moltoni & Vandoni had known by Arrigoni degli Oddi and later reported in Marotrelli's edition of 1931: "*Abbondante nello Stretto di Messina, ove è noto sotto il nome di fituseddu, disturba le pesche divorando in massa le esche oleose. Fino a pochi anni fa ne venivano catturati a sacchi coi retini da pesca. Basta un pezzetto di fegato dell'Orthagoriscus mola L., gettato in mare per attirare numerosi gli uccelli delle tempeste a pochi centimetri dalla barca, ed è così sopratutto che vanno a finire dentro il retino manovrato destramente dai pescatori.*"(1).

---

(1) "The Storm Petrel is abundant in the Strait of Messina, where it is called "fituseddu". Eating many oily baits it interfers with fishing. Until a few years ago, fishermen caught a lot of these birds with their fishing nets. It is sufficient to throw a piece of liver of O.m. in the sea to attract many Storm Petrels at few centimetres from the boat. In such a way they get caught by the fishermen with their nets."

Giglioli (1907) reported the Storm Petrel as a breeder in the Aeolian islands, where he found the fragments of its egg. Breeding in the islets surrounding Sicily was reported by Krapp (1970) at Marettimo, by Massa (1973) at Levanzo and Marettimo, and by Moltoni (1970) at Lampione, off Lampedusa. There is no evidence of breeding in Isola Pantelleria. Doderlein (1869-74) found that the Storm Petrel frequently followed the boat going to Pantelleria. Also Moltoni (1954), Steinbacher (1956) and Foschi (1968) observed some individuals off Pantelleria, but the Storm Petrel has not been found breeding till now.

The hypothesis that it could breed along the North-Western coast of Sicily (Orlando in Krampitz 1958) has never been confirmed and now we believe that it should be rejected.

In conclusion until the beginning of this century the Storm Petrel was widespread and abundant in Sicily and perhaps in the Mediterranean basin. On the contrary, from the beginning of this century until the 1950's it was less spread out but not rare. It seems that the peak of its decrease was during the last 30-40 years.

Frugis & Schenk (1981) include it among the species on the verge of extinction, but Brichetti & Cambi (1982) propose to include it among the species of an unknown status. In light of the reported data about the Storm Petrel's distribution, we think that it may be included among the vulnerable species as reported in Frugis & Schenk (1981).

The colony of the Isle of Marettimo

As it has been reported, the most important Mediterranean colonies are in the isles of Filfla and Marettimo. Thus, these islets are important at an international level for the conservation of this species in the Mediterranean. At Marettimo the Storm Petrel was recorded by Krampitz (1957) first, but breeding was confirmed by Krapp (1970) who found the nests inside the caves of the western coast of the islet as far as Punta Libeccio. He found 25 eggs in the cave called "Presepe" and some nests in another cave, called "Cammello". Massa (1973) found the Storm Petrel breeding in the same caves and also at Punta Bassana, a site later deserted because of the interference of Rats (Rattus rattus).

In 1984, some friends of ours (A. Ciaccio, T. La Mantia, F. Lo Valvo, G. Lo Verde and G. Rannisi) explored carefully the caves of the western coast of the islet, finding no site, among those quoted in the 1970's, occupied by the Storm Petrel except for a site up to that time unknown where many breeding pairs of Storm Petrels were found. In 1985, during three visits (26 July, 14 August and 21 September), we controlled again the western and eastern coasts of the islet, but we could not find any Storm Petrel in the caves where Krapp (1970) and Massa (1973) recorded it. The only colony was the one found in 1984. It was in a cave where access is not easy, but difficult to protect. In fact, throughout the summer tens of boats and rubber dinghies, loaded with tourists, cruise the coasts of the islet and visit the caves. There is even a big hydrofoil coming from Marsala visiting the Egadi islands. It goes into some of the caves such as the one of "Presepe".

The colony of Storm Petrels which still survives at Marettimo is inside a cave, whose access is a narrow tunnel accessible only by swimming. The tunnel is 60 metres long and completely dark. We observed that tourists did not go into this cave for three reasons. First the access of the cave was difficult; secondly there were large populations of annoying gnats; thirdly there was a pungent odour left by the Storm Petrels. Thus only some casual factors allow the survival of that important colony.

Data of breeding cycle

At Marettimo, Storm Petrels breed in some lateral cracks of the vault, in some inaccessible holes and spelunks, but especially in two rooms, one above each other, at the end of the main tunnel as well as in the narrow corridors at the end of the cave. We estimated the whole population between 500 and 1,000 pairs, and now we believe the second figure to be more reliable, since the numerous deserted eggs and breeding pairs of the holes, branches and lateral spelunks had been excluded from the estimate (Massa & Catalisano, in press). On the ground of our investigations, we confirmed that laying starts at the end of April and ends in the middle of July (as at Filfla: Sultana & Gauci, 1982), while fledging goes from August up to the middle of October. We collected some data about the weight and wing length of chicks. From the results given in fig. 2 we observe that the increase of weight is faster in the first life period and slower in the second. In the later period, preceding fledging, we observe a decrease of weight. Then it follows the general schema of all the Procellariiformes.

Conservation

We collected some biometrical data from ten adults at Marettimo and we found values very similar to those reported for the mediterranean population by Hemery & D'Elbée (1985). These authors observed that mediterranean Storm Petrels are statistically bigger than atlantic ones and they proposed to revaluate melitensis (Schembri, 1843) as a valid subspecies.

Thus the conservation of mediterranean populations, genetically isolated and differentiated from atlantic ones, is important. The only two large colonies are naturally protected from the predation of rats, but not from human interference. At Filfla the shooting at seabirds for collection is alarming (Sultana, 1986). At Marettimo the only true threat comes from human interference and precisely from boats and dinghies crossing the sea of the western coast in the summer, when the Storm

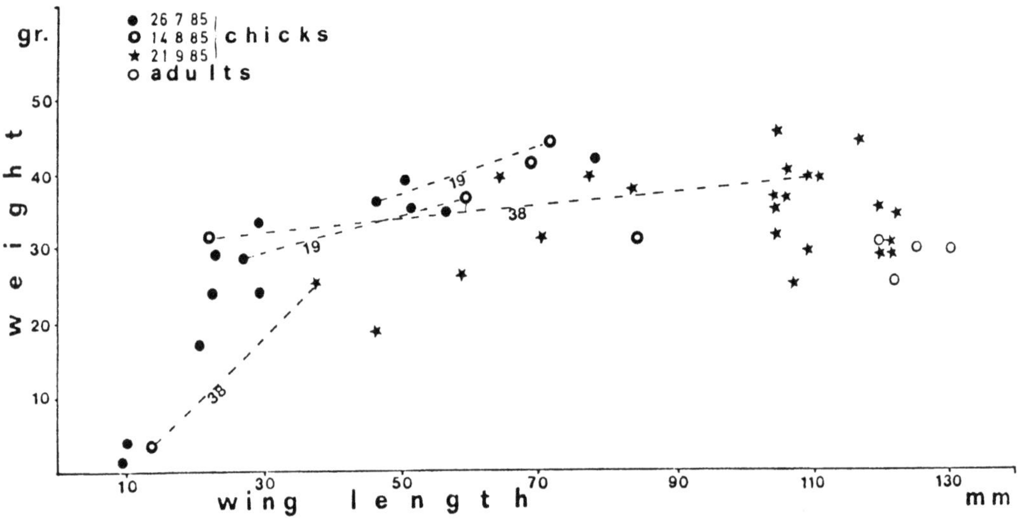

Fig. 2: Weight of the chicks of Storm Petrel correlated to wing length. Individuals measured in two different months are indicated with a dashed line. The figure under this line is the number of days passed between the two measurements.

Petrels are breeding. There are some proposals for the establishment of a nature reserve at Marettimo. In fact, some endemic invertebrates and some threatened and very rare species of vertebrates inhabit the islet. We hope that these proposals will be carried out soon.

Conservations measures include:

- a regulation of the movements of boats and tourists' dinghies along the western coast of the isle;
- a prohibition to enter the caves.

So we think that Storm Petrels will recolonize some caves as those of "Presepe" and "Cammello" nowadays deserted.

References

Akriotis, T. & Handrinos, G. (1986). First breeding case of Storm Petrel in Greece. in Medmaravis, Proceedings of 1st Medit. Seabird Symposium, Springer Verlag, Heidelberg (this volume).

Berthon, D. & Berthon, S. (1984). Compte rendu de l'expédition Balbuzard sur les côtes méditerranéennes du Maroc. L'Oiseau et Rev. Fr. Orn., 54: 201-213.

Brichetti, P. (1985). Guida degli Uccelli nidificanti in Italia. 2nd ed. F. lli Scalvi, Brescia.

Brichetti, P. & Cambi, D. (1982). Osservazioni sulla Lista Rossa degli Uccelli italiani. Avocetta, 6: 201-203.

De Juana, E. (1984). The status and conservation of seabirds in the Spanish Mediterranean. Pp. 347-361 in: Croxall, J.T., Evans, P.G.H. & Schreiber, R.W., Status and Conservation of the World's Seabirds. ICBP Technical Publ. n° 2.

De Juana, E., Varela, J. & Witt, H.H. (1980). Le Puffin cendré Calonectris diomedea nicheur aux îles Chaffarines. Alauda, 48: 27-31.

Doderlein, P. (1869-74). Avifauna del Modenese e della Sicilia. Giorn. Sc. nat. econom., 5-10: 381 pp.

Foschi, F. (1968). Monografia sugli Uccelli dell'isola di Pantelleria. Riv. ital. Orn., 38: 3-44.

Frugis, S. & Schenk, H. (1981). Red List of Italian Birds. Avocetta, 5: 133-141.

Giglioli, E.H. (1907). Avifauna italica. Secondo resoconto. S. Giuseppe ed., Florence.

Guyot, I., Launay, G. & Vidal, P. (1985). Oiseaux de mer nicheurs du Midi de la France et de Corse: évolution et importance des effectifs. Pp. 31-47 in: Oiseaux marins nicheurs du Midi et de la Corse. Annales du C.R.O.P., n° 2.

Handrinos, G. (1984) . Storm Petrel, in: European news. British Birds, 77: 586-591.

Hemery, G. & D'Elbée, E. (1985). Discrimination morphologique des populations Atlantique et Méditerranéenne de Petrel tempête Hydrobates pelagicus. Pp. 63-67 in: Oiseaux marins nicheurs du Midi et de la Corse. Annales du C.R.O.P., n° 2.

Jacob, J.P. & Courbet, B. (1980). Oiseaux de mer nicheurs sur la côte algérienne. Gerfaut, 70: 385-401.

James, P.C. (1984). The status and conservation of Seabirds in Mediterranean Sea. Pp. 371-375 in: Croxall, J.T., Evans, P.G.H. & Schreiber, R.W., Status and Conservation of the World's Seabirds. ICBP Technical Publ., n° 2.

Krampitz, H.E. (1957). Beobachtungen auf der Mittelmeerinsel Marettimo im Mai 1955. Vogelring, 26: 35-37.

Krampitz, H.E. (1958). Weiteres über die Brutvögel Siziliens. J. Orn., 99: 39-58.

Krapp, F. (1970). Vogelbeobachtungen während des Frühjahrzuges 1969 auf den Agadischen Inseln. Der Orn. Beobachter, 67: 280-294.

Ledant, J.P., Jacob, J.P., Jacobs, P., Malher, F., Ochando, B. & Roché, J. (1981). Mise à jour de l'avifaune algérienne. Gerfaut, 71: 295-398.

Martorelli, G. (1906). Gli Uccelli d'Italia. 1st ed. Rizzoli, Milan.

Martorelli, G. (1931). Gli Uccelli d'Italia. 2nd ed. by Moltoni, E. & Vandoni, C. Rizzoli, Milan.

Massa, B. (1973). L'avifauna estiva degli arcipelaghi delle Egadi e dello Stagnone (Trapani, Sicilia). Atti Accad. Gioenia Sc. nat. Catania, 5: 63-95.

Massa, B. & Catalisano, A. in press. Observations on the Mediterranean Storm Petrel Hydrobates pelagicus at Marettimo Isle. Avocetta.

Matveiev, S.D. & Vasic, V.F. (1973). Catalogus Faunae Jugoslaviae. Aves. Acad. Sc. Art. Slovenica, Ljubljana.

Moltoni, E. (1954). Gli Uccelli rinvenuti durante una escursione ornitologica all'isola di Pantelleria, Prov. di Trapani, nel giugno-luglio 1954 (29 giugno-21 luglio) con notizie su quelli noti per l'Isola. Riv. ital. Orn., 24: 1-41.

Moltoni, E. (1970). Gli Uccelli fino ad oggi rinvenuti o notati all'Isola di Pantelleria. Riv. ital. Orn., 43: 173-437.

Moltoni, E. & Frugis, S. (1967). Gli Uccelli delle Isole Eolie. Riv. ital. Orn., 37: 91-234.

Schembri, M. (1843). Catalogo ornitologico del gruppo di Malta. Anglo Maltese, Malta.

Schenk, H. & Torre, A. (1986). Breeding distribution, numbers and conservation of Seabirds in Sardinia, 1978-85. in Medmaravis,Proceedings of 1st Mediterranean Seabird Symposium, Springer Verlag, Heidelberg (this volume).

Steinbacher, J. (1956). Über den Herbst-Vogelzug auf Pantelleria. Vogelring, 25: 2-6.

Sultana, J. (1986). Seabird Conservation problems in the Maltese Islands. in Medmaravis, Proceedings of 1st Mediterranean Seabird Symposium, Springer Verlag, Heidelberg (this volume).

Sultana, J. & Gauci, C. (1982). A new guide to the birds of Malta. MOS, Malta.

Thomsen, P. & Jacobsen, P. (1979). The birds of Tunisia. Copenhagen.

Torre, A. (1983). Osservazioni di uccelli rari in Sardegna. Boll. Soc. Sarda Sc. nat., 22: 169-175.

THE STATUS OF BREEDING STORM PETRELS (HYDROBATES PELAGICUS) ON THE MEDITERRANEAN COAST OF FRANCE

John Walmsley, Fondation Tour du Valat,
Le Sambuc, 13200 Arles, France.

Introduction

The presence of the Storm petrel Hydrobates pelagicus on islands close by the large Mediterranean port of Marseille is surprising to say the least, and yet the Storm petrel was known as a breeding species on the "Archipel de Riou" during the mid-nineteenth century (Jaubert & Barthélémy-Lapommeraye, 1859). In 1915 a few pairs were reported breeding on the island of Riou (L'Hermitte 1915-16), and by 1918 large breeding colonies were found on "Ile Plane" and the two rocky islets: the "Grand and Petit Congloués" (Lavauden & Mourgue, 1918). In the spring of 1923 Heim de Balsac (1923) visited Riou, but gave no indication that the Storm petrel bred there. However, he did visit Ile Plane and found many petrels "apparently sleeping in recesses in the rocks". In early April 1924 he visited the "Grand Congloué" and found Cory's shearwater Calonectris diomedea and the Levantine race of the Manx shearwater Puffinus p. yelkouan occupying nest sites, but the Storm petrel was absent. Heim de Balsac & Mayaud (1932) again visited Ile Plane on 9 May 1931 and counted 40 petrels. The two fresh eggs found indicated that the breeding season had only just begun. Fifteen years later Storm petrels were still breeding on Ile Plane, some even nesting under dense low bushes (A. Rivoire pers. com.).

By 1950, visits to the island became more and more irregular, and that same year the Yellow-legged gull Larus cachinnans probably bred for the first time on Ile Plane, because in 1952 between 15 and 20 pairs were recorded (A. Rivoire pers. com.). Since that time the Storm petrel population is reported to have declined as the Yellow-legged gull population increased. Statements were even made that a breeding population no longer existed because of predation by Yellow-legged gulls on adult and young Storm petrels. During the period 1970-75 when the "Atlas of breeding birds of France" was compiled (Yeatman, 1976), the

NATO ASI Series, Vol. G 12
Mediterranean Marine Avifauna
Edited by MEDMARAVIS and X. Monbailliu

only observation of a Storm petrel was one bird found on Ile Plane in late July 1975 by Mr. Guillemont (A. Rivoire pers. com.). At that time the nearest Storm petrel breeding colonies were on the "Cerbicales and Lavezzi Islands" in Corsica, and "Cabrera" in the Balearic Islands in Spain.

Breeding was again confirmed in 1977 when a team (J.G. Walmsley & A.R. Johnson) from the Tour du Valat (Camargue) visited Ile Plane. This was followed by irregular visits during the period 1977-83 to census Storm petrels and determine their breeding distribution on the island. Different census methods were employed, and a capture and ringing programme was started.

The Study Areas

Of the 50 or more islands found off the Mediterranean coast of France, the "Archipel de Riou" is composed of 10 islands, the smallest: "Les Empereurs" (0.2 ha), and "La Moyade" (0.2 ha), and the largest "Riou" (90 ha) are composed chiefly of chalky limestone rock formation. Riou is by far the largest in the group, and rises to a height of 191 metres. In contrast to the vertical peaks and cliffs of Riou, Ile Plane is a relatively small and flat island with a total surface area of not more than 15 ha, and a maximum height of 22.8 metres. The vegetation is halophile, rarely above 80 centimetres in height, with short dense bushes dotted about the high plateau.

Potential predators of the Storm petrel are the Black rat *Rattus rattus* found on Ile Plane and the Grand and Petit Congloués (Lavauden & Mourgue, 1918), and more recently the Yellow-legged gull which is increasing alarmingly in the Mediterranean. The Rabbit *Oryctolagus cuniculus* may also be present on Ile Plane, but so far I have never seen one during my visits, nor have I witnessed any interference by rats or gulls on the Storm petrel population.

Census Methods

Our first visit to Ile Plane was at the end of May 1977. We assumed that breeding would be well under way, but had no idea where the Storm petrels were breeding on the island. We therefore prospected it thoroughly and identified the sector with potential nest sites. The question was then to decide which sites were occupied. This was done at night using three methods (tab. 1) involving sound and smell:

1. Listening to birds calling from the rocks (below ground),
2. Playing pre-recorded calls at the nest sites,
3. Noting whether the sites gave off the characteristic oily odour of the birds (particularly helpful during the day).

Year	Methods			Number of nests found	Number of man nights
	1	2	3		
1977	X			25	2
1978	X			23	5
1979	X	X		50	4
1980	X	X	X	40	5
1981	island not visited				
1982	X	X	X	13	1
1983	X	X		15	2

1 : Birds calling in nest
2 : replying to pre-recorded calls
3 : smell

Tab. 1: Storm petrel census techniques and results

The number of occupied sites is an index of the number of breeding pairs, but does not give an accurate estimate because it is well known that some immature birds occupy nest sites at least one year before they breed for the first time (Davis, 1957).

Results

Confirmation that the Storm petrel still breeds on the "Archipel de Riou" on Ile Plane, can be seen in the updated distribution map (fig. 1) showing the existing breeding colonies in the Western Mediterranean (after Cramp & Simmons, 1977; Guyot & Thibault, 1984; and Walmsley pers. obs.).

Fig. 2 shows the location of Ile Plane in relation to its neighbouring islands in the group. Former and present day breeding distribution of the Storm Petrel is shown (after Lavauden & Mourgue, 1918 & this study). However, due to difficulty of access the two Congloué's have rarely been visited in recent times, so it is not known whether the Storm petrel still breeds there.

Access to Ile Plane may also be difficult at times, but once established one can move about relatively easy on this flat island. The northern coastline has vertical cliffs while the southern side slopes down to the sea, from a vegetation-

Fig. 1: Breeding distribution of the Storm petrel Hydrobates pelagicus in the Western Mediterranean (after Cramp & Simmons, 1977; Guyot & Thibault, 1984; and Walmsley pers.obs.).

covered plateau to a boulder strewn shoreline. Yellow-legged gulls breed on the plateau, while the Storm petrels breed on the shoreline under large boulders, jumbled rocks, crevices and gullies, and under vegetation such as Salicornia sp., Crithmum maritimum, Inula crithmoides, Smilax aspera, and Brachypodium ramosum. Nests can be either clumped (up to four nests in close proximity to each other) or widely scattered, the colony extending some 700 metres along the southern shoreline.

Fig. 3 summarizes all the censuses made during the period 1977-83. The apparent population increase in 1979-80 was probably due to the different census techniques. In 1977-78 all nests were located by ear, whilst in 1979 besides locating nests by ear, a second method of pre-recorded Storm petrel calls was used (tab. 1). The island was not visited in 1981. In 1982 a strong south-westerly wind and rough seas prevented a complete census being made, and the low number of nests found in 1983 was probably because the island was visited late in the breeding season. These numbers are therefore minimum estimates. Nonetheless it is clear that the island could hold many more.

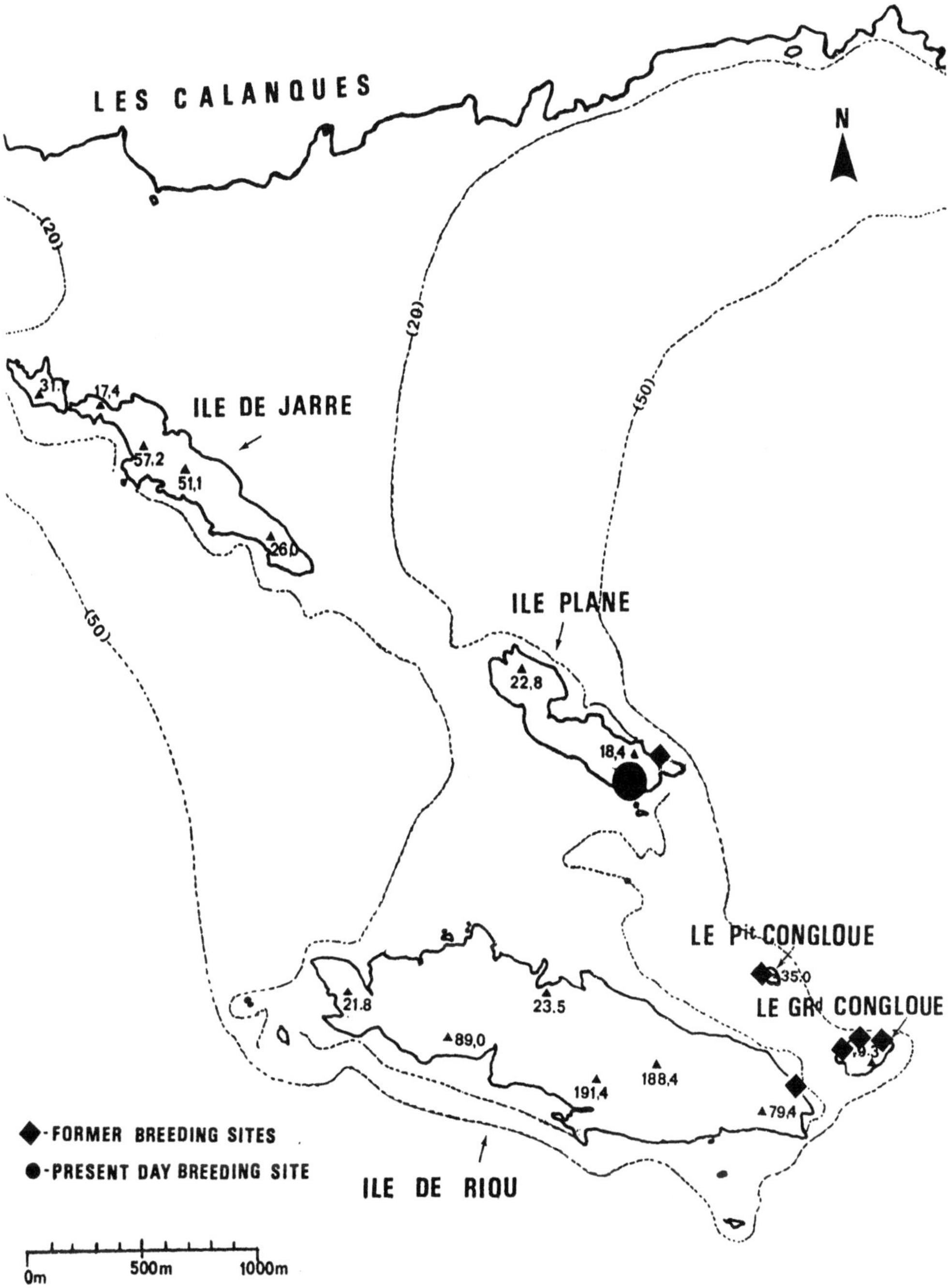

Fig. 2: The location of Ile Plane in relation to its neighbouring islands in the "Archipel de Riou". Former and present day breeding distribution of the Storm petrel is shown (after Lavauden & Mourgue, 1918; and Walmsley pers. obs.).

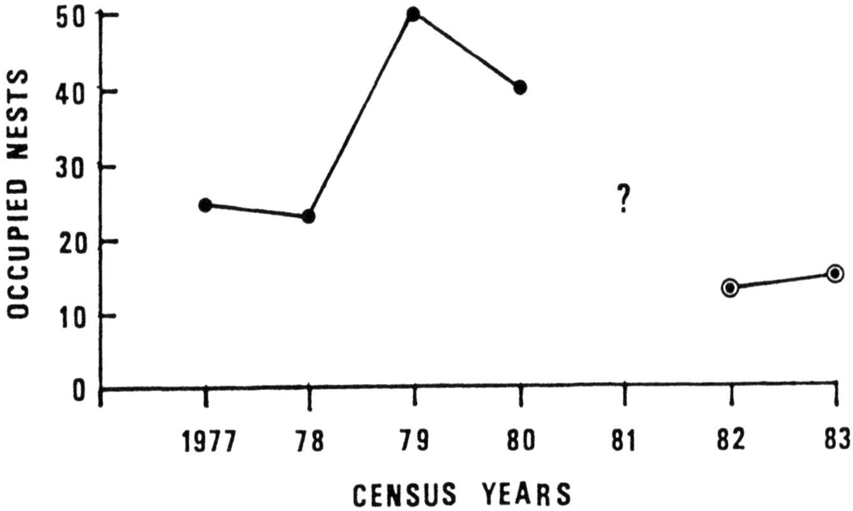

Fig. 3: Counts of occupied nests of Storm petrel on Ile Plane: 1977-1983.

Possible Factors Limiting Breeding Success

One of the principal factors which may limit the breeding success of Storm petrels on Ile Plane is the Yellow-legged gull. This species bred on Riou as early as the mid-nineteenth century (Jaubert & B-Lapommeray, 1859), and in 1918 was limited to the southern cliffs of Riou (Lavauden & Mourgue, 1918). The population was estimated at 100 breeding pairs in 1922-23 (Heim de Balsac, 1923-24). The first expansion of the breeding range within the island group was reported by Heim de Balsac & Mayaud (1932) who observed another colony of Yellow-legged gulls on the Grand Congloué. By 1949 the population had risen to about 1,500 pairs (A. Rivoire pers. com.) and had increased to nearly 3,000 pairs by 1963 (C.J.O. 1963).

Yellow-legged gulls probably first colonized Ile Plane in 1950, and by 1954 there were 12-15 breeding pairs (A. Rivoire pers. com.). In 1977 we estimated the breeding population at 500 pairs, and by 1980 had saturation level with 1,000 pairs. The population remained stable until 1983.

On visiting the island in 1982, it became apparent that some Yellow-legged gulls had overflowed from the vegetated plateau and were breeding on rocks under which the Storm petrels breed. Before 1982 only an occasional gull chick had been found sheltering in the rocks. It is well known that Yellow-legged gulls prey upon many species. Unsuspecting adult and young Storm petrels may become easy prey for the neighbouring Yellow-legged gull.

Visitors to the island in summer, especially family parties with dogs, and fishermen also present a danger to the breeding petrels. Most people stay well clear of the vegetated plateau where the Yellow-legged gulls breed and spend the day on the rocks on the southern part of the island. While human disturbance may be unimportant, dogs however present a greater danger. Perhaps a more worrying problem is the increase in the number of bird watchers to the island, since it became known that the Storm petrel still breeds there. They go there with the intention of seeing either a new species for their life list, a bird on the nest, or, better still, a bird in the hand. Disturbance of this kind is probably the greatest danger of all and may lead to birds abandoning breeding altogether for that year.

Conclusion

Since the Storm petrel was first discovered breeding on the "Archipel de Riou" in the mid-nineteenth century, the species has probably continued to breed until the present day, even though numbers and breeding colonies may have declined.

The census techniques developed for the kind of habitat occupied by the Storm petrel on Ile Plane have proved successful. This was particularly true when continuous play-back of Storm petrel calls was used.

Predation by Yellow-legged gulls and the increase in human and animal disturbance may be the factors limiting successful breeding. Regular "culling" would reduce the number of Yellow-legged gulls, while bird watchers should be informed of the danger, and, if possible, discouraged from visiting the island.

Acknowledgements

I would like to express my sincere thanks to all my colleagues at the Tour du Valat who assisted me in these nocturnal ventures: Alan Johnson, Tobias Salathé, Jean-Paul Taris and Marie-Louise Meyer. I would also like to thank Michel and Geneviève Pont and family. Without their valuable contribution of putting their private yacht at our disposal, and transporting the teams to the island and back safely, this paper would never have been written. Finally, my special thanks go to André Rivoire for the stimulating discussions we have had concerning Storm petrels, and to Dr. Hoffmann for allowing me the necessary time to carry out this work.

References

Centrale Ornithologique du C.J.O. (1963). Esquisse du statut des Laridés nicheurs en France. Oiseaux de France, 38: 1-19.

Cramp, S., and K.E.L. Simmons (eds) (1977). The Birds of the Western Palearctic, Vol. 1. Oxford University Press, Oxford, England.

Davis, P. (1957). The breeding of the Storm petrel. British Birds, 50: 85-101 & 371-384.

Guyot, I., and J.C. Thibault (1984). Les oiseaux marins nicheurs de Méditerranée occidentale: répartition, effectifs et recensement. Annales coll. Vert. Terre. Iles Med. Evisa, Oct. 1983.

Heim de Balsac, H. (1923). Les oiseaux de l'Ile de Riou. Rev. fr. Orn. 8: 103-111.

Heim de Balsac, H. (1924). Note complémentaire sur les oiseaux de Riou. Rev. fr. Orn. 8: 370-372.

Heim de Balsac, H., and N. Mayaud (1932). Nouvelles observations sur les oiseaux de l'Ile de Riou (Bouches-du-Rhône). Alauda 4: 85-88.

Jaubert, J.B., and Barthélémy-Lapommeray (1859). Richesses ornithologiques du Midi de la France. Marseille.

Lavauden, L., and M. Mourgue (1918). Contribution à l'étude du Thalassidrome tempête dans la Méditerranée. Rev. fr. Orn. 5: 305-309.

L'Hermite, J. (1915-16). Contribution à l'étude ornithologique de Provence. Rev. fr. Orn. 4: 331-357.

Yeatman, L. (1976). Atlas des oiseaux nicheurs de France. Société ornithologique de France, Paris, 282 p.

CHAPTER 2 : DATA BANKS AND CENSUS TECHNIQUES

Chairman: Regius Professor G.M. DUNNET. Department of Zoology. University of Aberdeen, Scotland.

DATA BANKS AND POPULATION MONITORING IN FRANCE

A contribution from the
"Groupe de Travail sur les Oiseaux Marins",
presented by Georges Hémery (*), Eric Pasquet (*),
and Pierre Yésou (**).

(*) Centre de Recherches sur la Biologie des Populations d'Oiseaux, Muséum National d'Histoire Naturelle, 55 rue Buffon, F-75005 Paris.

(**) Centre d'Etudes et de Recherches Appliquées sur les Oiseaux d'Eau, Office National de la Chasse, Champclou, F-85340 l'Ile d'Olonne.

Thanks to a limited number of pioneers French ornithology has, since the last century, contributed to the general knowledge of seabirds. Until recently however, interest in seabirds in France was almost restricted to reserve management and, for some enthusiasts, to sea-watching. Things rapidly evolved, mainly under the impulse of problems linked to the important development of Larus argentatus/cachinnans populations, sea-pollution and the dramatic decline of auks. Change in concerns became strongly evident in the mid-1970's. Research on seabirds has since spread out in many directions, including taxonomy, field-identification, reproductive biology, population dynamics, relations with fisheries, ethology.

The need soon arose for a structure to develop contacts and exchanges between seabird enthusiasts. The "Groupe de Travail sur les Oiseaux Marins" (G.T.O.M.) was thus created in 1980. It regroups all persons with an interest in seabirds, whether amateur or professional, member of an ornithology club, a society for nature protection or an independent observer, field-observation addict or searcher in a laboratory. Amongst other things, the G.T.O.M. co-ordinates enquiries on a nation-wide basis, two of which are presented here:

- the long-term monitoring of breeding populations,
- the study of bird distribution at sea.

NATO ASI Series, Vol. G 12
Mediterranean Marine Avifauna
Edited by MEDMARAVIS and X. Monbailliu

Apart from being activities in which G.T.O.M. members are fully engaged, these enquiries share two other characteristics. First, as they are of evident interest for the conservation of our natural heritage, they benefit from the support of the Ministry of the Environment. Secondly, they both use computerized data banks. Due to its previous experience in processing this type of data (Hémery et al. 1979), the "Centre de Recherches sur la Biologie des Populations d'Oiseaux" was given the responsibility for this task. The "Statistical Analysis System" is used for processing in both studies.

1. Breeding Seabirds Censuses Data Bank

1.1. Historical Background

Although of invaluable interest, pre-1950 testimonies on French seabird colonies are rather rare, sometimes of disputable accuracy, and always concern very restricted survey areas. Modern seabird censuses started in Brittany at the end of the 1950's: the creation of an important network of seabird reserves, managed mainly by the "Société pour l'Etude et la Protection de la Nature en Bretagne" (Henry & Monnat 1980), and the building-up of regional ornithological societies were followed by increasing field-work.

Simultaneously with the "Seafarer" survey run in the British Isles (Cramp et al. 1974), the first complete breeding seabird census was completed in Brittany in 1970 (Brien 1970). Meanwhile censuses were organized in Normandy (Braillon 1960-1970). Full coverage of the Channel and Atlantic coasts was carried out in 1977 and 1978. The results were compared to all former data available by Henry & Monnat (1981), giving the first historical synthesis of breeding seabird populations from the Belgian border to southern Biscay.

On the Mediterranean side, the Camargue is regularly surveyed since 1956 (Blondel & Isenmann 1981, Isenmann 1986), but most of the other breeding sites have only been sporadically visited, until the end of the 1970's. Intensive field-work was then organized under the aegis of the "Centre de Recherches Ornithologiques de Provence" and of environmental bodies - the "Parc National de Port Cros" and the "Parc Naturel Régional de Corse", This allowed a complete census of breeding colonies on the Mediterranean coasts and Corsica. Its published form also includes historical data (Guyot et al. 1985).

The status of French breeding seabirds is thus well known for the recent period (1978-1983). In many cases, their historical and geographical evolution is fairly well understood. It soon appeared, however, that problems concerning the precise analysis of previous records and the comparison of present results with those of forthcoming works, could arise. The difficulties expected can come from different sources, for example: the great amount of data to be taken into account, their accessibility, the bias introduced by non-standard census methods or -

not the least of problems - the various names given to a same site in different census reports.

A computerized data bank can greatly help to deal with these matters: proposals made in this way were accepted by the G.T.O.M. in 1982, and data processing was undertaken by the C.R.B.P.O. in 1983 (Pasquet 1984).

1.2. The data bank: its aims

The setting up of this data bank answers three main requirements:

(i) - it guarantees the safe-keeping and accessibility of information previously widely dispersed throughout numerous publications and observers and regional societies' files records. Some of this information was formerly difficult to obtain;

(ii) - by using different programs, data can be managed and analysed, either on a geographical basis or statistically;

(iii) - it is a strong basis for the organisation of future surveys and the development of standard data collection and presentation.

For the facilities it gives, the data bank is evidently of great interest to both biologists and conservation bodies. The registered data are of easy access to ornithologists studying, e.g., distribution shifts or population dynamics. Information can be acquired rapidly by those concerned by species or site conservation.

1.3. The species concerned

The species taken into account are those breeding along the Channel, Atlantic and Mediterranean coasts, including Corsica, i.e.: Fulmarus glacialis, Calonectris diomedea, Puffinus puffinus (with two subspecies treated separately: P.p. puffinus in Brittany, P.p. yelkouan in the Mediterranean), Hydrobates pelagicus (again two subspecies: H.p. pelagicus outside the Mediterranean, H.p. melitensis in the Mediterranean; cf. Hémery & Elbée 1985), Sula bassana, Phalacrocorax carbo, Ph. aristotelis (nominal subspecies in Normandy and Brittany, Ph. a. desmarestii in Corsica), Larus melanocephalus, L. genei, L. audouinii, L. canus, L. fuscus, L. cachinnans, L. argentatus, L. marinus, Rissa tridactyla, Sterna nilotica, Sterna sandviscensis, St. dougallii, St. hirundo, St. paradisea, St. albifrons, Alca torda, Uria aalge and Fratercula arctica.

A few other species, almost only coastal breeders in France (e.g. Tadorna tadorna, Haematopus ostralegus, Charadrius hiaticula, Ch. alexandrinus), were left out when the data bank was set up as they do not breed in colonies - or merely in diffuse ones - and are difficult to survey regularly.

NON-LINEAR LEAST SQUARES SUMMARY STATISTICS   DEPENDENT VARIABLE EFF

SOURCE	DF	SUM OF SQUARES	MEAN SQUARE
REGRESSION	2	442076.80168186	221038.40084093
RESIDUAL	16	5749.19831814	359.32489488
UNCORRECTED TOTAL	18	447826.00000000	
(CORRECTED TOTAL)	17	130487.11111111	

PARAMETER	ESTIMATE	ASYMPTOTIC STD. ERROR (2)	ASYMPTOTIC 95 % CONFIDENCE INTERVAL LOWER	UPPER
YO	23.68412724	3.65290713	15.94034853	31.42790594
A	0.10549407 (1)	0.00745524	0.08968974	0.12129840

ASYMPTOTIC CORRELATION MATRIX OF THE PARAMETERS

	YO	A
YO	1.000000	-0.982879
A	-0.982879	1.000000

(3)

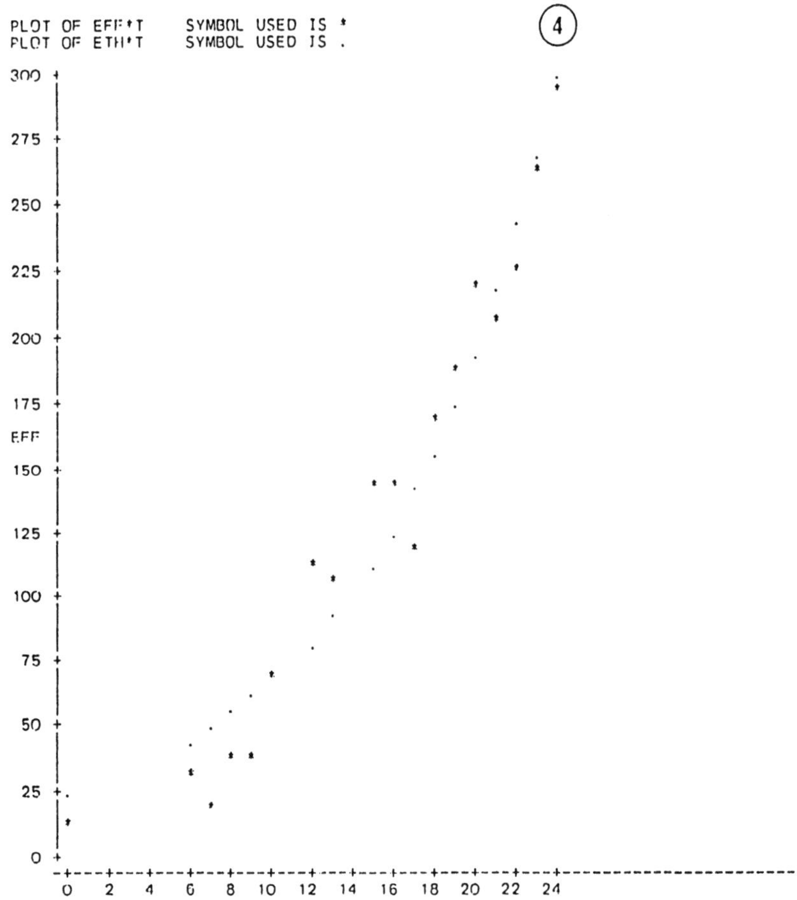

Fig. 1: Growth of a *Phalacrocorax carbo* colony: adjustment of census results to an exponential model.

On the other hand, some marine species also breed inland: Larus melanocephalus, L. canus, L. cachinnans, Sterna hirundo and St. albifrons. For these species, inland records are included in the data bank. Larus ridibundus, mainly an inland breeder, is not taken into account.

1.4. Methods

Two interactive index-files have been established: one for the census results, one for the localities surveyed. Each locality surveyed is given an alphanumeric code: it is the link between the two index-files.

The census file includes all the available census data. Data is understood as the number of breeding pairs of a given species at a precise locality at a given data. For each data, the following information is registered in the bank:

- alphanumeric code of the locality (see below)
- date (year, month, day)
- species (subspecies when necessary)
- name and surname of the observer)
- number of breeding pairs observed
- estimated number of breeding pairs (when available)
- census methodology (25 parameters can be taken into account).

It has been estimated that 7,000 to 8,000 data were available up to 1983. About 4,000 are already in the data bank and work is still in progress.

In the localities file each site surveyed is coded according to its geographical location. Sites can be grouped at three levels: sub-district, district and "département".

The département is an administrative entity, often covering over 100 kms of coast. In each département, districts are geographical or ecological homogenous areas. In a few cases, very large homogenous areas are divided in several districts. A district is divided in sub-districts, each of which regroups a certain number of sites: the census units. These units are easily defined when islets are concerned. They are more difficult to define in the case of large islands or coastal areas. Two rules have been adopted for such cases: respect of administrative and seabird reserve boundaries, respect of local census habits.

For each census locality, the following information is recorded:

- alphanumeric code of the locality (two figures for the département, one letter for the district, two figures for the census unit; a group of given consecutive units forms a sub-district);
- the census unit's precise name (possibly refering to its toponym or to its reference mark on hydrologic survey

maps from the French National Navy or other sources, according to availability);

- co-ordinates (degree-minute-second, Mercator's projection).

1.5. Results: two examples

The statistical analysis of population trends can easily be processed at different geographical levels (census unit, district(s), department(s), whole population) when sufficient data are available. This analysis adjusts the growth rate of a population to a mathematical model. The following models will be the ones most often used: in continuous form, exponential and logistic growth models; in discrete time, estimators of population trend index. In most cases, the adjustment can be obtained by the use of Gauss-Newton and Marquardt methods.

An example of such analysis is given by the evolution of *Phalacrocorax carbo* colony at St. Marcouf in Normandy (fig. 1). The shape of the graph of censused numbers shows that accurate results can be expected by use of an exponential model.

The mean growth ratio obtained for the period 1959-1983 (to = 1959) is:

$a = 0.1055$ (1); $\lambda = e^{2} = 1.111$ (11% increase per year)

Standard errors and confidence intervals of estimated parameters are given in (2), the correlation between parameters in (3). The graph (4) shows the superposition of the observed numbers curve (EFF) with the statistically adjusted one (ETH).

The automatic cartography traces a symbolic representation of colony size on a map background. Carthographical documents of different scales can be produced by the hierarchical organization of census units. Fig. 2 shows the distribution of censused pairs of *Larus argentatus* in Brittany and part of Normandy in 1969, data being grouped at the district scale.

2. Birds at Sea Data Bank

2.1. Historical Background

Until very recent years, data obtained by sea-watchers were the main source of information on seabirds present off French coasts. Such information often remains of limited interest, however, as sea-watching concerns inshore waters only, it is dependent on meteorological conditions and is rarely undertaken on a regular basis (Yésou 1982a). Stranded corpses collected during "beached bird surveys" can also give information on the birds' presence offshore and on how they are affected by chronic oil pollution (Thomas 1981). In the case of major pollution, such as in the wreck of the Amoco Cadiz, the mortality rate encountered is far greater. Biometrics and post-mortem exa-

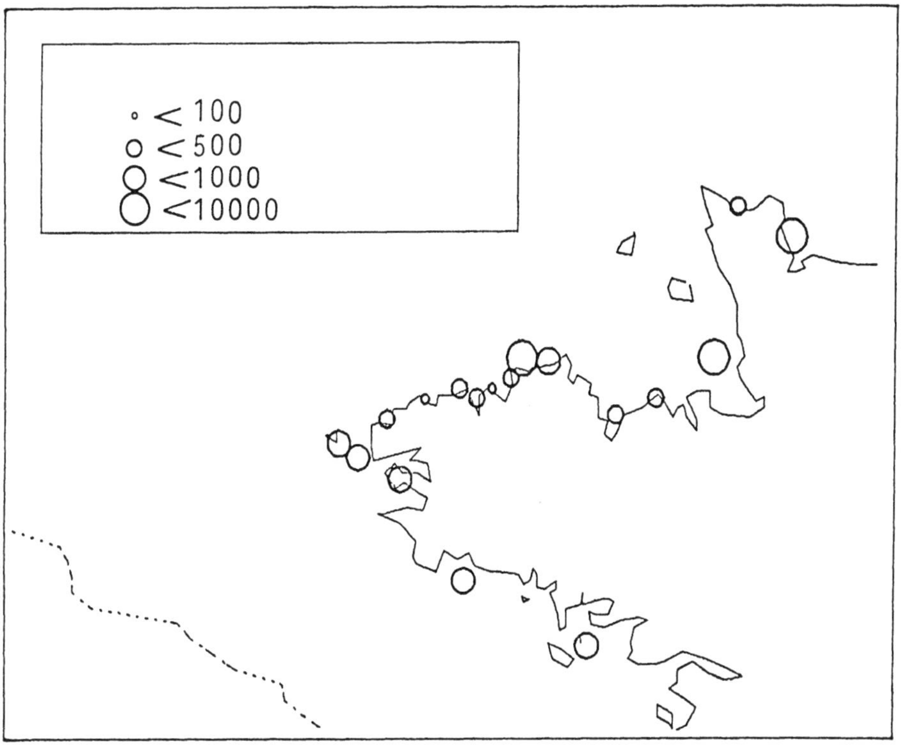

Fig. 2: Distribution by district of Larus argentatus breeding pairs censused in 1969 in Brittany and part of Normandy (note: some colonies were not censused that year).

minations of such casualities can give interesting information on population parameters: sex- and age-ratio, subspecific composition or geographical origins of the affected populations (Monnat 1978, Monnat & Guermeur 1979). From 1978 onwards, such damageable wrecks repeatedly occurred in Brittany, resulting in a high seabird mortality. Drastic measures concerning marine traffic in French waters were then taken. According to conservation bodies, such measures can still be improved to ensure a better protection of marine life - including seabirds.

In this perspective, accurate knowledge of bird distribution at sea is needed. A preliminary enquiry on the distribution of Gaviiformes and Podicipediformes in inshore waters was undertaken (Pasquet 1982). As available data on birds offshore was very limited (Prieur 1970, Brien 1971), an ambitious survey was organized: the "Programme C.R.B.P.O.-M.E.R.", managed since 1980 by the C.R.B.P.O. on a national basis, the field-work being carried out mostly by G.T.O.M. members (Hémery 1982, 1984). This program's organization benefits from previous experimental studies in southern Biscay, from 1975 to 1979 (Hémery, unpublished). The Birds at Sea Data Bank is linked to this program.

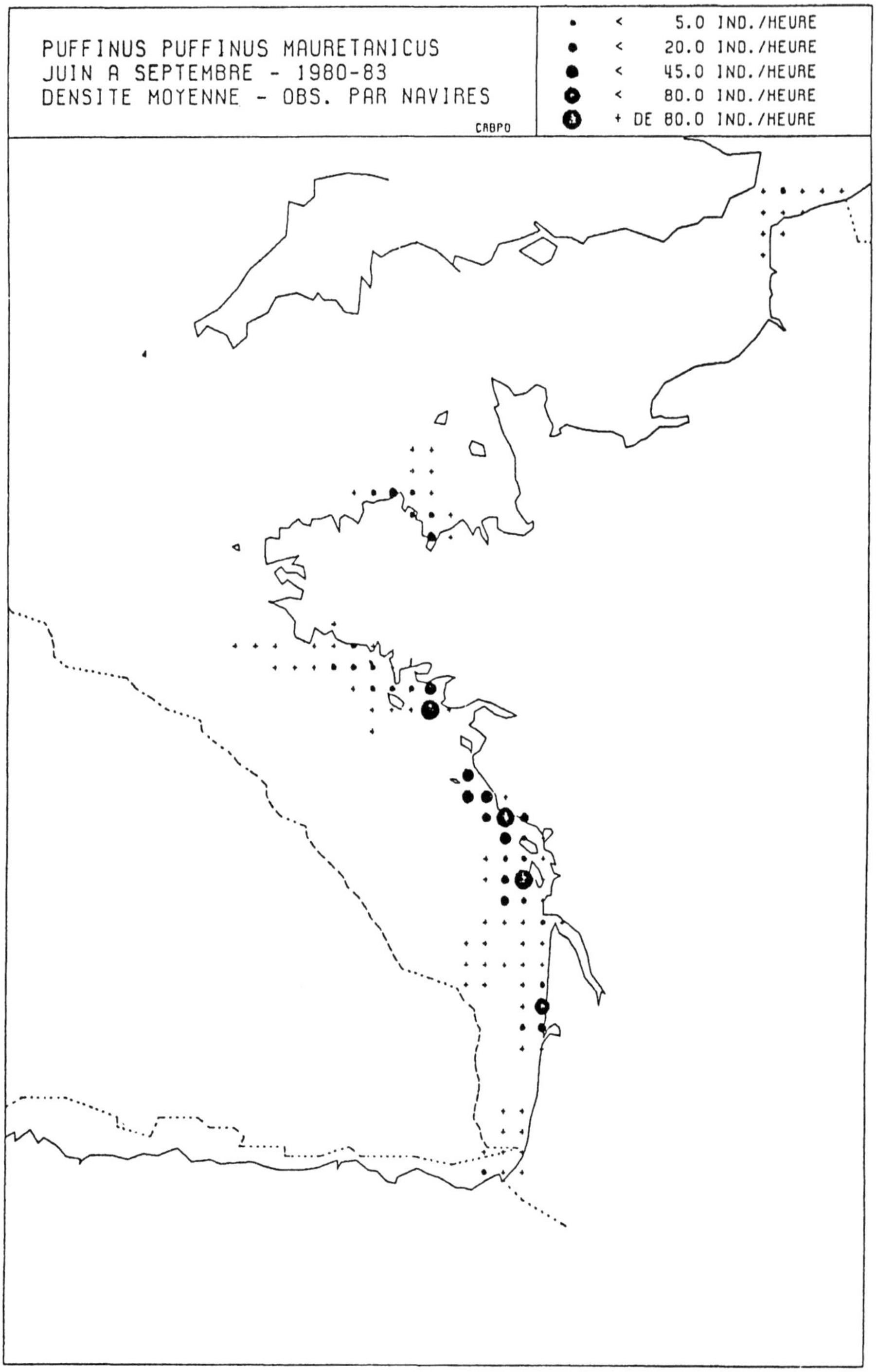

Fig. 3: Distribution of P. puffinus mauretanicus off western France, June to September: a provisional map. Crosses (+) denote the absence of birds in surveyed areas. Unsurveyed areas are left blank.

2.2. The survey: its aims

Current research has three main objects:

(i) - to define the relative density of each species according to time of year (variations calculated on a monthly basis) and place (differences between areas);

(ii) - at a lesser scale, the simultaneous study of oceanographic parameters which determine bird distribution at sea;

(iii) - the delimitation and typological classification of coastal areas according to their biological richness and their vulnerability to pollution.

2.3. Methods

The study area covers all the Channel, Atlantic and Mediterranean waters inside the French Extensive Economic Zone, i.e. for Atlantic, up to 200 nautical miles (360 km) off the coast.

The survey is based on line-transects made from French customs vessels and aircrafts. Vessels are by far the most used, but air transects can be of great interest for a rapid survey of large or distant areas, e.g. the Continental shelf. Rigourous standard methods of observation and recording are needed in order to obtain comparable data. This mainly concerns the following matters:

(i) - the characteristics of the vessel or aircraft: observer's elevation from sea-level; speed; general appearance of the vessel, which may influence bird detectability or bird behaviour;

(ii) - the conditions of observation: number of observers; visibility; weather and sea conditions; etc.;

(iii) - the nature of the observations made, and the way in which they are registered.

Standard conditions differ if the survey is conducted from ship, airplane or helicopter. Nevertheless, they are imperatively in agreement with conditions of observation currently in use for similar surveys in other countries (for a review, see Tasker et al. 1984).

The observer aboard a French customs vessel stands about 4 or 5 meters above sea-level. Covering a viewing arc of 360°, he looks for birds - either moving or settled on the sea - with the naked eye (binoculars are used for specific identification only). Observations are discontinued when good visibility and good sea conditions (under 4 Beaufort) are no longer available. The observer takes note of the birds seen, minute by minute (survey time unit) as well as cetaceans or trawlers encountered, etc., and any change in vessel speed (usually between 16 and 22 knots) or direction. Such changes are precisely noted on the vessel log, thus giving the precise location of the vessel at any moment.

Both sets of data (observer's records and vessel log) are fed into the data-bank, the time unit being the link between them. The use of a short unit of time and the simultaneous processing give very precise co-ordinates of each observation.

Automatic mapping and statistical analysis are then possible. Relative densities (number of birds per hour) can easily be calculated. Passage to true densities (number of birds per surface unit) is not as easy. Statistical adjustments need to take into account certain factors which are difficult to estimate: e.g. speed and direction of flying birds, bird behaviour in relation to the boat.

Information on oceanographic parameters comes from three sources:

(i) - analysis of infra-red images, obtained by meteorological satellites (U.S. NOOA series), that inform on the location and evolution of physical parameters such as upwelling and convergent front;

(ii) - contacts with laboratories specialized in oceanography, hydrology, ichthyology, etc.;

(iii) - to a limited extent, direct observation at sea during transect survey.

2.4. Results: some examples

Fig. 3 depicts the summer distribution of P. puffinus mauretanicus as appearing after 4 years of survey (1980-1983) in Western France. Fig. 4 shows monthly variations in the mean relative density for the same species. Although the information presented needs to be completed by forthcoming surveys, these examples clearly show the interest of such an enquiry:

(i) - the evolution of abundance throughout the annual cycle is rather well depicted;

(ii) - the distribution is far from homogeneous; this leads to interesting biological questions, and greatly helps the definition of the most sensitive areas for the species.

Satellite images clearly indicate the limit between contrasting warm and cold waters, illustrating the development of several physical parameters that strongly influence the nutrient richness of marine areas: true upwelling, tide-induced upwelling, convergent front, overheating area. The situation in the Atlantic at the end of August 1981 is given as an example in fig. 5. The superposition of such maps on maps of species density often shows strong correlations.

Here are examples of such correlations for the summer months: moulting P. puffinus mauretanicus are associated with the northern Biscay tidal convergent front; Sula bassana is the dominant species in the upwelling area running offshore from northern Biscay to the Celtic Sea; the overheating area of sou-

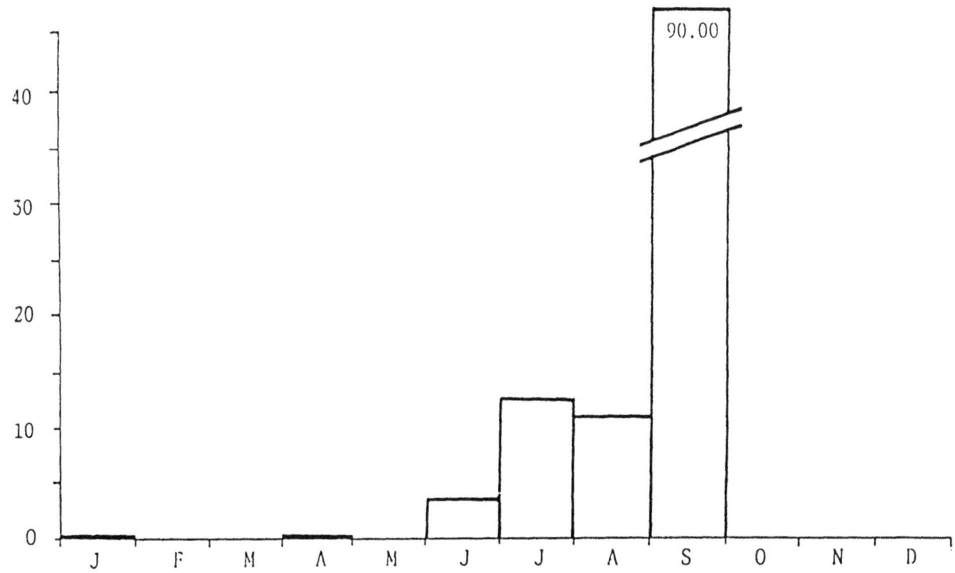

Fig. 4: Monthly variations in the mean density (number of birds per hour) of P. puffinus mauretanicus in western France: provisional data, 1980-1983. Note: insufficient data available for September and October result in a probably overestimated relative density in September, and a misleading absence in October.

thern Biscay seems characterized by a high density of Hydrobates pelagicus.

In a given area, specific densities can significantly change from year to year. Fig. 6 shows such changes for Hydrobates pelagicus and Sula bassana during a ten year period in southern Biscay. These species do not depend on the same prey items, one being piscivorous (S.b.), the other a plankton-eating species (H.p.). The similarity in their yearly and long term variations in density leads to the supposition that these changes are linked to physical and biotic characteristics of the marine environment.

Undoubtedly, the sustained accumulation of such data will, in a near future, tend to a much better understanding of birds' ecological preferences and biology at sea.

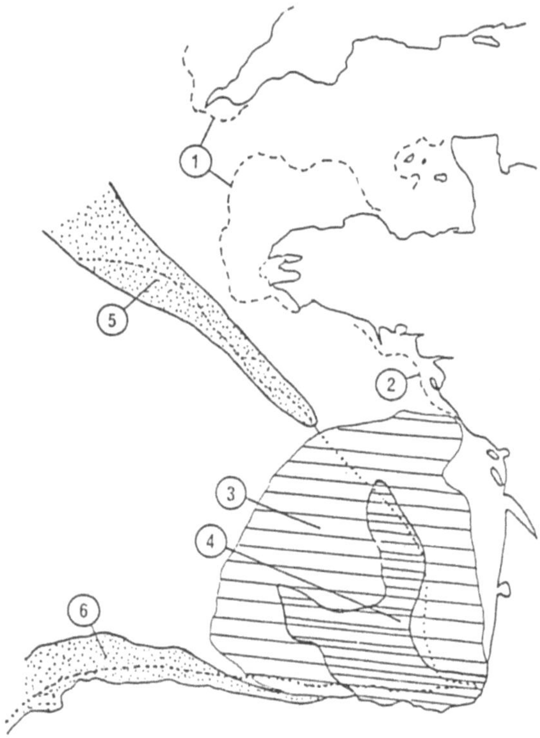

Fig. 5.: Simplified representation of oceanological parameters, drawn from a sattelite image taken on 26 August 1981: a typical summer situation in western Channel and Biscay.

1 : Channel convergent front
2 : northern Biscay tidal convergent front.
3 and 4: Biscay overheating area (warmer centre more densely shaded).
5 : tide-induced upwelling.
6 : true coastal upwelling.

A dotted line indicates the Continental shelf.

In other respects, it is worth mentioning that the "Programme C.R.B.P.O.-M.E.R." and seabird studies of other kinds are not exclusive. For example, a detailed description and discussion of an irruption of Calonectris diomeda in inshore waters from Biscay to the Celtic Sea during the summer of 1980 (Yésou 1982b), was obtained by combining data from this program, sea-watchers' records and information from breeding grounds.

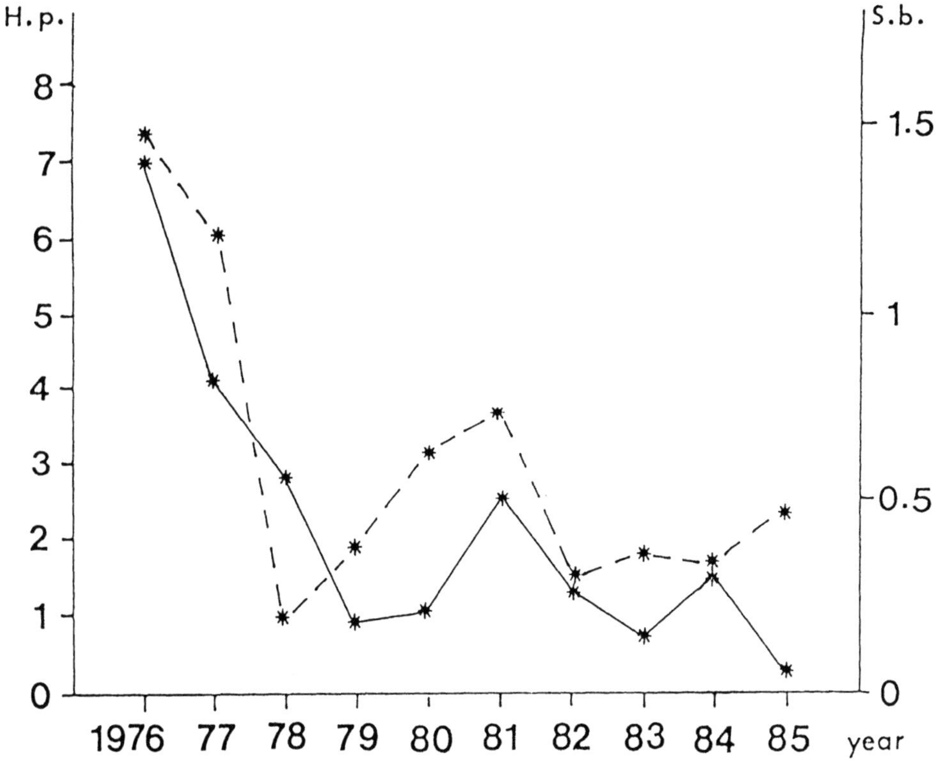

Fig. 6: Evolution of average density at sea (number of birds per hour) for Hydrobates pelagicus (solid line) and Sula bassana (dotted line), June to August, Fosse de Cap-Breton (southern Biscay).

Another example of how the work presented here and other studies are complementary can be seen in the case of the very high mortality of Rissa tridactyla in January-February 1984, when c. 10,000 corpses were collected in western France after severe gales (for preliminary reports, see Jouanin & G.T.O.M. 1984 and Cawthorne 1984). The age structure of dead birds was almost the same as the mean age structure observed at sea during the four previous winters: this strongly suggests that no difference occurred in the mortality affecting adults and immatures. Sea transects carried out just after this major wreck did not show any noticeable decrease in bird density. The density of surviving birds is one of the factors leading to the belief that concerned breeding populations are not threatened by the exceptional death of some tens of thousands individuals.

Lastly, there is a point of importance: even if this program is run, in some respects, on a quite different basis than in other European countries (Tasker et al. 1984, Evans 1986), it has the advantage of being fully standardized. Moreover, the geographical position of each record is located with greater precision than in any other scheme. Results can thus be compared to those from other countries and the data can, eventually, be integrated at international level.

References

Blondel, J. et Isenmann, P. (1981). Guide des Oiseaux de Camargue. Delachaux & Niestlé, Neuchâtel et Paris.

Braillon, B. (1969-1970). Les oiseaux marins nicheurs de Basse-Normandie: dénombrement de 1969 et récapitulation des données antérieures. Le Cormoran, 1: 42-64 et 100-101.

Brien, Y. (1970). Statut actuel des oiseaux marins nicheurs en Bretagne. VIII. Mise au point en 1970: visites récentes en état actuel des effectifs par localités. Ar Vran, 3: 167-275.

Brien, Y. (1972). Observations en milieu pélagique du golfe de Gascogne à l'Irlande au cours de l'été 1970. Ar Vran, 5: 109-123.

Cramp, S., Bourne, W.R.P. et Saunders, D.R. (1974). The Seabirds of Britain and Ireland. Collins, London.

Cawthorne, A. (1984). Storms hit seabirds. B.T.O. News, 131: 1.

Evans, P.G. (1986). Monitoring Techniques in the Atlantic. In Medmaravis: Proceedings 1st Mediterranean Seabird Symposium (this volume), Springer Verlag, Heidelberg.

Guyot, I., Launay,G. et Vidal,P. (1985). Oiseaux de mer nicheurs du Midi et de Corse: évolution et importance des effectifs. Annales C.R.O.P., 2: 31-47.

Hémery, G. (1982). Etude de la répartition géographique en mer des oiseaux marins. Compte-rendu C.R.B.P.O./M.E.R., Paris.

Hémery, G. (1984). Répartition géographique et écologie en mer des oiseaux marins. Rapport scientifique C.R.B.P.O., 1983-1984: 32-38.

Hémery, G. et Elbée, E. (d') (1985). Discrimination morphologique des populations atlantique et méditerranéenne de Pétrel tempête Hydrobates pelagicus. Annales C.R.O.P., 2: 63-69.

Hémery, G., Housta, F., Nicolau-Guillaumet, P. et Roux, F. (1979). Distribution géographique, importance et évolution numériques des effectifs d'Anatidés et de Foulques hivernant en France (janvier 1967 à 1976). Bull. O.N.C., Sp. Scien. Tech., Mai 1979: 5-91.

Henry, J. et Monnat, J.Y. (1980). Les réserves d'oiseaux de mer en Bretagne. Penn ar Bed, 12 (103): 381-396.

Henry, J. et Monnat, J.Y. (1981). Oiseaux marins de la façade atlantique française. Rapport de convention S.E.P.N.B./ M.E.R., Brest.

Isenmann, P. et al. (1986). Thirty years of fluctuations of the Laridae populations in the Camargue (1956-1985), in Medmaravis Proceedings 1st Mediterranean Seabird Symposium (this volume), Springer Verlag, Heidelberg.

Jouanin, C. et le G.T.O.M. (1984). A propos des échouages de mouettes. Courrier Nature, 89: 36.

Monnat, J.Y. (1978). Mortalités d'oiseaux à la suite du naufrage du pétrolier Amoco Cadiz. Penn ar Bed, 11 (93): 339-360.

Monnat, J.Y. et Guermeur, Y. (1979). L'Amoco Cadiz et les oiseaux. Rapport S.E.P.N.B./Ministère de l'Environnement. Brest.

Pasquet, E. (1982). Statut hivernal actuel des Plongeons, Grèbes et Grand Cormoran. Rapport C.R.B.P.O./M.E.R. Paris.

Pasquet, E. (1984). Création de la base de données relatives aux recensements d'oiseaux marins reproducteurs en France. Rapport C.R.B.P.O./S.R.E.T.I.E., Paris.

Prieur, D. (1970). Observations d'oiseaux pélagiques au large des côtes de Bretagne du 8 au 30 Septembre 1969. Ar Vran, 3: 42-50.

Tasker, M.L., Hope Jones, P., Dixon, T. et Blake, B.F. (1984). Counting seabirds at sea from ships: a review of methods employed and a suggestion for a standardized approach. The Auk, 101: 567-577.

Thomas, A. (1981). Recensements d'oiseaux marins échoués en Bretagne. Bilan de 1978 à 1980. Penn ar Bed, 13 (104): 15-29.

Yésou, P. (1982a). L'observation des mouvements en mer, ou "seawatching". Rapport au G.T.O.M. Ar Vran 29 S, 5: 29-33.

Yésou, P. (1982b). A propos de la présence remarquable du Puffin cendré Calonectris diomedea près des côtes du golfe de Gascogne et de la mer Celtique en 1980. L'Oiseau et R.F.O., 52: 197-217.

MONITORING SEABIRDS IN THE NORTH ATLANTIC

P.G.H. Evans, Edward Grey Institute, Zoology Dept.,
South Parks Road, Oxford OX1 3PS

Only in the last twenty years have ornithologists given serious attention to monitoring changes in bird populations. Seabirds, though sometimes presenting problems of access to their breeding sites, are unusual in generally offering opportunities to obtain estimates of their total population size. They usually display a clumped distribution often in sizeable colonies where there are stretches of suitable coastline or offshore islands, in contrast to most passerines that occur more or less evenly in a patchwork of suitable habitat. Despite these advantages there are very few seabird species (one example is the Northern Gannet Sula bassana) for which we can be confident of good estimates of their total population size. Instead we reply upon sampling populations over their natural range with the degree of effort varying greatly between regions (usually on a national basis) depending partly upon difficulties of access but mainly upon the resources available within that country. For most people, monitoring in this context means to follow changes in the numbers of birds or nests by frequent standardized counts. I should like to extend it to aim also for complete censuses at less frequent intervals (see below).

I shall first discuss the reasons for monitoring, suggesting priorities for attention, then identify our objectives, and finally examine ways of achieving them given resource limitations and sampling problems that exist for different seabird families. I shall consider only those species that breed primarily along the coast or upon islands (excluding marsh terns and some inland gulls such as Mediterranean Gull Larus melanocephalus and Little Gull Larus minutus) and whose range includes a good part of the European sector of the North Atlantic (equivalent to the Western Palearctic and including the North Sea, Barents and White Sea, Baltic and Mediterranean). I am therefore excluding tropical species with outlying populations north of the Tropic of Cancer (latitude 23°N) such as tropic-birds, boobies and frigate-birds in the Bahamas; and North American species such as Laughing Gull Larus atricilla and Iceland Gull La-

NATO ASI Series, Vol. G 12
Mediterranean Marine Avifauna
Edited by MEDMARAVIS and X. Monbailliu

(a) Species with entire range essentially within European Seas (including inland Seas)
Manx Shearwater Puffinus puffinus (if treat Pacific population as separate species)
British Storm Petrel Hydrobates pelagicus
Common Shag P. aristotelis
Great Skua Catharacta skua (if southern ocean populations are se-separate species)
Audouin's Gull L. audouinii
Yellow-legged Gull L. cachinnans
Mediterranean Gull L. melanocephalus (if including Black Sea and other inland Soviet seas)
(b) Other species with at least 50% their population size probably within European Seas (including inland Seas)
Cory's Shearwater Calonectris diomedea
Northern Gannet Sula bassana
Common Gull Larus canus
Lesser Black-backed Gull Larus fuscus
Great Black-backed Gull Larus marinus
Razorbill Alca torda
Slender-billed Gull L. genei (if including Black Sea and other inland Soviet seas)

Tab. 1: Geographic ranges of North Atlantic seabirds in a European context

(a) Species with entire population within Mediterranean Sea
Audouin's Gull Larus audouinii
(b) Other species with at least 50% their population size probably within Mediterranean Sea
Yellow-legged Gull Larus cachinnans
Little Tern Sterna albifrons (though outside North Atlantic region where subspecies occur)

Tab. 2: Geographic ranges of North Atlantic Seabirds in a Mediterranean context

rus glaucoides that do not breed regularly in the West Palearctic. I shall also not consider those populations of a species (e.g. Great Cormorant Phalacrocorax carbo sinensis) that occur in Europe on entirely enclosed seas such as the Black Sea, nor populations breeding along the Atlantic coast of islands in the Caribbean.

Reasons for monitoring, and criteria for choosing priority species/areas

Presently attention is devoted primarily to rare species such as the Audouin's Gull (Larus audouinii) or species we deem desirable but which we fear may be declining, such as the Roseate Tern (Sterna dougalli) and the Common Guillemot (Uria aalge). In some cases attention is also paid to troublesome species which are increasing, perhaps causing economic damage or presenting health problems. Usually, the latter are only those species such as the Herring Gull (Larus argentatus) that come into frequent contact with man. I suggest that we consider *all* species. Today's abundant seabird may be tomorrow's rarity and vice versa.

Clearly we do not have unlimited resources and so we need to set priorities, perhaps ensuring that we have adequate information from a small set of target species and then spreading the remainder of our resources upon the rest on an hierarchical basis. Criteria for choice should not be the ease with which that species may be counted although this will have to be considered. Most people would agree that priority should be given to those species with low population size or very restricted range, and those presently vulnerable or in marked decline. Since monitoring programmes are usually administered on a national basis, that country will concentrate upon its own seabird populations. However, these should be considered in an international context. It would not be the best use of resources to concentrate upon a small outlying population of Common Guillemots (Uria aalge) as exists for example on the Iberian Peninsula, unless there are good grounds for separating that population taxonomically. Only seven seabird species have their entire ranges within European Seas (tab. 1a) although another seven species have at least fifty percent of their population in that region (tab. 1b). Within the Mediterranean Sea, only the Audouin's Gull (Larus audouinii) has its entire population concentrated there although the majority of the Yellow-legged Gull (Larus cachinnans) population occurs within the region (tab. 2). The taxonomic status of different geographic populations of Little Tern Sterna albifrons is still uncertain. However, it is worth noting that 70% of the nominate race of the Little Tern breed within the Mediterranean. There are five species with endangered or vulnerable populations in European Seas (tab. 3a). Most of these are sub-tropical petrels with populations of very uncertain size. In the Mediterranean only the Audouin's Gull and Storm Petrel (Hydrobates pelagicus) may be considered vulnerable (tab. 3b); the former is probably increasing but the status of the latter is little known.

(a) Species with populations vulnerable or endangered in European Seas
Soft-plumaged Petrel *Pterodroma mollis*
Madeiran Storm Petrel *Oceanodroma leucorhoa*
Bulwer's Petrel *Bulweria bulwerii*
Audouin's Gull *Larus audouinii*
Roseate Tern *Sterna dougallii*
(b) Species with populations vulnerable or endangered in Mediterranean Sea
Storm Petrel *Hydrobates pelagicus*
Audouin's Gull *Larus audouinii*

Tab. 3: Vulnerability of North Atlantic seabirds in European and Mediterranean context.

Using our other conservation criterion of population change, most West European seabird populations are stable or actually increasing (tab. 4). This applies particularly to gull species, whereas various terns and some auk species are declining. There is still a large list of species for which we simply have too little data to assess their status. Many of these are petrels and shearwaters. Despite the difficulties they present for surveying colonies, we must give them priority attention.

The recent attention to population surveys allows us to make some crude estimates of the total population sizes within the North Atlantic region of these seabird species under consideration (tab. 5). These are derived from reviews by various authors in Croxall, Evans and Schreiber (1984), supplemented with recent data for particular species/areas (see footnote to tab. 5). It should be emphasized that they are very approximate although the accuracy of the estimates varies tremendously and ranges are very rarely given in the literature. This is a serious problem and everyone is urged to give their own estimates of the levels of uncertainty around their best estimates. At present a single estimate in the literature for a region, such as 'five thousand pairs' of Black Guillemots (*Capphus grylle*) for Franz Josef Land is treated as more accurate than an estimate of '15,000-20,000 pairs' in Scotland, yet the former almost certainly was not based on any detailed count and may be no more than a general impression. Furthermore, population estimates are necessarily based upon counts made often at very different points in time. This may not be serious if there has been little change in status but in some cases we know this is not the case. A good example is the Brunnich's Guillemot (*Uria*

Species with population generally			
increasing	declining	no marked overall change (1)	?
(a) NW EUROPE			
Northern Fulmar Northern Gannet Great Cormorant Common Shag Arctic Skua Great Skua (except Faeroes) Black-headed Gull Common Gull LBB Gull (except Finland) Herring Gull (except Denmark) GBB Gull Sandwich Tern	Roseate Tern Common Tern (except UK, West-Germany, Belgium) Little Tern (except Sweden, Finland, West-Germany) Common Guillemot (except UK, Scandinavia) Brunnich's Guillemot	Black-legged Kittiwake Caspian Tern Arctic Tern Atlantic Puffin (except France)	Manx Shearwater Brit. Storm Petrel Leach's Petrel Sabine's Gull Glaucous Gull Ross's Gull Ivory Gull Black Guillemot Little Auk
(b) SW EUROPE			
LBB Gull Herring Gull Black-legged Kittiwake	Cory's Shearwater Common Guillemot	Common Shag	British Storm Petrel Madeiran Storm Petrel Common Tern Little Tern
(c) MEDITERRANEAN			
Audouin's Gull Black-headed Gull Mediterranean Gull Slender-billed Gull Yellow-legged Gull	Common Shag Great Cormorant Common Tern	Sandwich Tern Little Tern	Cory's Shearwater Manx Shearwater British Storm Petrel

(1) includes species with populations some in decline and others increasing

Tab. 4: Present Status (during 1980's) of North Atlantic seabirds in a European and Mediterranean context

	Northern Fulmar Fulmarus glacialis	1.62 - 13.66 mill.
	Soft-plumaged Petrel Pterodroma mollis	100 - 1,600
	Bulwer's Petrel Bulweria bulwerii	10,000 - 112,000
*	Cory's Shearwater Calonectris diomedea	0.60 - 0.81 mill.
*	Manx Shearwater Puffinus puffinus	0.24 - 0.33 mill.
	Little Shearwater Puffinus assimilis	11,100 - 122,000
	White-faced Storm Petrel Pelagodroma marina	0.11 - 0.40 mill.
*	British Storm Petrel Hydrobates pelagicus	0.13 - 0.29 mill.
	Madeiran Storm Petrel Oceanodroma castro	11,000 - 110,000
	Leach's Storm Petrel Oceanodroma leucorhoa	0.78 - 1.13 mill.
*	Northern Gannet Sula bassana	0.225 - 0.275 mill.
	Great Cormorant Phalacrocorax carbo	57,500 - 61,000
*	Common Shag Phalacrocorax aristotelis	67,000 - 85,000
*	Great Skua Catharacta skua	13,500 - 14,000
	Pomarine Skua Stercorarius pomarinus	100 - 12,000
	Arctic Skua Stercorarius parasiticus	18,000 - 335,000
	Long-tailed Skua Stercorarius longicaudus	15,000 - 133,000
	Ivory Gull Pagophila eburnea	500 - 3,000
*	Audouin's Gull Larus audouinii	5,500 - 6,000
	Common Gull Larus canus	0.47 - 0.49 mill.
	Herring Gull Larus argentatus	1.07 - 1.12 mill.
	Yellow-legged Gull Larus cachinnans	114,500 - 131,500
*	Lesser Black-backed Gull Larus fuscus	0.17 - 0.75 mill.
*	Great Black-backed Gull Larus marinus	0.16 - 0.30 mill.
	Mediterranean Gull Larus melanocephalus	154,000 - 307,000
	Slender-billed Gull Larus genei	104,000 - 219,000
	Glaucous Gull Larus hyperboreus	29,000 - 525,000
	Black-headed Gull Larus ridibundus	1.16 - 1.32 mill.
	Ross's Gull Rhodostethia rosea	15 - 1,000
	Sabine's Gull Xema sabini	25 - 1,200
	Black-legged Kittiwake Rissa tridactyla	2.29 - 4.26 mill.
	Gull-billed Tern Gelochidon nilotica	9,000 - 11,500
	Caspian Tern Sterna caspia	5,500 - 6,600
	Common Tern Sterna hirundo	123,500 - 150,000
	Arctic Tern Sterna paradisaea	0.33 - 1.02 mill.
	Roseate Tern Sterna dougallii	6,000 - 7,400
	Little Tern Sterna albifrons	19,500 - 23,500
	Sandwich Tern Sterna sandvicensis	91,000 - 95,000
*	Little Auk Alle alle	8.76 - 27 mill.
*	Razorbill Alca torda	0.30 - 1.21 mill.
	Brunnich's Guillemot Uria lomvia	5.36 - 6.60 mill.
	Common Guillemot Uria aalge	3.01 - 5.03 mill.
	Black Guillemot Cepphus grylle	0.23 - 0.30 mill.
*	Atlantic Puffin Fratercula arctica	3.84 - 8.16 mill.

* Breeding range confined to North Atlantic region; rest extend to other areas.

NOTE: The above broad estimates are derived from the following sources: Brown & Nettleship, Buckley & Buckley, Clapp & Buckley, Sprunt, Evans (a,b), Golovkin, Barrett & Vader, Barcena et al., de Juana, de Juana et al., James, and Le Grand et al. in Status and Conservation of the World's Seabirds (1984), edited by J.P. Croxall, P.G.H. Evans, and R.W. Schreiber. ICBP Technical Publication No. 2, supplemented with additional recent data from Nettleship & Evans (1985), Thibault et al. (1985), Wanless (1986), contributors to this 1st Mediterranean Seabird Symposium, and personal communication with P.H. Becker, K. Kampp, A. Spaans, and S. Wanless.

For some regions (notably parts of the Soviet Union, and East European/North African Mediterranean countries) no population estimates at all exist for particular species. In those areas, I have made my own estimates usually as a range between two orders of magnitude, based upon qualitative comments of their distribution and abundance in the literature.

I have had to treat single estimates on an equivalent basis to a range of estimates published, although the former may not necessarily reflect greater accuracy.

Tab. 5: Approximate Population Sizes (breeding pairs) of Seabirds in North Atlantic region.

lomvia) which with its arctic distribution has colonies that are surveyed only occasionally. Populations of this species are in accelerated decline, particularly in Greenland (and possibly elsewhere). Most colonies within the region have been visited only three or four times this century. Estimates for those colonies last visited 30-40 years ago have to be considered alongside those made much more recently. Some adjustment may be made by calculating rates of population change at sample colonies and then applying these to the rest. This has the danger facing any extrapolation that a status change at one site/region may not mirror another. Thus Nettleship and Evans (1985) in their review of the distribution and status of Atlantic alcids gave the total Greenland population for the Brunnich's Guillemot (Uria lomvia) as between 465,000 and 475,000 pairs. These were using estimates for a variety of dates. When estimates were corrected using rates of decline from sample colonies, this gave a revised figure of 200,000-255,000 pairs (Evans, 1984a). This had assumed (by extrapolation) that the large colony of Sagdleq in Umanaq district had declined to 30,000-50,000 pairs by 1983. However, a recent visit (Kampp, pers.comm.) has revealed that colony to be extinct so the population estimate should be reduced at least to 170,000-205,000 pairs, values that are less than half of original estimates. It is very likely that population estimates of Brunnich's Guillemots (Uria lomvia) for areas such as Western Novaya Zemlya (one million pairs), Spitzbergen (one million pairs) and Franz Josef Land (100,000 pairs) will on further study reveal populations of much smaller size; they have not been surveyed in any detail and the Soviet islands have not been visited by ornithologists for over thirty years. We may find that this species has a total Atlantic population two or three million pairs lower than our present estimates.

This example illustrates another aspect that requires greater attention , that of the distribution of populations, the contribution each makes to the total estimate and the effect the range of estimates make to that total. Many seabird species have an epicentre where their populations are concentrated. For example, the Razorbill (Alca torda) has populations concentrated on the boundary between boreal and arctic waters (Nettleship and Evans, 1985). This means that nearly three-quarters of its population lies within Iceland (tab. 6). If we are concerned about monitoring that species from a conservation viewpoint, we should give high priority to obtaining a good estimate for the Icelandic population rather than paying most attention to other populations of trivial importance in terms of numbers. At present our estimate of the Icelandic population ranges from 100,000 to one million pairs (Petersen, 1982). This clearly has a large effect upon our estimate of the total population for the species. The same applies to an even greater extent for the Little Auk (Alle alle) where Greenland may account for over 90% of the world population but estimates there range from 8 to 25 million pairs. A detailed survey of the Greenland Little Auk population will obviously have a very important influence on our total estimate for the species.

(a)

Northern Fulmar	Iceland: 66%; Faeroes: 10%; Spitzbergen: 7%; Greenland: 6%; Canada: 5%; Britain/Ireland: 5%; Rest: 1%.
Soft-plumaged Petrel	Cape Verde: 65%;Madeira: 30%; Rest: 5%.
Bulwer's Petrel	Salvages: 90%; Madeira: 8%; Canaries: 1%; Azores: 1%; Rest: 0.5%.
Cory's Shearwater	Azores: 71%; Cape Verde: 8%; Canaries: 8%; Italy: 6%; Spain: 1.5%; Malta: 1.5%; Rest: 2%.
Manx Shearwater	Wales: 45%; Scotland: 31%; Ireland: 12%; Faeroes: 4%; Iceland: 2%; Spain: 1%; Rest: 5%.
Little Shearwater	Salvages: 78%; Cape Verde: 8%; Madeira: 7%; Canaries: 7%; Rest: 1%.
White-faced Storm Petrel'	Salvages: 78%; Cape Verde: 22%.
British Storm Petrel	Ireland: 36%; Faeroes: 36%; Scotland: 14%; Malta: 5%; Wales: 4%; Rest: 5%.
Madeiran Storm Petrel	Azores: 98%; Salvages: 1%; Rest: 1%.
Leach's Storm Petrel	Canada: 92%; Scotland: 6%; USA: 1%; Rest: 1%.
Northern Gannet	Scotland: 52%; Canada: 12%; Wales: 11%; Iceland: 10%; Ireland: 9%; Rest: 5%.
Great Cormorant	Norway: 27%; Holland: 17%; Denmark: 17%; Rumania: 7%; Scotland: 6%; Iceland: 6%; Ireland: 3%; Yugoslavia: 2.5%; Poland: 2.5%; Wales: 2.5%; E. Germany: 2%; Sweden: 2%; England: 2%; Rest: 9.5%.
Common Shag	Scotland: 34%; Norway: 19%; Faeroes: 9.5%; Iceland: 8%; Barents Sea: 7%; France: 5.5%; England: 4%; Ireland: 3%; Rest: 10%.
Great Skua	Scotland: 52%; Iceland: 44%; Faeroes: 2%; Rest: 2%.
Pomarine Skua	Canada: 83%; Barents Sea: 8.5%; W. Novaya Zemlya: 8.5%.
Arctic Skua	Canada: 29%; Iceland: 29%; Barents Sea: 29%; Norway: 4.5%; Greenland: 3%; Spitzbergen: 3%; Scotland: 2%; Rest: 0.5%.
Long-tailed Skua	Canada: 34%; Barents Sea: 34%; Sweden: 13.5%; Norway: 7%; Greenland: 7%; Rest: 4.5%.
Ivory Gull	Canada: 57%; Greenland: 29%; Spitzbergen: 14%.
Audouin's Gull	Spain: 78%; Algeria: 9%; Italy: 8%; France: 1%; Rest: 4%.
Common Gull	Norway: 32%; Sweden: 30%; Finland: 13%; Baltic Sea (USSR): 9%; Denmark: 8%; Rest: 8%.
Herring Gull	Norway: 24%; Scotland: 14%; USA: 8%; France: 7%; England: 7%; Sweden: 6%; Holland: 5%; Ireland: 5%; Wales: 5%; Denmark: 4.5%; Rest: 14.5%.
Yellow-legged Gull	Spain: 48%; Italy: 20%; France: 20%; Portugal: 7%; Rest: 5%.
Lesser Black-backed Gull	Barents Sea: 25%; Sweden: 13%; England: 10.5%; France: 9%; Holland: 8%; Norway: 8%; Finland: 6%; Scotland: 5%; Wales: 5%; Rest: 10.5%.
Great Black-backed Gull	Norway: 36%; Iceland: 18%; Barents Sea: 13%; USA: 8%; Scotland: 7%; Greenland: 5%; Canada: 4.5%; Rest: 8.5%.
Mediterranean Gull	USSR (mainly Black Sea): 97.5%; Greece: 2%; Rest: 0.5%.
Slender-billed Gull	USSR: 94%; Iraq: 3.5%; Mauritania: 1%; Italy: 0.5%; Rest: 1%.
Glaucous Gull	Greenland: 20%; Canada: 18%; Spitzbergen: 18%; Franz Josef Land: 18%; Barents Sea: 18%; Rest: 8%.
Black-headed Gull	Holland: 18%; Denmark: 16%; Sweden: 16%; Finland: 12%; Czechoslovakia: 6%; West Germany: 6%; East Germany: 5%; Baltic Sea (USSR): 4.5%; England: 4%; Rest: 12.5%.
Ross's Gull	Greenland: 98%; Canada: 1%; Rest: 1%.
Sabine's Gull	Greenland: 88%; Canada: 10%; Rest: 2%.
Black-legged Kittiwake	Greenland: 20%; Spitzbergen: 17%; Norway: 15%; Scotland: 12%; Iceland: 12%; Canada: 6%; Bear Is.: 6%; England: 3%; Faeroes: 2%; Rest: 7%.
Gull-billed Tern	USA: 44%; Mauritania: 16%; Spain: 16%; Greece: 5.5%; Turkey: 5.5%; Algeria: 4.5%; Baltic Sea (USSR): 4.5%; Rest: 4%.
Caspian Tern	Mauritania: 24%; USA: 20%; Sweden: 18%; Finland: 17%; Baltic Sea: 8.5%; Rest: 12.5%.
Common Tern	USA: 23%; Sweden: 16%; Holland: 9%; Norway: 8.5%; Baltic Sea (USSR): 6%; Finland: 5%; West Germany: 4%; Turkey: 4%; England: 3.5%; Italy: 3%; France: 2%; Spain: 2%; Ireland: 2%; Poland: 2%; Azores: 2%; Scotland: 2%; Rest: 8%.
Arctic Tern	Iceland: 44%; Scotland: 10%; Canada: 8%; Greenland: 7%; Norway: 5.5%; Baltic Sea (USSR): 4%; White Sea (USSR): 4%; Sweden: 4%; Finland: 2%; Faeroes: 1.5%; Barents Sea: 1.5%; Spitzbergen: 1.5%; Rest: 5%.
Roseate Tern	USA: 52%; Bahamas: 22%; Ireland: 10%; Azores: 9%; Rest: 7%.
Little Tern	Italy:28%; Greece: 14%; Egypt: 10%; Spain: 9.5%; England: 7.5%; France: 5%; Poland: 4%; Holland: 3%; Tunisia: 3%; Sweden: 2%; West Germany: 2%; Rest: 9%.
Sandwich Tern	USA: 48.5%; England: 11%; Holland: 21%; West Germany: 7.5%; France: 7%; Sweden: 4%; Denmark: 4%; Rest: 7%.
Little Auk	Greenland: 92%; Spitzbergen: 6%; Franz Josef Land: 1%; Rest: 1%.
Razorbill	Iceland: 73%; Scotland: 11%; Ireland: 7%; Canada: 2%; Rest: 7%.
Brunnich's Guillemot	Iceland: 24%; Canada: 24%; West Novaya Zemlya: 17%, Spitzbergen: 17%; Bear Is.: 13%; Greenland: 2%; Rest: 3%.
Common Guillemot	Iceland: 29%; Bear Is.: 18%; Scotland: 16%; Canada: 14.5%; Faeroes: 7%; Norway: 3%; Ireland: 2%; Rest: 10.5%.
Black Guillemot	Canada: 26%; Iceland: 19%; Greenland: 16%; Scotland: 7%; Spitzbergen: 6%; Franz Josef Land: 6%; Norway: 6%; Sweden: 4%; Rest: 10%.
Atlantic Puffin	Iceland: 50%; Norway: 22%; Scotland: 11%; Faeroes: 10%; Canada: 6%; Rest: 1%.

(b) The Mediterranean contains the following approximate proportions of the total North Atlantic population: Cory's Shearwater 11.3%; Manx Shearwater: 4.9%; British Storm Petrel: 5.8%; Madeiran Storm Petrel: 0.2%; Great Cormorant: 6.1%; Mediterranean Gull: 2.4%; Slender-billed Gull: 4.8%; Common Shag: 8.2%; Audouin's Gull: 100%; Yellow-legged Gull: 98%; Black-headed Gull: 2%; Caspian Tern: 13.6%; Gull-billed Tern: 26%; Common Tern: 15.2%; Little Tern: 70%; Sandwich Tern: 6.4%.

NOTE: These percentages are calculated from the mean of the range of estimates provided from the sources listed in tab. 5. It should be emphasized that in a number of cases the estimates have a very broad range of uncertainty around them and this could significantly alter the relative importance of a population for a particular country in relation to the total population size. Those values where the range of estimates could significantly affect the total are underlined with solid lines; those which would not do so but where the estimates have a wide range of uncertainty, are underlined with hatched lines.

Tab. 6: North Atlantic Seabird Populations - Country Totals as % of Total N. Atlantic Population Sizes.

In an age when many expeditions are visiting remote parts of the world, we should think internationally and put some of our resources into helping one another to survey populations where it matters most. An obvious example would be intensive surveys of seabirds in the archipelagos of Azores, Madeira, Salvages, Canary and Cape Verde. A number of shearwater and petrel species have populations concentrated here yet in most cases we have little more than informed guesses of their population sizes. The eastern Mediterranean and parts of the North African coasts likewise require surveying in some detail for their Procellariform and tern populations. A third priority would be surveys in Iceland (particularly for auks, the Northern Fulmar and arctic tern), and arctic regions such as Greenland (all seabirds but particularly Little Auk, Ross's and Sabine's Gulls, and Black-legged Kittiwake) and some islands such as Novoya Zemlya (all seabirds but particularly auks). A fourth priority would be greater attention to censusing particular seabirds, notably Shearwaters and petrels, skuas and arctic gulls everywhere in the North Atlantic region.

Objectives of monitoring

Having identified broadly how resources may be partitioned into priority species and areas, we should give some attention to our aims. These should be carefully and unambiguously defined at the outset and they may vary between studies. Many monitoring schemes have been established without thought to whether the data collected are interpretable. There are two issues to consider:

Firstly, monitoring aims to detect changes in numbers over time by replicated counts. We must therefore ensure that our methods do indeed detect any changes that take place. This requires attention to sampling problems, the levels of detection practicable to us and the time scale over which we are measuring changes.

Secondly, we should not only consider population changes but provide some evidence that can be used to assess causes for those changes. This requires monitoring also certain parameters, notably fecundity (reproductive success) and survival.

These two aspects are considered below:

- How to detect population changes

Although objectives (and the resources to carry them out) do vary at least in the short term, where possible we should independently set ourselves some longterm aims. These may be arranged hierarchically. As noted above, I would suggest first that ornithologists should cooperate internationally to arrive at better total population estimates for all seabird species, using baseline surveys. This will not only allow us to assess more accurately which species are particularly vulnerable but also will identify better where the populations of a particular species are concentrated. Our second objective (which of course may proceed at the same time) should be to use a sampling method to monitor changes in populations over the entire range of each species. This should best be performed on a national basis, as it is at present, but attention should be given to those regions (for example remote Arctic islands and the Macaronesian archipelagoes) which might otherwise receive little treatment. The latter will require international and sensitive cooperation. Ideally complete censuses should be carried out for every species at intervals of perhaps fifteen years, preferably with different countries undertaking national censuses in the same year. This overcomes the problems of range expansion or contraction which might not otherwise be detected by regular counts at sample sites. Obviously a fifteen year census of some species, for example the Northern Gannet (*Sula bassana*), is more feasible than others, for example the Herring Gull (*Larus argentatus*), and if it turns out to be impractical then an alternative might be to lengthen the time interval between counts (to 20 or perhaps even 25 years). However, there is a danger that by the time a population change has been detected it may be too late to do much about it. Sample censuses at fixed sites will help overcome this problem (at least if the species is in decline) but not if the species alters its behaviour (for example large colonies of Black-legged Kittiwakes *Rissa tridactyla* might decline but the species alternatively form new small-sized colonies).

There is still much discussion over the minimum time interval for sample counts required to detect population changes in time to respond to them with action. The shorter the time interval the better, and in United Kingdom a number of species are monitored annually (Stowe, 1982). Where resources allow this level of intensity then we should not discourage it (so long as it does not divert resources from other species requiring attention). If, as is the case for most regions this is not feasible, then replicate counts at sample sites might best be made every five years (every two or three years if those populations of a species are immediately threatened or in obvious decline).

Considerable effort has been devoted to counting methodology and sampling problems. At times I think the desire for precision has been overdone. Small-scale annual variations in numbers at sample sites may not have any more than academic interest even if they can be distinguished from counting error, the effects of weather, and inter-site movement. If a population declines at 5% per annum then counts every five years need to detect a 22% reduction in that population. Most of our present counting techniques probably allow us to do this. Obviously we should seek accuracy but not be unduly deterred by the many practical difficulties we face.

Perhaps insufficient attention has been paid to the problems of sampling. This also will probably be determined at least partly by local resources available. A resident warden has the opportunity to survey seabird colonies on his/her reserve whereas another site might require special visits. Nevertheless we should aim to count sample areas through the range of a species and within those areas select sample colonies or parts of colonies. This presents another problem which has not received the attention it should. On a regional basis we have little idea where to draw the limits around the population we assume we are monitoring. This is particularly relevant when we make the often artificial demarcation of the population into colonies. Many observed changes in numbers are probably simply reflecting movement between segments of the population, either temporary ones of non-breeders or more permanent ones that involve the exchange of genes (see Evans, 1985, for further discussion of this topic). Both are of importance if appropriate conservation action is to be taken, although they have different implications. Individual seabirds do not observe regional or national boundaries, and gene flow, though sadly little studied in seabirds, is probably fairly extensive at least for some species. For those reasons it is probably best to ensure that at least a few widely dispersed areas are sampled rather than many smaller ones.

Temporal shifts of segments of the population may be an important feature with Cormorants, Gulls (including Kittiwakes) and terns. When population surveys are being undertaken, nations would be strongly advised to integrate their activities so that surveys were carried out in the same year by all countries.

- Determining causes of population changes

Assuming we can detect changes in population size of a species, we should next consider what are the possible causes of those observed changes. By the nature of their life style, seabirds generally spend a significant portion of their life away from land where they are relatively inaccessible for study. We should obviously attempt to overcome these problems and encouragingly a number of North Atlantic countries are now doing so. However, population parameters are probably best monitored at the breeding colony. Detailed information on reproductive success requires intensive study and few countries will have the resources to do this on anything more than a small scale. However, there are a number of indices of reproductive success that can be taken. These will vary between species. Where species have clutch sizes exceeding one, counting a sample of 50-200 nest sites and recording their contents will give a measure of productivity. This is best done two to three times through the entire breeding season at the study colony, with each nest site marked in some way (on photographs if the colony is not directly accessible or is vulnerable to disturbance). Marked sites allow one to follow the fortunes of individual nests. Such information could be further supplemented by measuring the weight and wing length of a sample (minimum 25, preferably 50+) of chicks spanning a range of ages to obtain a generalized growth curve. To obtain such information may require 2-3 visits to the colony. At least some of these colony visits could be combined with visits to record the contents of marked nests. Obviously care must be taken to keep disturbance to a minimum and some sites will not be amenable to this technique (for example narrow ledge cliff sites occupied by Common Guillemots Uria aalge). Nestling growth curves should give an indication of the feeding conditions prevailing in the area that season. Further ways to obtain such information are to carry out timed (e.g. 100 minutes) watches at a selection of nests to record the number of feeds/nest and identify where possible the prey type, estimating size of prey with reference to the bill length of the bird. Care should be taken to adjust feeding rates to the number of young per nest and to carry out watches at comparable times of day (in most species, the first 2-3 hours after dawn, or time of lowest light intensity in arctic regions, represents the period of highest feeding intensity). A minimum of five 100-minute feeding watches is recommended per size.

Most censusing tries to estimate the number of pairs breeding at a colony although often this is not possible and head counts include breeders and non-breeders alike. Much of the variation in numbers counted at a colony during the season arises from the behaviour of the non-breeding element of the population. However, we should not ignore this element. It is probably almost as important to have a measure of the proportion of non-breeders at the colony as of breeders. This will require taking sample areas and observing in greater detail the number of nest sites actually occupied then relating this to the total number of birds associated with that area. This results in a series of 'k' values, calculated by dividing the number of breeding pairs in sample areas (N_p) by the total count of numbers of individuals (N_i). Maximum values of N_i provide an estimate of the proportion of non-breeders. It is important to remember that many species have extended laying periods so that counts should be designed to span the entire period during which each site contains an egg or chick. Obviously the timing of counts also affects the results one obtains since a proportion of nests will fail and some of these will be abandoned. 'k' values are best calculated from counts of the total number of pairs that laid eggs but this involves following the fortunes of marked nests.

Although many temperate and arctic seabirds spend the majority of the year at sea away from their breeding sites and causes of mortality are difficult to examine, an overall measure of survival is possible by intensive ringing of birds on an annual basis at selected sites. This is an aspect of monitoring that has received far too little attention to date. In United Kingdom, considerable numbers of Storm Petrels (*Hydrobates pelagicus*) are ringed each year but many of the data arising are of marginal value. Many of the birds captured are non-breeders, of unknown age. There is an urgent need with this species and other seabirds, to follow birds of known age. This requires large scale ringing of nestlings at select sites and in those species where it is possible to follow the fortunes of individuals by observation at the colony, it is important that they be colour-marked (with a hatching year code and if possible a site code). By so doing we should be able to arrive at estimates of survival for different age cohorts and a determination of the age structure for different sample populations. Much of the information we can derive from those activities rely upon mark-recapture (or resighting) analytic techniques (see Morgan and North, 1985, for a recent review of the scope for such analyses) but if birds are also given a birth site code, we can study the extent of intersite movement that occurs; together with the recoveries of metal rings from dead birds, we can also gain much information of the wintering areas of different populations. Resources will dictate how many sites receive intensive ringing effort and whether all species may be covered. It is recommended that priorities are drawn up on the criteria noted in the previous section.

Finally although not usually directly within our province, we should not forget a variety of environmental factors that could usefully be monitored alongside that of seabird populations and aspects of their population biology. These will include climatic factors that are probably already being measured by other interested parties, but should also include those factors that might be detrimental (or advantageous) to seabirds such as populations of fish, squid or planktonic prey, pollutants, predators, habitat changes and human disturbance.

Counting methods and units

Ideally the total population of a species would be monitored by sampling throughout its range. However, it is likely that coverage will have to vary greatly between countries and such decisions will need to be made by the nations concerned. On a more regional basis, monitoring areas should be selected to give reasonable geographical coverage; within those areas, whether counts are made of whole colonies or sample plots within colonies will depend upon the species, its abundance and the dispersion of breeding birds. Likewise, the feasibility of counting one particular element of the population (for example breeders only) will also vary between species. These aspects are considered in more detail below.

The method for selection of sample plots is a matter for debate. On statistical grounds, it would be best to select plots at random from a list drawn up of all countable plots in the area under consideration (see Harris et al., 1983). This depends upon the distribution of colonies remaining the same, but this may not be so. To take account of the formation of new colonies, it is recommended that the initial delimiting of colonies and plotting onto a map is repeated at intervals (say, every five years) to provide a new selection of plots chosen from the new visible population. Some continuity with the previous scheme would be an advantage and this could be achieved by replacement of only a proportion of plots. However, the statistical arguments in favour of this sampling method should be balanced against other considerations; it may not be practical to operate in this manner and site selection may need to combine an element of opportunism with biological intuition (hopefully of an informed nature!).

Those sites used for regular surveillance should be marked permanently, either by fixed stakes (bearing in mind removal/ loss of insecure stakes), or photographs with boundaries of study plots accurately and unambiguously marked out. When counts are made of individuals these should be repeated until two counts are within 2% of one another and the last of these used.

Counting units (pairs, apparently occupied nest sites, individuals) should be defined at the outset and clearly reported in the literature. With whole colony counts, there has in the past been a tendency to use pairs and individuals interchangably on the assumption that one member of the pair will be away feeding; others take a head count and divide by two. Both should be discouraged. If head counts are being used they are best presented as individuals. If it is necessary to convert these to an estimate of breeding pairs, then the appropriate 'k' conversion values (determined for that site or region) can be applied but *must be referred to*.

Counting methods and the units to use vary between species and clearly depend upon the resources available. Below I present those methods recommended for different North Atlantic seabird families, arranged for different species hierarchically from the best method given sufficient resources, to methods one could adopt with few resources. These recommendations reflect the outcome of a monitoring workshop held in England in February 1985 by the UK Seabird Group, organized by M.P. Harris and myself. The methods have been developed for seabirds in Northern Europe (and North America) but the principles should apply to other regions such as the Mediterranean. Local researchers should obviously use their discretion and make modifications where appropriate.

- Procellariformes: Nocturnal petrels and shearwaters.

Unit: apparently occupied nest site
Date: shortly after completion of egg laying
Time of day: daylight hours (combined with night-time visits).

Define limits of colony and mark on a map. Some colonies may be checked directly and total counts made of apparently occupied nest sites. For other colonies, superimpose a grid of a map and with table of random numbers determine a number of random points. Locate these by compass on the ground and place circular quadrats (square quadrats in wooden terrain) at these points, regardless of habitat and whether nest sites are present or not. Use circular quadrats of 10 sq. m (petrels at high density), 20 sq. m (petrels at low density or shearwaters at high density), or 30 sq. m (shearwaters at low density) with a rope of 1.78 m (= 10 sq. m), 2.52 m (= 20 sq. m) or 3.09 m (= 30 sq. m) around a fixed stake. Count number of occupied and unoccupied burrows/nest sites (distinguishing between these if possible - see below). Carry out c. 30 quadrats, with stratified sampling if the colony is clearly divided into areas of markedly different burrow density (i.e. select areas of two or three obviously different density and then place 10-20 quadrats in each, for a minimum of thirty). Reference: Harris & Norman (1981).

If possible, occupied burrows/nest sites should be distinguished from unoccupied ones. This can be checked directly by use of an optical fibrescope or alternatively by playback recordings (these are useful for initial surveys of potential colonies) taking account of response rates for different species and differences in response rate between the sexes. Such values exist in the literature for Manx Shearwater (Puffinus puffinus) but it is wise to establish this information by opening up burrows within a sample area. References James & Robertson (1985); also Brooke (1978a, b).

Non-breeders can be monitored by netting in areas where the breeding population has been sampled, and scoring the proportion of birds without obvious brood patches. This preferably should be done during the early or mid-nesting period to ensure that all breeders do have incubation patches by that time. The method is not entirely satisfactory because non-breeders move around more than breeders and so will be sampled more frequently, but if methods are standardized they probably allow comparison between years and between sites.

If sites are completely inaccessible, alternative though less satisfactory methods include counts of shearwaters offshore in the late afternoon either prior to egg laying or during the incubation period. Petrels may be mist-netted (as indicated above) at the nearest adjacent sites which are accessible. Neither of these methods can provide an estimate of the breeding population size, but an overall population estimate which may serve as an index.

- Northern Fulmar (Fulmarus glacialis)

  Unit: apparently occupied nest site
  Date: late incubation - early nestling period
  Time of day: daylight hours.

If whole colony counts cannot be made, delimit all lengths of cliff where the population to be monitored is nesting. Walk the entire length and plot on a map all the places where one can safely view numbers of birds. Photographs of the cliffs are then taken from these places and early in the season all counting areas containing preferably 100-200 birds are delimited on these photographs, using natural features such as fault lines where appropriate. Plots may be continguous but each is treated as a separate unit. Each plot is numbered and a proportion of these are selected using a table of random numbers. The actual number of plots to be counted will depend upon the total number of plots there are and the time available for counting, but a minimum of five and preferably 20% of the total number available is recommended. Each study plot is photographed and the boundary of the counting area marked on a photographic print. Observation points should be marked out with permanent stakes and counts always made from this position. Count the total number of apparently occupied nest sites. Occupied sites are only counted when an individual is sitting tightly on a reasonably horizontal area large enough to hold an egg. Two individuals sitting beside one another and apparently paired count as one site as do single sitting individuals. At least two counts are recommended, one during the main incubation period and one in the early - mid nestling period. Large grassy colonies in some areas may best be photographed and numbers counted directly from the photographs (photographs are recommended anyway as a permanent record of birds on counting plots).

Although the above method is designed to minimize the number of non-breeding birds included in the count, inevitably failed breeders and some non-breeders will be included, and that proportion will be higher than for a species that builds a substantial nest. If the problem needs to be overcome, this will require more detailed observations on a smaller sample area with counts of eggs and/or chicks through the season. A measure of the non-breeding element is provided by total head counts preferably repeated two or three times between the mid-incubation and mid-nestling periods. These counts may conveniently be made at the same time as the counts of apparently occupied sites.

- Laridae: Black-legged Kittiwake (*Rissa tridactyla*)

Unit: apparently occupied nest site
Date: early incubation - early nestling period (the earlier the better)
Time of day: daylight hours.

Use the same sampling method outlined above for Fulmar, with the exception that counting areas should preferably comprise c. 200 nests (obviously these will have to be smaller in regions with low population sizes), and one will-timed count is probably sufficient. Count all substantial or well constructed nests capable of holding 2-3 eggs. Count nests whether or not adults are associated with them, although a separate count of nests occupied by at least one adult may be useful.

Since the nests of Kittiwakes are fairly clearly defined, such counts should more closely approximate to the breeding element of the population. However, the timing of the counts will affect what proportion of failed breeders are included, and the only way to overcome this is to follow through the occupancy of each nest, marked onto photographs. Non-breeders often aggregate on adjacent rocks or ledges lacking nests and can usually readily be counted separately. Reference: Heubeck and Richardson (1985).

- Other coastal Larids

Unit: apparently occupied nest site
Date: mid incubation period (preferably when most nests have complete clutches or some young have hatched)
Time of day: daylight hours.

Complete counts of the colony are best and in many cases should be feasible. In some cases it will be possible to count marked nests by placing numbered tags at each nest, if necessary during 2-3 well-timed visits. However, care must be taken to reduce disturbance and this may necessitate alternative, less accurate methods such as head counts from a suitable vantage point, or counts from photographs. In those latter cases, the observer must make a judgment of which birds are actually breeding.

Where complete counts are not possible, define limits of the colony and mark on a map. Superimpose a grid and with a table of random numbers determine a number of random points on this map. Locate these by compass on the ground and place quadrats at these points, regardless of habitat and whether nest sites are present or not. The size of the quadrats will depend upon the size and density of the colony, but suggested sizes are 30 m x 30 m, 50 m x 50 m, or 100 m x 50 m in order of decreasing density. The total number of nests and the number containing eggs are then recorded for every quadrat. The number of quadrats will also be dictated by overall colony size but a minimum of thirty is recommended. References: Wanless & Langslow (1983); Wanless & Harris (1984).

As with the Kittiwake, clearly defined nests should be the unit that is counted and this should reasonably closely approximate to the breeding population. However, as before, the timing of the count will affect the proportion of failed breeders included and so the stage of breeding should be checked prior to counting. Since larids have precocial young, some nests may no longer be occupied whilst others still contain eggs or may not yet have had eggs laid in them. This is partly taken into account by recording both occupied and unoccupied reasonably fresh (i.e. the current season's) nests. Otherwise, it requires monitoring those quadrats over an extended time period. A total head count (including satellite groups of birds) should give a measure of the non-breeding element in relation to the number of occupied nests counted.

• Sternidae Terns

Unit: apparently occupied nest site
Date: mid incubation period to early nestling period
Time of day: daylight hours.

Use the same counting techniques as described above for coastal larids, attempting complete colony counts where possible. As with gulls, care must be taken not to cause undue disturbance to the colony and this may limit the time spent in the colony and the number of quadrats counted together with whether counts are repeated. It may be feasible to count used nests after the birds have left but this should be verified for that site/species in the first season. In many cases, one may have to resort the head counts only.

Terns differ from many other seabird families in being more susceptible to change of breeding site either within a season or between seasons; the same probably applies to some gull species and to cormorants. This may mean that it is advisable to undertake monitoring counts at all selected colonies in the same season, preferably not only on a national basis but also internationally. It also emphasizes the need to review the location of all colonies at intervals of say, five years.

- Alcidae Common Guillemot (*Uria aalge*) and Brunnich's Guillemot (*Uria lomvia*)

  Unit: individual
  Date: incubation to early nestling period
  Time of day: preferably 0700 - 1500 GMT except in arctic regions for Brunnich's Guillemot where it may be any time of day (but pattern of attendance should be determined first for each colony)
  Weather: if possible, avoid winds greater than force four, rain/snow or fog.

Delimit all lengths of cliff where the population to be monitored is nesting. Walk the entire length and plot on a map all the places where one can safely view a concentration of birds. Photographs of the cliffs are then taken from these places and early in the season all counting areas containing preferably c. 200 birds are delimited on these photographs. Where possible each plot should run vertically up and down including the tops of cliffs so as to cover a range of habitats. Plots may be contiguous but each is treated as a separate unit. Each plot is then numbered and a proportion of these are selected using a table of random numbers. The actual number of plots to be counted will depend upon the total number of plots and time available, but a minimum of five and preferably 20% of the total number available is recommended. Each study plot is photographed and the boundary of the counting area marked on a photographic print. The number of individuals within these plots should be counted on at least five days spanning the entire counting period (avoiding runs of consecutive days). All plots should be counted on each date, keeping to a consistent order within each day. Reference: Harris et al. (1983); Harris et al. (1985).

As noted earlier, the above method may need to be modified by circumstances - local resources, access, etc. and counters will need to use their discretion.

Since Guillemots do not build nests, it is very difficult to distinguish breeders from non-breeders. The above method will therefore inevitably include a proportion of non-breeders which is likely to be higher than for many other seabird species. This can only be overcome by more detailed studies of sample ledges, with 'k' values (N_p/N_i) generated. The proportion of non-breeders is estimated from maximum values of N_i (total number of individuals per sample plot). Generally, the number of individuals attending the colony and in turn, the proportion of non-breeders in auk species is lowest around dawn and dusk and early in the breeding season; it is highest around midday and during the early mid nestling period. 'k' values should obviously be made at the same times, stages of the season within sites and between years and between sites, perhaps taking times when the proportion of non-breeders are likely to be at both a maximum and a minimum. If time is limited, such information can be obtained using automatic cameras. Reference: Olsen (1985).

- Razorbill (Alca torda)

  Unit: individual (unless occupied nest site feasible)
  Date: incubation to early nestling period
  Time of day: preferably 0700 - 1500h GMT in temperate regions and 0500-1500h GMT in the arctic (but check diurnal pattern of attendance for each monitoring site)
  Weather: if possible, avoid winds greater than force four, rain/snow or fog.

For both razorbills and guillemots, the diurnal pattern of attendance is often characteristic of the area, although it may be affected by the proportion of non-breeders and the feeding conditions for that particular year. Where possible check this for each monitoring site, at least on a regional basis. Try to keep the time of the count constant.

It is unlikely that in more than a very few cases the sampling method proposed for guillemots will be practical for the razorbill. This is because the species rarely forms discrete concentrations of breeding individuals. It may be possible to divide up lengths of cliff where the population to be monitored is nesting, and number plots that are both safe and easy to view (as outlined for the Fulmar). Whether it is feasible to adopt a random sampling programme or necessary to subjectively select plots, it is recommended that counting areas should comprise 100-200 individuals. The number of individuals within these plots should be counted on at least five days spanning the entire counting period (avoiding runs of consecutive days). All plots should be counted on each date. References: Evans (1980), Harris et al. (1983).

The number of plots recommended to count depends upon the size of the population, its density and the time available, but at least five plots should be monitored.

The difficulties of distinguishing breeders from non-breeders noted for the guillemot apply even more so for the razorbill. This is particularly the case because nest sites are generally out of view and may be inaccessible. Where colonies are amongst boulder scree, and these are accessible, it is best to count separately individuals in each of five discrete sub-colonies (aggregations) and then immediately to move carefully into the area counting the actual number of individuals and the number of nest sites (chicks/eggs) associated with each sub-colony, bearing in mind that some may not have laid and others may have lost eggs. Where possible, such sub-colonies should be distributed over the cliff and not confined to areas close to the cliff top. The resulting census should give an estimate of the proportion of occupied sites associated with the numbers of individuals counted.

Some types of colony (for both razorbill and guillemots) will not lend themselves to a monitoring scheme. This applies, for example, to cave colonies, whilst flat top colonies of guillemots can only be monitored by photographing the colony from above, marking on the boundaries and with a grid and table of random numbers, selecting a number of points at which 1 sq.m or 5 sq.m quadrats may be laid on the ground and all eggs counted. A minimum of thirty such quadrats are recommended but the need to minimize undue disturbance may necessitate a compromise. In almost all cases, it is likely that visiting such colonies would cause too much disturbance,so it is probably best to confine monitoring to aerian photography, if necessary using 'k' values derived from sites in other habitats. Reference: Birkhead & Nettleship (1980).

For some very large mixed auk colonies on high cliffs (e.g. in Iceland), an alternative procedure has been to count the number of Razorbills and Guillemots on the sea below the colony and using the ratio to relate to total counts (of both species combined where it is not easy to differentiate them) on the cliffs in sample plots (Gardarsson, 1985). The same approach may be applied to mixed colonies of Brunnich's and Common Guillemots. Care should be taken to ensure that counts on the sea account for any bias in the distance offshore that different species may occupy. If possible, counts of each species should be made in sample plots on or flying to the cliff itself. Alternatively, some measure of breeding numbers may be obtained before birds return to the cliffs, at the start of the season, when they are sitting on the sea at the foot of the colony (Gardarsson, 1985).

- Black Guillemot (Cepphus grylle)

  Unit: individual (in adult plumage)
  Date: pre-laying period (preferably)
  Time of day: between dawn and 0800h (but check diurnal pattern of attendance for each monitoring site)
  Weather: avoid winds greater than force four, rain/snow or fog.

This species presents further problems because not only are nest sites generally out of view, but nests are at lower densities than for either Razorbill or the two Guillemot species. There is no entirely satisfactory method of censusing this species short of mapping out nest sites in sample areas following protracted concealed watches (from sunrise for four hours duration in Northern Britain, West Greenland and Iceland, although this needs checking first for each region) at different stages of the season. However, a more practical though less precise method is to count the total number of potential breeders (individuals in adult plumage) by walking slowly along the coast soon after dawn, flushing birds off the cliff and counting all those individuals offshore. These counts are best repeated 3-5 times on separate days during the recommended period, preferably choosing calm sea conditions. The sampling area will obviously depend somewhat on the size of the region to be monitored, the density of nesting birds and the time available. However, it is recommended that at least 20% of coastline with nesting birds should be monitored and a random sampling technique may be used to select coastline lengths. If regions to be monitored are large, it may mean that the species should be treated in the same way as Gannet, Cormorant and Shag with total counts of the population at longer intervals (five or ten years). Reference: Ewins (1985).

Counts of potential breeders are likely to include a proportion of birds that, though capable of breeding, do not attempt to do so; others may breed but fail. To distinguish these elements of the population requires the more intensive method described earlier, with protracted concealed watches and nest site occupancy checked at intervals through the breeding season.

• Atlantic Puffin (*Fratercula arctica*), Little Auk (*Alle alle*)

Unit: apparently occupied nest site
Date: preferably before and during laying period (Puffin); early to mid nestling period (Little Auk)
Time of day: if burrow counts of either species are made during the nestling period, they should preferably be between 1000 - 1600h GMT to minimize effects of disturbance.

The circular quadrat method described for nocturnal Procellariformes should be used, with the only modification being one of the methods for determining occupancy. In the case of the Puffin, where nest burrows occur in soil, they may be determined by signs such as fresh digging, droppings, and during the nestling period by hatched eggshells or fish in the entrance. Rabbit burrows are usually larger with much soil outside, often bare earth, and pellets in the entrance. However, it is very difficult to separate burrows of Manx Shearwater from those of Puffin, and occupied burrows amongst boulders can only be determined by observation of parents flying into the nest sites (preferably in 3 hour period after dawn, from a concealed position, at least for Little Auks). Where possible, occupancy should be confirmed by checking the actual contents of nests. This can be carried out towards the end of the breeding season. Reference: Harris (1984).

Little Auks generally nest among boulders and accurate counts of occupied sites can rarely be made until hatching (because of the difficulty of locating nests during the incubation period when adults are very secretive and change-overs at the nest infrequent). In this case, parents returning with food to their young are best watched during the three hour period after first light. Reference: Evans (unpubl. data).

If the use of quadrats is not practical, an alternative though less satisfactory method for both species is to make head counts of the entire colony or sample areas. Peak attendance in the Puffin is usually in the evening, 1-2 hours before dusk, and in the Little Auk between midnight and 1000h GMT (although these times need to be checked first for each monitoring site since they may vary). If the nest entrances are visible, the following method may be useful during the chick-rearing period. Select a sample area in which the number of sites containing chicks can be determined. This sample site can then be used to determine the number of chicks in a larger area, by comparing the number of birds carrying food (fish in the bills of Puffins; extended throat pouches in the case of Little Auks) to chicks in the area during a given time (best in early morning when feeding frequency is highest in both species). The ratio of the number of birds with food counted in the larger area and in the sample area will be the same as the ratio of the number of chicks in each area, provided the two areas are counted at comparable times and the young of both areas are of comparable age. This of course only gives a measure of nests with chicks and does not allow for breeding failure. That requires calculating the proportion of sample sites that are occupied but do not have chicks. A similar ratio method can also be used on head counts of individuals for sample areas and then applied to the total area counted. Reference: Evans (1980).

Some seabird species are probably best monitored by infrequent surveys and complete censuses at intervals of 5-10 days. This is either because the total population comprises a relatively small number of colonies which may be counted in their entirety or because the dispersion of nests and the monitoring problems associated are so great that counts at more frequent intervals are unlikely to be feasible.

• Sulidae Northern Gannet (Sula bassana)

Unit: apparently occupied nest site
Date: early - mid nestling period
Time of day: preferably between 0800 - 1500h GMT in temperate latitudes; probably a similar period further north though this has yet to be established
Weather: avoid rain, fog or high winds.

Count directly of from good quality photographs (in some cases best from the air) the number of nest sites occupied by one or two adults. Include all sites irrespective of whether or not any nest material is present so long as the site appears suitable for breeding. Inevitably this will include a proportion of non-breeders, but most of the latter will be concentrated in clubs either on separate rocks often at the base of the colony or on ledge sites with little or no nest material. These can be counted separately to obtain a measure of the non-breeding element of the population.

- Phalacrocoracidae Great Cormorant (*Phalacrocorax carbo*) and Common Shag (*Phalacrocorax aristotelis*)

Unit: apparently occupied nest site
Date: early to mid nestling period (bearing in mind the fact that the species has a protracted breeding season with some nests containing well fledged young when others may not yet have eggs)
Time of day: daylight hours (unless colonies vulnerable to predation).

Count all substantial/well constructed nests occupied by at least one individual (in some cases this may best be done photographically). In some areas, nest sites may be inaccessible and out of view, so that accurate counts are not possible. There is no entirely satisfactory way to overcome this problem and it may necessitate restricting the population to be counted to that element which is visible from a safe position. Head counts of birds associated with the sample population (including those on headlands and offshore rocks) will enable a measure of the proportion of non-breeders, but note that birds are very mobile and may shift around the coast. The same applies to the breeding population so that general surveys are necessary at intervals to locate new colonies.

- Stercorariidae: All skua species

Unit: apparently occupied territory
Date: mid incubation to mid nestling period (preferably when most nests have complete clutches)
Time of day: daylight hours.

For discrete colonies or those at reasonably high densities, count all occupied territories from a suitable vantage point scanning from a distance with binoculars. Beware of counting members of a pair, standing apart, as two single territory holders, and of overlooking individuals that blend against similarly coloured backgrounds. Preferably repeat counts three times spanning the counting period with territories marked on a map. This will help to reduce the error arising from some sites not yet containing eggs when others may be empty (and thus unattended) as precocial young begin to wander. For these reasons the technique will slightly underestimate the total breeding population. Total head counts as for larid gulls and terns, when related to the above counts of occupied territories, will provide an estimate of the number of non-breeders present. Reference: Furness (1982).

Monitoring breeding success

Counts of occupied nests or sample plots early and later in the breeding season will give one measure of breeding success. For more detailed information, it is necessary to count eggs and chicks per nest (sampling time once in mid-incubation and once in late nestling period) or measure growth parameters of young, as outlined in an earlier section. Both methods may be applied to gannets, shags and cormorants, kittiwakes, larid gulls, skuas and black guillemots. The latter method is useful for species laying single eggs such as petrels and shearwaters and the rest of the auks. For those species, such as ledge nesting guillemots and razorbills, where sites may be inaccessible or very susceptible to breeding failure from disturbance, it may be possible to obtain a measure of growth rates by catching young soon after fledging and obtaining fledging weights. Supplementary information on the conditions affecting breeding success include feeding rates, and the sizes and type of prey taken, and these will necessitate timed watches, preferably from hides.

Conclusions

Much of the information presented here will doubtless be modified, perhaps considerably, in the future. Recommended methods have generally been worked out for a small number of areas (particularly United Kingdom and Canada) and for a few well-worked species (such as auks). It is important that European workers recognize this and do not take all that is said literally. They may find that local conditions (or resources) necessitate alternative approaches. However, hopefully they form a basis from which to work. The same may be said of the population estimates for North Atlantic seabirds. They provide some idea of the order of magnitude of populations of each species but perhaps more importantly they draw attention to our ignorance of some of the largest populations that exist in our region.

Acknowledgements

The population data derive primarily from contributions to the ICBP book on the Status and Conservation of the World's Seabirds and the recommended monitoring methods from the workshop on this subject held at Uttoxeter, England by the UK Seabird Group in February 1985. I thank in particular Dr. M.P. Harris for his extremely valuable input to the latter. He, together with Drs. T.R. Birkhead, E.K. Dunn, A.J. Gaston and S. Wanless kindly commented upon a draft of this paper at short notice. Recent unpublished population estimates were provided by Drs. P.H. Becker, A. Bermejo, M. Fasola, E. de Juana, K. Kampp, B. Massa, J. Mayol, P. Meininger, A. Spaans, A. Teixeira, S. Wanless and P. Yesou and I thank these.

References

Barcena, F.; Teixeira, A.M. & Bermejo, A. (1984). Breeding Seabird Populations in the Atlantic Sector of the Iberian Peninsula. Pp. 335-346. In Croxall, J.P.; Evans, P.G.H. & Schreiber, R.W. (editors) Status and Conservation of the World's Seabirds. ICBP Technical Public. No. 2. Cambridge.

Barrett, R.T. & Vader, W. (1984). The Status and Conservation of Breeding Seabirds in Norway. Pp. 323-334. In Croxall, J.P., Evans, P.G.H. & Schreiber, R.W. (editors) Status and Conservation of the World's Seabirds. ICBP Technical Public. No. 2. Cambridge.

Birkhead, T.R. & Nettleship, D.N. (1980). Census methods for murres Uria species - a unified approach. Canadian Wildlife Service Occas. Paper No. 43: 1-25.

Brooke, M. de L. (1978a). Some factors affecting the laying date, incubation and breeding success of the Manx Shearwater Puffinus puffinus. J. Anim. Ecol. 47: 477-495.

Brooke, M. de L. (1978b). Sexual differences in the voice and individual vocal recognition in the Manx Shearwater Puffinus puffinus. Anim. Behav. 26: 622-629.

Brown, R.G.B. & Nettleship, D.N. (1984). The Seabirds of Northeastern North America: Their present status and conservation requirements. Pp. 85-100. In Croxall, J.P., Evans, P.G.H. & Schreiber, R.W. (editors). Status and Conservation of the World's Seabirds. ICBP Technical Public. No. 2. Cambridge.

Buckley, P.A. & Buckley, F.G. (1984). Seabirds of the North and Middle Atlantic Coasts of the United States: Their status and conservation. Pp. 101-134. In Croxall, J.P., Evans, P.G.H. & Schreiber, R.W. (Editors). Status and Conservation of the World's Seabirds. ICBP Technical Public. No. 2. Cambridge.

Clapp, R.B. & Buckley, P.A. (1984). Status and Conservation of Seabirds in the Southeastern United States. Pp. 135-156. In Croxall, J.P., Evans, P.G.H. & Schreiber, R.W. (Editors). Status and Conservation of the World's Seabirds. ICBP Technical Public. No. 2. Cambridge.

Cramp, S. (editor) (1985). Handbook of the birds of Europe, the Middle East, and North Africa: the birds of the Western Palearctic, Vol. IV, Terns to Woodpeckers. Oxford.

Croxall, J.P., Evans, P.G.H. & Schreiber, R.W. (editors) (1984). Status and Conservation of the World's Seabirds. ICBP Technical Public. No. 2. Cambridge, 778 pp.

Evans, P.G.H. (editor) (1980). Auk Censusing Manual. British Seabird Group. 13 pp.

Evans, P.G.H. (1984a). The Seabirds of Greenland: Their status and Conservation. Pp. 49-84. In Croxall, J.P., Evans, P.G.H. & Schreiber, R.W. (editors). Status and Conservation of the World's Seabirds. ICBP Technical Public. No. 2. Cambridge.

Evans, P.G.H. (1984b). Status and Conservation of Seabirds in Northwest Europe (excluding Norway and the U.S.S.R.). Pp. 293-322. In Croxall, J.P.; Evans, P.G.H. & Schreiber, R.W. (editors). Status and Conservation of the World's Seabirds. ICBP Technical Public. No. 2.Cambridge.

Evans, P.G.H. (1985). Genetic Structure of Seabird Populations. Pp. 5-6. In Tasker, M.L. (editor). Population and Monitoring Studies of Seabirds; Proceedings of Second International Conference of the Seabird Group. 36 pp.

Ewins, P.J. (1985). Colony attendance and censusing of Black Guillemots Cepphus grylle in Shetland. Bird Study 32: 176-185.

Furness, R.W. (1982). Methods used to census skua colonies. Seabird Report 1977-1981. 6: 44-47.

Gardarsson, A. (1985). Methods used in a survey of seabird cliffs in Iceland. P. 5. In Tasker, M.L. (editor). Population and Monitoring Studies of Seabirds; Proceedings of Second International Conference of the Seabird Group. 36 pp.

Golovkin, A.N. (1984). Seabirds Nesting in the U.S.S.R.: The Status and Protection of Populations. Pp. 473-486. In Croxall, J.P.; Evans, P.G.H. & Schreiber, R.W. (editors). Status and Conservation of the World's Seabirds. ICBP Technical Public. No. 2. Cambridge.

Le Grand, G.; Emmerson, K. & Martin, A. (1984). The Status and Conservation of Seabirds in the Macaronesian Islands. Pp. 377-392. In Croxall, J.P.; Evans, P.G.H. & Schreiber, R.W. (editors). Status and Conservation of the World's Seabirds. ICBP Technical Public. No. 2. Cambridge.

Harris, M.P. & Norman, F.I. (1981). Distribution and status of coastal colonies of seabirds in Victoria. Memoirs Nat. Mus. Vict. 42: 89-106.

Harris, M.P.; Wanless, S. & Rothery, P. (1983). Assessing changes in the numbers of Guillemots Uria aalge. Bird Study 30: 57-66.

Harris, M.P. (1984). The Puffin.T. & A. Poyser, Staffs.

Harris, M.P.; Wanless, S. & Rothery, P. (1985). Review of British and Irish Guillemot monitoring studies. Pp. 21-22. In Tasker, M.L. (editor). Population and Monitoring Studies of Seabirds; Proceedings at Second International Conference of the Seabird Group. 36 pp.

Heubeck, M. and Richardsen, M.G. (1985). Kittiwake monitoring: methods and interpretation. P. 20. In Tasker, M.L.

(editor). Populations and Monitoring Studies of Seabirds; Proceedings of Second International Conference of the Seabird Group. 36 pp.

James, P.C. (1984). The Status and Conservation of Seabirds in the Mediterranean Sea. Pp. 371-376. In Croxall, J.P.; Evans, P.G.H. & Schreiber, R.W. (editors). Status and Conservation of the World's Seabirds. ICBP Technical Public. No. 2. Cambridge.

James, P.C. & Robertson, H.A. (1985). The use of playback recordings to detect and census nocturnal burrowing seabirds. Seabird. 8: 18-20.

De Juana, E.A. (1984). The Status and Conservation of Seabirds in the Spanish Mediterranean. Pp. 347-362. In Croxall, J.M.; Evans, P.G.H. & Schreiber, R.W. (editors). Status and Conservation of the World's Seabirds. ICBP Technical Public. No. 2. Cambridge.

De Juana, E.A.; Varela, J. & Witt, H.-H. (1984). The Conservation of Seabirds at the Chafarinas Islands. Pp. 363-370. In Croxall, J.P.; Evans, P.G.H. & Schreiber, R.W. (editors). Status and Conservation of the World's Seabirds. ICBP Technical Public. No. 2. Cambridge.

Morgan, B.J.T. & North, P.M. (editors) (1985). Statistics in Ornithology. Lecture Notes in Statistics, n° 29, 418 pp., Springer Verlag, Berlin.

Nettleship, D.N. & Evans, P.G.H. (1985). Distribution and Status of the Atlantic Alcidae. Pp. 54-155. In Nettleship, D.N. & Birkhead, T.R. (editors). The Atlantic Alcidae. Academic Press, London & New York.

Olsen, B. (1985). Photographic Surveillance of Guillemots on study photos. P. 18. In Tasker, M.L. (editor). Population and Monitoring Studies of Seabirds: Proceedings of Second International Conference of the Seabird Group, 36 pp.

Petersen, A. (1982). Sjofuglar. In Gardarsson, A. (editor). Fuglar. Pp. 15-60. Rit Landverndar 8, Reykjavik.

Sprunt, IV, A. (1984). The Status and Conservation of Seabirds of the Bahama Islands. Pp. 157-168. In Croxall, J.P.; Evans, P.G.H. & Schreiber, R.W. (editors). Status and Conservation of the World's Seabirds. ICBP Technical Public. No. 2. Cambridge.

Stowe, T.J. (1982). Recent population trends in cliff-breeding seabirds in Britain and Ireland. Ibis 124: 502-510.

Thibault, J.C.; Guyot, I. & Cheylan, G. (editors) (1985). Oiseaux marins nicheurs du Midi et de la Corse. Annales du Centre de Recherches Ornithologiques de Provence, No. 2, 88 pp.

Wanless, S. & Langslow, D.R. (1983). The effects of culling on the Abbeystead and Mallowdale gullery. Bird Study 30: 17-23.

Wanless, S. & Harris, M.P. (1984). Effect of date on counts of nests of Herring and Lesser Black-backed Gulls. Ornis Scand. 15: 89-94.

Wanless, S. (1986). A survey of the numbers and breeding distribution of the North Atlantic Gannet *Sula bassana* and an assessment of the changes which have occurred since Operation Seafarer 1969/70. NCC Report on Contract ESS 50/F3B/S12. Nature Conservatory Council, Petersborough.

SEABIRD CENSUS AND STUDY TECHNIQUES

Elizabeth A. Schreiber and Ralph W. Schreiber
Los Angeles County Museum of Natural History
900 Exposition Blvd.
Los Angeles, California 90007 USA

In order to carry out an effective conservation programme, or to model bird population dynamics, accurate data on the size and dynamics of avian populations are required.

Problems inherent in censusing

The necessity for and difficulty of accurately censusing bird populations have received considerable attention (reviewed in Ralph and Scott 1981, Verner 1985) and seabirds have received some special consideration (Dunnet 1982; Croxall et al. 1984). Atlas projects have been accomplished and are especially useful in providing information on locations of colonies and ranges of species (i.e. Sowls et al. 1980). The next step is to determine population sizes and monitor any changes through time. Considerable effort has been expended on censuses of birds at sea (references in Gould et al. 1982, Vermeer et al. 1983, Briggs et al. 1985a, b). So we will not discuss the specific problems involved in those efforts here.

During the past two decades we have encountered difficulties in making counts of birds in island colonies and we hope our discussion here will be useful to persons just starting such projects.G.M. Dunnet (1982) and his students (Ollason and Dunnet 1983, and references therein) have previously discussed some of those problems. We wish to add to his points in hopes of assisting others in avoiding the pitfalls in census techniques, both in designing a study and in interpreting data.

Dunnet has stressed that monitoring studies must take into account problems of life cycle, annual cycle, variation due to time and weather, site tenacity of birds, observer consistency, sampling methods, and statistical manipulation. Weins et al. (1986) emphasized the spatial and temporal scales in ecological

NATO ASI Series, Vol. G 12
Mediterranean Marine Avifauna
Edited by MEDMARAVIS and X. Monbailliu

studies. Jared Verner (1985) discussed various techniques of estimating and counting populations, pointing out the difficulty of making comparisons between counts done by various people using different techniques. There are basically 5 categories of bias in estimating numbers of birds: 1) observer, 2) habitat, 3) birds' behaviour, 4) study design, and 5) weather (Verner, 1985).

Our major point is that estimating a seabird population, even in just one small, confined, island colony is difficult and subject to tremendous error. This makes discussing population fluctuations extremely difficult. Just counting nests and number of birds present is insufficient. We have found populations on our Pacific island study colonies to have 10% to 500% more adults using a given colony/island than are present and breeding on any one day.

Long term population cycles occur naturally, so that no "normal" population size exists. Historically the numbers present have varied considerably, and perhaps especially since different persons made separate counts. Until we witnessed the 1982-83 ENSO in the central Pacific (Schreiber and Schreiber 1984) we thought the various counts of birds on Christmas Island represented a steady state population and any fluctuation in numbers was due to the timing of counts or the difficulty of accurately censusing such a tremendous area.

During 1982 and 1983, 80 to 90% of the Christmas Island avian population died. Since then we have documented only a gradual increase in the number of birds present and breeding. This has led us to believe that the "normal" situation of the bird populations is to be always increasing, until a natural crash or major decline occurs. Then the population begins to increase again (see Schreiber and Schreiber, this volume).

Attempting to make estimates of the numbers of birds using an area is difficult: Are we making censuses just after a crash or are the birds at their peak numbers? Were the counts made during the height of the nesting season, which can vary from year to year, or during the start or end of the season? Are nest sites used twice during the season? Long term consistent data, on the order of 10 to 20 years, taken throughout each breeding season are essential to truly understanding seabird population dynamics.

- Banding Studies

During 1984 and 1985 we banded 400 White Terns (*Gygis alba*) on Johnston Atoll. In each year we estimated 20 nests were present, and the maximum number of birds we counted on the Atoll at any one time was 85-100. By October 1985, in random samples we found 1 in 5-7 birds banded. This means there are potentially 2,800 White Terns using the Atoll. We were astonished to learn this in light of the low number of nesting birds. We suspect the reason for the low number of nests is related to a lack of suitable habitat. It may be that there are a large number of White

Terns moving through the Pacific and that they simply "rest" here now and then.

• Problems with single surveys

Another common error in making population estimates comes from basing counts on one visit to a colony. We have found approximately 2% to 10% error in nest counts between successive surveys of the same colony. Brown Boobies (Sula leucogaster) nest in open areas ans are easy to see. Between successive nest checks we find that we miss 1-3 nests out of 100, and those are all individually marked nests. Red-tailed Tropicbirds (Phaethon rubricauda) nest under bushes and in our study colonies where all nests are marked we miss 1-5 nests out of 100 on any given survey. Masked Boobies (Sula dactylatra) nest over large, spread out areas in vegetation 5-40 cm. tall. Their nests are easily missed and to check 100 nests we walk about 2 miles. We tend to miss 5-10% of the nests on each survey.

In a study of Wedge-tailed Shearwaters (Puffinus pacificus) on Mutton Bird Island, New South Wales, Floyd and Swanson (1983) estimated 12,400 breeding pairs in 1983, + 6,127! Then they estimated the number of fledglings at 6,270 to 18,530. This number was considered to be a 125% increase over the 5-6,000 pairs estimated in 1976 (Swanson 1976). It seems to us that one should not discuss or indicate a 125% increase in a population when in 1980 the number of breeding pairs could be anywhere from 6,000 to 18,000: a range from no increase at all to a 300% increase.

• Time of Day Counts are Made

The time of day when counts are done is also important to consider when comparisons are to be made with other data. Our data for roost counts on Johnston Atoll (Schreiber and Chovan, in prep.) show order of magnitude changes between noon and midnight, and a 25% to 50% increase in a population from sunset to midnight (fig. 1). Clearly censuses made at different times of day are not comparable.

Growth Rate Analysis

Often environmental stress does not cause direct mortality, but insidious, difficult to observe factors such as low asymptotic weights and decreased growth rates (Zach and Mayoh 1984). Chicks may never reach normal adult body size (Tyler and Stearner 1966, Tyler et al. 1967), and have shortened life spans (Stearner and Christian 1972). Effects such as these cannot be measured when a study consists of only counts of birds present. Individuals must be banded and growth increments measured in order to truly determine the status of a population.

Monitoring the health of a population by an analysis of growth rates and asymptotic weights of chicks has proven to be a very useful and relatively simple technique (Ricklefs 1984). Samples of both large and small chicks, 15-20 individuals of

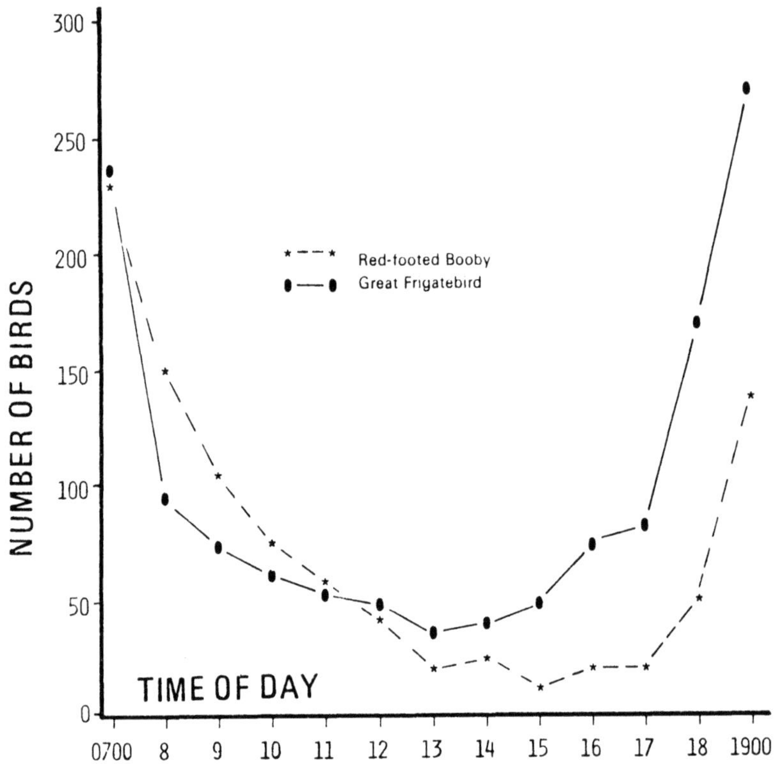

Fig. 1: The number of Red-footed Boobies and Great Frigate-birds roosting on Sand Island, Johnston Atoll, Central Pacific Ocean in relation to time of day.

each, are required for accurate monitoring. A change in growth rates is often the first indication of a pertubation in the environment (Ricklefs and White 1975, Zach and Mayoh 1984). It is well documented that intraspecific comparisons of growth rates and asymptotic weights of chicks are an index to the nutritional state of the nesting population and also can reflect stress (Zach and Mayoh 1984) or variation in feeding conditions, both seasonal and annual (Ricklefs and White 1975).

In order to measure growth rates we record measurements of culmen, wing, tarsus, tail and weight, on a sample of the same chicks 4 to 7 days apart. Using the Gompertz growth equation we obtain a growth rate and asymptotic weight for the population (Ricklefs 1984). The initial weights taken in the increments are adjusted for variation in wing length, then a regression analysis is performed to determine if the adjusted initial weights (RXWT) differ between seasons or years. This tells if birds of the same wing length had the same body weights from year to year. Since the rate of wing growth changes as a chick gets older we separate large and small chicks and do the regression on each separately. Then we determine if the adjusted (for wing length) weight increment between measurements (RDWT) differed between years or seasons, again by sizes large and small. The Duncan Multiple Range test of the General Linear Model procedure is used to determine if a significant difference exists in RXWT and RDWT between seasons, years and islands.

Study Techniques

We believe it worthwhile here to present a few techniques that we use in the field while collecting data. Many of these have evolved through trial and error over several years.

On each visit to a colony we weigh and measure a series of 20 to 30 adult birds in order to track seasonal, year to year and interisland variation in size. This is most useful if the same birds can be caught in succeeding years. The various Pesola spring scales are reliable and easy to use for recording weight. Birds are placed in a pillow case or tubular cardboard food container (such as a Pringles Potato Chip can) with the scale attached. The scale should be tared frequently while in use in case a bird has left something in the bag after being weighed, or the spring connections have become loose.

Length measurements are taken with a plastic dial caliper or with 1/2 of a meter stick. In specific study areas, each newly banded adult is measured (culmen, wing, tarsus and tail). These same measurements and weights are taken on a series of chicks for growth increment analysis. We also weigh and measure a sample of 20-33 eggs on each survey.

We record molt patterns of the primaries and rectrices for about 30 adults of each species on each visit. We try to use the same area from year to year with the hope of catching the same birds in order to discuss individual molt patterns as well

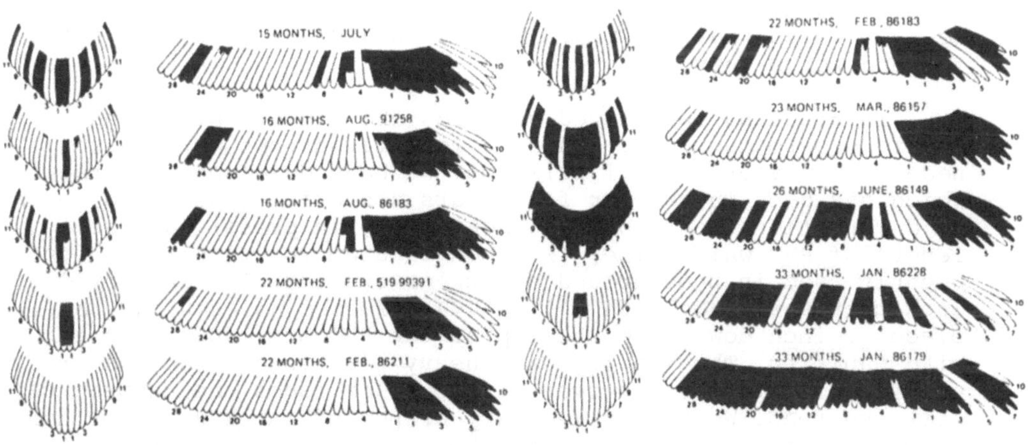

Fig. 2: "Molt maps" of Brown Pelicans (Pelecanus occidentalis) from Florida showing the stage of growth of individual feathers. Black indicates the feather is new. Numbers to right of date are the Los Angeles County Museum of Natural History specimen number of a band number.

as patterns for the species as a whole. It is very instructive to draw molt maps (fig. 2) to easily identify patterns and aberrations. Length of each feather (missing, 1/4, 1/3, 1/2 etc.) in the primaries and secondaries and from left to right on the rectrices is recorded. If the feather is full, its age is recorded, old or new. An asterisk is used to indicate the center of the tail since number of rectrices varies in some seabird species (Schreiber et al. in prep.).

The United States Fish and Wildlife Service, Bird Banding Laboratory provides individually numbered bands to licensed banders and keeps records of all banding activity. We have found that the aluminium alloy of the bands provided does not last nearly the life of the bird. We are working with Laysan Albatross (Diomedea immutabilis) on Midway Island and have found that bands put on in the late 1950's through mid 1960's are now illegible and many have already been lost. We can etch the illigible bands with an acid bath (Ludwig 1942) and read most of them but there is no way to replace the information lost with the lost bands.

Lambournes (Coleman House, Station Rd., Knowle, West Midlands, Great Britain) makes a special alloy band that lasts longer but these must be purchased by the bander and cost approximately $ 170.00 per thousand. Considering that thousands of albatros must be banded in order to have some remaining in the population after 30 to 40 years, this can be prohibitively expensive. And yet with birds that live 30 to 50 years there is no way to follow an individual's life history without bands which last for their entire lifespan.

Depending on the species and the habitat a catching device is frequently needed to make capture of individuals efficient. We have found that a 2 meter pole with a study fine mesh net on the end works very well.

In studying long-lived seabirds who exhibit site tenacity and form permanent pair bonds, a long-lived nest marker is a necessity. After several failures, we have developed a marker which has been in use for four years and still appears to be in good condition (Schreiber and Schreiber, in press). We use a length of PVC white pipe, 1/2 inch in diameter such as is frequently used in underground sprinkler systems. A 3 inch by 4 inch cattle ear tag with a 1/2 inch diameter hole at the top is placed 1 inch down from the top of the PVC pipe and glued on with caulking. The ear tags are heavy plastic and the numbers are etched into them so even if the color fades the numbers can still be read. We have tried many brands of caulk and some disintegrate after a few weeks in the tropical sun. Generally the outdoor type which is guaranteed for 20 years seems good. The white pipe is surprisingly visible but should be long enough to stand above grassy vegetation. On coral reefs or hard surfaces we use a 5 pound sledge hammer and an iron spike to make the hole for the marker. It is useful to place the marker on a consistent side of all nests for reference in future visits.

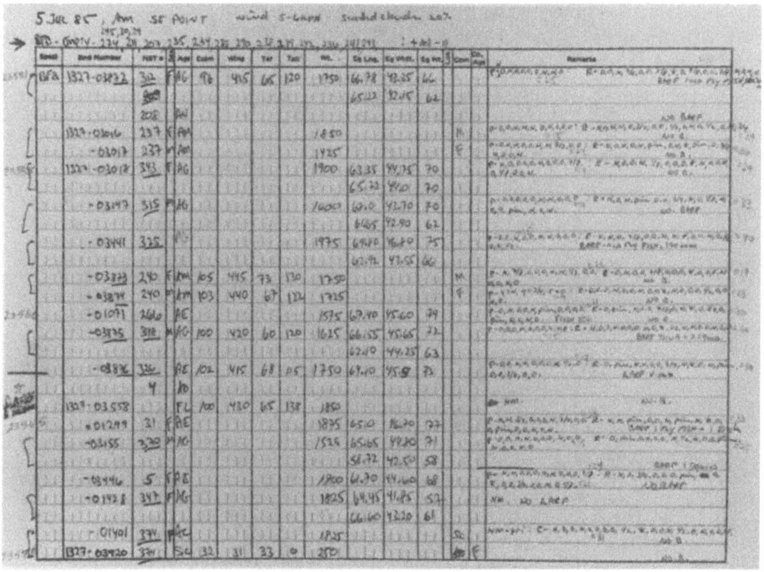

Fig. 3: Field data sheet showing species (BFB = Sula dactylatra) and other information collected from birds on Christmas Island, Pacific Ocean. An underlined NST is a newly placed nest number and an underlined band number is a new banding. The numbers on the far right side are the calculated molt score, based on the molt data in the remarks section.

We always "field mark" adult birds that we handle with spray paint or permanent felt tip pen so we know we have handled that individual when next visiting the colony (if the visit is within 2 months). Then, according to what we are trying to accomplish, we can either catch the individual again or ignore it. This marking has saved us considerable time.

We have devised a data sheet for recording information in the field (fig. 3). The sheet is arranged for easy recording of data in the field and our computer entry system is arranged so data can be entered directly from these sheets. Many of the data are coded but we have tried to keep codes very simple i.e. AE = adult on egg, AC = adult with chick, etc.

Our data management system, called NESTS, is extremely versatile and we can format data as needed and sort them by any category. By sorting on species and nest number we can follow what has happened at each nest over the years (fig. 4). These figures are for Red-tailed Tropicbirds (Phaethon rubricauda) for which we have the best data. We left out measurements of adults and eggs in fig. 4 for simplicity. Note that in nest 25 in 1982, 1983 and 1985 we only caught one member of the pair and in 1983 the first egg failed and the pair relaid. In nest 7 the same pair was together in 1982 through 1985. To find out what happened to the pair that was in nest 7 in 1980 we do a sort of the data by band number and follow each individual.

Sample of data produced by NESTS software program.

Red-tailed Tropicbird (Phaethon rubricauda)

Data for Nest 25

Band Number	Date	Age and Status
785-03077	25 Aug 79	AE
785-02512	28 Jun 80	AE
785-02569	13 Jul 80	AE
785-02512	16 Jul 80	AE
785-02512	18 Jul 80	AE
785-02512	20 Jul 80	AE
785-02646	22 Jun 82	AE
785-02646	28 Jun 82	AE
	27 Oct 82	E
785-02733	25 Jun 83	AE
785-02733	28 Jun 83	AE
	04 Jul 83	E
785-02733	29 Jul 83	AE
785-02823	21 Oct 83	MC
785-02823	30 Oct 83	LC
785-02733	06 Jul 84	AE
615-86067	11 Jul 84	AE
785-02733	17 Jul 84	AE
615-86067	03 Jul 85	AE
	07 Jul 85	2 Day chick.

Data for Nest 07

Band Number	Date	Age and Status
785-03092	27 Jun 80	AE
785-03092	01 Jul 80	AE
785-02510	04 Jul 80	AE
785-02510	08 Jul 80	AE
785-02510	09 Jul 80	AE
785-03092	10 Jul 80	AE
785-03092	13 Jul 80	AE
785-03092	16 Jul 80	AE
785-02510	18 Jul 80	AE
785-02604	16 Jun 82	AE
785-02604	22 Jun 82	AE
785-03075	25 Jun 82	AE
785-03075	28 Jun 82	AE
785-03075	25 Jun 83	AE
785-02604	28 Jun 83	AE
785-02604	01 Jul 83	AE
785-03075	04 Jul 83	AE
785-02604	26 Jul 83	AC (small chick)
785-03075	30 Jul 83	AC
785-02810	03 Oct 83	LC
785-03075	11 Jul 84	AE
785-02604	15 Jul 84	AE
615-86384	06 Oct 84	LC
785-02604	03 Jul 85	AE
785-03075	10 Jul 85	AE

Fig. 4: Print out data from individual nest sort from NESTS software.

Sample of data produced by NESTS software program.

Red-tailed Tropicbird (Phaethon rubricauda)

Data for Band Number 785-03501

Date	Age	Co-Occp.	Nest #	Egg Length mm	Egg Width mm	Egg Weight gms	Bird Culmen mm	Bird Wing mm	Bird Weight gms
19 Aug 1979	AC	JH	063						725
28 Jun 1980	AE		014	63.75	46.40		59		740
01 Jul 1980	AD		014						660
04 Jul 1980	AE		014				60		600
16 Jul 1980	AC	CK	014						725
18 Jul 1980	AC	CK	014						690
22 Jun 1982	AE		014				59		650
25 Jun 1982	AE		014	66.10	45.94	67	58		600
28 Jun 1982	AE		014			66			550
28 Jul 1983	AE		137						700
31 Jul 1983	AE		137	60.00	44.75	60			600
06 Jul 1984	AE		137	62.60	47.00	70			725
12 Jul 1985	AE	SC	419						650

Fig. 5: Print out of data from individual band number sort from NESTS software.

By sorting on band number we find how many times we have caught an individual and what stage of the nesting cycle it was in (fig. 5). We compare this band number sort to the nest number sort and find that Red-tailed Tropicbird number 785-03501 got a new mate between 1980 and 1982 but was still in the same nest. The bird kept the same mate from 1982 through 1985 but in 1983 they moved to a bush 20 meters away and in 1985 moved 1.5 meters to the other side of the same bush. Some birds have had the same mate for 5 years and nested in the same site. We believe when a pair moves their nest site it is due to a change in the vegetation. The nesting habitat of the Red-tailed Tropicbird on Christmas Island is discussed by Clark et al. (1983).

In fig. 6 we illustrate the recaptures of Red-tailed Tropicbirds banded in 1979 in our study area on Christmas Island. Thirty adults were originally banded and 13 (43%) were recaptured the next year. We did not visit the colony in 1982 but our data for later years in this figure illustrate the variability in returns through time. In 1984 we recaptured 12 (40%) and in 1985 9 (30%). We are presently analyzing these data by the dynamic or composite life table method to calculate mortality rates.

Banded as adult with chick	= 5
Banded as adult with egg	= 15
Banded as adult alone	= 10
Banded as chicks	= 24

1979 +	1980						= 7
1979 +	1980 -	+ 1982	+ 1983	+ 1984	+ 1985		= 4
1979 +	1980 -	+ 1982	+	+ 1984			= 2
1979 +	1980 -	+ 1982	+ 1983	+	+ 1985		= 1
1979 +	-	+ 1982	+	+ 1984	+ 1985		= 1
1979 +	-	+	+ 1983	+ 1984			= 1
1979 +	-	+	+ 1983				= 2
1979 +	-	+	+	+ 1984			= 1
1979 +	-	+	+	+ 1984	+ 1985		= 3
						Total	= 22

None captured in 1981.
First chick recaptured in October 1983 on an egg.

Fig. 6: Red-tailed Tropicbirds banded in 1979 and subsequently recaptured.

Conclusion

We believe that only through detailed studies of individual birds and their nests can we make valid statements regarding breeding biology parameters of a species. Only with such data can a valid population estimate be obtained, and the dynamics of the population understood.

There is obviously no substitute for being present in the study area throughout the annual cycle, and especially during the breeding season. This is one of the drawbacks of studying seabirds whose nesting season can last 6 to 8 months or longer. Being present during a 1 to 3 week period once or twice during a season is very frustrating. We can get data on egg sizes, adult sizes, chick growth rates and some pair bonds, but there is no way to accurately estimate reproductive success.

We hope this brief outline is of value and we welcome discussion of the points we have attempted to illustrate. We are more than willing to correspond with individuals regarding our techniques and analysis methods.

References

Briggs, K.T., W.B. Tyler, and D.B. Lewis (1985a). Comparison of ship and aerial surveys of birds at sea. J. Wildl. Manage. 49: 405-411.

Briggs, K.T., W.B. Tyler, and D.B. Lewis (1986b). Aerial surveys for seabirds: methodological experiments. J. Wildl. Manage. 49: 412-417.

Clark, L., R.E. Ricklefs and R.W. Schreiber (1983). Nest-site selection by the Red-tailed Tropicbird. Auk 100: 953-959.

Croxall, J.P., P.G.H. Evans and R.W. Schreiber, Eds. (1984). Status and conservation of the world's seabirds. ICBP Technical Publication No. 2.

Dunnet , G.M. (1982). Ecology and everyman. J. of Animal Ecology 51: 1-14.

Floyd, R.B., and N.M. Swanson (1983). Wedge-tailed Shearwaters on Muttonbird Island: an estimate of the breeding success and the breeding population. Emu 82: 244-250.

Gould, P.J., D.J. Forsell, and C.J. Lensink (1982). Pelagic distribution and abundance of seabirds in the gulf of Alaska and eastern Bering Sea. U.S. Dept. of Interior, Fish and Wildlife Service, Biol. Services Programme, FWS/OBS 82/84. 294 p.

Ludwig, F.E. (1942). An improved method of determining band numbers on illegible bands. Bird Banding 13: 103-104.

Ollason, J.C., and G.M. Dunnett (1983). Modelling annual changes in numbers of breeding fulmars, Fulmarus glacialis, at a colony in Orkney, J. of Animal Ecology 52: 185-198.

Ralph, C.J., and J.M. Scott, Eds. (1981). Estimating numbers of terrestrial birds. Studies in Avian Biology No. 6, Cooper Ornithological Society, 630 p.

Ricklefs, R.E., and S.C. White (1975). A method for constructing nestling growth curves from brief visits to seabird colonies. Bird-Banding 46: 135-140.

Ricklefs, R.E. (1984). The optimization of growth rate in altricial birds. Ecology 65: 1602-1616.

Schreiber, R.W., and E.A. Schreiber (1984). Central Pacific Seabirds and the El Nino Southern Oscillation: 1982-1983 Perspectives. Science 225: 713-716.

Schreiber, E.A., and R.W. Schreiber. In press. Permanent nest markers for ground nesting birds. Journ. of Field Ornithology.

Schreiber, R.W., and E.A. Schreiber. Christmas Island (Pacific Ocean) seabirds and the El Nino Southern Oscillation (ENSO): 1984 perspectives. In Medmaravis, Proceedings first Mediterranean Seabird Symposium (this volume), Springer Verlag Heidelberg.

Schreiber, R.W., D.W. Anderson, E.A. Schreiber and D. Bradley. In prep. Plumage and molt of the Brown Pelican.

Schreiber, R.W., and J. Chovan. In prep. Roosting by pelagic seabirds: energetic, populational and social considerations.

Sowls, A.L., A.R. DeGange, J.W. Nelson, and G.S. Lester (1980). Catalog of California seabird colonies. U.S. Dept. of Interior, Fish and Wildlife Service, Biol. Services Program. FWS/OBS 37/80. 371 p.

Stearner, S.P., and E.J.B. Christian (1972). Late effects of ionizing radiation in the chicken: survival, body weight, and pathology. Radiation Research 52: 179-196.

Swanson, N.M. (1976). Seabird islands No. 32, Mutton Bird Island, New South Wales. Aust. Bird Bander 14: 88-91.

Tyler, S.A., and S.P. Stearner (1966). Growth in the chicken after high-rate and low-rate cobalt-60 gamma irradiation. Radiation Research 29: 257-266.

Tyler, S.A., S.P. Stearner, and H.D. Muller (1967). Radiation effect on growth of the chicken x-irradiated during embryonic development. Growth 31: 105-117.

Vermeer, K., I. Robertson, R.W. Campbell, G. Kaiser, and M. Lemon (1983). Distribution and densities of marine birds on the Canadian west coast. Environment Canada, 73 p.

Verner, J. (1985). Assessment of counting techniques. Current Ornithology Vol. 2, pp. 247-302. R.F. Johnston, Ed. Plenum Publ. Corp.

Wiens, J.A., J.F. Addicott, T.J. Case, and J. Diamond (1986). Overview: the importance of spatial and temporal scale in ecological investigations. Chapt. 8, pp. 146-153 in Diamond, J. and T.J. Case, Eds., Community Ecology. Harper and Row, New York.

Zach, R., and K.R. Mayoh (1984). Gamma radiation effects on nestling tree swallows. Ecology 65: 1641-1647.

CHAPTER 3 : POPULATION CONTROL OF INVADING SPECIES

Chairman: Prof. E. de JUANA. Departamento de Zoologia.
Universidad Autonoma de Madrid.

THE BREEDING BIOLOGY OF ADOUIN'S GULL

ON THE CHAFARINAS ISLANDS

by Patricia Bradley
International Centre for Conservation Education
Guiting Power, Cheltenham, U.K.

The breeding biology of Audouin's Gull on the Chafarinas Islands

According to a recent ICBP International Newsletter, Audouin's gull, *Larus audouinii* Payraudeau 1826, is the Mediterranean's most important bird. It is the Mediterranean's only endemic breeding seabird. In 1977 Hans H. Witt estimated the world population size to be between 600-800 pairs. IUCN/WWF, in 1978, initiated a programme to promote the conservation of Audouin's gull throughout its range. Largely as a result of the more detailed surveys which insued, current estimates suggest that there are at least 4,000 pairs in the breeding population. As part of this project a three year study of the factors affecting the reproductive output of this rare gull species has been undertaken at, what is numerically, the most important breeding site. In this paper the breeding success or failure of Audouin's gull is discussed and the factors which influence reproductive success are described. A management plan is also suggested which would safeguard and perhaps improve the productivity at the colony in future seasons.

The largest colony is to be found on the Chafarinas Islands, 3 km off the Moroccan Coast. In 1985 2,020 pairs of Audouin's gulls bred there. This represents almost 50% of the world's total breeding population. The presence of Audouin's gull, at this site, in such large numbers can be understood in terms of the facts that firstly: only one island of the three island group is inhabited. Audouin's gulls avoid humans and have in the past suffered heavy losses due to egg stealing. Locally, Audouin's gull eggs are famous as being good for pastry-making. The colony has until recently enjoyed the inadvertent protection afforded by the presence of a detachment of Spanish Regular soldiers on the inhabited island of the Chafarinas Group. Secondly, recent oceanographic surveys have given further weight to the fame

NATO ASI Series, Vol. G 12
Mediterranean Marine Avifauna
Edited by MEDMARAVIS and X. Monbailliu

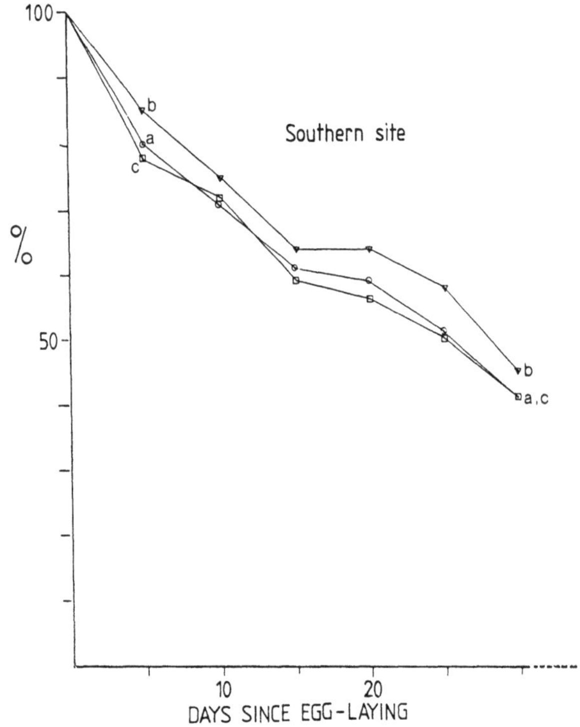

Fig. 3: Survivorship of chicks, expressed as the percentage surviving beyond a certain age.

the two sites. The most common clutch size was three. These three eggs generally hatch in the same sequence as they were laid. Other studies have recorded differential mortality between the siblings with the third chick suffering highest mortality rates (Parsons 1970). In fig. 3 the survival of the first (a), second (b) and third (c) chicks to hatch are plotted for the southern site in 1983 in the same manner as before. No significant difference was found between their survival rates. At the northern site, as illustrated in fig. 4, the c or third chick to hatch survived less frequently than the a or the b. This was due to many c chick deaths during the first five days after hatching. The third egg is on average smaller, by 7%, than the first egg laid and it has been shown in other gull studies that large chicks develop from large eggs and that these large chicks have greater chances of survival (Parsons 1972). In this study egg size in itself was not found to have a significant effect on chick survival to an age of five days.

Data collected during hidewatches and from the records made during visits to the sites, indicate that chicks at sites which lack vegetation leave the nests significantly earlier than those at vegetated. They leave the nests to find cover, shelter from the heat and from predation. Densities of Audouin's gulls are high at vegetated sites. The trips to cover, which are longer at less vegetated sites, may involve crossing another territory which may be defended by other adult Audouin's gulls. The third

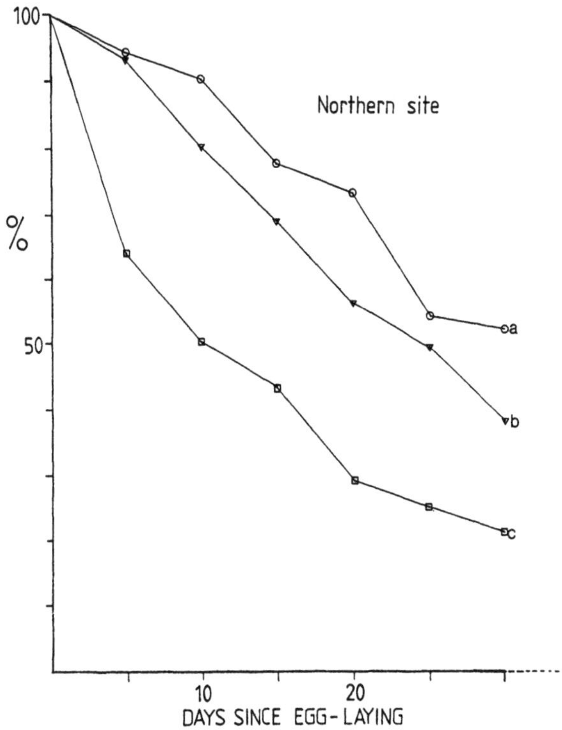

Fig. 4: Survivorship of chicks, expressed as the percentage surviving beyond a certain age.

chick being younger is less mobile than its siblings and either takes longer to make the trip or is left behind in the nest and is therefore vulnerable to attacks by other adult Audouin's gulls, should it stray onto another territory, attacks by Yellow-legged gulls or fatal temperature changes. In 1984 artifical cover was provided to examine the importance of cover for both the chicks and the incubating, often apparently heat-stressed, incubating adults. Estimation of the value of artifical cover was, however, frustrated by another factor having a much greater impact.

• The Yellow-legged Gull Problem...

During the late 1970s numbers of Audouin's gulls breeding on the Chafarinas increased, however, since the turn of the decade the breeding population size has remained stable. Meanwhile the numbers of Yellowlegs have increased. Of particular relevance has been the expansion of the Yellowleg subcolony on the island of Rey, to which the Audouin's gulls have confined their nest sites during this study (for details of both species nesting distributions see Varela et al. in this volume). In fact between 1984 and 1985 the Yellowleg breeding population on Rey increased by 49%. This increase may be due to the apparently limitless food supplies available to the Yellowlegs and a lack of competition. Yellow-legged Gulls feed on the refuse from the military garrison; the waste associated with the neighbouring Moroccan fishing and market village; and they take offal and discards from the numerous fishing boats in the area. Audouin's gulls, in contrast, persist in fishing, except when winds are high and seas rough in which case they take insects and migratory birds (Witt et al. 1979). Yellow-legged gulls, as elsewhere in the Mediterranean, begin to breed approximately three weeks earlier than do Audouin's gulls and therefore have the pick of the nesting sites. The Yellow-legged gull is physically much larger than Audouin's gull and is more aggressive, capable and prone to predating upon the smaller species' eggs, chicks and adults.

...and its effects

In the breeding season of 1984 Yellow-legged gull numbers increased in the immediate vicinity of the southern site where the Audouin's gull nests were well spaced around the vegetation. In the north, the tightly-packed nesting Audouin's gulls successfully deterred potential predators by mobbing approaching Yellow-legged gulls. In the south at no time were enough Audouin's gull nests threatened at any one time for the numbers of defending adults to be sufficient to fend off an attack. Yellow-legged gulls were observed taking eggs, and later, chicks at the southern site, reducing the productivity at that site to zero. At one of the control sites where a small group of Audouin's gulls were breeding in close proximity to Yellow-legged gulls, breeding success was similarly reduced. As this control site was visited only infrequently it would suggest that the presence of the observer was not a major factor influencing the success of failure at a site.

In 1985 Audouin's gulls did not recolonize the sites where breeding had failed in 1984, but bred in other areas, some of which were obviously unsuitable, for example being within the splash zone during storms. Again in 1985 the southern site suffered heavy predation. At the first study site established in 1985 in the south none of the eggs in a study area, which contained fifty nests, lasted for more than thirteen days. The southern control site of 1985 was then adopted as a substitute study site but there also the predation rate was high and no chicks survived longer than five days.

At the northern site nesting densities of Audouin's gulls were higher than previously recorded, nevertheless a greater percentage of the chicks which were killed by Yellow-legged gulls (see tab. 1). This may be due to greater effort, on my part in 1985, being made to recover chick rings from Yellow-legged gull food pellets. The percentage of chicks which were killed by conspecifics was also higher in 1985 than in 1984 (see tab. 1). This killing of chicks by adults was only observed at the northern site (see tab. 2). In tab. 2 the fates of the chicks at both sites in 1984 are compared. At the northern site nesting densities were high and the frequency with which chicks trespassed into the territories of neighbours was much higher than at the southern site. It was on these occasions that adults attacked chicks. The chicks were never eaten. These attacks may be an adaptation to prevent adoptions of neighbours' offspring which might compete with the adult's own chicks.

CAUSES OF AUDOUIN'S GULL CHICK DEATHS
(expressed as % of the total number of the chicks known to die at the northern sites in 1984 and 1985).

	NORTH 1984	NORTH 1985
% killed by Yellow-legged gulls	2	33
% killed by Audouin's gulls	28	34
% impaled on bushes	10	7
% died hatching	18	7
% causes unknown	40	22
(sample size)	(50)	(55)

Tab. 1

CAUSES OF AUDOUIN'S GULL CHICK DEATHS
(expressed as % of the total number of the chicks known to die at the study sites in 1984).

	NORTH 1984	SOUTH 1984
% killed by Yellow-legged gulls	2	64
% killed by Audouin's gulls	28	0
% impaled on bushes	10	0
% died on hatching	18	11
% causes unknown	40	24
(sample size)	(50)	(58)

Tab. 2

In fig. 5 and 6 the survival of eggs and chicks are shown for the three years at the southern and the northern sites respectively. Please note that as it was not possible each time to mark all the chicks (due to time restrictions as only one hour was spent working within view of the gulls per session) there is an artificial drop between the percentage of the eggs from which chicks successfully hatch and the number of chicks which were marked and therefore included in further estimations of survival.

Tab. 3 shows the reproductive success at the two sites during the three years. The number of fledglings per egg at the southern site in 1983 compares favorably with values recorded for other gull species. It is, however, apparent that the reproductive success of Audouin's gull on the Chafarinas Islands has

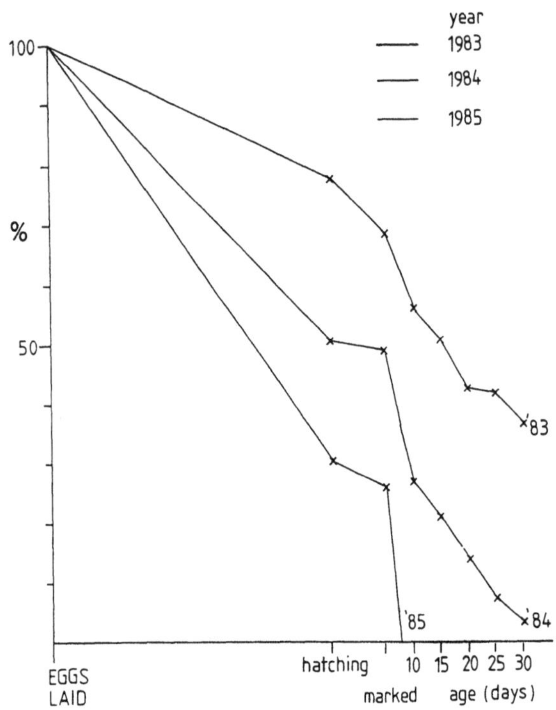

Fig. 5: The percentage of eggs/chicks surviving to various stages at southern sites.

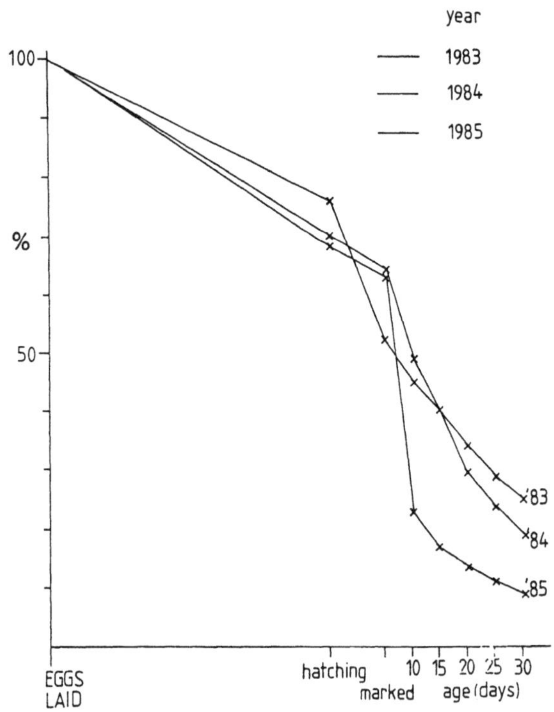

Fig. 6: The percentage of eggs/chicks surviving to various stages at northern sites.

THE REPRODUCTIVE SUCCESS OF THE TWO SITES DURING THE THREE YEARS

	1983		1984		1985	
	north	south	north	south	north	south
Number of eggs laid	229	197	236	233	267	134
Mean number of chicks hatching per egg	0.79	0.79	0.68	0.68	0.69	0.32
% of chicks known to die	13.7	17.3	25.4	25.1	29.7	37.2
% of chicks known to survive > 20 days	39.3	55.8	36.0	37.7	21.6	0
Mean number of fledglings/egg	0.31	0.44	0.24	0.26	0.15	0

Tab. 3

declined. There is no indication of any limit to Yellow-legged gull potential for expansion on the Chafarinas, given the abundance of food and absence of competition, until both the uninhabited islands are fully colonized. Schemes must be adopted now to manage the gull populations on the Chafarinas. The expansion of the Yellow-legged gull colony on Rey must be curbed and suggestions have been made as to how this may be achieved (Bradley 1984). A warden is necessary to implement this management programme, to increase the environmental awareness of the Moroccan and Spanish neighbouring villagers and to prevent the disturbance of the colony during the breeding season. Experimental trials are needed to ascertain the value of artificial cover in the regulation of nesting densities and the reduction of heat stress and predation. Finally, in view of recent developments in the type of military activity on the islands, the islands must be recognized as in International Nature/Marine Reserve (not only because of the Audouin's gulls but also the Cory's Shearwaters, the Mediterranean Monk Seal, the Ospreys) to indicate the value of this tiny archipelago and as a means of conserving the richness and variety of these islands' fauna, once typical of the Mediterranean as a whole.

References

Bradley, P. (1984). Proposals for the protection of the breeding colony of Audouin's Gulls on the Chafarinas Islands. Report to ICBP and ICONA.

Parsons, J. (1970). Relationship between egg size and post-hatching chick mortality in Herring Gull (L. argentatus). Nature 228: 1221-1222.

Parsons, J. (1972). Egg size, laying date and incubation period in the Herring Gull. Ibis 114: 536-541.

Witt, H.H. (1977). Zur Biologie der Korallenmöwe L. audouinii - Brut und Ernährung. Journ. f. Orn. 118: 134-155.

Witt, H.H., Crespo, J. de Juana, E. and Varela, J. (1981). Comparative feeding ecology of Audouin's Gull and Herring Gull. Ibis 123: 519-523.

Acknowledgements

This project was supported by The Department of Education for Northern Ireland, The International Council for Bird Preservation, The British Ornithological Union and The Fauna and Flora Preservation Society. Its successful completion has only been possible with the advice and practical assistance of Dr. P. Monaghan, of Glasgow University, E. de Juana and J. Varela, of Autonoma University, P. Goriup and N. Collar of ICBP and J.M. Cabo from Melilla. I am also grateful to the staff of ICCE for their encouragement.

THE LARUS CACHINNANS MICHAHELLIS COLONY OF CHAFARINAS ISLANDS

Juan M. Varela, Eduardo De Juana
Dept. de Zool. C-XV, Facultad de Ciencias,
Universidad Autonoma, 28034 Madrid, Spain

Introduction

The literature on the Yellow-legged Gull (Larus cachinnans) is relatively abundant (see Goethe in Glutz & Bauer, 1982). Nevertheless most information refers to the Northern or European coasts of the Mediterranean sea, the North African side being relatively undiscovered (Jacob & Courbet, 1980; Thevenot et al., 1981; Mayaud, 1983). This paper presents and discusses the data obtained at a breeding colony of L.c. michahellis, off the Mediterranean Moroccan coast (Chafarinas Islands) during the period of 1976 to 1985. Some of these data have been preliminary reported elsewhere (de Juana et al., 1979, 1984; Witt et al., 1981; Carrera et al., 1985; Bermejo & Carrera, in press). The information hereby presented could prove of value in establishing the geographical pattern of some biological characteristics of the species as a way to support taxonomical differences, and in facilitating population management in areas where this gull could be posing problems to the survival of other species, as is the case of the Audouin's Gull (Larus audouinii) in the study area (see Bradley in this same volume).

Method

The Chafarinas Islands were visited in 1976 and every year from 1979 to 1985 within the period April-June (most years 2-3 visits of 1-2 weeks each). As our main scope was to study the breeding colony of Audouin's Gull, the information on Yellow-legged Gull was gathered in some irregular way, except during two years when 100 nests (1979) and 40 nests (1984) were marked and subsequently followed. A general description of the Chafarinas Islands is to be found in de Juana, Varela & Witt (1984) while the vegetation is described in Asensi & Salvo (1980).

NATO ASI Series, Vol. G 12
Mediterranean Marine Avifauna
Edited by MEDMARAVIS and X. Monbailliu

The colony

The Yellow-legged Gull breeds on the islands of Congreso (max. dimensions: 980 m x 520 m, 137 m height) and Rey (870 m x 200 m, 37 m height). The first one is also the breeding site of a great colony of Cory's Shearwater (Calonectris diomedea, up to 3,600 birds censused at dusk, de Juana, Varela & Witt, 1984), whereas the second one holds the Audouin's Gull colony (1,000 to 2,200 pairs during the study period).

Accurate census counts of adults or nests have been made in Rey every year, showing an increase from some 250 pairs in 1976 to 850 pairs in 1985. The colony of Congreso being much bigger was only thoroughly estimated in 1983, when various complementary methods were used (Nettleship, 1976): counts of adults both directly (from neighbouring Isabel II Island or from a boat) or from photographic enlargements and 10 x 10 m quadrat samples (n = 16) made at random in different vegetation areas. This gave for 1983 an estimation of 3,000 pairs at Congreso, or roughtly some 3,500 pairs in the Chafarinas.

Quadrat samples were also carried out in 1980 (n = 11) and 1985 (n = 13), the first case being too late in the season to allow valid comparisons. From 1983 to 1985 the mean density in Congreso raised from 4.5 nests/100 sq.m. to 5.6 n/100 sq.m., what would imply an increase from 3,000 pairs to some 3,650 pairs, or some 4,500 pairs in the whole colony. From these data we can deduce a population increase of 28%. Besides methodological difficulties, we estimate a present population of 4,000 $\pm$ 500 breeding pairs.

Brosset (1957) was the first to report the breeding of this species in Chafarinas Islands. He observed "hundreds of birds" at Congreso, where only 23 nests were accessible, the majority being situated on cliffs. In a further visit, Brosset & Olier (1966) found the L. audouinii colony at Rey where not a single pair of Yellow-legged Gull bred.

In 1976, during our first visit, this species was already installed in both Congreso (a probably too conservative estimate of 600 pairs) and Rey (250 pairs). Mayol (1978) briefly visited Congreso at the end of June finding just 500 pairs, when many birds already left the colony. For an unknown period of time, until 1977, the southernmost area of both islands was occupied by a small Army detachment, which clearly limited the distribution of colonies and perhaps the number of gulls.

Further censuses on Rey gave numbers of 465 (1983), 570 (1984) and 850 pairs (1985). Fig. 1 and 2 show the numerical and spatial evolution of this colony leading to a progressive invasion of the western and southernmost areas from their original settlements at the steep north-eastern coast. The increase rate (82.8% in just two years, 1983-1985), can only be explained by an important contribution of first-time breeders, probably most of them born in Congreso. The recovery of two adults in Rey ringed as pulli in Congreso may support this hy-

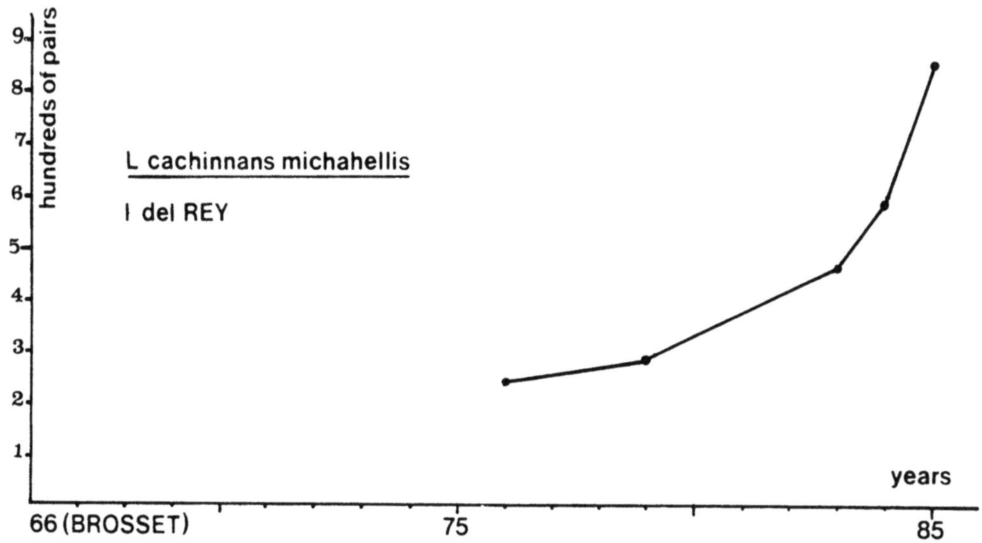

Fig. 1: Population trend of the Rey Island colony

pothesis. All the same, a slight delay in the clutches of Rey compared to those of Congreso could suggest a higher proportion of young breeders in the former (see Launay, 1985, for differences in breeding advance related to age in this species).

Following our quadrat samples, the nest density in Congreso clearly declines from North to South, both in 1983 and 1985 (mean density for the two years: 5.2 nests/100 sq.m. Max-min: 9-1 n/100 sq.m.). Fig. 3 compares the mean densities to the approximate distribution of vegetation cover. Some authors (Isenmann, 1976; Jacob & Courbet, 1980; Carrera & Vilagrasa, 1984; Launay, 1985) suggest a preference of the species to nest in areas of medium-low height of relatively disperse vegetation. This is not the case in Congreso where the highest densities are found in nearly unvegetated areas, apparently less suitable for nesting but far apart from the army barracks. These densities are not comparable to those given by the afore mentioned authors because of differences in census techniques. As an example, it should be noted that some nests are situated only 50 cm apart as measured between edges.

Nests

Nests are preferably installed close to the shelter of bushes that may partially cover them. In just one occasion we found a nest on the top of a bush some 40 cm tall. Nests can also be placed in unvegetated areas or on rocks; only a small percentage of nests are built on cliffs as is also the case in

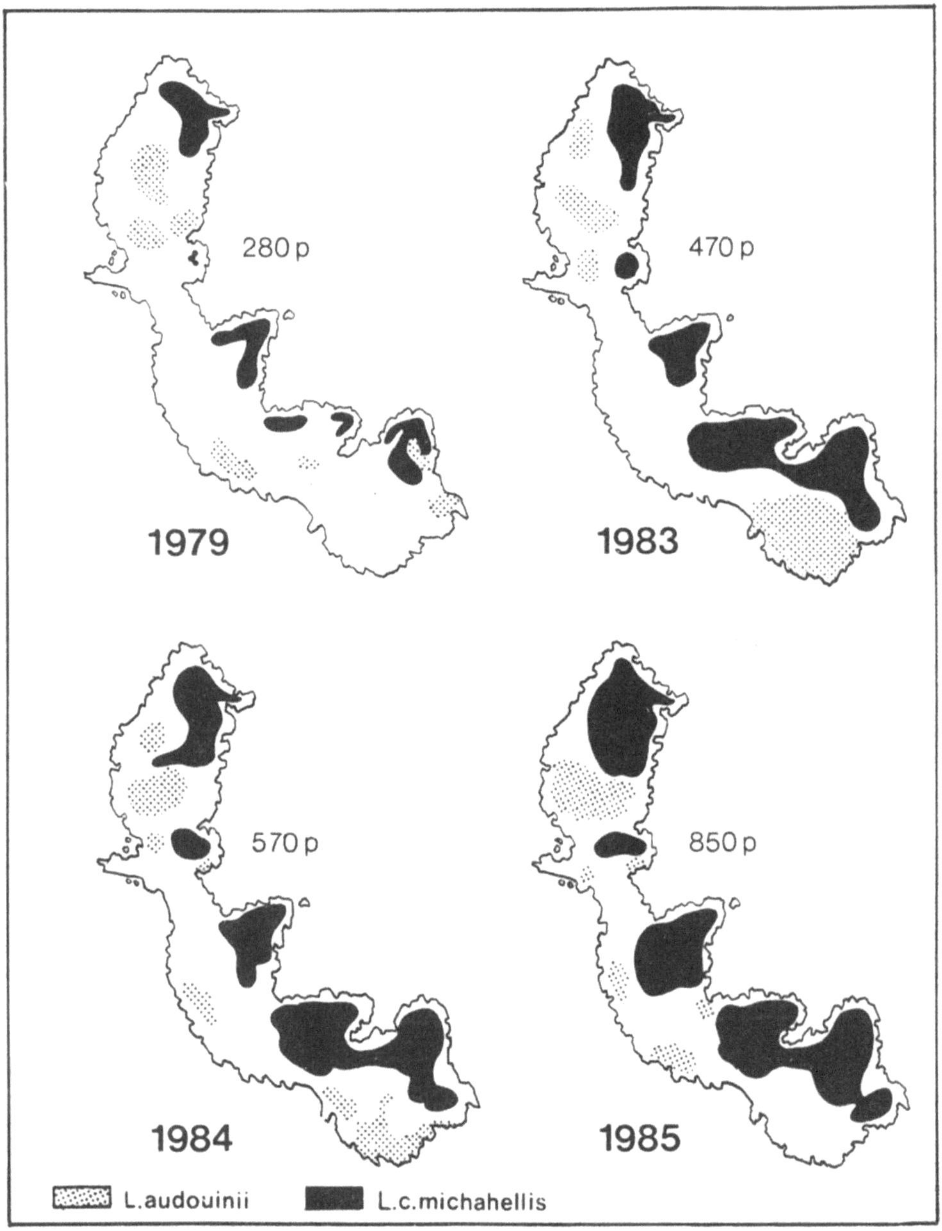

Fig. 2: Spatial evolution of the Rey Island colony.

Algerian colonies (Jacob & Courbet, 1980; Fernandez, 1973) instead found the highest percentage of nests on cliffs at the Tajo de Barbate colonies in southern Spain). Nest material is mainly collected from nearby vegetation, but feathers, garbage and even chick corpses from previous years are also incorporated. As already reported by Isenmann (1976) and Lambertini & Bassi (1983), the more elaborated nests are found in richly vegetated areas.

Mean sizes of nests are: external diameter: $\bar{x}$ = 33.96 cm (49-28), (σ= 3.36); inner diameter, $\bar{x}$ = 21.45 cm (24-18) (σ= 2.25); depth, $\bar{x}$ = 6.7 cm (9-4.5) (σ= 1.19); n = 29 (de Juana et al., 1979). We have not found a statistical correlation between nest size and mean egg size (see also Lambertini & Bessi, 1983).

Eggs

The mean measures of eggs were obtained from complete clutches. Only freshly laid eggs were weighed.

Length : $\bar{x}$ = 70.3 mm (78.8-63.0); σ = 2.73; n = 260
Width : $\bar{x}$ = 49.0 mm (51.9-43.8); σ = 1.36; n = 260
Weight : $\bar{x}$ = 91.4 g (104-70); σ = 6.20; n = 131

A dwarf egg measured 48.0 x 37.6 mm, 37 g (this data was not taken into account for statistical purposes).

The mean egg sizes corresponding to different years are:

1976 : $\bar{x}$ = 71.0 mm (σ = 2.88) x 49.4 mm (σ = 1.41); n = 40
1979 : $\bar{x}$ = 69.5 mm (σ = 2.89) x 48.96mm (σ = 1.24); n = 42
1984 : $\bar{x}$ = 69.8 mm (σ = 2.35) x 49.22mm (σ = 1.23); n = 72

Though no significant differences were found, the comparison of these interannual variations can be of interest if compared to other colonies cited in other papers (Lambertini & Bessi, 1983; Carrera & Vilagrasa, 1984). The mean weight of eggs from the Chafarinas colony is lighter than the one from the Camargue (Isenmann,1976: $\bar{x}$ = 96.65 g, σ = 7.61, n = 123).

Compared to the Iberian colonies, Chafarinas' gulls produce eggs similar in size to those from Catalonia (Carrera & Vilagrasa, 1984: 69.88 x 48.23) and Gibraltar (Cortês, 1981: 69.5 x 48.3), but probably bigger than those from Galicia (NW Spain) (Bermejo in Glutz & Bauer, 1982: 68.0 x 47.3). These variations should be considered in relation to the biometrical studies (Carrera et al., 1985) that have clearly separated the North Western Iberian populations from the colonies in Catalonia and Chafarinas which are similar in size, thus giving supplementary support to the hypothesis of the existance of different subspecies.

As happens in other Laridae species, the third egg in a clutch is smaller than the first one. Isenmann (1976) found

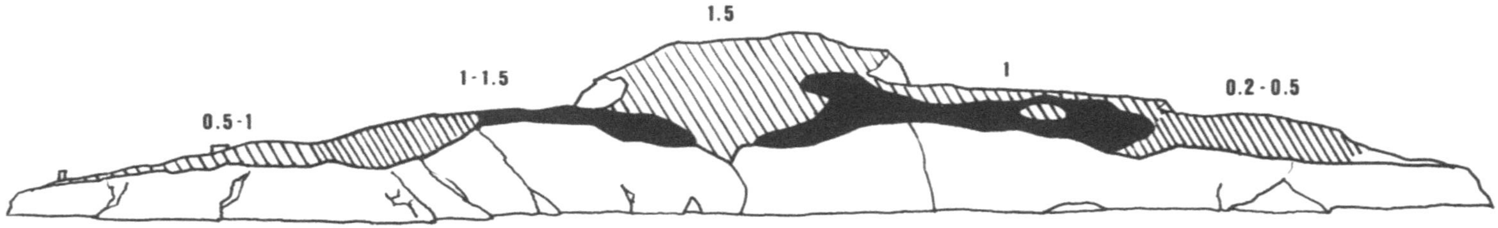

Vegetation density and height (meters)

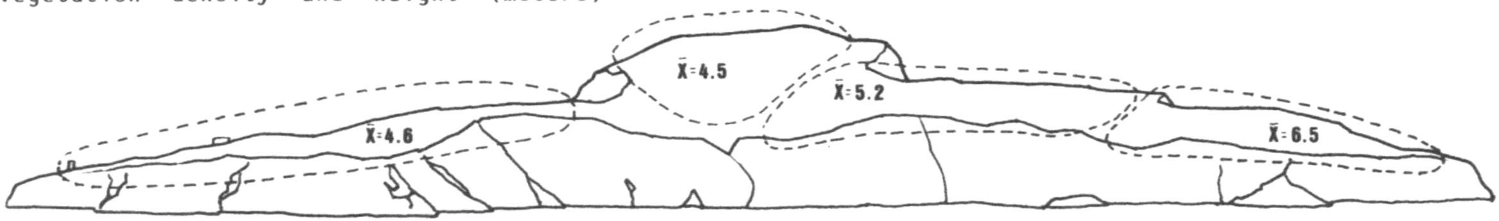

Mean densities of nests / 100 m^2 (years 1983 and 1985)

Figure 3.

View of Congreso Island as seen from Isabel II Island (white circle : barracks of the Army).

these differences significant in Camargue. All the same we have found the third egg smaller and lighter than the other two. The second egg instead weighs less but is similar in size ot the first one. In the following table, where I is length x width/100 we compare both dimensions (in mm and g).

$\bar{x}_1$ = 71.21 x 49.26; I = 35.08, (σ = 1.96); $\bar{W}$ = 93.13, (σ = 5.37); n = 45
$\bar{x}_2$ = 70.21 x 49.18; I = 34.59, (σ = 2.59); $\bar{W}$ = 92.26, (σ = 6.1); n = 48
$\bar{x}_3$ = 69.23 x 48.61; I = 33.67, (σ = 2.89); $\bar{W}$ = 88.36, (σ = 6.2); n = 38

Laying and incubation

The mean clutch size is 2.69 eggs per nest. This is obtained from marked nests in which no variation in number of eggs was observed within 4 days time (1979: $\bar{x}$ = 2.63; n = 93; 1984: $\bar{x}$ = 2.73; n = 38) (see Glutz & Bauer, 1982, for variations in clutch size). The corresponding percentage is: 71.4% (nests with 3 eggs), 25.5% (2 eggs) and 2.9% (1 egg). Only one nest contained 4 eggs, a very low rate compared to other localities (Spitzer, 1978; Launay, 1985).

The earliest laying date can be deduced from a nest containing two chicks and a chipped egg, found on 18.4.1985. For an incubation period calculated at 27 days (see below) laying in this nest would have commenced on 18.3. Only Muntaner & Congost (1979) and Carrera & Vilagrasa (1984), cite an earlier laying: on 13.3. The normal onset of laying could take place around 25.3., as deduced from the incubation period, marked nests and first fledged birds that can be already found from 30.5. on (using a fledging period of 35 days as given by Isenmann, 1976, in Camargue).

The laying progress in marked nests is shown in fig. 4. From this we infer a mean date of 9.-10.4., very similar to that one reported in Camargue (Isenmann, 1976). Our data come from a certain area of Rey Island, chosen for being a traditional nesting site but not representative for the breeding progress in all other areas.

Launay (1985) reports a delay in laying by "unexperienced" breeders. Something similar may be deduced from our laying progress samples. (Percentage over the clutch size: 2.7 = 100%)

Phenology of egg clutch completion:

A	Congreso (Total)	Rey (Total)	
1983 (10.4.)	82.9%	(10.4.)	71.8%
1984 (5.4.)	40.0%	(5.4.)	38.38%
	-	(10.4.)	48.1%
1985 (8.4.)	44.7%		-

B	North	Centre	South
Congreso 1983 (10.4.)	80.3%	-	67.0%
Rey 1983 (10.4.)	-	61.1%	82.5%
1984 (10.4.)	43.7%	-	52.5%

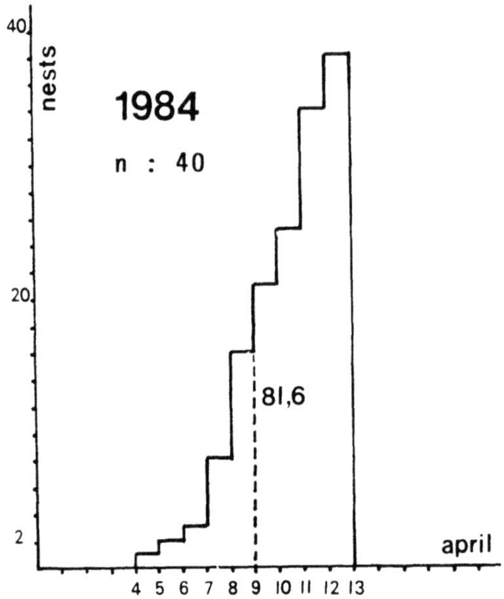

Fig. 4: Cumulative histogram giving the number of nests that reach the final clutch up to a given date (81.6% over the mean clutch size: 2.7 = 100%).

Such variations are produced both between colonies and between years although Congreso seems to be in advance as compared to Rey. Very few data are available for late broods. Fernandez Palacios (in lit.) found two "half feathered chicks" on 23.7.83 (the approximate laying date would be 8.6.) but this is obviously a very extreme case. The laying progress samples indicate that around the last week of April most of the colonial nuclei have reached a partial clutch size of 2.5, that is more than a 90% of the definitive clutch size (2.7). Although annual variations may exist the most probable extent of the laying period ranges from the last week of March to the end of April, thus more or less coinciding with dates provided for other Mediterranean colonies (see Glutz & Bauer 1982 for comparisons) but well in advance compared to the Atlantic ones (Galicia and Portugal: beginning of laying in late April. Bermejo & Teixeira, pers. com.) with the exception of southern Spain (Alonso, 1983).

The laying interval of marked nests is 2 days between subsequent eggs ($\bar{x}$ = 2.03, σ = 0.72, n = 26); ($\bar{x}$ = 2.15, σ = 0.37, n = 19) and 4 days between the first and third eggs ($\bar{x}$ = 4.0, σ = 0.76, n = 18). The incubation starts with the second egg. Temperature samples made early in the morning (n = 100) indicate that 37.1% of nests with one egg showed a warm egg temperature. This rate rises to 85.7% in the observations of nests containing 2 eggs and to 100% in 3 eggs nests. This coincides with the observations of Isenmann (1976). However Launay (1985) accepts the incubation onset from the first egg.

The mean incubation period takes nearly 27 days: ($\bar{x}$ = 26.96); ($\bar{x}_1$ = 28.1, σ = 0.83, n = 11); ($\bar{x}_2$ = 26.8, σ = 0.6, n = 9); ($\bar{x}_3$ = 26, σ = 0.63, n = 6). A slightly longer period (28 days) was calculated by Isenmann (1976) for Camargue.

The first hatchings of marked nests took place on 30.4. (though we have found chicks already on 28.4.). On 3.5., 40% of these nests had got at least one chick. In Camargue Isenmann (1976) found a mean date for third egg hatchings around 6.5. On 5.5., 41.3% of marked nests had third egg hatchings and a 20.6% of the third eggs were chipped.

Hatching success

We managed to check the development of 95 eggs out of a sample of 108. Our early departure was preventing us to follow the development of the other broods. The enclosed table compares the results obtained through a similar method in other colonies.

Colony	Hatching rate	Number of eggs					
		Rolled	Broken	Disap.	Sterile	NC	T
Chafarinas	75.9%	0	7	6	?	13	108
Marseille (1)	73.4%	1	18	16	5	0	184
	78.2%	0	7	15	2	0	86
Camargue (2)	94.0%	0	1	1	6	0	135

(1) Launay, 1985; (2) Isenmann, 1976.

The hatching rate of the 95 eggs was 86.3%. The remaining 13 ones close to finish the incubation period, would likely enlarge the number of hatchings or sterile eggs. Anyway the rate of infecund eggs is difficult to determine if we do not know the productivity of those disappeared. Twenty of the 40 clutches had a 100% hatching rate, 6 of them had to put up with the loss of all or some of the eggs (15.3%) and the rest (33.5%) had eggs or chicks born on the date of our departure.

Little can be said about development of chicks. At the end of June many of them are able to fly. During our ringing activities of Larus audouinii (last week of June, first weeks of July) we took the chance to ring some of the small percentage of Yellow-legged chicks that were still in the pre-fledging stage. From these accidental observations we suppose that the fledging time and later on the abandonment of the colony do not differ too much from other Mediterranean areas.

Feeding

Many pellets were collected within the colonies. The result of its analysis will be the subject of a separate paper, although a preliminary study has already been published (Witt et al., 1981). The following table summarizes the study of regurgitated food obtained during ringing work.

Adults (n = 13)		Chicks (n = 30)	
5	Garbage	25	Fish
3	Fish	3	Molluscs
2	Insects	4	Insects
2	Molluscs	1	Garbage
1	Lizzard		

Despite the small sample the results seem to follow the scheme described for other Mediterranean colonies (Isenmann, 1976; Araujo et al., 1977; Carrera & Vilagrasa, 1984; Launay, 1985). In this respect, it is interesting to compare our results with those obtained in the same colony through the analysis of pellets (n = 65) (Witt et al., 1981). In the latter study the main left overs found in pellets of adults are fish and insects, followed by molluscs and mice. This difference is likely brought about by the different decomposition process of left overs: soft diet items (rice, meat, fruit) found in garbage disappear rapidly.

The main source of adult's food seems to be human waste followed by fish, obtained both from direct fishing or from discards of trawl boats. The three subsequent items contain several species of the Island's micro-fauna captured, to some extent, in or around the colony site. Chicks however could favour fish followed by invertebrates and garbage.

Predation and cleptoparasitic behaviour have already been cited by several authors (Blondel, 1963; Araujo et al., 1977; de Juana et al., 1979; Varela et al., 1980; Carrera & Vilagrasa, 1984; Launay, 1985). Accordingly this gull successfully captures small to medium sized preys as well as eggs and chicks of neighbouring breeding birds.

A notorious increase of these habits has been observed at Chafarinas islands. Audouin's gulls were often chased by groups of up to 15 Yellow-legged Gulls forcing them to regurgitate food. Even the stronger species Pandion haliaetus is also treated in this way (see also Araujo et al., 1977). Besides, a high percentage of eggs and chicks are robbed from Audouin's nests (see Bradley in this same volume) and some adults are also killed and eaten. The earlier breeding period of L.c. michahellis poses two problems to Audouins: a) the lack of nesting sites or at least of the best nesting places, and b) the higher nutritional requirements of Yellow-legged chicks coincide with the incubation and hatching periods of Audouin's Gull, thus making of their clutches and broods an additional supply of food.

Movements

A total of 1,053 Yellow-legged Gulls were ringed at Chafarinas:994 of them at the stage of pulli and the remaining 59 as adults mostly captured in trap cages after the Mills & Ryder (1979) method. Moreover 400 of them got a red plastic band on the tarsus.

Fig. 5 shows the distribution of the 13 bird recoveries all ringed as pulli. Most recoveries are within an area of 50 km from the ringing place and only two, corresponding to first winter birds were found further (up to 130 km) in Alhucemas bay. Two adults recovered at Rey island deserve our attention, being 6 years old birds ringed as pulli at Congreso island. The group of colour ringed gulls has produced just one observation: one

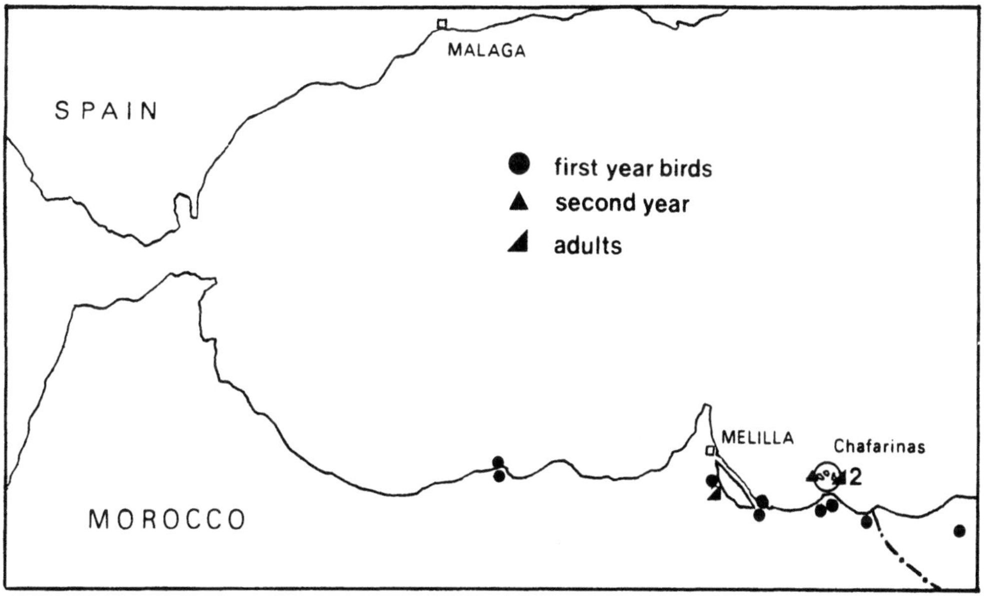

Fig. 5: Recoveries of Larus cachinnans michahellis ringed at Chafarinas Islands.

"immature bird" (Garrido et al., 1985) was seen in Malaga, the only one in Europe and at some 250 km distance from Chafarinas.

Despite the few data available, we may come to believe that the Yellow-legged Gull population of the Chafarinas shows a higher degree of sedentarity than North Mediterranean gulls as reported by Isenmann (1973) and Carrera et al. (1981). Tunisian gulls also have often been recorded far away from their breeding colonies: France, northern Italy and Sardinia (Glutz & Bauer, 1982).

Acknowledgements

This study was made possible thanks to funds of IUCN/WWF (project 1413 "Conservation of Audouin's Gull") and ICONA (Instituto para la Conservacion de la Naturaleza).

Several colleagues gave their help during field work seasons among them, Jesus Crespo, Patricia Monaghan, Andrew Paterson and Hans Witt. Patricia Bradley and John Dogherty kept on the tracks of events during our absences from the islands.

The Ministry of Defence gave permits for visiting the Chafarinas. The military garnison of the islands, as well as the officers and crew of some Navy ships, provided us with food, comfort, transportation and friendship. We are grateful to all of them.

References

Alonso, J.A. (1983). Propuesta de proteccion para la pajarera de Garcilla Bueyera del Tajo de Barbate (Cadiz). Alytes, 1: 139-150.

Araujo, J.; Munoz Cobo, J. & Purroy, F.J. (1977). Las Rapaces y Aves Marinas del archipielago de Cabrera. Naturalia Hispanica 12, Icona, Madrid.

Asensi, A. & Salvo, A.E. (1980). La vegetacion de las Islas Chafarinas, Jabega, 23: 55-59. Diputacion Provincial, Malaga.

Bermejo, A. & Carrera, E. (in press). Premières données sur les reprises de Goéland leucophée bagués en Péninsule Iberique, Baléares et Chafarinas. Réunion Nationale G.T.O.M. La Rochelle,France, 1983.

Blondel, J. (1963). Le problème du contrôle des effectifs du Goéland Argenté (Larus argentatus michahellis) en Camargue. Terre et Vie 17: 301-315.

Brosset, A. (1957). Excursion aux Iles Chafarines. Alauda 25: 293-295.

Brosset, A. & Olier, A. (1966). Les Iles Chafarines lieu de reproduction d'une importante colonie de Goélands d'Audouin (Larus audouinii). Alauda 34: 187-189.

Carrera, E.; Nebot, M.R. & Vilagrasa, X. (1981). Comentaris sobre els desplaçaments erràtics de la poblacio catalana de Gavià argentat (Larus argentatus michahellis). Buttll. Inst. Cat. Hist. Nat.,Sec. Zool. 47: 143-153.

Carrera, E. & Vilagrasa, X. (1984). La colonia de Gavià argentat (Larus argentatus michahellis) de les Illes Medes. In: Els Sistems Naturals de les Illes Medes, pp. 291-298. I.E.C. Barcelona.

Carrera, E.; Trias, J.; Bermejo, A.; de Juana, E. & Varela, J. (1985). Contribution à l'étude de la biométrie de la population ibérique et nord-africaine du Goéland leucophée (Larus cachinnans michahellis). Montpellier, Novembre 1985, Colloque sur l'avifaune méditerranéenne; in press: l'Oiseau et la R.F.O.

Fernandez, J.A. (1973). La colonia de Gaviotas plateadas mas meridional de Europa. Vida Silvestre, 3(3): 166-175.

Garrido, M.; Alba, E. & Gonzalez, J.M. (1985). Aves de las zonas humedas malaguenas. Diputacion Provincial, Malaga.

Glutz von Blotzheim, U.N. & Bauer, K.M. (eds.). Handbuch der Vögel Mitteleuropas Band 8/1, (Charadriiformes, 3). Akademische Verlagsgesellschaft, Wiesbaden.

Isenmann, P. (1973). Données sur les déplacements erratiques des Goélands argentés à pieds jaunes (Larus argentatus michahellis) nés en Méditerranée. L'Oiseau et R.F.O., 43 (3): 187-195.

Isenmann, P. (1976). Contribution à l'étude de la biologie de la reproduction et de l'écologie du Goéland argenté à pieds jaunes (Larus argentatus michahellis) en Camargue. Terre et Vie 20: 551-563.

Jacob, J.P. & Courbet, B. (1980). Oiseaux de mer nicheurs sur la côte algérienne. Le Gerfaut 70: 385-401.

De Juana, E., Bueno, J.M.; Carbonell, M.; Pérez, V. & Varela, J. (1979). Aspectos de la alimentacion y biologia de reproduccuion de Larus audouinii Payr. en su gran colonia de cria de las Islas Chafarinas (ano 1976). Bol. Estacion Central de Ecol. 8 (16): 53-65. Icona, Madrid.

De Juana, E.; Varela, J. & Witt, H.H. (1984). The conservation of seabirds at the Chafarinas Islands. ICBP Technical Publication 2: 363-369.

Launay, G. (1985). Nouvelles données sur la biologie du Goéland leucophée (Larus cachinnans michahellis) dans le Midi de la France. In Oiseaux Marins Nicheurs du Midi et de la Corse, pp. 77-81. Annales du C.R.O.P. No. 2, Aix-en-Provence.

Lambertini, M. & Bessi, M. (1983). Alcune note sulla biologia riproduttiva del Gabbiano reale (Larus argentatus michahellis) all'isola di Capraia (Li). Quaderni Mus. St. Nat. Livorno 4: 131-141.

Mayaud, N. (1983). Les oiseaux du Nord-Ouest de l'Afrique: notes complémentaires. Alauda 51 (4): 271-301.

Mills, J.A. & Ryder, J.P. (1979). Trap for capturing shore and seabirds. Bird Banding 50 (2): 121-123.

Muntaner, J. & Congost, J. (1979). Avifauna de Menorca. Treb. Mus. Zool. Barcelona, 1. Barcelona.

Nettleship, D.N. (1976). Census techniques for seabirds of arctic and Eastern Canada. Canadian Wildlife Serv. Occas. Paper 25. Ottawa.

Thevenot, M.; Beaubrun, P.; Baouab, R.E. & Bergier, P. (1982). Compte rendu d'ornithologie marocaine. Année 1981. Institut Scientifique, Rabat.

Spitzer, G. (1978). Zur Reproduktionsrate der Mittelmeer-Silbermöwe (Larus argentatus michahellis). Die Vogelwarte, 29: 272-275.

Varela, J.; Pérez, V. & de Juana, E. (1980). Nuevos datos sobre la etologia de la Gaviota de Audouin (Larus audouinii Payr.) Ardeola 25: 71-92.

Witt, H.H. (1974). Zur Nahrungsökologie der Mittelmeer-Silbermöwe (Larus argentatus michahellis) an einem Brutplatz auf Sardinien, Die Vogelwelt 95 (4): 148-150.

Witt, H.H.; Crespo, J.; de Juana, E. & Varela, J. (1981). Comparative feeding ecology of Audouin's Gull (Larus audouinii) and Herring Gull (Larus argentatus) in the Mediterranean. Ibis, 123: 519-526.

NEST-SITE SELECTION AND INTERACTION OF YELLOW-LEGGED AND AUDOUIN'S GULLS AT ISOLA DELL'ASINARA

Xaver Monbailliu, 20 rue St. Martin, 75004 Paris
Antonio Torre, 96 via S. Satta, Alghero

Introduction: Description of the study area

The island of Asinara (51 sq. km) is situated 2 km north of the most north-westerly point of Sardinia (Capo del Falcone). A small island of approximately 1 sq. km, Isola Piana, forms a "land-bridge" between Capo del Falcone and the southern coast of Asinara (fig. 1). Vegetation is composed of low xerophile scrubs (annual rainfall is less than 500 mm) such as Tree Spurge (*Euphorbia dendroides*), lentisc (*Pistacia lentiscus*), Cist (*Cistus Monspeliensis*), *Thymelaea hirsuta* and the endemic *Centaurea horrida*. The highest vegetation in the gull colonies is made up of the Phoenician Juniper (*Juniperus phoenicea*), *Malva silvestris*, *Calycotome spinosa* and *Ferula communis*. In the cliff colonies of Punta Tumbarino and Romasino, Sea Stock (*Matthiola sinuata*) dominates while wild leeks (*Allium porrum*), Sea Purslane (*Obione portulacoides*) and Glassworth (*Salicornia fruticosa*) cover the islets in Cala Scombro di Dentro.

Granite and micaschist rock outcrop provides shelter for gulls and shags, the main breeding species in Cala Scombro di Dentro. At Asinara the main Yellow-legged Gull colonies are situated on the 200 m high cliffs of Punta Tumbarino (400 breedings pairs were estimated there in 1984), at Cala Romasino (230 pairs) and in Cala Scombro di Dentro where 328 nests were counted in 1984 all located on islets (fig. 2). Two other colonies are situated in the northern part of the island: 50 pairs breed at Punta Porco and another 5 south of Cala d'Oliva. The nearest gull colony in mainland Sardinia is situated on Isolotto dei Porri, 16 km south of Cala Romasino, where some 200 pairs of Yellow-legged Gulls (*Larus cachinnans michahellis*) breed as well as a dozen pairs of Shags (*Phalacrocorax aristotelis*) and occasionally a pair of Audouin's Gull (*Larus audouinii*).

Besides Yellow-legged Gulls the colony at Cala Scombro di Dentro also shelters a flourishing flock of shags (100 breeding

NATO ASI Series, Vol. G 12
Mediterranean Marine Avifauna
Edited by MEDMARAVIS and X. Monbailliu

Fig. 1: Situation of the study area, Cala Scombro di Dentro, at Isola dell'Asinara.

pairs) and a colony of Audouin's Gulls. The latter is currently threatened by the population increase of Yellow-legged Gulls. Occasionally a few pairs of Common Tern (Sterna hirundo) try to breed amidst the Yellow-legged Gulls. Two pairs of Mallards (Anas platyrhynchos) breed in the dense Yellowleg colony of Isolotto Anatre.

Apart from some passerines (Passer hispaniolensis, Sylvia melanocephala, Saxicola torquata) no other bird species breed in the Gull colonies of Cala Scombro di Dentro. The study area, Cala Scombro di Dentro, is composed of 11 islets where Yellow-legged Gulls bred in 1985. The smallest one, Isolotto Lentisco, is not larger than 800 sq. m. Isola Scombro, the largest islet, covers some 2.8 ha. The deepest zone in Cala Scombro di Dentro is situated north of Isola Scombro (- 11 m). The seabed is covered with a dense Posidonia vegetation. Due to the special penitentiary status of Isola dell'Asinara, birds are relatively undisturbed in the islets of Cala Scombro. Occasionally fishermen, sometimes accompanied by a dog, may visit the study area. All islets where Yellow-legged Gulls bred in 1985 are inaccessible to stray cattle. Cows and sheep are serious limiting factors for the reproduction of gulls and terns on the Asinara main coast. This also applies to an islet (Isolotto Capretti) east of

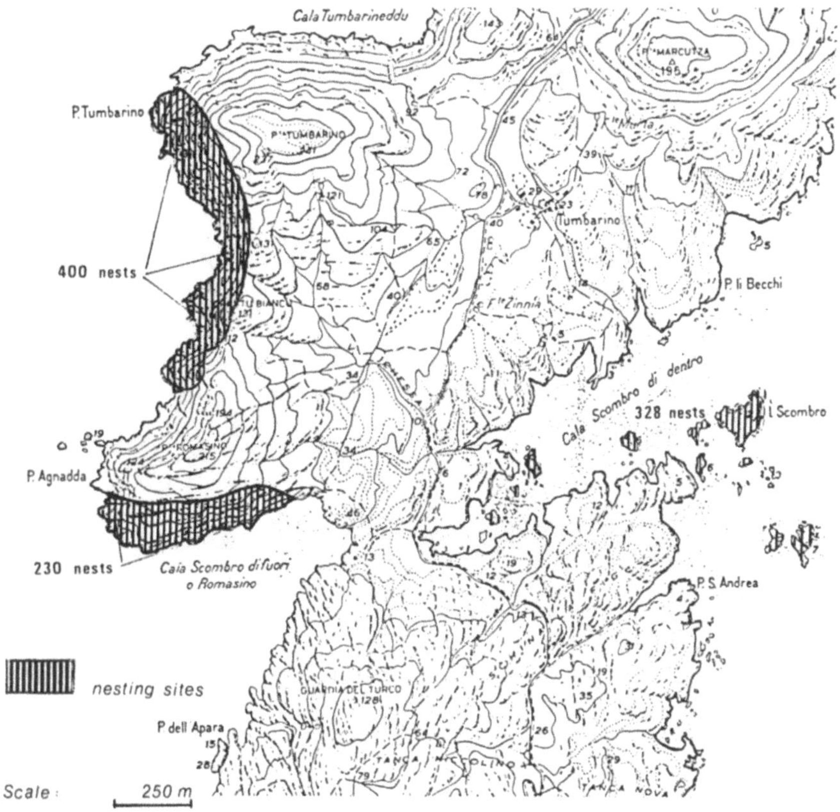

Fig. 2: The main nesting sites of Yellow-legged Gulls at Asinara with their 1984 population numbers: Punta Tumbarino: 400 nests; Punta Romasino: 230 nests; Cala Scombro di Dentro: 328 nests.

Isola Piana where Little Terns (Sterna albifrons) and Common Terns (Sterna hirundo) try to breed every year but are disturbed by cows crossing a shallow sea strait. May we refer to Monbailliu (1985) and Torre (1979) for further data on the avifauna of Isola dell'Asinara and Nurra (North-West Sardinia) respectively.

The Audouin's Gull colony at Asinara

Audouin's Gulls were discovered breeding at Asinara during spring 1983. No earlier observations of this gull species have been registered. On May 19th 1983 at least 158 nests with eggs or freshly born chicks were found on an islet only 20 m off the mainland coast of Asinara. The Audouin's Gulls occupied, in 1983 and 1984, the flat southern part of the islet not larger than 230 sq. m. This high density, an average of 1 nest per 1.45 sq. m., is important moreover as the available nesting site is in reality much smaller since dense vegetation of both Glasswort and Sea Purslane impedes nest building on part of the islet. Indeed, many nests were found only a few feet apart reaching a density of approximately 1 nest per sq. m. Bradley (1986) reports in this publication a maximum population density of

The Audouin's enigma: some adults, for unknown reasons, show a double black stripe on both mandibules.

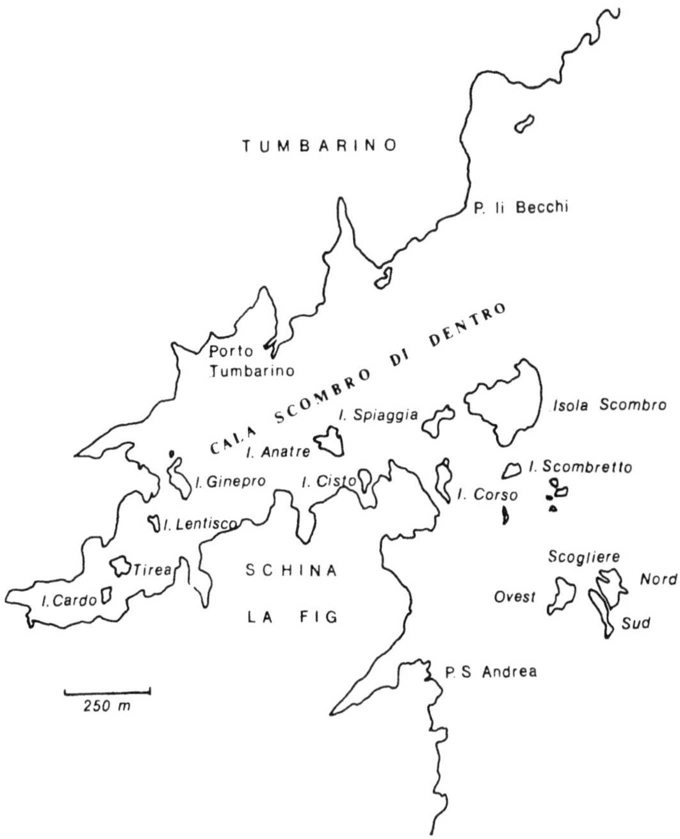

Fig. 3: The islet colonies in Cala Scombro di Dentro

1 nest per 3 sq. m. in the Audouin's Gull colony at the Chafarinas.

The very high density of nests of Audouin's Gulls at Asinara can be explained by the shortage of suitable islets for breeding. In Cala Scombro di Dentro virtually all islets are occupied from February onwards by Yellow-legged Gulls. Audouin's prefer to breed on small islands where man and animals (dogs, foxes, boars, moufflons, cattle) are absent. They look for the company of the much larger and agressive Yellow-legged Gull but this sympatric behaviour often leads to dramatic predation by territorial Yellowlegs on adult Audouins during nest making and especially on chicks of Audouins at the end of May and the beginning of June. Since in dense colonies Audouins may kill congenerous chicks if they trespass neighbouring nests, we decided not to continue our research in the colony islet. Moreover, once disturbed by man, Audouins temporarily leave their colony which enables Yellowlegs, breeding in small numbers at the northern part of the islet, to predate on chicks.

In 1984, only 115 pairs of Audouins built their nests on the same islet (average brood size on 26th May: 2.49). A few nests contained 4 eggs. In 1985, due to a 20% population increase of breeding Yellow-legged Gulls, the southern part of the Audouin islet was occupied by Yellowlegs. Since they start breeding up to 5 weeks earlier than Audouins, Yellow-legged Gulls dispose of full control for nest-site selection. Consequently, the Audouins did not manage in 1985 to recolonize the islet hitherto normally "reserved" for them. The Audouins finally settled down, with 7 days delay, on a peninsula north of Cala Scombro di Dentro. Their average brood size on the 23rd May was only 1.9 eggs. On mainland Asinara the abundance of stray cattle (cows, sheep, etc.) is the major limiting factor for coastal breeders. On May 31st the Audouin colony was visited by cows. Only a few chicks survived. Despite the construction of a metallic fence by prisoners to protect the remaining Audouin nests, all Audouins left their colony by June 3rd.

Objectives and Methodology

In order to find out why Yellow-legged Gulls occupied the Audouin's islet we have studied several parameters related to nest-site selection by this invading gull species such as:

- occupancy rate of islets during the pre-nuptial period,
- breeding advance in April,
- nest density,
- biometrics of eggs.

These criteria have been analyzed for a sample of nests situated on all islets in Cala Scombro di Dentro as well as in the cliff colony of Punta Romasino. By comparing these data in the various colonies we hope to get a deeper insight into the choice made by Yellow-legged Gulls for breeding sites. A better understanding of nest-site selection helps to determine optimum population numbers of this invading species enabling other Larids

Audouin's Gulls at Isolotto Corso in 1984.

4 nest clutches of Audouin's Gull at Isolotto Corso. The nests are less than 1 m apart. Note the 8 cm long caliper rule in the middle of the photo.

(Audouin's Gull, Common Tern) to co-habit in the same breeding area. Working conditions at Asinara are difficult since an official permit is required for each study period. Working conditions are also subject to the availability of transport (motor boat, dinghies), weather and control by prison guards. They are responsible for some inconsistencies in data gathering.

Results

A. Occupancy rate of colony islets during nuptial period

Immature Yellow-legged Gulls may spend the winter inland on the continent. Recently, Dr. P. Géroudet (Switzerland) observed an immature gull at Lake Geneva on 15th December 1985 and two others on 22nd February 1986, all colour-ringed by us at Isola Scombro on the 27th of May 1985 (pers. comm.). On the other hand, adult Yellow-legged Gulls stay over winter in or close to their colonies. In the Gulf of Asinara they are present in the colonies or occupy the domestic waste dump of Porto Torres. They also roost in fields close to the dump or in the port of Porto Torres.

Since Yellowlegs already display territorial behaviour during winter in the colonies we have studied the occupancy rate of each islet during the last week of February 1985. In the western Mediterranean this corresponds to the period of prenuptial behaviour and copulation. Tinbergen already indicated in 1953 that Herring Gulls (*Larus argentatus*) display a high grade of site tenacity. From biometrics of egg clutches at Asinara we also know that the smallest islets and the more prominent sites in the larger islets are occupied first and this by gulls producing the largest clutches (tab. 2). If the larger clutches are laid by larger and dominating birds we might expect that the strongest (= largest and dominant) pairs occupy the same but first choice nest sites every year or compete for them with intruding newcomers. In order to compete for the best nest sites, Yellow-legged Gulls need to occupy their selected territories in February or earlier. The other adult gulls not present yet in the colonies occupy the remaining nest site possibilities later on in the beginning of March or take possession of new nest sites (for ex. another islet) if a population increase implies such a choice.

Tab. 1 summarizes the occupancy rate of colony islets by Yellow-legged Gulls during the last week of February 1985 as compared to the number of breeding pairs in 1984 and 1985. The colonies are ranked according to the size of the islets (largest islets as well as the mainland cliff colony of Punta Romasino on the left hand side). These bird counts in winter are relatively easy since the gulls while being approached by boat fly up from their islet and regroup themselves nearby on the sea surface. However we were unable to get exact gull counts for the three Scogliere islets (Nord, Ovest, Sud) since Yellowlegs nesting there gather together in one single group at sea.

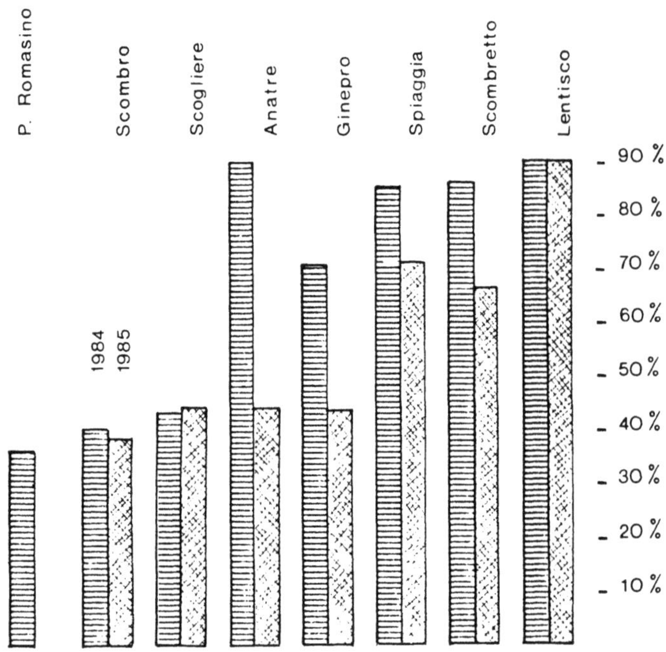

Tab. 1: Occupancy rate of colonies of Yellow-legged Gulls at Cala Scombro during the last week of February 1985 as compared to the population numbers in 1984 and 1985. The colonies are ranked according to their size.

From tab. 1 we may deduce that the largest islets (Scombro, Scogliere) as well as the Romasino cliff colony have a moderate occupancy rate in winter as compared to the 1984 and 1985 population numbers. For example at Isola Scombro only 34% (as compared to 1985) or 40.3% (as compared to the breeding population of 1984) of the nesting population occupied their nesting site at the end of February 1985. The same might have applied to the islets of Anatre and Ginepro if their population numbers had remained stable in 1985. However in Cala Scombro di Dentro the breeding population of Yellow-legged Gulls increased by 20% in spring 1985 and the medium-size islets (Anatre, Ginepro and to a lesser extent Spiaggia and Scombretto) registered the largest population increase. Isolotto Anatre even doubled its population in one year. Thus, the medium-sized islets seem to attract the greatest numbers of newcomers well ahead of the breeding season while the smallest islets (I. Lentisco) are just too little to accommodate more breeding pairs.

If medium-sized islets attract, in February or earlier, the largest percentage of newcomers and to some extent also the strongest pairs, we may infer that the islets of 0.3 to 0.4 ha. correspond to first choice breeding sites. The islet where the Audouin's Gull bred till 1984 belongs to the same group (0.3 ha) but was obviously left aside as a secondary choice since the islet is only separated from the mainland by a 20 m wide strait.

• Factors affecting nest-site selection

Apart from island size at least three other parameters determine nest-site selection: distance to the mainland, geo-morphology (relief, slope, rock outcrop) and vegetation cover. For safety reasons (stray cattle, dogs, man) seabirds breed on inaccessible cliffs or on small islets uninhabited by man. At Cala Scombro di Dentro some islets are situated just too close to the Asinara mainland coast and thus represent a secondary choice for breeding. This applies to Isolotto Corso where, apart from the Audouins until 1984, only a few Yellowlegs breed. This applies equally to I. Tiria where for the first time, due to the population increase in 1985, two pairs of Yellow-legged Gulls bred and to I. Cardo situated in the innermost part of Cala Scombro di Dentro and thus too close to the mainland coasts (fig. 3). Another islet, Isolotto Cisto, becomes part of the mainland during low tide and is not suitable for breeding since horses, moufflons and stray cattle frequent the peninsula of Schina la Fig. All other islets in Cala Scombro are considered too small for breeding. They are constituted of outcrop of granite, are unvegetated and are only a few meters above sea level.

Geomorphology can also be a limiting factor but at Cala Scombro di Dentro all colony islets share roughly the same characteristics: they are all relatively flat offering similar nesting potential. However the surface of mineral unvegetated parts is extensive on some islets (Scogliere Ovest, Tiria) and the absence of vegetation or of protruding rock is obviously a serious limiting factor there.

Because of conservation interests we did not analyze the nature and coverage of vegetation on the islets since plants are growing very fast during March and April (thus during the breeding season) and disintegrate partly in summer (Graminaceae, Allium, Euphorbia, Ferula). Such phyto-ecological surveys can seriously disturb the breeding shag and gull population and may result in a negative impact for Audouin's Gulls. Burger and Gochfeld (1981), studying nest site selection by Kelp Gulls (Larus dominicans) in South Africa and Namibia, researched the influences habitat and vegetation may have on the process of choosing breeding territories. They studied more particularly the percentage of cover, species, growth form and height of plants in relation to the position of nests. Kelp Gulls breed in various habitats, differing in substrate and cover but they seem to avoid vegetation cover exceeding 70% of an area.

At Cala Scombro di Dentro our impression is that Yellow-legged Gulls display, as in other colonies, a high degree of elasticity as far as breeding site selection is concerned in relation to vegetation cover and are not discouraged by vegetation cover exceeding 70%. Isolotto Anatre has perhaps the densest low scrub vegetation (lentisc, obione, glassworth but also leeks,

Asphodelus and Ferula) but registered the highest population density of Yellowlegs. Some gulls even breed on top of lentisc scrubs. Thus, despite the lack of detailed vegetation mapping, we infer that vegetation cover at Cala Scombro is not a limiting factor for nest site selection by Yellow-legged Gulls since all major colony islets dispose of a similar canopy and plant mosaic. However this may not apply to Audouin's Gulls since this smaller Larid might prefer a mineral soil or low herbaceous plants (grasses, Sea Stock etc.) situated in close range of higher vegetative cover (Glassworth, Sea Purslane, Lentisc) or of rock outcrop (crevices, little coves, shaded zones).

B. Breeding advance in April

At Asinara egg laying by Yellow-legged Gulls starts during the third week of March. Hatching starts mid-April (on the 17th April in 1984). This is slightly one week earlier as compared to the Porquerolles colony (Provence) studied by Launay (1985) and roughly 12 days earlier than in the Camargue (see Isenmann, 1976). In order to find out which islets enjoy the highest percentage of early breeders we counted nest clutches during 12-21 April 1984 and 3-6 April 1985. Tab. 2 demonstrates that breeding advance during the 17-21 April 1984 period is similar for Isola Scombro and the medium-sized islets such as Anatre, Ginepro and Spiaggia. The 7 clutches at Scombretto were all about to hatch during the third week of April. Only the three Scogliere islets (66%) as well as Lentisco (60%) registered a slight delay in breeding advance as compared to the other islets for which we have no explanation since these islets scored a normal breeding advance during the following breeding season of 1985.

In 1985 the number of nests at Cala Scombro di Dentro augmented from 328 (in 1984) to 394 which represents a population increase of 20%. Isolotto Anatre more than doubled its population of breeding adults. The number of pairs increased more moderately at Scombro, Ginepro and Spiaggia. Scogliera Sud was the only islet loosing some breeding pairs (7 instead of 10 nests). Perhaps this islet group is situated off the centre of gravity which is Isola Scombro. The smallest islets remained stable (Lentisco) or hosted a few additional couples (Scombretto). During the breeding season of 1985 the Yellow-legged Gull also occupied, for the first time, Isolotto Tiria (two nests) and the southern part of I. Corso where the Audouins used to breed until 1984. The total number of breeding pairs (8) at I. Corso is only an estimate since we destroyed repeatedly, to no avail, the nests of Yellowlegs in order to lure back the Audouins to their former breeding site.

Breeding advance in 1985 (3-6 April) was equally high on all islets except at Isola Scombro (58%). This relatively low level of breeding advance at Scombro can be explained by an influx of newcomers and first time breeders. Indeed the occupancy rate at Scombro during February 1985 was very low (38%) taking into account that it is the principal colony in Cala Scombro di Dentro. At the end of February 1985 some 200 adult Yellow-legged

The study site Cala Scombro di Dentro. In the centre Isola Scombro as seen from Schina la Fig.

Gulls were still seen flocking together on a roost in a field close to the dump of Porto Torres. Most likely the bulk of this group joined the various gull colonies at Asinara a few weeks later.

But why should the main colony islet host the highest percentage of newcomers? Probably four different reasons help to explain why this contingent of new but late breeders settled down at Scombro with up to two weeks delay.

1. The "strongest" Yellow-legged Gulls select their breeding habitat in the medium-sized islets most likely because of safety (the smallest islets have less chance to be inhabited or to be infested by rats).
2. Some 80 pairs of shags breed at Isola Scombro from January onwards. They occupy the most prominent rocks and coves which are potentially also the best breeding sites for gulls.

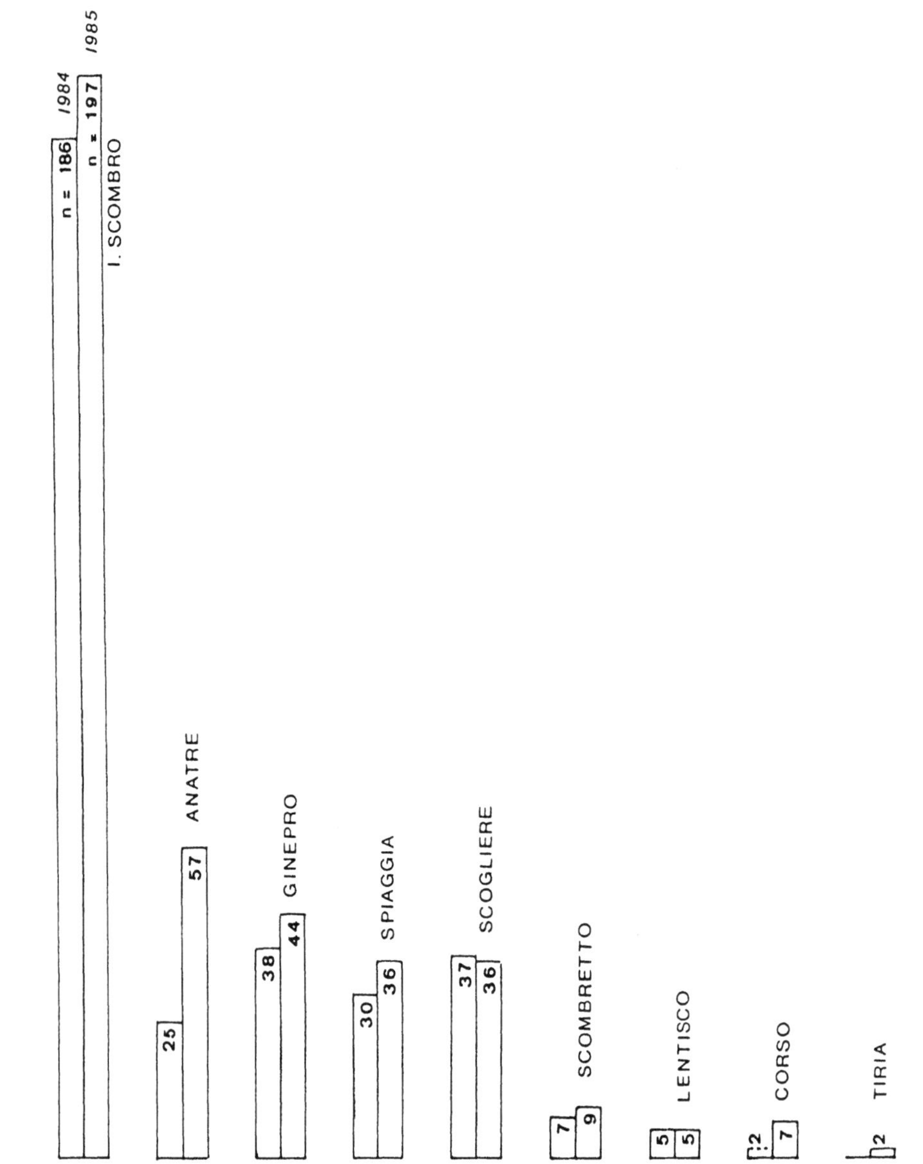

	I. Scombro	Anatre	Ginepro	Spiaggia	Scogliere	Scombretto	Lentisco	Corso	Tiria
1984:	81 %	80 %	81 %	76 %	66 %	100 %	60 %	-	-
1985:	58 %	85 %	88 %	83 %	80 %	78 %	100 %	-	100 %

Tab. 2: Breeding advance of Yellow-legged Gulls at Cala Scombro di Dentro (1984 and 1985). Breeding advance in 1984 is based upon the percentage of broods containing at least 3 eggs or chicks during the period 17-21 April 1984; of 1985 upon broods containing 2 or 3 eggs during the period 3-6 April 1985 (9th April for Isolotto Tiria). The total number of counted nests is given for each islet on top of the histograms.

3. The Black Rat (Rattus rattus) present at Scombro predates on eggs and chicks of shags in winter and of gulls in spring. They also thrive on vegetation and especially on olive pits regurgitated by gulls. Rats split the pits open and eat the softer inner core.
4. The sheer size of Isola Scombro makes it possible to host the new breeding population of Yellow-legged Gulls without obtruding too much on the territories around the nests of shags and formerly established gulls.

The gulls breeding for the first time install themselves on the flatter grass covered areas such as on the eastern part of the island as well as on marginal areas such as on lentisc and along the shore line. Thus if gulls start nesting at the same time on the medium-sized islets we observe that the breeding pattern at Isola Scombro is asynchronous: the older birds occupy the best available breeding sites in winter and start egg laying in mid-March. The second contingent, most likely young or weaker adults, starts to breed with up to two weeks delay.

If this hypothesis proves to be correct, why did Isola Scombro only attract 11 more pairs as compared to 1984? Was there a population shift already in February of older former Scombro breeders to the surrounding islets such as Anatre, Ginepro and Spiaggia because 20 more pairs of shags bred in 1985 occupying the best available nesting sites at Isola Scombro? Did gulls move because of increased predation by rats? Or did we underestimate the 1985 Yellow-legged Gull population at Scombro since it was the only islet colony counted by a third observer? All three arguments are worth considering.

In February 1985 some 200 Audouins occupied rocks very close to the islets of Anatre and Ginepro where they obviously aroused aggressivity among Yellowlegs. One adult Audouin was killed there by a Yellow-legged Gull. Later in April the Audouins occupied the north-western shore of Cala Scombro di Dentro. Sometimes they were seen flying very high as they were leaving Asinara. A few dozens settled down temporarily at Punta Sabina, north of Asinara, but without breeding success. Later in April the Audouins must have occupied the only peninsula along the northern coast of Cala Scombro, Porto Tumbarino, where they were found breeding in May.

C. Nest Density

A third parameter enabling us to assess preference for breeding territories is obviously nest density: the higher the density of a colony of Yellow-legged Gulls, the greater the preference for a breeding site. This theory is valid for Yellowlegs at Asinara since this invading species has virtually no competitors (shags however compete to some extent since they are sympatric at some islets) nor predators apart from Black Rats on the islets of Cala Scombro and probably Brown Rats (Rattus norvegicus) in the cliff colonies. Nest-site selection is also made pos-

Ranked following size of island

SIZE		DENSITY 1984	DENSITY 1985
Isola Scombro:	28,000 m2	150.5 m2 or 66.4 nests/ha	142.2 m2 or 70.4 nests/ha
Scoglie Nord :	4,200 m2 *	100 m2 or 100 nests/ha	91.3 m2 or 109.5 nests/ha
Isola Anatre :	4,000 m2	160 m2 or 62.5 nests/ha	70.2 m2 or 142.5 nests/ha
Isola Ginepro:	3,400 m2	111.8 m2 or 89.5 nests/ha	77.3 m2 or 129.4 nests/ha
Scoglie Ovest:	3,200 m2 *	266.6 m2 or 37.5 nests/ha	idem
Isola Spiaggia:	3,200 m2	106.7 m2 or 93.7 nests/ha	88.9 m2 or 112.5 nests/ha
Isola Corsi :	3,000 m2 *	no data	187.5 m2 or 53.3 nests/ha
Scoglie Sud :	2,700 m2 *	135 m2 or 74 nests/ha	193 m2 of 51.8 nests/ha
Isola Tirea :	1,600 m2 *	not occupied	400 m2 or 25 nests/ha
Isola Scombretto:	1,100 m2	157.1 m2 or 63.6 nests/ha	122.2 m2 or 81.8 nests/ha
Isola Lentisco .	800 m2	160 m2 of 62.5 nests/ha	idem

The density of islets marked with a * is based on 50% of the islet size since these territories dispose only of breeding potential on half the size of the islet.

Tab. 3: Average area of territories and breeding density of Yellow-legged Gulls at Cala Scombro di Dentro.

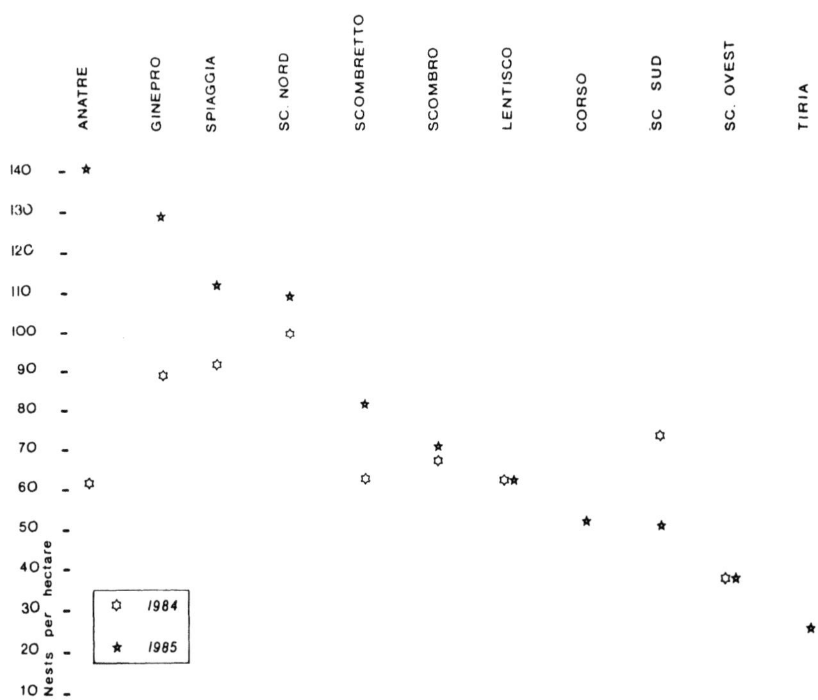

Tab. 4: Scatter diagram of nest density of Yellow-legged Gulls in islet colonies at Cala Scombro di Dentro, 1984 and 1985. The islets are ranked according to density of colonies in 1985.

sible in areas with sufficient possibilities for nesting. Since nesting density at Isola Scombro is only half as much as at Isolotto Anatre (1985) we may suppose that Yellow-legged Gulls have sufficient breeding possibilities at Cala Scombro di Dentro. This choice is even more important on mainland Asinara where the western coast provides ample breeding potential on cliffs inaccessible to wild boars, moufflons and cattle. Since the population density increased in 1985 (20% more nests in 1985 at Cala Scombro), nesting densities differ in the more successful colonies (tab. 3). Breeding density is considered high at Anatre where the territory of one pair is limited to max. 70 sq.m. At Ginepro this vital breeding territory is 77 sq.m. and 85 sq.m. at I. Spiaggia. As a matter of comparison the average area per breeding pair in the densiest Provençal colony situated on a similar but larger islet and studied by Launay (1985) is approximately 100 sq.m. (Ilôt du Grand Saranier).

The diagram tab. 4 summarizes the breeding densities in 1984 and 1985 and shows, as already discussed, that the smallest islets, which are colonized by only a few pairs, did not attract more breeding couples in 1985 (Lentisco, Scogliera Ovest). Isolotto Scombretto is an exception since breeding territories were limited in 1985 to 122 sq.m. per pair, less than half the territory of a couple at Scogliera Ovest. The proximity of Isola Scombro, centre of gravity of the Cala Scombro di Dentro colony, can explain this relatively high population density on Scombretto. Scogliera Sud is, for unknown reasons, the only islet having had less breeding pairs in 1985.

If the five densest colonies can host so many more breeding pairs, why did they also bother occupying the southern part of the Audouin islet hitherto a marginal choice in nest-site selection of Yellow-legged Gulls? The biometrical analysis of egg clutches might provide some clarification.

D. Biometrics of Clutches

A sample of nest clutches was measured during the first week of April 1985 (length, width, weight). Correlation between weight and size was found to be significant suggesting synchronous breeding and a similar egg content rate at Punta Romasino and in the islets of Cala Scombro di Dentro. Tab. 5 shows that the largest (and thus also the heaviest) clutches are to be found at the islet of Corso as well as at Scogliere Nord, Sud and Ovest. The medium sized islets enjoying the largest population densities approach the mean clutch size found at Cala Scombro di Dentro. The clutches (all containing 3 eggs) at Scombro and Scombretto are ranging from -9.5 to -13 grams, as compared to the mean clutch weight.

All eggs were measured in the smallest colonies (Scombretto, Scogliere Sud & Ovest, Corso). A random sample was taken choosing 6 nests containing three eggs in the colonies of Anatre, Spiaggia, Scogliera Nord and Ginepro. At Scombro we measured clutches of 9 nests while we selected 22 clutches at Punta Ro-

L. cachinnans BIOMETRICS OF EGGS ISOLA DELL'ASINARA

Mean Values of Weight of Eggs

1st egg		2nd egg		3rd egg		Total Brood	
Scogl. Sud	90.2 gram	Corso	87	Scogl. Nord	84.5	Corso	257
Corso	89.2	Scogl. Sud	85.7	Scogl. Ovest	81	Nord	256.2
Ginepro	89.1	Scogl. Nord	84.3	Corso	80.8	Sud	254.4
Scogl. Ovest	88.2	Scogl. Ovest	83.8	Scogl. Sud	78.5	Ovest	253
Cala Scombro	88.2	Cala Scombro	83.5	Cala Scombro	78	Cala Scombro	249.7
Spiaggia	88.1	P. Romasino	83.1	P. Romasino	77.9	Ginepro	249.6
P. Romasino	87.4	Scombro	83.1	Ginepro	77.7	P. Romasino	248.4
Scogl. Nord	87.4	Ginepro	82.8	Spiaggia	76.8	Spiaggia	247.2
Anatre	87.2	Anatre	82.8	Anatre	76.7	Anatre	246.7
Scombro	86	Spiaggia	82.3	Scombretto	72.9	Scombro	240.1
Scombretto	83.8	Scombretto	79.7	Scombro	71	Scombretto	236.4

Tab. 5: Mean weight of egg clutches of Yellow-legged Gulls at Cala Scombro di Dentro and Punta Romasino (1-6 April 1985). Data are ranked per colony and for each egg composing the clutch.

Cala Scombro		MEAN VALUES		RANGE OF SIZE	
1st egg	88.2 gram	(sample: 38)	69.5 x 48.7 mm	(79-64 x 53-46.5)	(sample: 54)
2nd egg	83.5	(sample: 38)	68.7 x 48.1 mm	(78-61.5 x 50.5-45)	(sample: 51)
3rd egg	78	(sample: 38)	66.5 x 46.9 mm	(78-59 x 50-43.5)	(sample: 38)
All eggs	83.2 gram	(sample: 114)	68.5 x 48 mm	(79-59 x 53-43.5)	(sample: 143)
Punta Romasino					
1st egg	87.4 gram	(sample: 22)	71.2 x 49 mm	(77-65 x 52.5-45)	(sample: 22)
2nd egg	83.1	(sample: 22)	69.8 x 48.3 mm	(77-64 x 51-45)	(sample: 22)
3rd egg	77.9	(sample: 22)	67.1 x 47.3 mm	(76-60 x 51-44)	(sample: 22)
All eggs	82.8 gram	(sample: 66)	69.2 x 48.2 mm	(77-60 x 52.5-44)	(sample: 66)

Tab. 6: Biometrics (weight, length, width) of egg clutches at Cala Scombro di Dentro and Punta Romasino (April 1985). Data are given for each egg (largest, medium-sized and smallest egg per nest).

masino. Data of tab. 5 are to be assessed realizing that the clutches of the biggest colonies were randomly selected thus deliberately including nests appearing relatively small. On the other hand practically all egg clutches were measured at the islet of Corso, Scogliere Sud and Ovest as well as on Scombretto. If we may assume that large clutches are produced by large gulls (and thus by strong and dominant individuals) we can conclude that all the smaller islets are occupied by the largest gulls of Cala Scombro. Scombretto is an exception since this islet is a mere extension of I. Scombro. Indeed one doesn't need a dinghy to visit Scombretto since several rocks between both islands form a land bridge there.

Biometrics of eggs indicate that clutches at Asinara are smaller than those measured in some other West Mediterranean colonies. Tab. 6 demonstrates that the mean length and width in Cala Scombro (n = 143) are 68.5 x 48 cm and 69.4 x 48.2 at Punta Romasino (n = 66) (tab. 6). Isenmann (1976) reports a mean size of 71.4 x 49.9 in the Camargue (n = 123) while de Juana et al. (1979) gives an account of 70.8 x 49.3 (n = 68) at Islas Chafarinas. Our data resemble more those of Lambertini (1982) who studied clutch sizes at Capria (68.0 x 47.3, n = 90) and of Mayol (1978) who measured eggs in the Balearics (69.3 x 49.7 cm, n = 50).

In order to find out if there might be a relationship between clutch size and the nature of nest-sites (covered partly by vegetation or rocks and uncovered nests), we measured 38 complete clutches in Cala Scombro and another 22 at Punta Romasino. However, unlike Lambertini (1983) at Isola Capraia, we did not find a significant correlation. Most nests at Asinara are uncovered (without scrubs or rocks shading the nest). Large clutches can be found both in the shadow of vegetation cover and on top of protruding rocks. Yet, at Cala Scombro di Dentro where vegetation is more abundant, the mean weight of 38 clutches containing all 3 eggs during the first week of April 1985 was slightly heavier in covered nests (251.2 gr.) than in uncovered nests (248.4 gr.). This was not the case in the cliff colony of Romasino at April 8th 1985 where we registered a mean weight of 232 gr. in covered nests and 252 gr. for uncovered clutches! The mean brood size at Cala Scombro di Dentro is to be considered among the highest in the Mediterranean, notably 2.7 eggs for a sample of 281 nests measured between 17th and 21st of April 1984.

Conclusion: Scenario of Nest-Site Selection

From the above data related to occupancy rate of colonies in winter, breeding advance in spring, nest density and biometrics of clutches, we can reconstruct the course of the following scenario. In February 1985 all small, medium-sized islets were occupied by 70 to 90% of their 1984 breeding population. Only the largest island, Scombro, failed to attract less than 50% of its 1984 breeding population as was the case at Punta Romasino. All colonies, including the smallest ones, had dominant birds occupying the best breeding territories and nesting sites: being not too close to the coastal edge, on a prominent position such as a rock and close to vegetation cover producing shade for chicks. Four weeks later, breeding advance was more or less identical in the islet colonies except at Scombro where we observed asynchronous breeding since, due to a 20% population increase, a group of newcomers or first-time breeders joined the island with some 10 days delay. This might have also been the case in the prime choice colonies of Anatre, Spiaggia and Ginepro registering the highest population increase, if they did not score such high synchronous population densities. From this we may deduce that the strongest pairs were able to select what they judged to be the best breeding habitats (medium-sized islets of approximately 0.4 ha situated at least 40 m off the mainland coast) and the best nest sites as described above. Dominant birds installed themselves also on hitherto unoccupied territories of apparently second choice notably at I. Tiria and in the southern part of I. Corso where the Audouins used to breed. Scombro and Scombretto attract a population producing slightly smaller egg clutches. This may be due to the influx of first time breeders.

If this scenario proves to be exact we can predict what will happen if we reduce the Yellow-legged Gull population at Asinara. If the breeding pairs culled will not be replaced during the same season by others (sub-adults for ex.), then the islets of secondary choice will be abandoned which will restore the breeding opportunities for other species such as the Audouin's Gull and, to a lesser extent, the Common Tern. We also may expect that culling on the smaller islets, which are first choice for nest-site selection, is going to be more effective in keeping the Yellowlegs off Isolotto Corso, than reducing them on Isola Scombro.

In most colonies, Audouins are breeding in sympatry with Yellow-legged Gulls. The patchy distribution of their main food source, fish, could explain why Audouins seem to like to breed in the vicinity of Yellowlegs although the latter have modified their diet to chicken bones and pastaciutta from refuse dumps! Both species have probably always bred together. However the current population boom of the "commensal" Yellow-legged Gull needs to be monitored urgently if we want to protect the much rarer and only endemic seabird of the Mediterranean.

Acknowledgements

This study has been carried out thanks to a special permit from the Italian Ministry of Grazia e Giustizia and the Prison Authorities at Cala d'Oliva, Asinara. We would like to thank the commune of Port Torres for the necessary logistical aid such as accommodation and transport over sea. May we also take this opportunity to thank the skippers and fishermen of Stintino who braved out the Mistral and Levante winds during crossings in the Gulf of Asinara. We are also grateful for the financial contribution of the European Economic Communities to help undertake this research.

References

Bradley, Patricia (1986). "The Breeding Biology of Audouin's Gull on the Chafarinas Islands", in MEDMARAVIS, Proceedings 1st Mediterranean Seabird Symposium (this volume), Springer Verlag, Heidelberg.

Burger and Gochfeld (1981). "Nest site selection by Kelp Gulls in Southern Africa", Condor 83: 243-251.

De Juana et al. (1979). "Aspectos de la alimentacion y biologia de reproduction de Larus audouinii Payr. en su gran coloni de cria de las islas Chafarinas (ano 1976). Boletin de la Estacion Central de Ecologia 16.

Isenmann, P. (1976). "Contribution à l'étude de la biologie de reproduction et de l'écologie du Goéland à pieds jaunes en Camargue", La Terre et la Vie, 30.

Lambertini, M. (1983). "Alcune note sulla biologia riproduttiva del Gabbiano reale all'Isola di Capraia (LI)", Quaderni Museo Storia Naturale Livorno, 4.

Launay, G. (1985). "Nouvelles données sur la Biologie du Goéland Leucophée dans le midi de la France", in Oiseaux Marins Nicheurs du Midi et de la Corse, Annales du C.R.O.P., n° 2, Aix-en-Provence.

Mayol Serra, J. (1978). "Observaciones sobre la gaviota de Audouin, Larus audouinii Payr., en el Mediterraneo occidental (primavera de 1978). Naturalia Hispanica, numero 20.

Monbailliu, X. (1985). "Importanza Ornitologica dell'Isola dell' Asinara", in Atti del Convegno sui Parchi Naturali, Regionali e Nazionali, Porto Torres 1984.

Tinbergen, Niko (1953). "The Herring Gull's World", William Collins & Co. Ltd., London.

Torre, A. (1979). "Osservazioni sulla avifauna della Nurra", Bollettino della Società Sarda di Scienze Naturali, vol. 19.

CHAPTER 4 : DEMOGRAPHY AND POPULATION FLUCTUATIONS

Chairman: Dr. Ralph W. SCHREIBER, Chairman I.C.B.P.
Seabird Group.

BREEDING ECOLOGY OF LARIDAE AND STERNIDAE IN THE COMACCHIO WETLANDS (PO DELTA) AND RECENT POPULATION TRENDS

Pierandrea Brichetti, Gran Museo Civico Scienze Naturali,
via Ozanam, 4 - Brescia
Ugo F. Foschi, Museo Ornitologico "F. Foschi",
via Pedriali, 12 - Forli

The Northern Adriatic coastal band (Valli di Comacchio, Po Delta, the Venice and Friuli Lagoons) is the most important wetland in Italy for waterbirds and one of the most important in southern Europe. This aquatic zone is not only an important area for wintering nordic populations, but the southernmost part ("Valli di Comacchio") hosts the largest populations of Laridae and Sternidae breeding in Italy. 8 of the 11 species constantly present in Italy reproduce in the 10,200 hectares saved from reclamation (the reclamation of the approx. 18,000 hectares in the western part of the Valli was terminated in 1962). An additional two species (Sterna caspia, Sterna bengalensis) have been occasionally found breeding there.

The Valli di Comacchio represents today a vast brackish area (witness of the former wetlands formed by the Po delta). They are characterized by a large surface area of open water interrupted in the middle by a series of islets or remains of an old dune complex prevalently aligned from north to south. The mean depth varies from 60 to 80 cm, reaching a maximum of 150-200 cm. The percentage of salinity varies seasonally and increases from south to north (18-48‰). The "Valli" communicate with the sea by way of several canals whereof the flow can be controlled. The salt-work and fishfarming basins, which are situated in the north-eastern part of the Valli have formed a specific habitat.

The islets which host the colonies of gulls and terns emerge generally only about 20 to 40 cm. They are covered by a halophyte vegetation known as Salsoletum sodae on sandy soils enriched by

NATO ASI Series, Vol. G 12
Mediterranean Marine Avifauna
Edited by MEDMARAVIS and X. Monbailliu

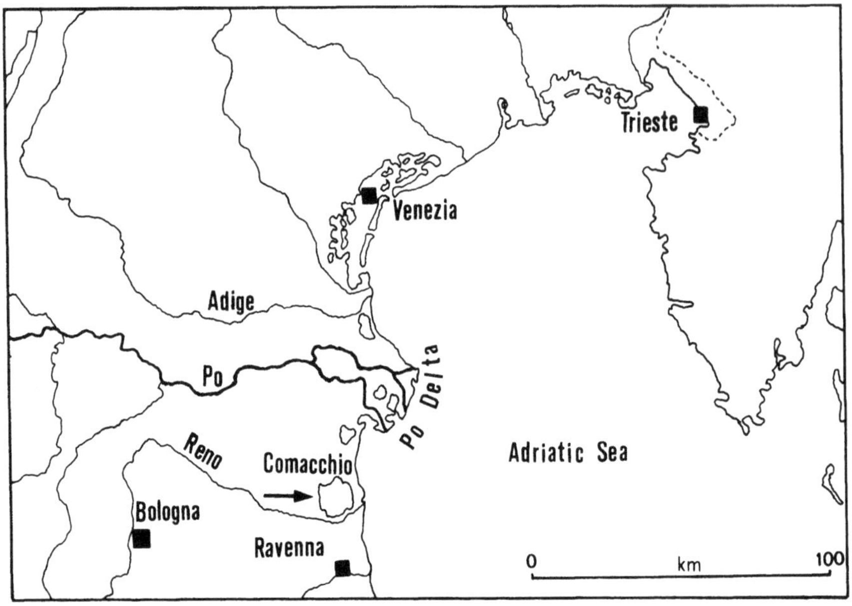

Fig. 1: Geographical situation of the Comacchio wetlands.

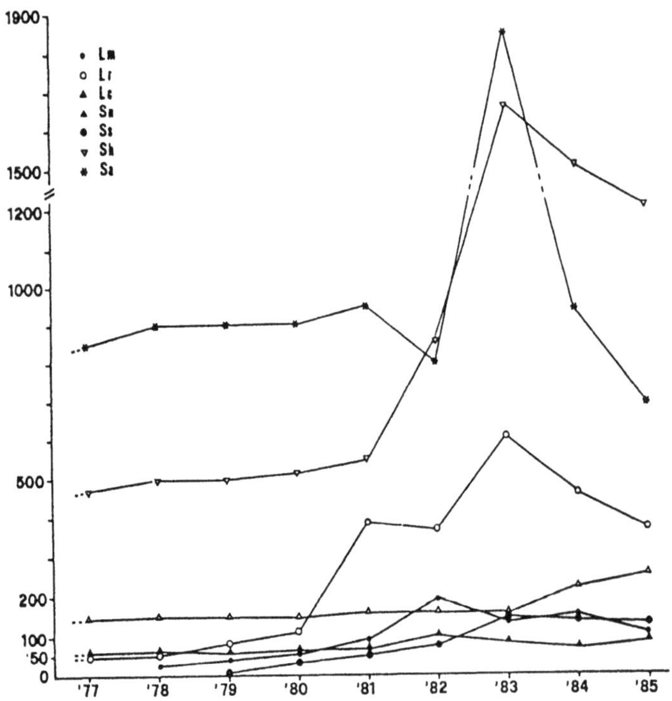

Fig. 2: Number of breeding pairs of Laridae and Sternidae in the Comacchio wetlands from 1977 to 1985 (Larus genei excluded).

organic substances, Salicornietum fructicosae on wet clay and Salicornietum herbaceae in the areas where the waters are calm. Large areas of some islets lack vegetation and are covered completely with shell debris.

The structure and dynamics of breeding populations

The scarcity of past available ornithological data seems to indicate that the Valli di Comacchio have become increasingly more important in recent years. The species found breeding here in the first half of this century were: Larus cachinnans, Sterna hirundo and Sterna albifrons. Around the 1950's Sterna nilotica was found nesting, and in 1960 Larus ridibundus bred.

There was an increase of the existing populations of S. hirundo and S. albifrons. At the end of the 1970's three other species formed colonies: Larus melanocephalus (1978), Larus genei (1978) and Sterna sandvicensis (1979). This was accompanied by a general increase of the population of other species and the creation of new colonies, some of them in neighbouring wetlands (Valle Bertuzzi). Additionally, in 1978, the occasional breeding of Sterna caspia was noted (1-2? pairs). One pair of Sterna bengalensis bred in 1985. The detailed numerical evolution and spatial distribution can be summed up as follows:

- The population of Larus melanocephalus increased from approximately 25 pairs in 1978 to 103 in 1985 (with a maximum of 197 in 1982). In 1982 two pairs were found colonizing the neighbouring Valle Bertuzzi. They breed mainly in areas covered with low halophyte vegetation (Obione portulacoides) and secondarily in Salicornietum of medium height, preferably in areas close to both L. ridibundus and S. hirundo.

- The population of Larus ridibundus has increased from the three nests found in 1960 to 30-50 pairs during the seventies with a sizable increase since 1980 passing from 115 to 372 pairs in 1985 (with a maximum of 614 in 1982). The species clearly prefers the high and thick Salicornietum fructicosae and associates predominantly with S. hirundo and L. melanocephalus but only marginally with S. sandvicensis.

- Larus genei seems to have failed to establish a stable breeding pattern after having bred in 1978 (2 pairs). From a maximum of 10 to 13 pairs in 1983-84 the population dropped to 4 pairs in 1985. The nests are found mostly in areas with low sparse vegetation, bordering the colonies of S. nilotica, S. hirundo, L. melanocephalus and L. ridibundus.

- Larus cachinnans: the population dynamics of this species, which showed difficulties in establishing itself in the past, seemed stable during the 1970's-early 1980's (150-168 pairs) and has shown an increase in recent years (230-260 pairs in 1984-85). The Yellow-legged gull breeds in scattered pairs or in monospecific colonies both on islets with and without halophyte vegetation as well as on wooded artificial banks.

Larus genei, Rondine di mare maggiore _Hydroprogne tschegrava_. Riv. ital. Orn. 48: 215-233.

Brichetti, P. (1979). Sulla nidificazione in Italia di _Sterna sandvicensis_ Latham (Beccapesci). Riv. ital. Orn. 49: 147-207.

Brichetti, P. & Foschi, U. (1982). Valli di Comacchio: situazione _Laridae_ e _Sternidae_ 1982. Riv. ital. Orn. 52: 207.

Brichetti, P. & Foschi, U. (in press). Status of _Laridae_ and _Sternidae_ breeding in the Valli di Comacchio (Italy). Proceedings Colonial Waterbirds and Pacific Seabirds Symposium, San Francisco, 1985.

Brichetti, P. & Isenmann, P. (1981). Studio preliminare sulla evoluzione degli effettivi nidificanti di _Laridae_ e _Sternidae_ nelle Valli di Comacchio (Italia) e nella Camargue (Francia). Riv. ital. Orn. 51: 133-161.

Fasola, M. (ed.) (1982/1983). Rapporti preliminari _Laridae_ - Italia. INBS e Dip. Biol. Anim., Pavia.

Foschi, U. (1979). Accertamento di nidificazione in Italia del Beccapesci (_Sterna sandvicensis_). Uccelli d'Italia 4: 271-272.

Foschi, U. & Teodorani, G. (1978). Accertamento di nidificazione di Gabbiano corallino (_Larus melanocephalus_) e Gabbiano roseo (_Larus genei_). Uccelli d'Italia 3: 144-148.

Toschi, A. (1955). Sulla nidificazione di alcuni uccelli nelle Valli di Comacchio. Ric. zool. appl. caccia Bologna n. 26.

Toschi, A. (1960). La nidificazione in Italia della Cicogna bianca, del Mignattaio e del Gabbiano comune. Ric. zool. appl. caccia, Bologna n. 32.

Warncke, K. (1962). Die avifaunische Sonderstellung der Valli di Comacchio für Italien. Vogelwelt 83: 129-139.

FLUCTUATIONS OF THE GULL AND TERN POPULATIONS IN THE EBRO DELTA, NORTH-EAST SPAIN (1960-85)

Xavier Ferrer and Albert Martinez-Vilalta
Departament de Zoologia, Facultat de Biologia,
Universitat de Barcelona. C/Diagonal 645
08028 - Barcelona, Spain

Introduction

The Ebro delta is a 320 km2 area situated in North-East Spain (fig. 1). It is mainly constituted of cultivated land: ricefields (46%), market gardens and orchards (20%). The natural area is thus very small: about 7,300 Ha (23%) (fig. 1). Compared to other deltas, this small surface contains a high number of sea and marsh birds. Indeed, the Ebro delta is the most important Iberian site quantitatively as well as qualitatively, for Common, Little and Sandwich Terns. The total number of terns nesting is about 3,400 pairs (1980-85) consisting of 5 regular species: Sterna nilotica, S. sandvicensis, S. hirundo, S. albifrons and S. bengalensis, and 1 accidental: S. dougallii. The total breeding population of gulls is about 2,300 pairs (1980-85) composed of 5 species: Larus ridibundus, L. genei, L. audouinii, L. fuscus graellsii and L. cachinnans.

Medium and long term censuses of avian populations are quite rare in the Mediterranean basin and are usually dispersed in local papers. The Ebro delta is not an exception and despite its international importance and the regular field work during the last twenty-five years (especially in the last decade) quite a lot of ornithological data remain to be collected, as we can see from the recently published palearctic handbooks (Cramp & Simmons, 1983; Glutz & Bauer, 1982).

The present paper analyzes the censuses made since 1960. It summarizes the situation in the breeding populations of gulls and terns and suggests some possible causes for their fluctuations. We also try to demonstrate, as much as possible, the international importance of the site in the Western Mediterranean, and thus the urgent need of its conservation.

NATO ASI Series, Vol. G 12
Mediterranean Marine Avifauna
Edited by MEDMARAVIS and X. Monbailliu

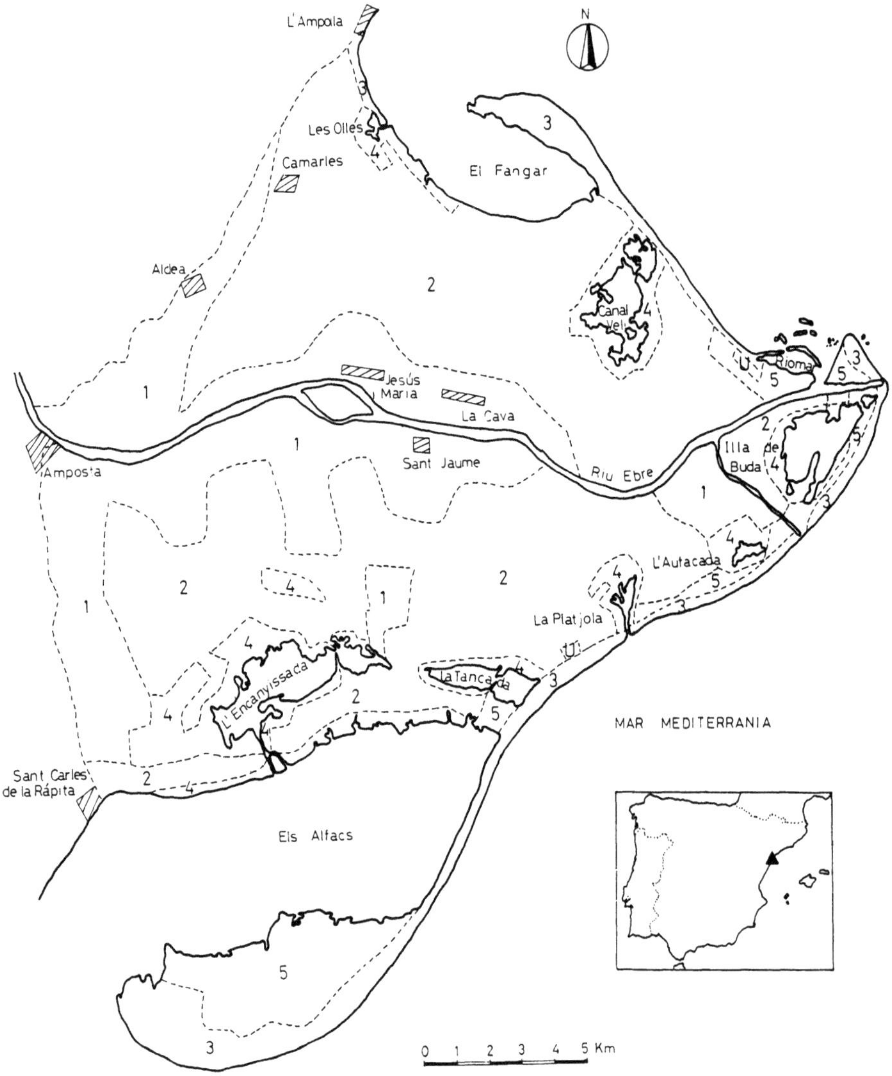

Fig. 1: The landscapes of the Ebro delta, adapted by F. Comin and X. Ferrer from an R.M. Masalle's original.

1 - Market gardens and orchards
2 - Paddy fields
3 - Psammophilous vegetation
4 - Helophitic vegetation
5 - Halophilous vegetation
U - Urban soil

Methods and Materials

The censuses we are presenting here are based on sampling of nests. In some problematic areas, they have been complemented with counts of territorial adults and of pulli.

Checking of nesting populations of gulls and terns started in the late fifties. Sampling however was irregular as far as time, species and colonies are concerned. In fact the species that breed in reduced, concentrated colonies were censused thoroughly, as with L. audouinii, L. fuscus , L. genei, L. ridibundus and S. sandvicensis. Species such as S. hirundo, S. nilotica, S. albifrons and L. cachinnans that nest scattered and/or in very wide areas were hardly covered by total censuses. The long nesting period, from the first clutches of L. cachinnans in mid-April to the last broods of S. albifrons in mid-July made it also difficult to obtain complete counts for breeding seabirds. In fact, the most complete censuses were those of 1976 and 1980. Few complete data are available between 1960 and 1972, where, although there was in intensive sampling, censusing was concentrated in some localized geographic sectors. Thus there are no total censuses and those published for some species, should be taken into account carefully, like in the case of the Common Tern (Sterna hirundo), which we will refer to later.

The material used (see tab. 1) was taken from our own censuses and from the works of: Maluquer (1960, 1971), Maluquer & Pons (1961), Mestre & Ferrer (1974), Ferrer (1977), Muntaner et al. (1984) and Ferrer et al. (1986).

The Cramp & Simmons (1983) and Glutz & Bauer (1982) handbooks are presenting the censuses compiled in the Ebro delta. These either display lower totals than the ones here presented, or are misinterpreted from some of the publications mentioned above, as is the case with L. genei in Cramp's handbook.

Results

The results are shown in tab. 1 and in fig. 2-6. Here below we will treat each species individually.

• Black-headed Gull (Larus ridibundus)

The first appearance as a breeding species in the Ebro delta dates back to approximately 1960 or the late 1950's. In any case, its population remained wavering during the 60's with a single colony of 10 to 30 couples. This however disappeared in 1968 due to sea erosion. In 1970 the colony reappeared in a saline area with a less fluctuating water level. From this moment on, the colony started to grow slowly, but constantly. In 1979, the population increased in a totally linear way (see fig. 2) and this year, it recolonized its first site in the Delta. This increase may be due to the change of breeding size. The new site, in the saltworks is much better located than the old one. It is also important to remember that the establishment and the later growth of the new colony coincides chronologically with the consolidation of this species as breeder in Spain during the late 1960's and early 1970's. In other words, it seems that the factors that caused the growth are extrinsic to the Delta and partially intrinsic to it (improvement in the colony sites). There does not seem that a favorable change in food conditions occurred, like for example a large extension of ricefields, as was the case in the Camargue (Blondel & Isenmann, 1981).

	1960	61	62	63	64	65	66	67	68	69	70	71	72	73	74	75	76	77	78	79	80	81	82	83	84	85
Larus ridibundus Black-headed Gull	25+	18	12	12	10	15-20	?	12	6+	?	2	?	0	1	3	7	16	20	38	86	146	+	+	+	420	+
Larus genei Slender-billed Gull																12	+	8	6	10	24	12+	+	+	280	+
Larus audouinii Audouin's Gull																						36	200	546+	1200	1200
Larus fuscus graellsii Lesser black-backed Gull																						2	2	3	?	13
Larus cachinnans Yellow-legged Gull	+	+	+	+	+	+	500-600	+	+	+	+	+	+	+	+	+	+	+	+	+	230	+	+	+	+	500
Sterna nilotica Gull-billed Tern	+	+	+	50+	70+	100+	+	+	+	+	+	+	30+	15+	+	25+	50+	20+	+	+	54	+	+	+	100+	+
Sterna sandvicensis Sandwich Tern		5									?	4+	10	37	36	45	93	70	198	255	231	+	+	+	330	+
Sterna hirundo Common Tern	+	├──	──	──	──	700 ++	──	──	──┤	+	+	+	+	900+	+	800+	1000-1500	1000+	650+	+	2256	+	+	+	+	+
Sterna albifrons Little Tern	+	├──	──	──	──	650 - 700	──	──	──┤	+	+	+	+	120+	+	+	325+	+	+	500+	652	+	+	+	+	+
Sterna bengalensis Lesser crested Tern																			?	1	?	1	?	?	2	+
Sterna dougallii Roseate Tern		1																								

Tab. 1: Census of breeding couples of gull and terns in the Ebro delta. The counts with + refer to absence of sampling of some colonies or partial counting. References: Maluquer (1960, 1971), Maluquer & Pons (1961), Mestre & Ferrer (1974), Ferrer (1977), Muntaner et al. (1984), Ferrer et al. (1986) and personal data.

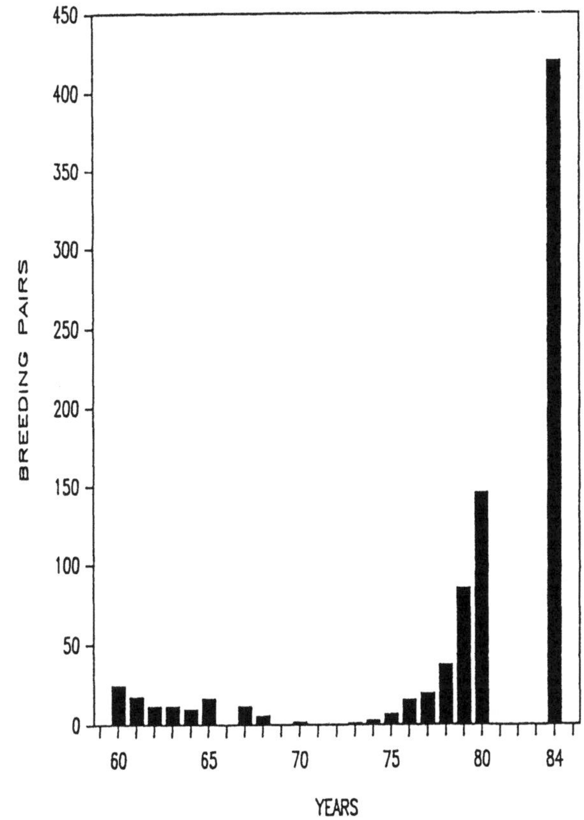

Fig. 2: Total of Black-headed gull (Larus ridibundus) breeding pairs in the Ebro delta from 1960 to 1984. Breeds every year except in 1972 and perhaps in 1969 and 1971. See tab. 1.

• Slender-billed Gull (Larus genei)

Bred for the first time in 1975 in the Ebro Delta, where 12 couples were reported. Its continuous increase (fig. 3) starts in 1979, totaling 280 couples in 1984. This growth matches with the establishment of a protected area in its breeding area. In our opinion however, the reasons for the increase lie outside the Delta: perhaps in the flourishing colonies of Sardinia or in Banc d'Arguin (Mauritania), and/or in the deplacement of gulls from the diminishing colonies (Marismas del Guadalquivir, Fuentedepiedra, Camargue, etc.).

• Audouin's Gull (Larus audouinii)

The establishment of L. audouinii as a breeding species in the Delta was perfectly documented in 1981, although more or less regular observations existed since 1979, when the species

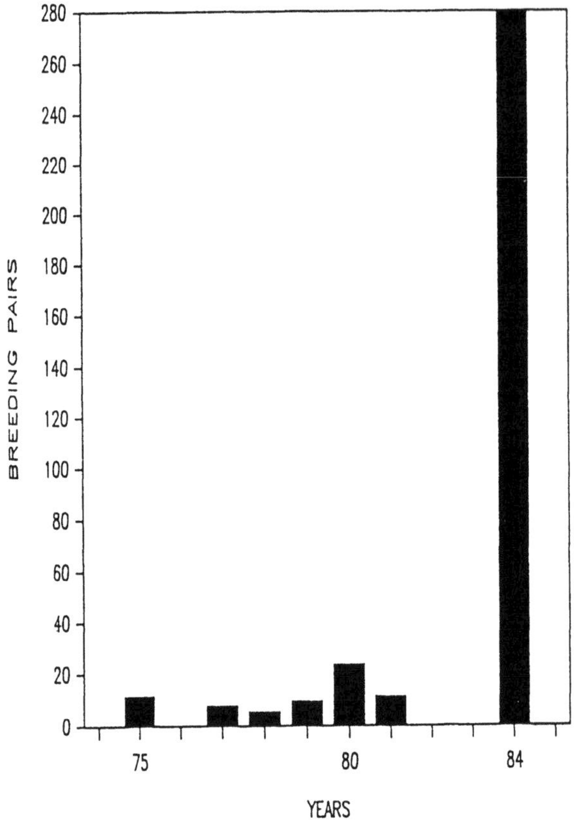

Fig. 3: Total Slender-billed Gull (Larus genei) breeding pairs in the Ebro delta since the colonization in 1975. See tab. 1.

was rather exceptional in the area (Martinez & Motis, 1982). The establishment and growth of the colony coincide with the protection of the breeding area. It seems more likely that the growth of the colony is the result of a general growth of the western mediterranean population (Carrera, E. in Muntaner et al., 1984).

• Lesser black-backed Gull (Larus fuscus graellsii)

This very rare breeder in the western Mediterranean bred for the first time in 1981 (2 couples). In 1984, as much as 13 territorial pairs were counted. Regularly, some immatures moult in the Ebro Delta and stay in spring-summer (Carrera, E. in Muntaner et al., 1984).

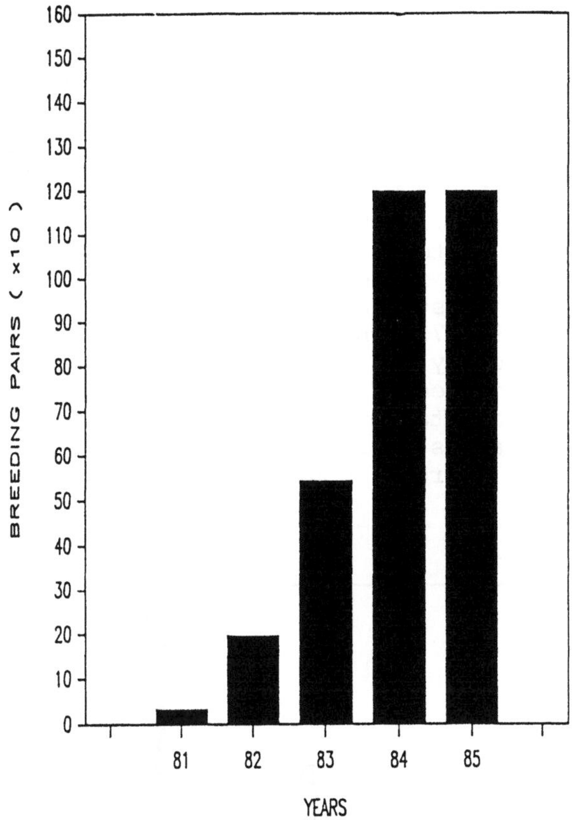

Fig. 4: Total Audouin's Gull (Larus audouinii) breeding pairs in the Ebro delta since the colonization in 1981. Bred every year. See tab. 1.

• Yellow-legged Gull (Larus cachinnans)

There have been some problems with the censuses and none of them are exact. To judge from some controlled colonies and from the 1980 census, it seems that the number of individuals decreased during the 1970's, whereas some increase has been observed in the 1980's (it has recolonized some old, deserted breeding areas of the 1970's). As an example, there is the case of Buda island colony: 1965 (100 pairs), 1973 (22), 1975 (13), and 1980 (71). The censuses are based on Maluquer (1971), Ferrer (1977) and our own data. Unfortunately, this trend is only tentative because the sample is small and apart from 1980, there exist only two other total censuses (1966 and 1985). Although sampling of these surveys is not comparable they give a similar order of magnitude: about 500 couples.

• Gull-billed Tern (Sterna nilotica)

This is a very scattered breeder occupying a vast area. Evaluation of its population is difficult since there is only one complete census (1980) and another one nearly complete (1976). However one can postulate that during the 1960's the total population oscillated between 100 and 200 couples. There was an important decrease in the early 1980's and during the 1980's the total amounts to between 50 and 100 pairs (Muntaner et al., 1984). A possible reason for this decrease is the massive aerial use of the pesticide Queletox over the rice fields of the Delta during 1968-73. This affected directly the adult breeders which food-supply was based mainly in the rice fields. However, one can not disregard other more general factors such as the intrinsic unpredictable population fluctuations already observed in other areas as in Camargue (Blondel & Isenmann, 1981) and the general decreasing pattern of its European population.

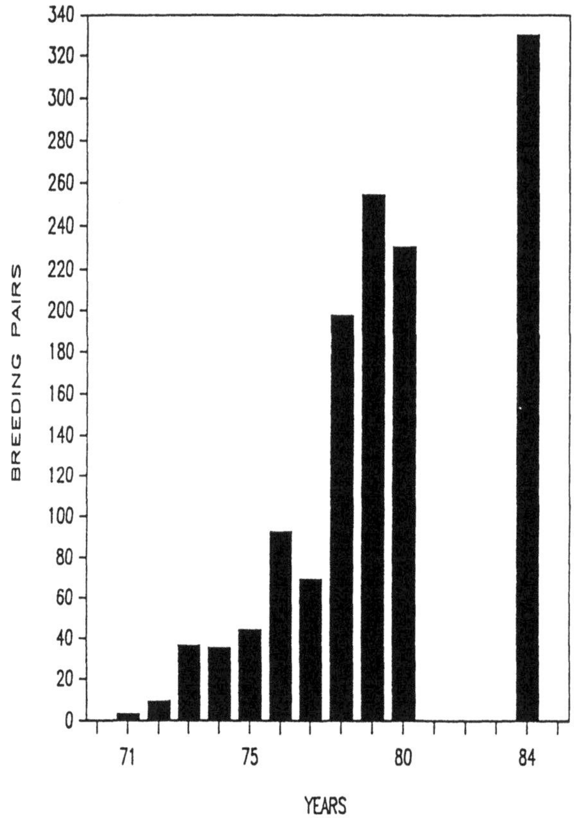

Fig. 5: Total Sandwich tern (Sterna sandvicensis) breeding pairs in the Ebro delta since the recolonization in 1971. Previously, 5 pairs bred only once: in 1961 on another site. See tab. 1.

• Sandwich Tern (*Sterna sandvicensis*)

5 couples bred apart by themselves in 1961. The colony was in a more suitable area in 1971, or maybe already in 1970. In this new area a stable population was maintained during the first 5 years, followed by a considerable, almost continuous growth. It is not absolutely sure that this is a new breeder, since the Sandwich tern was already reported in the Ebro Delta during the spring of 1870. However, identification is regarded as uncertain (Muntaner et al., 1984). The last census in 1984 indicates as many as 330 couples.

• Common Tern (*Sterna hirundo*)

This is a typical breeder in this Delta. Its colonies are multiple and some are difficult to evaluate such as the existing ones in the spits. Absolutely total censuses were not given until 1976. In the 1960's the total of 700 pairs given by Maluquer (1971) seems to be low since the observers at that time did not sample the most important area which holds a mean population of 800 pairs. As a consequence, it is probable that the total number was above 1,500 couples. Just as with *S. nilotica* and *S. albifrons*, and according to some partial data, it seems that there was a decrease of the population during the early 1970's. It is hard however, to obtain exact numbers with this species since it can experience important annual oscillations. Based on the evolution of the most important and best known breeding nucleus (fig. 6) it looks like from the first exhaustive control in 1973, that the total population is in expansion. This has been substanciated by the colonization after 1978 of some spanish mediterranean areas (Santa Pola and Albufera de Valencia) and probably of South France (Salses and Canet in 1984) (Ferrer et al., 1986).

• Little Tern (*Sterna albifrons*)

With a similar situation to that of *S. hirundo*, the Little Tern has experienced some oscillations. Between 1968 and 1973 it directly suffered from the effects of aerial fumigations (Vaughan, 1971). Nonetheless, one must take into account the general fluctuating behaviour of the species in other mediterranean breeding quarters (Blondel & Isenmann, 1981). Thus in the most complete censuses of 1979 and 1980, the breeding population was of about the same order of magnitude as in the first half of the 1960's (about 650 couples), when the Ebro Delta was said to hold one of the most dense european populations of Little Tern (Maluquer, 1971).

• Lesser crested Tern (*Sterna bengalensis*)

This species was first observed in Ebro Delta in August of 1978. Its reproduction was reported for the first time in 1979. It also bred in June of 1981 and in spring of 1984. According to tab. 1, it seems probable that this species is a regular breeder in the region since its reproduction was recorded during the years in which the sampling of the colonies was exhaustive, whereas it was not reported during other years.

ciclopèdia Catalana, Barcelona.

Glutz Von Blotzheim, U.N. & Bauer, K.M. (1982). Handbuch der Vögel Mitteleuropas. Vol. 8/1. Akademische Verlagsgesellschaft. Wiesbaden.

Maluquer, S. (1960). La avifauna primaveral de la isla de Buda. Ardeola 6: 151-210. Madrid.

Maluquer, S. (1971). La avifauna del Delta del Ebro en primavera-verano. Ardeola vol. esp. 1971: 191-334. Madrid.

Maluquer, S. & Pons, J.R. (1961). La avifauna primaveral de la isla de Buda en primavera-verano de 1961. Ardeola 7: 79-111. Madrid.

Martinez, A. & Motis, A. (1982). Quelques observations sur la présence du Goéland d'Audouin (Larus audouinii) pendant la période
6: 158-161.

Mestre, P. & Ferrer, X.(1974). Censo de algunas aves coloniales del Delta del Ebro en 1972-73. Misc. Zool. 3: 143-149. Barcelona.

Muntaner, J.; Ferrer, X. & Martinez-Vilalta, A. (eds.) (1984). Atlas dels ocells nidificants de Catalunya i Andorra. Ketres ed. Barcelona.

Vaughan, R. (1971). Tragedy of the Ebro Delta birds. Country Life Feb. 18: 356-357.

FLUCTUATIONS OF THE LARIDAE OF THE RHONE DELTA OVER THE PAST 30 YEARS (1956-1985)

Paul Isenmann, Centre Louis Emberger (C.N.R.S.), B.P. 5051, F - 34033 MONTPELLIER (France)

Alan Johnson & John Walmsley, Fondation Tour du Valat, Le Sambuc, F - 13200 ARLES (France)

The gulls and terns (Laridae) together with the herons (Ardeidae) and flamingos (Phoenicopteridae) undoubtedly take pride of place amongst the aquatic bird species breeding in the Camargue (southern France). Most species of Laridae colonize small islands in the lagoons of the southern part of the Rhône delta. About two-thirds of this area have for many years now been integrated into two extensive salines (Salin de Giraud, Salin d'Aigues-Mortes). As a group this community exploits diverse feeding habitats in this region; the sea, salt-pans, brackish lagoons, fresh-water marshes, agricultural land and even refuse tips (Lévêque 1957, Isenmann 1979, Blondel & Isenmann 1981). Eight species presently breed, four belonging to the genus Larus and four to the genus Sterna (this paper ignores the Whiskered Tern Chlidonias hybrida an irregular breeder in fresh-water marshes). The first more or less complete survey was made in 1937 by Lomont (1938) and the first complete census in 1956 (Lévêque, 1957). Since then complete censuses have been carried out by the Station Biologique de la Tour du Valat every three years in collaboration with the C.N.R.S. (Centre National de la Recherche Scientifique), the "Réserve Nationale de Camargue" and the "Parc Ornithologique du Pont de Gau" (Blondel 1964, Johnson & Isenmann 1971, Blondel & Isenmann 1981, Hafner et al. 1980, 1985). This paper is based on the results of the ten censuses carried out between 1956 and 1985.

Results

The results of these censuses are summarized in tab. 1, from which it can be seen that only five species were recorded breeding in 1937: Black-headed Gull (Larus ridibundus), Yellow-

NATO ASI Series, Vol. G 12
Mediterranean Marine Avifauna
Edited by MEDMARAVIS and X. Monbailliu

	1956	1962	1967	1969	1970	1973	1976	1979	1982	1985
LARUS RIDIBUNDUS	2900	6000	4500	7600	10000	6550	8000	7000	9920	5350
LARUS GENEI	1	-	1	-	-	13	12	13	26	17
LARUS MELANOCEPHALUS	-	-	-	2	-	1	7	7	13	23
LARUS CACHINNANS	300	600	725	680	630	1150	2750	2600	2480	1900
STERNA ALBIFRONS	400	350	496	290	395	352	450	250	340	325
STERNA HIRUNDO	2650	2000	1100	1000	1900	1360	1500	1100	1800	1600
STERNA SANDVICENSIS	15	95	287	440	453	310	1050	700	1285	1000
STERNA NILOTICA	250	262	166	210	73	140	200	200	258	285

Tab. 1: Results of 10 censuses of breeding pairs of Laridae in the Camargue.

legged Gull (L. cachinnans), Little Tern (Sterna albifrons), Common Tern (Sterna hirundo) and Gull-billed Tern (Sterna nilotica). Two others were added in 1956, Slender-billed Gull (Larus genei) and Sandwich Tern (Sterna sandvicensis) and finally the Mediterranean Gull (Larus melanocephalus) in 1965. During these thirty years three species have increased markedly, the Black-headed Gull, the Yellow-legged Gull and the Sandwich Tern.

The main increase of the Black-headed Gull population seems to have occurred after the second world war; 395 pairs in 1937, 2,900 pairs in 1956, 10,000 pairs in 1970. Since this peak numbers have fluctuated at a lower lever (fig. 1).

This dramatic increase is consistant with a general increase in numbers throughout western Europe this century (Isenmann, 1976/77), Glutz von Blotzheim & Bauer 1982) but there are now signs that the population is levelling off (see Reichholf 1983). In the Mediterranean region, the Camargue seems to have been the first area colonized by this species which probably later spread to localities in Spain and Italy. This increase seems to have been brought about by reduced winter mortality of adults combined with favourable agricultural changes (development of rice-fields) and good breeding success (see Lebreton & Isenmann 1967). During season rice-fields provide important feeding areas.

The Yellow-legged Gull population has undergone a similar increase from just a few scattered breeding pairs in the 1930's, to 300 pairs in 1956 and 2,750 pairs in 1976. Refuse tips in the area provide important feeding areas for this species. These predatory gulls constitute a serious threat to other breeding birds in the Camargue. Culling of these gulls in spring began in 1961 (Blondel 1963) and has been carried out annually since then. It is planned to continue this action.

The third species which has increased markedly in numbers is the Sandwich Tern (fig. 2). This species was first recorded breeding in the Camargue in 1948 though it may have occurred earlier (Isenmann 1972). Up to 1962, the population level remained low, the main increase taking place between 1967 and 1982 when a peak of 1,285 pairs was reached. It is not known why this increase occurred but it is thought that tern numbers fluctuate according to the abundance of the principal food items, in this case the Pilchard (Sardina pilchardus) as Lévêque (1957) suggested in the case of the Common Tern.

Two other species have increased recently, the Slender-billed Gull and the Mediterranean Gull. The former was an irregular and scarce breeder until 1971 (Isenmann 1976). Since 1972 some pairs have attempted to breed most years with a maximum of 26 in 1982. Breeding success varies greatly from year to year, from complete failure some years to a maximum of 60 chicks in 1983 (Hafner et al. 1985). A similar tendency to increase has been noted during the same period in Sardinia and N.E. Spain (Isenmann 1976, Glutz v. Blotzheim & Bauer 1982, Ferrer pers. com.).

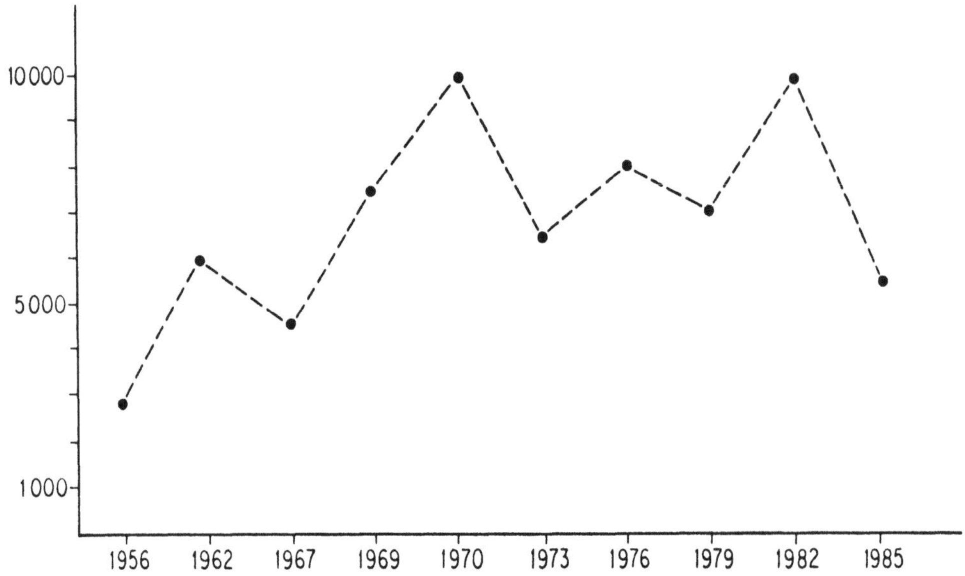

Fig. 1: Numbers of breeding pairs of the Black-headed Gull over the past 30 years (1956-1985).

Fig. 2: Numbers of breeding pairs of the Sandwich Tern over the past 30 years (1956-1985).

As for the Mediterranean Gull, numbers and breeding success over the past twenty years varied much less, this species showing signs of taking a slow but sure foot-hold in the Rhône delta. It has also colonized other parts of western Europe and increased in number within its traditional breeding range (Cramp et al. 1982).

The populations of the remaining three tern species are relatively stable but fluctuate up to 50% and more on either side of the mean.

The Rhône delta hosts 45% of the French breeding population of Common Terns and 50% of Little Terns (see Muselet 1983) whilst two of the eight Laridae of the region (Slender-billed Gull, Gull-billed Tern) breed nowhere else in the country.

Discussion

In the Rhône delta population trends of most species have fluctuated over the past 30 years in a manner similar to that shown throughout the major part of the breeding range of each species. In the case of the Black-headed Gull, its abundance allows it to respond quickly to new situations which develop and which are favourable for breeding and/or feeding. Common and Sandwich Terns may fluctuate in numbers according to prey abundance and availability, whilst other species such as the Slender-billed Gull and to a lesser extent the Gull-billed Tern exploit feeding and breeding habitats which are temporary and unpredictable.

In the Camargue, a series of fairly recent developments have undoubtedly enhanced the quality of the area for breeding gulls and terns (see Blondel & Isenmann 1973, 1981). These developments have affected both the security of the breeding sites, and the food supply. The areas in which the majority of the colonies are located are now protected (Réserve Nationale de Camargue, Réserve des Impériaux, Salins du Midi (C.S.M.E.)). The reduction in disturbance is likely to have led to a higher breeding success. The considerable extensions of the salt exploitations brought about an increase in the area of lagoons whose water levels are permanent and stable each year throughout the breeding season. This is where most colonies are established on islands or dykes free from disturbance from terrestrial predators.

Formerly a much larger area of lagoons underwent a seasonal drying out, to such an extent that reproduction was not possible at all some years (Lomont 1938). Breeding islands appear not to be a limiting factor to population size. There are many suitable islets and not all are colonized in any one year.

As far as the food supply is concerned, agricultural developments and notably rice-growing have favoured the Black-headed Gull and perhaps also the Gull-billed Tern. In many cases these fields, rich at times in invertebrates, are in the neighbourhood of the breeding lagoons.

The development of refuse tips around the Camargue has undoubtedly favoured both the Black-headed Gull and the Yellow-legged Gull which frequent these areas in large numbers throughout the year. These species may even feed their chicks on food items collected from the tips. The latter species has also benefitted from an increase of coastal fishing activity. Finally, the extension of the salinas not only offers added protection to colonies but also provides an abundant supply of food for those species which feed in shallow waters.

Being one of the richest breeding areas for Laridae not only in the western Mediterranean but in the whole of the Western Palearctic it is to be hoped that the present abundance and diversity of species can be maintained in the future. This depends to a large extent on the continued good-will and understanding of the salt company (C.S.M.E.) towards this important bird community.

References

Blondel, J. (1963). Le problème du contrôle des effectifs du Goéland argenté (Larus argentatus michahellis) en Camargue. Terre et Vie, 17: 301-315.

Blondel, J. (1964). L'avifaune nidificatrice des eaux saumâtres camarguaises en 1962 et 1963. Terre et Vie, 18: 309-330.

Blondel, J. & P. Isenmann (1973). L'évolution de la structure des peuplements de Laro-limicoles nicheurs de Camargue. Terre et Vie, 27: 62-84.

Blondel, J. & P. Isenmann (1981). Guide des Oiseaux de Camargue. Delachaux & Niestlé, Neuchâtel-Paris.

Cramp, S. & K.E.L. Simmons (eds.) (1982). The Birds of the Western Palearctic. Vol. III. Oxford University Press, Oxford.

Glutz von Blotzheim, U. & K. Bauer (1982). Handbuch der Vögel Mitteleuropas. Band 8/I. Akademische Verlagsgesellschaft, Wiesbaden.

Hafner, H., Johnson, A. & J. Walmsley (1980). Compte rendu ornithologique camarguais pour les années 1978 et 1979. Rev. Ecol. (Terre et Vie) 34: 621-647.

Hafner, H., Johnson, A. & J. Walmsley (1985). Compte rendu ornithologique camarguais pour les années 1982 et 1983. Rev. Ecol. (Terre Vie) 40: 87-112.

Isenmann, P. (1972). Aire de répartition de la Sterne caugek (Sterna sandvicensis) en Méditerranée et données sur sa biologie en Camargue. Nos Oiseaux, 31: 150-162.

Isenmann, P. (1976). Contribution à l'étude de la biologie de la reproduction et de l'étho-écologie du Goéland railleur (Larus genei). Ardea, 64: 48-61.

Isenmann, P. (1976/1977). L'essor démographique et spatial de la Mouette rieuse (Larus ridibundus) en Europe. L'Oiseau et R.F.O., 46: 337-366 et 47: 25-40.

Isenmann, P. (1979). Le partage des biotopes de Camargue par les Laridés nicheurs. L'Oiseau et R.F.O., 49: 91-103.

Johnson, A. & P. Isenmann (1971). L'évolution récente des effectifs nicheurs de Laro-limicoles en zone saumâtre de Camargue. Alauda, 39: 29-36.

Lebreton, J.D. & P. Isenmann (1976). Dynamique de la population camarguaise de Mouettes rieuses (Larus ridibundus): Un modèle mathématique. Terre et Vie, 30: 529-549.

Lévêque, R. (1957). L'avifaune nidificatrice des eaux saumâtres camarguaises en 1956, essai de recensement suivi d'une première esquisse écologique. Terre et Vie, 11: 150-178.

Lomont, H. (1938). Rapport ornithologique. Actes de la Réserve de Camargue, 21: 13-20.

Muselet, D. (1983). Répartition et effectif de Sterna hirundo et de Sterna albifrons nicheuses en France pour l'année 1982. L'Oiseau et R.F.O., 53: 309-322.

Reichholf, J. (1983). Bestandstendenzen bei der Lachmöwe (Larus ridibundus). Anz. Orn. Ges. Bayern, 22: 211-217.

BIOMETRICAL AND BIOLOGICAL CONSIDERATIONS ON THE CORY'S SHEARWATER CALONECTRIS DIOMEDEA

Bruno Massa & Mario Lo Valvo, Istituto di Zoologia
dell'Università di Palermo, 18 Via Archirafi, 90123 Palermo

Introduction

Today the Mediterranean distribution of Cory's Shearwaters is believed to be well known. New colonies may be discovered, but certainly the largest ones are known. Many data were published during the last ten years and we believe to have a satisfactory knowledge of their biology, though some problems, as the winter movements of each population, the first breeding age and the ratio immatures/adults are still to be examined carefully.

In this paper we examine only the following aspects:

1) the estimate of the Mediterranean populations;
2) a biometrical comparison between some populations of the Mediterranean basin and the Atlantic ones (ssp. borealis);
3) a comparison of the breeding cycle of previous populations.

Material and methods

The estimate of the populations and the biometrics have been drawn from the available literature. Some authors did not recognize the males and the females and their data probably are the mean of both sexes. We made the average of the males and females of all the populations in order to compare these data with those of the authors who did not recognize the sexes. Therefore we did not test the small homogeneous values and we used the t of Student only to compare populations of the same sex. We also took some measurements on skins collected along the coasts of Tuscany and Liguria (Tyrrhenian sea), probably coming from colonies of Corsica or Sardinia, and preserved in the Museum of Natural History of Milan. We also measured some

NATO ASI Series, Vol. G 12
Mediterranean Marine Avifauna
Edited by MEDMARAVIS and X. Monbailliu

skins from I. Selvagen (Atlantic) (2 males and 1 female), I. Cabrera (Balearic Islands) (3 males and 2 females) and I. Dragonada (Crete) (2 males and 3 females), all taken in order to research chemical pollutants (Renzoni et al., in prep.). Original data on the breeding cycle were collected in the I. Linosa (Sicilian Channel) during research between 1981 and 1985.

• Numbers and distribution in the Mediterranean Sea

Fig. 1 shows all the colonies, according to contemporary authors.

Aegean Sea. Kumerloeve (1972) published a map in which he reported 12 confirmed colonies and 7 possible ones, but he did not estimate the population. Vaughan (1980) found a small unknown colony of about 50 pairs in the Aegean; he also stated that all the Aegean colonies were small, but those of Crete. Wink et al. (1982), who studied the colony of Paximada (Dionisiades, Crete), think that the largest colonies of the Southern Aegean exist in these isles holding about 1,600 pairs. Round & Swann (1977) had found 63 nests of Cory's Shearwaters in the I. Dragonada (Dionisiades) and probably that colony is constituted by some hundreds of pairs (Massa, unpubl.). James (1984) estimates 5,000 pairs all over Greece.

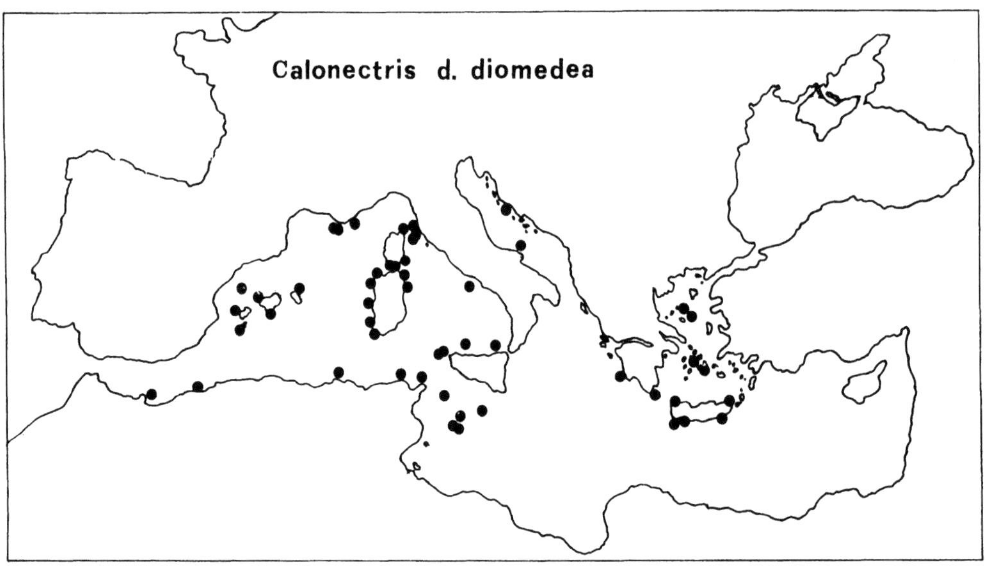

Fig. 1: Mediterranean distribution of the Cory's Shearwater.

Adriatic Sea. According to Matveiev & Vasic (1973) the Cory's Shearwater breeds certainly along the dalmatian coasts, where, according to Krpan (1962-64 in Vaughan,1980) there should be only small colonies. The only Italian breeding site in the Adriatic is the I. Tremiti, from which the name "Diomedea" of this species came (Vaurie, 1965). Neither Di Carlo (1963) nor Spina (1982) estimate that population, but the latter makes us understand that it consists of hundreds of pairs. Census is difficult because of inaccessibility of the breeding site.

Sicilian Channel. Here we can find the largest colonies of the whole Mediterranean. Cachia Zammit & Sultana (pers.comm.) estimate about 8-10,000 pairs in the Maltese Islands. We have no data for the I. Pantelleria, but some observations at dusk in 1978, 1982 and 1985 make us think that at least 1,000 pairs breed there. In the I. Linosa (Pelagic Isles), where Vaughan (1980) estimated from 1,000 to 5,000 pairs, we censused at least 10,000 pairs with the capture-recapture method (see Appendix 1). It seems that a few pairs breed in the I. Lampedusa (Pelagic), whereas in Lampione, an islet off Lampedusa, at least 300-400 pairs still breed (Massa, unpubl.). Massa (1973) recorded it as abundant in the I. Egadi. The largest colony of these islets breeds on I. Marettimo; The colony is accessible, though with difficulties, and consists of hundreds of pairs. The largest colony of the Sicilian Channel lives on I. Zembra. Gaultier (1981) estimates this colony between 20 and 25,000 pairs.

Tyrrhenian Sea. The islets along the northern coast of Sicily are breeding sites of the Cory's Shearwater. It breeds on I. Ustica and I. Aeolian (small populations). There are no numbers for the populations of the islets of Tuscany and Latium (Brichetti, 1979), but it seems that while the Cory's Shearwater is plentiful in the Sicilian Channel, in the Tyrrhenian sea the Manx Shearwater (Puffinus puffinus) is more abundant.

Sardinia and Corsica. Schenk & Torre (1986) estimate 2,500-4,000 pairs in the islets and along Sardinian coasts. Guyot et al. (1985) estimate the whole Corsican population between 350 and 490 pairs.

Mediterranean French coasts. Breeding has been confirmed only in the islets of Marseille and Hyères, where Guyot et al. (1985) estimate a population between 460 and 570 pairs.

Mediterranean Spanish coasts. According to de Juana (1984) there is no evidence of breeding along the Spanish coasts. Therefore the only confirmed colonies are in the I. Columbretes (some hundreds of pairs) and in the Balearic isles (5,000-10,000 pairs, including those of Cabrera, where Araujo et al. (1976) censused about 300 pairs in 1974).

Coasts of Algeria and Morocco. Jacob & Courbet (1980) found two colonies, one in some islets near Collo and another in the I. Habibas near Oran. The former consists of about 20 pairs, while the latter is larger, but it has not been censused. Along the moroccan coasts Berthon & Berthon (1984) report it only as

		SELVAGES	CHAFFARINAS	CABRERA	TYRRHENIAN SEA	ZEMBRA	LINOSA	MALTA	PAXIMADA
WEIGHT grams	males + females	945 n=52	664 n=29	641 (63.3) n=63	n.a.	639 n=121	623 n=375	606 n=43	552 n=211
	males	1014 (55.9) n=26	717 (65.6) n=13	629 n=3	n.a.	703 n=73	675 (59.4) n=187	n.a.	586 (58.7) n=113
	females	877 (61.6) n=26	622 (34.7) n=16	510 n=2	n.a.	575 n=49	577 (48.5) n=178	n.a.	514 (64) n=98
WING mm	males + females	361 n=31	355 n=27	351 (8.5) n=66	344 (9.2) n=19	349 (6.5) n=218	351 n=114	346 n=39	337 n=127
	males	363 n=20	361 (7.5) n=11	345 n=3	352 (3.4) n=8	355 (6.1) n=109	356 (9.3) n=59	n.a.	342 (7.6) n=65
	females	358 (4.5) n=11	351 (7.1) n=16	340 n=2	339 (8.4) n=11	343 (6.9) n=109	345 (9.6) n=55	n.a.	333 (8.4) n=62
TARSUS mm	males + females	55.7 n=113	55 n=21	51.9 n=66	47.7 (1.8) n=18	n.a.	53.2 n=114	54.6 n=34	52.7 n=73
	males	57 (1.2) n=52	56.8 (1.3) n=7	n.a.	48.4 (1.1) n=8	n.a.	53.6 (3.2) n=59	n.a.	53.4 (1.7) n=35
	females	54.6 (1.3) n=61	54.1 (1.1) n=14	n.a.	47.2 (2.1) n=10	n.a.	52.8 (2.1) n=55	n.a.	52 (1.2) n=38
BILL LENGTH mm	males + females	54 n=112	50.1 n=29	50.6 n=66	50.4 (2.1) n=18	48.8 (1.6) n=218	52.1 n=114	49.5 n=34	47.9 n=79
	males	55.5 (1.7) n=52	52.2 (2) n=12	50.8 n=3	52.3 (1.7) n=7	50.7 (1.5) n=109	53.6 (1.8) n=59	n.a.	49.5 (1.4) n=40
	females	52.8 (1.8) n=60	48.7 (1.8) n=17	47.9 n=2	49.2 (1.3) n=11	46.9 (1.6) n=109	50.6 (1.6) n=55	n.a.	46.2 (1.2) n=39
BILL DEPTH mm	males	n.a.	14.7 (1) n=19	13.3 n=3	12.9 (.9) n=7	13.8 (.6) n=109	13.7 (.7) n=58	n.a.	12.9 (.7) n=65
	females	n.a.	12.1 (.6) n=17	11.4 n=2	11.7 (.6) n=11	12.1 (.5) n=109	12.0 (.5) n=51	n.a.	11.6 (.4) n=58

Tab. 1: Biometrics of Cory's Shearwater from Mediterranean and Atlantic Isles (Selvagen).
In brackets: s.d.
n.a. = figures not available
Selvagen: Zino, 1971; Chafarinas: Witt et al., 1984; Cabrera: Araujo et al., 1977; Tyrrhenian Sea (skins): original; Zembra: Gaultier, 1981; Linosa: Iapichino et al., 1983 and original; Malta: Sultana, pers.comm.; Paximada: Ristow & Wink, 1980.

possible, but they could not search in the islets at night. The only moroccan known colony to day is that of I. Chafarinas, quoted by de Juana et al. (1980). In 1984 de Juana et al. estimate it in the order of thousands of pairs. They observed 3,600 individuals at dusk on 6 June 1981. Witt observed in April about 9,000 individuals (de Juana, pers.comm.).

Some considerations about the Mediterranean population

In the Mediterranean basin there are 50-60 colonies generally localized on islets. The population numbers of only 27 colonies have been estimated. In these 27 colonies at least 60,000 pairs are known to breed, whereof 40,000 pairs in the islets of Zembra and Linosa. Since the colonies not censused are small, we think that less than 100,000 pairs would breed in the Mediterranean. We also have to consider the immatures which occur in the colonies during the breeding cycle. According to Mougin & Stahl (1982) in the I. Selvagen the immatures are slightly more numerous than the adults, whereas according to Araujo et al. (1976) in I. Cabrera they may be three times as many immatures as the number of adults. We were not able to estimate the ratio immatures/adults in the colony of Linosa, but we consider a ratio of 0.5/1 reliable (see Appendix 2). In this case the total mediterranean population (including immatures) could be over 250-300,000 individuals. For the time being it is not possible to estimate them more accurately.

It is useful to remember that between 22nd October and 15th November 1977 Telleria (1980) observed 33,355 Cory's Shearwaters to leave the Mediterranean through the Strait of Gibraltar, and estimated a passage of over 150,000 individuals by the means of sample counts. It is noteworthy that: first Cory's Shearwater also migrate during the night, secondly Telleria observed only individuals along the Spanish coast and thirdly when he left the observatory place, migration was still going on. On the ground of that we believe that many more than 150,000 individuals migrated through the Strait in that period.

Telleria (1980) observed two different peaks, one between 23 and 30 October and another between 5 and 11 November, probably corresponding respectively to adults and juveniles. If we consider that the breeding pairs lay every year one egg, with a breeding success of 0.8 (0.64-0.93: Araujo et al., 1977; Gaultier, 1981; Thibault, 1985; Wink et al., 1982), the juveniles of the second peak should be at least 50,000.

In conclusion by assessing the above data and considering also that some individuals winter in Sicilian waters (Brichetti, 1979; Lo Valvo & Massa, unpubl.) and in the Balearic sea (Muntaner et al., 1984), it seems that the mediterranean population may be over 300,000 individuals.

- <u>Biometrics of the population</u>

Tab. 1 and fig. 2 show the values of the Mediterranean and Atlantic populations. Tab. 2 gives the level of significance of statistical difference between the pairs of populations.

<u>Wing</u>. The largest individuals live in the I. Selvagen. The individuals of the I. Chafarinas have the longest wing of the Mediterranean, whereas those of Aegean (Dionisiades) have the shortest one. Other populations have intermediate measurements, statistically different from each other. Only Zembra and Linosa have populations statistically not distinguishable by the wing length.

	WING		TARSUS		BILL LENGTH		BILL DEPTH		WEIGHT	
	males	females	males	females	males	females	males	females	males	females
SELVAGEN CHAFARINAS	-	P=.012	N.S.	N.S.	P>.001	P>.001	-	-	P>.001	P>.001
SELVAGEN TYRRHENIAN	-	P>.001	P>.001	P>.001	P>.001	P>.001	-	-	-	-
SELVAGEN ZEMBRA	-	P>.001	-	-	P>.001	P>.001	-	-	-	-
SELVAGEN LINOSA	-	P>.001	P>.001	P>.001	P>.001	P>.001	-	-	P>.001	P>.001
SELVAGEN PAXIMADA	-	P>.001	P>.001	P>.001	P>.001	P>.001	-	-	P>.001	P>.001
CHAFARINAS TYRRHENIAN	-	P>.001	P>.001	P>.001	N.S.	N.S.	P>.001	N.S.	-	-
CHAFARINAS ZEMBRA	P=.008	P>.001	-	-	P=.002	P>.001	P>.001	N.S.	-	-
CHAFARINAS LINOSA	N.S.	P=.018	P=.011	P=.027	P=.02	P>.001	P>.001	N.S.	N.S.	P=.001
CHAFARINAS PAXIMADA	P>.001	P>.001	P>.001	P>.001	P>.001	P>.001	P>.001	P>.001	P>.001	P>.001
TYRRHENIAN ZEMBRA	N.S.	N.S.	-	-	P>.001	P>.001	P>.001	P=.014	-	-
TYRRHENIAN LINOSA	N.S.	N.S.	P>.001	P>.001	N.S.	P=.008	P=.007	P=.008	-	-
TYRRHENIAN PAXIMADA	P>.001	P>.001	P>.001	P>.001	P>.001	P>.001	N.S.	N.S.	-	-
ZEMBRA LINOSA	N.S.	N.S.	-	-	P>.001	P>.001	N.S.	N.S.	-	-
ZEMBRA PAXIMADA	P>.001	P>.001	-	-	P=.02	P=.008	P>.001	P>.001	-	-
LINOSA PAXIMADA	P>.001	P>.001	N.S.	P=.034	P>.001	P>.001	P>.001	P>.001	P>.001	P>.001

Tab. 2: Levels of statistically significant probability among different populations of Cory's Shearwaters (test t of Student).

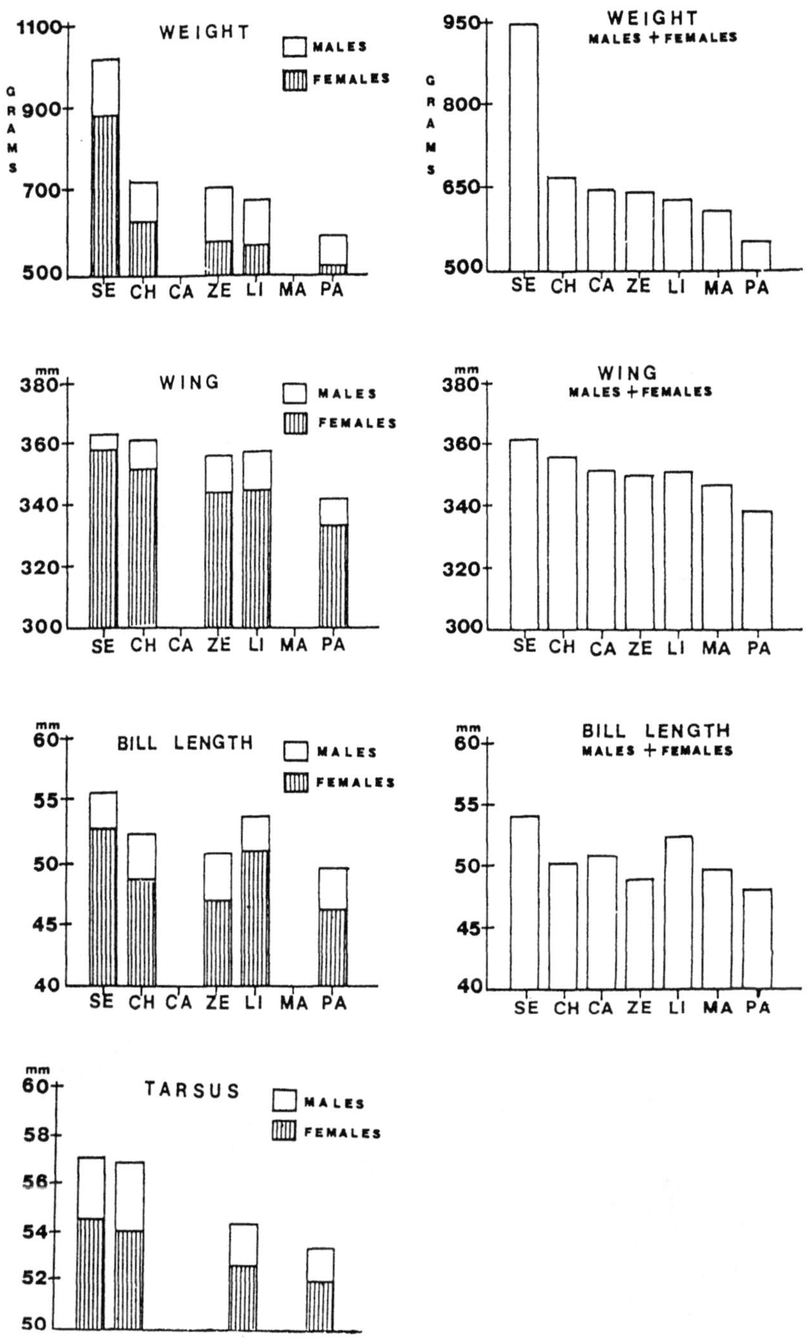

Fig. 2: Geographical trend of some biometrical parameters of the Cory's Shearwater.
SE = Selvagen; CH = Chafarinas; CA = Cabrera; ZE = Zembra; LI = Linosa; MA = Malta; PA = Paximada.

Tarsus. The individuals of the I. Selvagen and Chafarinas have no statistically different tarsi, but they are longer than those of other populations. The specimens of the Tyrrhenian sea, probably coming from colonies of Corsica or Sardinia, have the shortest tarsi. Their measurements were taken from skins and the results may be due to this reason. Similarly the individuals measured coming from Cabrera have also quite short tarsi.

Bill. The total bill length (b.l.1 of fig. 3) proved to be statistically different in all populations, excluding the comparison between those of Chafarinas and the skins of Tyrrhenian origin and also between the individuals of Linosa and the same skins (males only).

It is interesting to observe that this was the only statistically different parameter between Zembra and Linosa. The bill of individuals of Linosa resulted to be rather long and one could think that there were some errors while taking the measurements. However we compared two samples of measurements taken independently by each of us at Linosa, and we did not find any appreciable difference (0.6 mm). According to Busse (1983) a possible error in measuring is in the order of 1 mm. We can therefore conclude, unless both of us made a mistake, that the bill of individuals of Linosa is longer than the one of Shearwaters of Zembra.

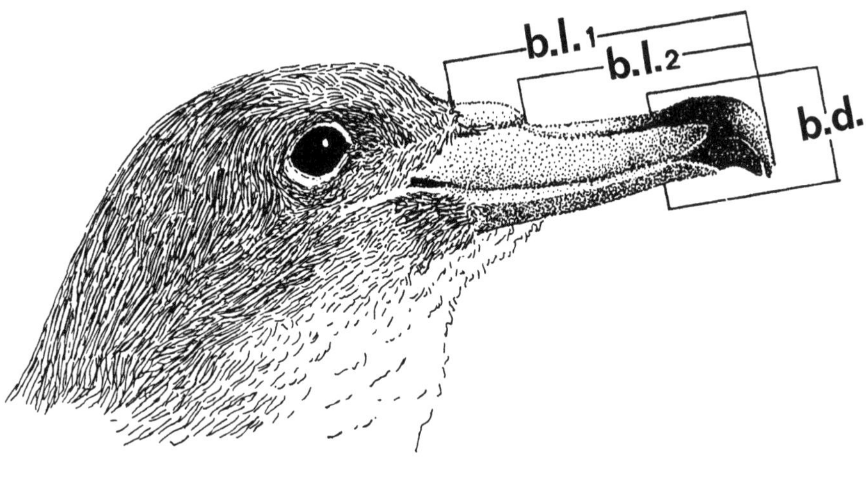

Fig. 3: Bill measurements of the Cory's Shearwater. b.l.1 = total bill length; b.l.2 = bill length from the nostril to the tip; b.d. = bill depth.

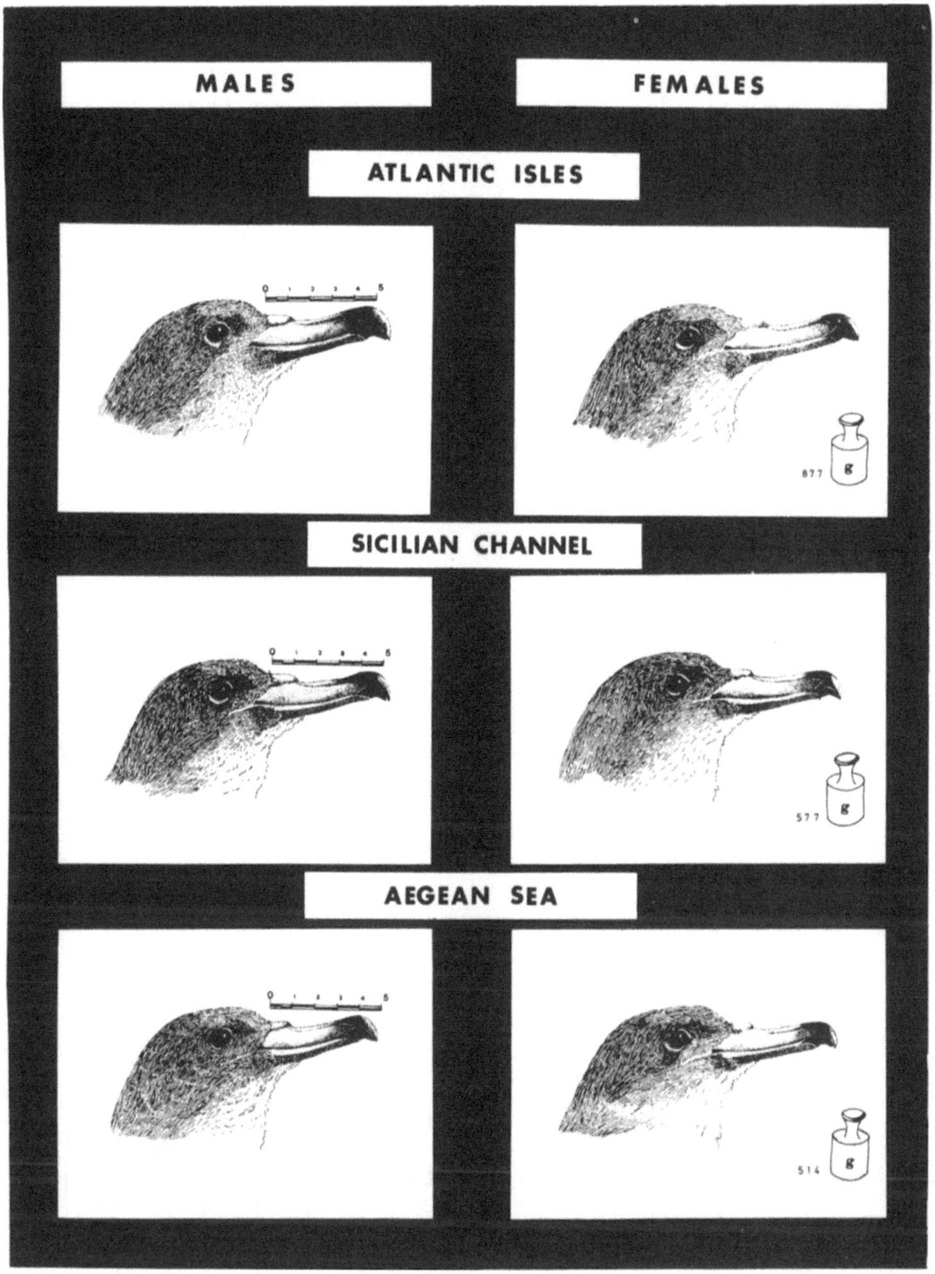

Fig. 4: Bill proportions of the Atlantic (Selvagen) and Mediterranean (Linosa and Dragonada) Cory's Shearwaters. See also tab. 1.

The second bill length (b.l.2 of fig. 3) was measured of only the individuals at Linosa and Paximada (Iapichino et al., 1983; Wink et al., 1982) as well as of the Tyrrhenian skins (40.8 ± 1.3 and 38.4 ± 1.2 respectively for the males and the females of Linosa; 38.6 ± 1.2 and 36.5 ± 0.9 for those of Paximada; 39.6 ± 0.6 and 37.8 ± 0.7 for the skins of the Tyrrhenian). We found a statistically significant difference between the values of Linosa and Paximada ($P < 0.001$ for both sexes), between those of Linosa and Tyrrhenia ($P < 0.05$ for the males and $P < 0.1$ for the females) and between those of Paximada and Tyrrhenia ($P < 0.05$ for the males and $P < 0.001$ for the females).

The bill depth (b.d. of fig. 3) of only some populations was taken and it proved to be statistically different. The skins of the Tyrrhenian sea resulted to have a bill less thick. A statistical difference between these and the individuals of Paximada was not found. Sexual dimorphism in bill depth is not evident. Thus statistical difference was only tested for males from Zembra, Linosa and the Tyrrhenian sea.

In fig. 4 the heads of males and females of the Atlantic isles (Selvagen), Sicilian Channel (Linosa) and Aegean Sea (Dragonada) are drawn in order to show the proportional differences of bill measurements.

The product bill length x depth is thought to be a good index to separate the sexes (Ristow & Wink, 1980; Wink et al., 1982; Witt et al., 1984). In fact, as shown in fig. 5 and tab. 3,

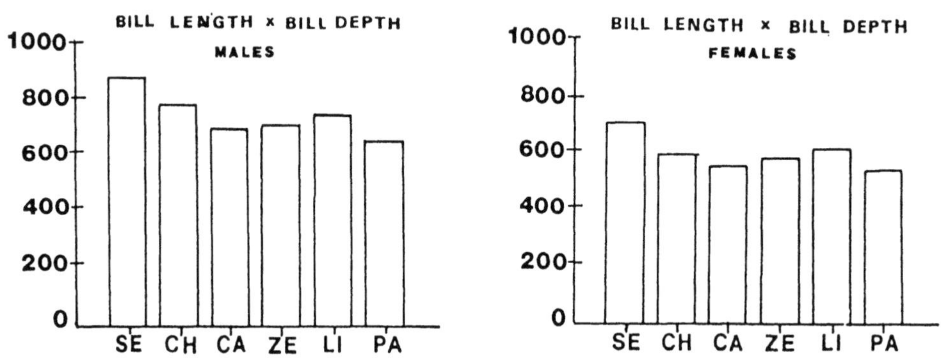

Fig. 5: Bill length x depth of some Cory's Shearwater populations. SE = Selvagen; CH = Chafarinas; CA = Cabrera; ZE = Zembra; LI = Linosa; MA = Malta; PA = Paximada.

this product allows both to separate the sexes and the populations. Data from Cabrera are very similar to those of Paximada, but the sample of Cabrera is too small for any statistical analysis. It is interesting to observe that also the skins of the Tyrrhenian males score low values. The trend is different for the females, excluding the specimens of Cabrera and Paximada, which show the smallest values.

	SELVAGEN	CHAFARINAS	CABRERA	TYRRHENIAN SEA	ZEMBRA	LINOSA	PAXIMADA
Males	866 (791-949)	767 (688-851)	676 (633-718	675 (608-745)	701 (652-751)	734 (673-798)	638 (587-692)
Females	700 (672-721)	589 (539-655	548 (515-581)	576 (532-621)	570 (529-613)	607 (563-652)	536 (504-569)

Tab. 3: Values of bill length x bill depth of different populations of Cory's Shearwaters (mean, minimum and maximum).

Weight. Data of tab. 1 are statistically different for all the pairs taken into account, except for those concerning the males of the Chafarinas and Linosa (tab. 2). Yet the weight values fluctuate from year to year and throughout the year.

At Zembra Gaultier (1981) observed the following trend during 1978 and 1979: in May he found the maximum weight dimorphism of the sexes (185.6 g), probably due to the weight decrease of the female for egg laying; the female weight increased in July and the sexual dimorphism diminished to 111.6 g. At last, in October, both sexes lost weight and sexual dimorphism lowered to 86.7 g.

In I. Linosa we observed three different trends in three different years (see also tab. 4 and fig. 6).

1982. In April we found a weight dimorphism of the sexes of 118 g, which increased in May to 139 g, because of the weight decrease of the females. In August the dimorphism was only 65 g: the females had put on weight, while the males got lighter.

1983. The weight dimorphism was 94 g in April, and 102 g in the first days of September. In comparison to the same period of 1982, in April the males were 6% heavier and the females 10%, while in September males and females were respectively 9% and 5% heavier than in August 1982.

1984. In April the weight dimorphism was still of 94 g, as the previous year, but the weight of both sexes was very low as com-

		1982	1983	1984
APRIL	males	705	747	636
	females	587	653	542
MAY-JUNE	males	694	-	666
	females	555	-	559
AUGUST-SEPTEMBER	males	678	747	701
	females	613	645	579

Tab. 4: Trend of weight of the Cory's Shearwater at Linosa (Sicilian Channel) in 1982, 1983 and 1984.
1982: 25-30 April; 21-29 May; 20-30 August.
1983: 10-15 April; 10-14 September.
1984: 20-30 April; 4-6 June; 20-30 August.

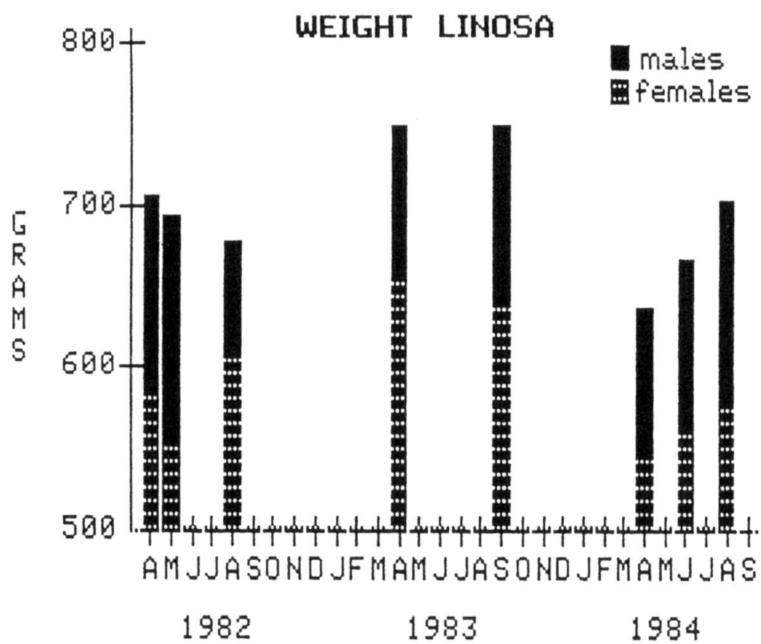

Fig. 6: Trend of weight (males and females) in different years, taken on the population of Linosa.

pared to 1983 (10% less than in 1983 for the males and 8% for the females). In the first days of June weight increased and the dimorphism was 107 g. In August, weight still increased especially among males and the dimorphism was found to be 122 g, the highest observed in this period at Linosa.

We can conclude that, since weight depends on the availability of food in the wintering and breeding areas, this statistical comparison may be pointless. It is probable that, especially in some difficult years, the long period of courtships preceding laying effects the increase of fat needed for the metabolic activity that Cory's Shearwaters are undergoing during the following months.

Ratio egg weight/female weight

Jouanin (1976) thought that there was no difference about this ratio between the population of the Selvagen and that of Zembra. On the contrary the data today available seem to confirm the difference on the ratio between the populations of the Selvagen and the Mediterranean (tab. 5). Jouventin & Mougin (1981) give the following linear regression between the logarithm of the egg weight and that of the adult for the Procellariiformes: $y = 0.774\ x^{0.722}$. The values for the Cory's Shearwaters give us the following regression: $y = 0.415\ x^{0.55}$ ($r = 0.87$; $P < 0.05$). Generally the ratio egg weight/incubating female weight of mediterranean populations is more similar to that of some Procellariiformes as Puffinus and Fulmarus than the ratio of Atlantic Cory's Shearwaters (see the Appendix of Jouventin & Mougin, 1981). This characterizes even more the smaller body size of Mediterranean individuals in respect to Atlantic ones.

SELVAGEN	11.8	(Zino, 1971)
ZEMBRA	14.6	(Gaultier, 1981)
LINOSA	14.5	(Lo Valvo & Massa, unpubl.)
PAXIMADA	14.9	(Ristow & Wink, 1981)

Tab. 5: Ratio egg weight - female weight x 100

Concluding considerations about biometrics

Iapichino et al. (1983) hypothesized a clinal variation from Atlantic isles towards the Eastern Mediterranean. Some data here discussed seem to confirm this hypothesis. However there are some populations that do not follow this rule, e.g. those of the Balearic Islands and perhaps of Corsica or Sardinia (Tyrrhenian skins of tab. 1) which seem to be smaller than the other Western Mediterranean.

Statistical differences in biometrics should show that every population is genetically closed, probably without any relationship or exchange with other populations. Moreover, the breeding site fidelity of juveniles and adults, observed in the I. Selvagen (Jouanin et al., 1977) and the breeding site fidelity of adults in I. Linosa (Lo Valvo & Massa unpubl.) seem to be good reasons for considering at least the largest colonies closed or isolated, probably as isolated as Puffinus p. yelkouan compared to P. p. mauretanicus.

• Fishing areas of the Mediterranean

Probably the size of colonies and the biometrics of populations are correlated to food availability during the breeding cycle. From FAO (1983) we come to know the annual catches of fishing in mediterranean areas for a period of ten years. Tab. 6 gives the average values. There are no differences between the data of the Western and Central Mediterranean, whereas there are remarkable differences with the Eastern Mediterranean. If these annual fishing values can explain the relatively low number of the Cory's Shearwater in the Aegean Sea, they do not explain

		MEDITERRANEAN			
		Western	Central	Eastern	Black Sea
All the species	$\bar{x}$	108,675	157,238	45,087	496,510
	s.d.	54,501	25,283	18,970	115,173
Clupeidae	$\bar{x}$	34,450	30,847	9,329	99,869
	s.d.	25,360	12,475	3,744	38,279
Engraulis encrasicholus	$\bar{x}$	16,212	22,369	7,898	299,040
	s.d.	10,406	13,301	2,013	85,559
Cephalopoda	$\bar{x}$	6,339	12,832	1,274	-
	s.d.	3,043	1,297	642	-
All coastal pelagics	$\bar{x}$	64,508	62,595	19,867	447,446
	s.d.	40,162	24,288	10,752	107,783

Tab. 6: Annual catches by major groups and by fishing areas in metric tons for a period of ten years (from FAO, 1983). Western Mediterranean = from the Strait of Gibraltar to Cape Bon, Northern Sicily and Italy. Central Mediterranean = Sicilian Channel, Ionian Sea and Adriatic Sea to Greece, excluding Aegean Sea. Eastern Mediterranean = Aegean Sea and Levant Sea.

the particular concentration of colonies in the Sicilian Channel. Besides it seems that a great percentage of caught fish is thrown back to the sea. In the Sicilian Channel at least 50,000 tons of fish are thrown back from the boats of Mazara del Vallo (F. Andaloro, pers. comm.). These figures are not included in the official statistical data.

In the Sicilian Channel the populations of Zembra (Gaultier, 1981) and Linosa (Sarà, 1983) are feeding especially on superficial pelagic fishes with slender forms, weighing no more than 80 g (Sarà, 1983). In September-October 1982 Sarà (1983) found the largest concentrations of Cory's Shearwaters in the Sicilian Channel at about 60km (Sarà, 1983, was able to establish that they came from Linosa, finding among them an individual ringed in that isle) and even 250 km southwards Linosa. According to Gaultier (1981) one adult ringed while incubating on 30 May 1974 in the I. Zembra, was captured in the month of September of the same year (while it had to attend the chick) off the Capo Murro di Porco (Siracusa), 250 km from Zembra.

These distances from the breeding sites are rather below those that can be covered by a Cory's Shearwater. It has been observed that Mediterranean Shearwaters feed their chicks on the average once every two days (Thibault, 1985; Lo Valvo & Massa, unpubl.). Since a chick is being fed in turns by male and female, each adult has four days to go to the feeding areas and back with the food concentrated on its oil. Mougin et al. (1984), by Pennycuick's (1969) formulas, estimate 30.7 km/h the minimum-power speed (= Vmp) (that is the speed at which energy is used most slowly) of the Selvagen Cory's Shearwaters. Mediterranean Cory's Shearwaters are smaller both in weight and wing surface, and their Vmp, according to Pennycuick's (1969) formulas, results in only 26.9 km/h. If a Shearwater flies for 12 hours per day, it may cover 650 km every two days. As we observed earlier, the real covered distance is smaller, probably because of the great availability of food in the Sicilian Channel.

Tab. 6 also shows some fishing data in the Black Sea. This area is particularly fertile, as it results from the catches of *Clupeidae* and *Engraulis encrasicholus*. Many Manx Shearwaters coming from the Aegean Sea have been observed in the Black Sea during the spring-summer months (15-20,000, but also up to 30,000: Van Impe, 1975), whereas the Cory's Shearwater was rarely observed there (see Kumerloeve, 1972, note 2). This could show the proportion of the two Shearwater populations in the Aegean Sea.

• Notes about the breeding cycle

The Cory's Shearwater is among the species of Procellariiformes with a highly synchronous breeding cycle. In the Mediterranean basin laying begins on 19-20 May and ends in the first days of June. In the I. Selvagen, and probably in the Azores and Canaries, laying is a week later. In tab. 7 we report the available data about the laying and hatching periods and the

	Laying period	Incubation days	Hatching period	References
SELVAGEN	27 May - 6 June	53.8	17 - 19 July	(Zino, 1971)
CABRERA	25 May - 6 June ?	55 ?	18 - 29 July ?	(Araujo et al., 1977)
CORSICA	20 May - 2 June $\bar{x}$ = 26 May	51.3	14 - 26 July	(Thibault, 1985)
ZEMBRA	19 May - 2 June $\bar{x}$ = 27 May	50.8	10 - 24 July	(Gaultier, 1981)
LINOSA	20 May - 3 June $\bar{x}$ = 27 May	51	10 - 25 July	(Lo Valvo & Massa, unpubl.)
CRETA	20 May ? - 1 June	51 ?	13 - 22 July	(Wink et al., 1982) (Round & Swann, 1977)

Tab. 7: Dates of laying and hatching and days of incubation for six different populations of Cory's Shearwater. ? = figures not certain.

time required for incubation. Only figures of Cabrera are not within the Mediterranean range, but since Araujo et al. (1977) controlled the laying of the colony of Cabrera only from 22 to 27 May 1974, their data do not have to be considered definitive.

In literature we found only one date disagreeing with those reported in tab. 7: Giglioli (1907) wrote to have found on the 10th August 1882 some adults incubating on the islet of Lampione. Is this a mistaken date?

In the Mediterranean Sea the number of incubation days is lower than in the I. Selvagen (51.3 ± 1.12 and 53.8 ± 0.83). This difference proves to be statistically significant (*t* of Student; $P < 0.001$).

We remember that generally there is a significant correlation between the weight of the adult and the number of incubation days; for the Procellariiformes $r = 0.83$, $P < 0.001$: Jouventin & Mougin (1981). We also found a good correlation for the Atlantic and Mediterranean Cory's Shearwaters ($r = 0.98$, $P < 0.001$).

Appendix 1 - Census of the Cory's Shearwaters in I. Linosa (Sicilian Channel) by capture-recapture method per sample-area

• Method

Census was performed in 1982 in I. Linosa. 855 adults of Cory's Shearwater have been ringed near their nests in an area estimated at least 1/10 of the total area of the colony (excluding some smaller colonies located in different areas of the isle). During ringing, along a pre-arranged route, all the ringed and recovered individuals were recorded (200 in total). Afterwards the elaboration method of Petersen-Lincoln (Begon, 1979) was applied. It consists of taking from a population, whose total amount N is unknown, a sample of M individuals. These are ringed and released. Subsequently in the same population n individuals are recaptured, but only m of them ringed and $n-m$ unringed. If all the individuals have the same probability to be randomly captured, the proportion of the ringed individuals of the second sample is an estimate of the proportion of the ringed individuals in the total population (Meunier & Solari, 1979). Then $n/m = N/M$, from which $N = Mn/m$ (index of Petersen-Lincoln).

If we used the capture-recapture method in different nights, it is possible to obtain a mean of different estimates. In this case the sample i ($i = 1,\ldots\ldots,K$), constituted by n_i individuals, contains m_i ringed individuals. Every night the unringed n_i-m_i are ringed and released. Before the i-th sample, population contains N individuals (total number), whose M_i are ringed. Calculation may be performed by the mean of estimates of the population for each capture-recapture sample. Formulas have been obtained by Chapman (1951 in Meunier & Solari, 1979) and Seber (1973 in Begon, 1979) from Petersen-Lincoln method.

• Results

The two formulas of the tab. 8 give respectively the values of 1,622 and 1,466 individuals, that is 700-800 pairs in the sample-area. If we consider that sample-area is at least one tenth of the total area of the colony, the total number of pairs may be estimated of at least 7-8,000 pairs. Since the Cory's Shearwater breeds in other areas of Linosa, an estimate of 10,000 pairs for the whole islet is not excessive.

As it is custom of Linosa inhabitants to take the eggs of the Cory's Shearwater, we can calculate figures for some areas of the islet. So, at least 400 eggs have been collected in the sample-area (the accessible part) and 500 in another colony outside the study area. Thus our estimate of 10,000 pairs for the whole islet should result reliable.

Appendix 2 - Estimate of the immatures within a colony

It is $P = Eka/1-b$, where P is the total number of the population, k is the rate of survival of immatures in the first life year, E is the number of laid eggs by the considered popu-

Jacob, J.P. & Courbet, B. (1980). Oiseaux de mer nicheurs sur la côte algérienne. Gerfaut, 70: 385-401.

James, P.C. (1984). The status and conservation of Seabirds in Mediterranean Sea. Pp. 371-375 in: Croxall, J.T.; Evans, P.G.H. & Schreiber, R.W., Status and conservation of the World's Seabirds. ICBP Technical Publ. n°2.

Jouanin, C. (1976). Note sur la biométrie des Puffins cendrés de Tunisie. Oiseau et Rev. fr. Orn., 46: 97-102.

Jouanin, C.; Roux, F. & Zino, A. (1977). Sur les premiers résultats du baguage des Puffins cendrés aux îles Selvagen. Oiseau et Rev. fr. Orn., 47: 351-358.

Jouventin, P. & Mougin, J.L. (1981). Les stratégies adaptatives des oiseaux de mer. Rev. Ecol., 35: 217-272.

Kumerloeve, H. (1972). Le statut du Puffin cendré Procellaria diomedea Scopoli dans le golfe saronique (Grèce). Alauda, 40: 102-104.

Massa, B. (1973). L'avifauna estiva degli arcipelaghi delle Egadi e dello Stagnone (Trapani, Sicilia). Atti Accad. Gioenia sc. nat. Catania, 5: 63-95.

Matveiev, S.D. & Vasic, V.F. (1973). Catalogus Faunae Jugoslaviae. Aves. Acad. Sc. Art. Slovenica, Ljubljana.

Meunier, M. & Solari, A. (1979). Estimation de la densité de population à partir des captures-recaptures: application au campagnol des champs. Mammalia, 43: 1-24.

Mougin, J.L.; Roux, F.; Stahl, J.C. & Jouanin, C. (1984). L'évolution des effectifs des Puffins cendrés Calonectris diomedea borealis de l'île Selvagen Grande (30°09' N, 15°52' W) de 1980 à 1983. Bocagiana, 75: 1-8.

Mougin, J.L. & Stahl, J.C. (1982). Essai de dénombrement des Puffins cendrés Calonectris diomedea borealis de l'île Selvagen Grande. Bocagiana, 63: 1-17.

Muntaner, J.; Escandell, A.; Ramos, E. & Orfila, G. (1984). Adicion y revision faunistica de las especies de la publicacion "Avifauna de Menorca". Pp. 177-205 in: Muntaner, J. & Congost, J., Avifauna de Menorca, 2nd ed. Treb. Mus. Zool., 1: 1-205.

Pennycuick, C.J. (1969). The mechanics of bird migration. Ibis, 111: 525-556.

Ristow, D. & Wink, M. (1980). Sexual dimorphism of Cory's Shearwater. Il Merill, 21: 9-12.

Round, P.D. & Swann, R.L. (1977). Aspects of the breeding of Cory's Shearwater Calonectris diomedea in Crete. Ibis, 119: 350-353.

Sarà, M. (1983). Osservazioni sulla consistenza numerica e sull'alimentazione della Berta maggiore (Calonectris diomedea) nel Canale di Sicilia. Riv. ital. Orn., 53: 183-193.

Schenk, H. & Torre, A. (1986). Breeding distribution, numbers and conservation of Seabirds in Sardinia, 1978-85. in Medmaravis, Proceedings of 1st Mediterranean Seabird Symposium, Springer Verlag, Heidelberg (this volume).

Spina, F. (1982). Note ornitologiche dalle isole Tremiti (Mare Adriatico). Riv. ital. Orn. 52: 110-118.

Sultana, J. & Gauci, C. (1982). A new guide to the birds of Malta. MOS, Malta.

Telleria, J.L. (1980). Autumn migration of Cory's Shearwater through the Straits of Gibraltar. Bird Study, 27: 21-26.

Thibault, J.C. (1985). La réproduction du Puffin cendré Calonectris diomedea en Corse. Pp. 49-55 in: Oiseaux marins nicheurs du Midi et de la Corse. Annales du C.R.O.P., n°2.

Van Impe, J. (1975). Sur les mouvements du Puffinus p. yelkouan en Mer Noire. Alauda, 43: 185-187.

Vaughan, R. (1980). Notes on Cory's Shearwater (Calonectris diomedea) and some other birds on Linosa, Pelagic Isles. Riv. ital. Orn., 50: 143-154.

Vaurie, C. (1965). The Birds of the Paleartic Fauna.Vol. 2 (Non Passeriformes). Whiterby, London.

Wink, M.; Wink, C. & Ristow, D. (1982). Brutbiologie mediterraner Gelbschnabelsturmtaucher (Calonectris d. diomedea). Seevogel, 127-135.

Witt, H.H.; De Juana, E. & Varela, J.M. (1984). Geschlechtsspezifische Grössenunterschiede in einer westmediterranen Population des Gelbschnabelsturmtauchers (Calonectris diomedea). Die Vogelwarte, 32: 234-236.

THE IMPORTANCE OF THE IBERIAN MEDITERRANEAN COAST AS A WINTERING AREA FOR GULLS AND TERNS

Enric Carrera i Gallissa, Consell de Cent 341, 4rt 1a.
08007-Barcelona
Jordi Garcia i Petit, Magdalenes 6, 2on 1a
08002-Barcelona
(Grup Català d'Anellament - Catalan Ringing Group)

Introduction

Until recently little was known of the importance of the Mediterranean coast of the Iberian Peninsula, as a wintering area for gulls and terns. Almost all ideas were based on rather generalized studies and on the results of ringing (Bernis, 1967). However in recent years various people have carried out censuses of varying depth, extent and quality, concentrating mainly on the east coast of Spain, especially Catalonia and the Valencia region (Isenmann, 1972; Isenmann, 1976; Carrera et al., 1981; Carrera, 1983; Vilagrasa et al., 1982; Carrera et al., 1984; Bermejo et al., 1985; Pardo, 1984). Of particular interest is the 1984 overall census of gulls and terns wintering in the Iberian Peninsula, which was carried out in fair depth (Bermejo et al. in press.). This gives us our first general picture of these birds' winter stay in the Peninsula.

The present article aims to present an overall view of the wintering of gulls and terns on the Mediterranean coast of the Iberian Peninsula. It is based on the results of the published censuses of 1977, 1979, 1981 and 1984, and an unpublished one of January 1985, which includes the Balearic Islands for the first time.

Material and Methods

Censuses were carried out at roosts, as the birds move about so much during the day. Although this is the most effective method, it must be emphasized that it is extremely diffi-

NATO ASI Series, Vol. G 12
Mediterranean Marine Avifauna
Edited by MEDMARAVIS and X. Monbailliu

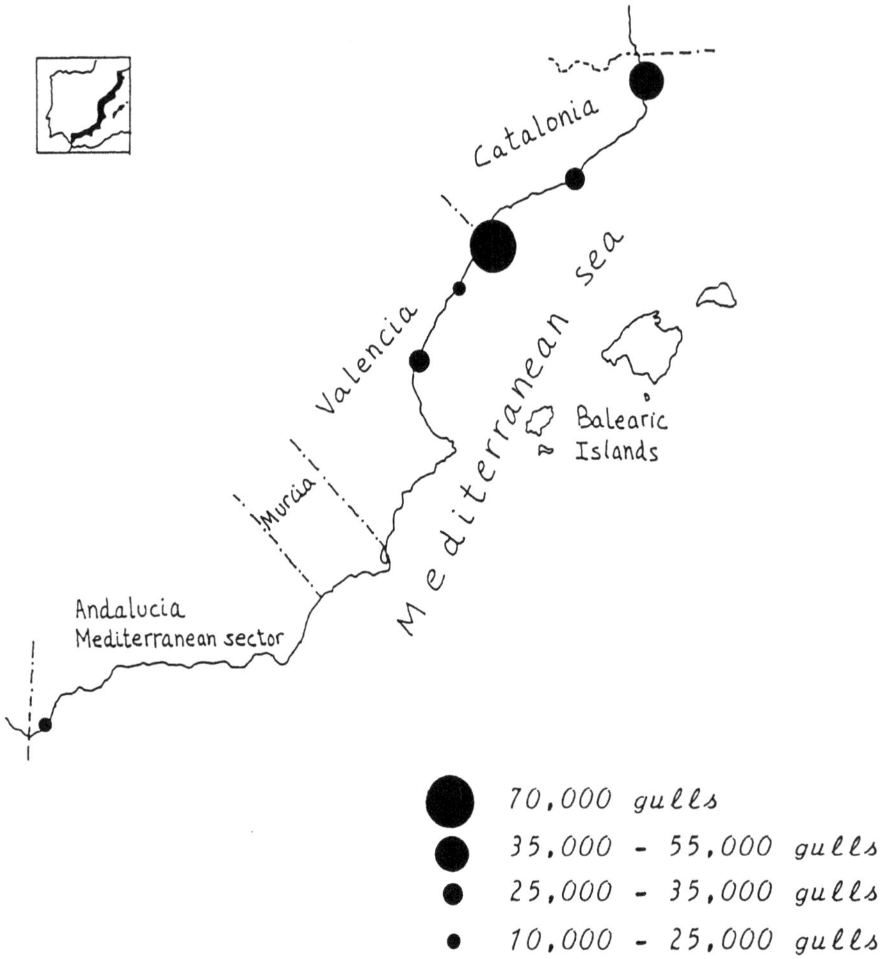

MAIN WINTERING CONCENTRATIONS OF GULLS

cult to locate and census the infinite number of small-scale roosts up and down the coast, especially in the case of a species as common as Larus ridibundus. Thus the increase in numbers, demonstrated by tab. 1, indicates improved methods rather than annual fluctuations in the population size. Again censusing standards greatly influence results in places where a large number of birds are involved (Medes Islands, Ebro Delta, Albufera de Valencia, etc.).

The censuses were performed using telescopes 20-60 x 60 and binoculars 8 x 40 and 10 x 50.

Year	Province	Number of localities	Number of gulls	Number of species
1977	Gi, B, T, Cs	11	60,509	6
1978	Gi, B, T, Cs, V	42	74,637	7
1980	Gi, B, T, Cs, V	43	108,585	6
1981	Gi, B, T, Cs, V	53	95,093	6
1984	Gi, B, T, Cs, V, A Al, Mu, Ma, C	100	213,154	13
1985	Gi, B, T, Cs, V, A, Ba	106	216,811	10

Gi : Girona
B : Barcelona
T : Tarragona
Cs : Castellon
V : Valencia
A : Alicante
Al : Almeria
Mu : Murcia
Ma : Malaga
C : Cadiz
Ba : Balearic Islands

Tab. 1

Overall Results

The maximum and minimum annual totals of gulls and terns counted were 216,811 and 60,509 respectively. However it must be pointed out that no complete census has been carried out; the maximum figure, obtained in 1985, included only Catalonia, the Valencia Region and the Balearic Islands, thus leaving almost half of the coast to be covered.

The total number of wintering gulls and terns has been estimated at a minimum of some 330,000 birds, of which about 60% are Larus ridibundus, 18% Larus cachinnans, 15% Larus melanocephalus, 4.5% Larus fuscus, 0.3% Larus minutus, 0.2% Sterna sandvicensis, 0.3% Larus audouinii; the remainder including

Larus genei, Larus canus, Rissa tridactyla, Chlidonias hybridus, Sterna hirundo, Sterna albifrons and Larus sabini.

The main concentrations are in the following areas: Ametlla de Mar-Ebro Delta-Les Cases d'Alcanar (70,000 minimum); Medes Islands-Roses (55,000); Albufera de Valencia (35,000); the Port of Barcelona and environs (26,000); Mediterranean coast of Cadiz (25,000); and Benicarlo-Peniscola (15,000).

Species by Species Account

● Mediterranean Gull *Larus melanocephalus*

Probably the most interesting species which winters on this coast. Isenmann (1972) discovered the importance of this coast as winter quarters for this gull, counting 9,000 in 1975 (Isenmann, 1975). Later censuses boosted this total to 15,000 - 18,000 birds, though in 1985 (without a count for Malaga) a total of 43,755 was reached, thanks to the discovery of a new roost of some 13,000 birds (confirmed January 1986) at Les Cases d'Alcanar (Tarragona), and a very high total for Castellon.

This species is found almost exclusively in the Provinces of Barcelona, Tarragona and Castellon. The greatest concentrations occur at l'Ametlla de Mar, Les Cases d'Alcanar, the Port of Barcelona and Vinaroz-Peniscola. These four sites are responsible for about 95% of the Mediterranean Gulls of the study area. All contain important fishing ports and are set in areas where there is extensive cultivation of olives, which represent an important source of food for this species. In the province of Malaga, only 49 were counted in 1984, in contrast with a figure for November 1975 of almost 5,000 (Isenmann, 1976), and the large number of ringing recoveries for this province (Bernis, 1967).

The census results demonstrate the great importance of the study area for the wintering of Mediterranean Gulls, with an estimated minimum of 48,000 - 50,000 birds.

● Little Gull *Larus minutus*

Out of the breeding season, this species favours the open sea and thus censuses carried out at roosts (as in most cases in the present article) are bound to give poor results, normally referring to single birds.

However, in the first four censuses notable concentrations (up to 1,270 birds in 1977) were recorded at the Llobregat Delta (Barcelona). These constitute the highest winter totals for the Mediterranean, which Erard (1960) presumed to be the main overwintering area for this species. Absent from Balearic Islands.

● Sabine's Gull *Larus sabini*

One observation only: the Straits of Gibraltar, 1st January 1984. A pelagic species with very few records from the Iberian Peninsula.

● Black-Headed Gull *Larus ridibundus*

By far the commonest and most widespread gull of Spain's east coast. A minimum of some 200,000, which represents 60% of all the gulls and terns censused.

Although it is found all along the coast, about 80% of its numbers are concentrated in Catalonia and the Valencia region (83% of the ringing recoveries in Bernis (1967)). The main concentrations occur near marshes, industrial zones and fishing ports (Ebro Delta, Roses Bay, Albufera de Valencia, Barcelona-Llobregat Delta).

The only species, apart from Yellow-legged gull, found inland, where it is relatively frequent on most rivers, lakes and reservoirs.

● Slender-Billed Gull *Larus genei*

No winter records existed until 1983, 1984 and 1985, when 34, 16 and 80 birds respectively were seen at the Ebro Delta. This coincides with a great boost in the breeding population of this site (Ferrer, 1986), which suggests that part of the Spanish population may be sedentary.

● Audouin's Gull *Larus audouinii*

In our study area several hundred (minimum estimate 400) spend the winter in the vicinity of breeding grounds and post-breeding moult areas. The discovery of a small overwintering population at the Ebro Delta (confirmed January 1986) seems to coincide with the great increase in the breeding population of the same site (Ferrer op.cit.).

● Common Gull *Larus canus*

The mouth of the Garonne (France) represents the southern limit for wintering Common gulls in any great number. However in very cold weather they may move down to the Iberian Peninsula and the Mediterranean (Cramp & Simmons, 1985). For this reason it passes the winter here very rarely and in very small numbers. Only 16 birds have been recorded in the 6 censuses, the greatest concentration being 6 in Catalonia in 1985. Absent from Balearic Islands.

● Lesser Black-backed Gull *Larus fuscus*

Abundant on the Atlantic coast of the Peninsula, with a minimum of some 90,000 (Bermejo et al. in press.). It decreases steadily from Straits of Gibraltar up to the Mediterranean coast being present in some numbers only where there are few Yellow-legged gulls (Tarragona, Castellon, Valencia and the Mediterranean coast of Andalusia).

Its northern limit is in the province of Barcelona, where the Yellow-legged gull's great zone of influence begins (on the south coast of France and the province of Girona, some 35,000 pairs of Yellow-legged gulls breed, which explains why only 5 Lesser Black-backs have been seen in Girona in the 5 censuses and in 1979 only 12 were recorded on the French coast (Isenmann, 1980)). It is rare also in the Balearic Islands. The estimated minimum for the study area is 15,000 birds.

• Yellow-legged Gull Larus cachinnans michahellis

The second most numerous species in the area in winter. Although it can be found all along the coast, it is especially abundant in the immediate vicinity of its breeding grounds, thus 60% of its numbers occur in Catalonia, which boasts the largest colony in the western Mediterranean (Carrera & Vilagrasa, 1985). Nonetheless censuses in Catalonia tend to produce very variable results, well below the real figure, due basically to the difficulties posed by a census of the Medes Islands, where at least 26,000 birds roost. The only census carried out in the Balearic Islands was incomplete and probably well below the true figure. We have always been surprised by the abundance of the species in the Valencia region, especially in the province of Alicante, given the low numbers breeding here (Bermejo & Carrera, 1983).

Yellow-legged gulls may move inland, following rivers and they can be found in lagoons and reservoirs of coastal provinces and also those of one inland province (Lerida). The winter population is estimated at a minimum of 61,500 birds.

• Kittiwake Rissa tridactyla

Tends to favour the open sea, so censuses carried out at roosts are unlikely to detect many. A total of 13 have been recorded in the 6 censuses. Some counts have been carried out at sea (Sargatal & Llinas, 1978; Llorente & Montori, 1982; V. Estrada & J. Garcia,pers.com.) and these suggest that Kittiwakes regularly winter in small numbers around the northern half of the Spanish Mediterranean coast. It is, however, difficult to estimate numbers.

• Sandwich Tern Sterna sandvicensis

Present along the whole coast, although never in large groups. The largest concentrations are at the Ebro Delta, the Llobregat Delta, the Albufera de Valencia, the Mar Menor, Roses Bay and the Costa del Sol, all which are characterized by sandy beaches near wetlands. Virtually never occurs along rocky coasts, a tendency also noted in Algeria (Jacob, 1978).

Not detected in Murcia in 1984, due to inefficient censusing, but has regularly been recorded there in winter (Macivor

& Navarro Medina, 1972). The total population is estimated at a minimum of about 1,000.

• Common Tern Sterna hirundo

The Peninsula represents the European population's northern limit in winter (Muselet, 1982). In the 6 censuses, 2 Common terns were recorded in the Albufera de Valencia and 7 at Almenara (Castellon). However there are previous records of small numbers at the Albufera de Valencia (Erard & Viellard, 1975) and the Ebro Delta (Ferrer, 1977).

• Little Tern Sterna albifrons

Only two records; 1 bird at El Saler (Valencia) and 1 at Calpe (Alicante), both in 1984. No previous records known.

• Whiskered Tern Chlidonias hybridus

Though not included in the censuses, it can be considered a regular in winter, though in small numbers. 15 observations of a total of 83 birds were recorded in the Ebro Delta and the Albufera de Valencia between 1978 and 1983 (Carrera et al., 1983); previous records also exist for these localities (Erard & Viellard, 1966; Isenmann, 1972).

Conclusion

As a wintering zone for gulls and terns, the Mediterranean coast of Spain presents the following characteristics:

- numbers are relatively lower than those of neighbouring countries (France and Italy), but higher than those of the North African coast (Algeria and Marrocco);
- the greatest density occurs in the northern half of the zone (Catalonia and the Valencia Region);
- it is the most important wintering area of the Mediterranean Gull in the western Mediterranean;
- the greatest winter concentrations of Little Gulls so far recorded in the Mediterranean belong to the zone;
- certain species, such as Larus genei, Larus audouinii, Sterna hirundo and Chlidonias hybridus are on the edge of their wintering areas here and are regularly sighted in small numbers.

Six years' experience has demonstrated how difficult it is to perform an efficient census of these birds in winter. As the birds are scattered and move about a lot, a large, well-equipped team is needed and this, in turn, reduces the quality and consistency of the census. We consider that a good knowledge of the zone and the main roosts is essential. For all these reasons, we realize that it is very dangerous to attempt to estimate overall populations of a country or region, merely on the strenght of a sample from one or two years.

Larus melanocephalus
13,000 ~ 15,000 gulls
2,000 ~ 4,000 gulls
500 ~ 2,000 gulls
100 ~ 500 gulls

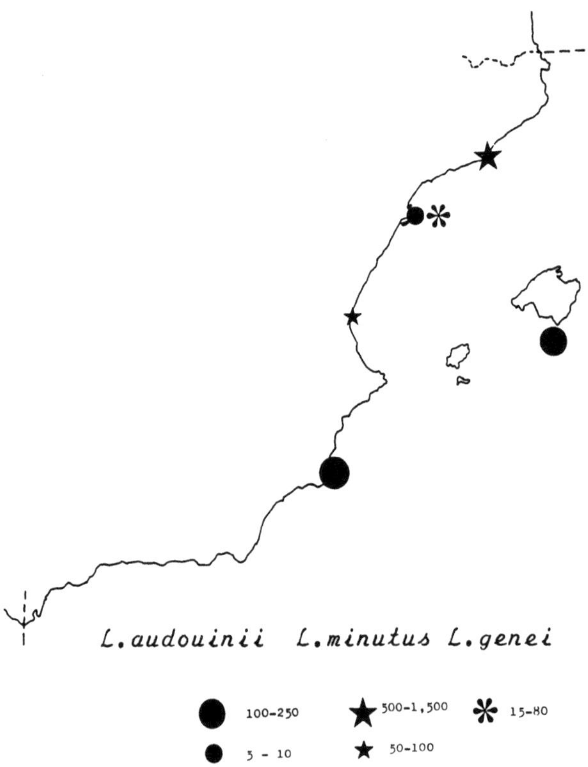
L. audouinii L. minutus L. genei
100-250
500-1,500
15-80
5 - 10
50-100

Census Results

• Mediterranean Gull *Larus melanocephalus*

	1977	1979	1980	1981	1984	1985	Estimated minimum
Catalonia	7,890	10,006	9,198	5,384	13,888	33,030	35,000
Valencia	1,400	150	6,395	7,187	4,156	10,725	13,000
Balearics	-	-	-	-	-	0	0
Murcia	-	-	-	-	0	-	0
Andalucia Mediterranean sector	-	-	-	-	58	-	?
TOTAL	9,290	10,156	15,593	12,571	18,102	43,755	48,000 50,000

• Little Gull *Larus minutus*

	1977	1979	1980	1981	1984	1985
Catalonia	1,270	601	501	852	14	18
Valencia	0	3	0	0	6	60
Balearics	-	-	-	-	-	0
Murcia	-	-	-	-	0	-
Andalucia Mediterranean sector	-	-	-	-	0	-
TOTAL	1,270	604	501	852	20	78

• Black Headed Gull Larus ridibundus

	1977	1979	1980	1981	1984	1985	Estimated minimum
Catalonia	44,885	36,006	39,303	28,279	84,543	70,636	100,000
Valencia	150	14,900	20,018	26,682	44,865	52,677	60,000
Balearics	-	-	-	-	-	4,210	15,000
Murcia	-	-	-	-	745	-	2,000
Andalucia Mediterranean sector	-	-	-	-	13,104	-	20,000
TOTAL	45,035	50,906	59,321	54,961	143,257	127,523	200,000

• Slender-Billed Gull Larus genei

	1977	1979	1980	1981	1984	1985
Catalonia	0	0	0	0	16	80
Valencia	0	0	0	0	0	0
Balearics	-	-	-	-	0	-
Murcia	-	-	-	-	0	-
Andalucia Mediterranean sector	-	-	-	-	0	-
TOTAL	0	0	0	0	16	80

• Audouin's Gull Larus audouinii

	1977	1979	1980	1981	1984	1985
Catalonia	0	0	0	0	0	5
Valencia	0	0	0	0	0	0
Balearics	-	-	-	-	-	216
Murcia	-	-	-	-	111	-
Andalucia Mediterranean sector	-	-	-	-	0	-
TOTAL	0	0	0	0	111	221

• Common Gull Larus canus

	1977	1979	1980	1981	1984	1985
Catalonia	2	2	0	0	1	6
Valencia	0	0	0	0	5	0
Balearics	-	-	-	-	-	0
Murcia	-	-	-	-	0	-
Andalucia Mediterranean sector	-	-	-	-	0	-
TOTAL	2	2	0	0	6	6

L. ridibundus
20,000 - 40,000 gulls
15,000 - 20,000 gulls
10,000 - 15,000 gulls
5,000 - 10,000 gulls

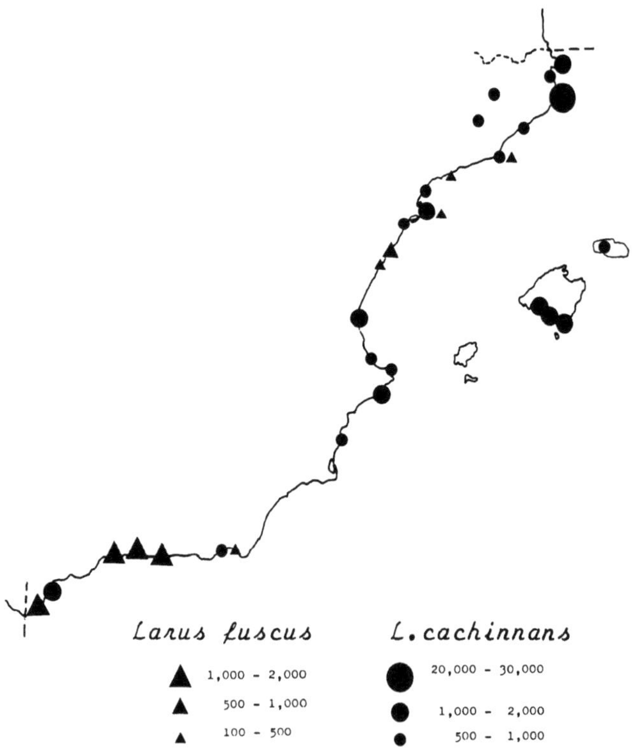
Larus fuscus
1,000 - 2,000
500 - 1,000
100 - 500
L. cachinnans
20,000 - 30,000
1,000 - 2,000
500 - 1,000

• Lesser Black-backed Gull *Larus fuscus*

	1977	1979	1980	1981	1984	1985	Estimated minimum
Catalonia	467	1,063	2,093	1,307	1,566	1,370	3,000
Valencia	128	326	3,303	3,500	1,680	3,070	4,000
Balearics	-	-	-	-	-	5	50
Murcia	-	-	-	-	71	-	150
Andalucia Mediterranean sector	-	-	-	-	5,562	-	7,000
TOTAL	595	1,389	5,401	4,807	8,879	4,445	14,000 15,000

• Yellow-legged Gull *Larus cachinnans michahellis*

	1977	1979	1980	1981	1984	1985	Estimated minimum
Catalonia	4,675	9,136	24,943	18,769	30,220	21,013	35,000
Valencia	22	2,045	2,676	2,485	5,841	7,667	8,000
Balearics	-	-	-	-	-	6,983	15,000
Murcia	-	-	-	-	473	-	500
Andalucia Mediterranean sector	-	-	-	-	2,501	-	3,000
TOTAL	4,697	11,181	27,619	21,254	39,035	35,663	61,500

Sterna sandvicensis
250 terns
15 - 50 terns
1 - 5 terns

• Kittiwake *Rissa tridactyla*

	1977	1979	1980	1981	1984	1985
Catalonia	0	2	0	0	2	3
Valencia	0	0	0	0	3	0
Balearics	-	-	-	-	-	0
Murcia	-	-	-	-	0	-
Andalucia Mediterranean sector	-	-	-	-	3	-
TOTAL	0	2	0	0	8	3

• Sandwich Tern *Sterna sandvicensis*

	1980	1981	1984	1985	Estimated minimum
Catalonia	153	273	204	141	400
Valencia	-	39	160	-	200
Balearics	-	-	-	22	70
Murcia	-	-	0	-	150
Andalucia Mediterranean sector	-	-	70	-	150
TOTAL	153	312	434	141	1,000

Acknowledgements

We would like to thank everybody, who has taken part in the censuses.

References

Bermejo, A. & Carrera, E. (1983). Premières données sur les reprises de Goélands leucophées bagués dans la Péninsule ibérique, aux Baléares et aux Chafarinas. 3ème réun. G.T.O.M. La Rochelle, sept. 1983.

Bermejo, A., Carrera, E., de Juana, E. & Texeira, A.M. (1984). Primer censo de Laridos invernantes en la Peninsula Ibérica, Islas Canarias y Ceuta. La Garcilla 64: 30-32.

Bermejo, A., Carrera, E., de Juana, E. & Texeira, A.M. Primer censo general de gaviotas y charranes (Aves: Laridae) invernantes en la Peninsula Ibérica (enero 1984). in press.

Bernis, F. (1967). Aves Migradoras Ibéricas. Publicaciones especiales de la S.E.O. Madrid 1967.

Carrera, E., Ferrer, X., Martinez, A. & Muntaner, J. (1981). Invernada de Laridos en el litoral mediterraneo catalan y levantino. Ardeola 28: 35-50.

Carrera, E. (1983). Hivernada de larids (Aves, Laridae) al litoral del Principat y Pais Valencia (1980 i 1981). Butll. Inst. Cat. Hist. Nat. 49 (secc. zool.): 143-150.

Carrera, E., Martinez, A., Motis, A. & Pardo, R. (1981) (1983). Nota sobre la invernada del Chlidonias hybrida (Pallas) (Aves, Sternidae) en el Mediterraneo ibérico. Misc. Zool. 7: 222-224.

Carrera, E. & Vilagrasa, X. (1983). La colonia de Gavia argentat Larus argentus michahellis de les illes Medes. in: Els sistemes naturals de les illes Medes. Edited by: J. Ros, I. Olivella & J.M. Gili. Arxius de la seccio de ciències num. 73. I.E.C. Barcelona 1984.

Cramp, S. Ch, Ed. (1983). Handbook of the Birds of Europe in the Middle East and North Africa. Oxford University Press.

Erard, C. (1960). Sur l'aire de reproduction, zones d'hivernage et les migrations de la Mouette pygmée Larus minutus. Alauda 28: 196-228.

Erard, C. & Viellard, J. (1965). Comentarios sobre la avifauna invernal en el oriente espanol. Ardeola 11: 95-100.

Ferrer, X. (1977). Introduccio ornitologica al Delta de l'Ebre. Treb. Inst. Cat. Hist. Nat. 8: 227-302.

Ferrer, X. (1986). Fluctuations of the Gull and Tern populations in the Ebro Delta, NE. Spain (1960-1985). in Medmaravis: Population Studies and Conservation of the Mediterranean Marine Avifauna. 1st Mediterranean Seabird Symposium. Alghero (Sardinia) 26-30 March 1986, Springer Verlag, Heidelberg (this volume).

Isenmann, P. (1972). Notas sobre algunas especias de aves acuaticas en las costas mediterraneas espanolas (enero 1982). Ardeola 16: 242-245.

Isenmann, P. (1975). Contribution à l'étude de la biologie de reproduction et d'écologie de la Mouette mélanocephale Larus melanocephalus. Nos Oiseaux 33: 66-73.

Isenmann, P. (1976). Note sur le stationnement hivernal de laridés sur la côte méditerranéenne d'Espagne. L'Oiseau et R.F.O. 46: 135-142.

Isenmann, P. (1980). Résultats d'un premier recensement de Laridés hivernants sur le litoral méditerranéen français (février 1979). L'Oiseau et R.F.O. 50 (2): 161-163.

Jacob, J.P. (1979). Résultats d'un recensement hivernal de Laridés en Algérie. Gerfaut 69: 425-436.

Llorente, G. & Montori, A. (1982). Algunas observaciones de Rissa tridactyla (Linn.) (Aves Laridae) y Stercorarius skua (Brunnich) (Aves, Stercoraridae) en el litoral barcelones. Publicaciones del Departamento de Zoologia 7: 81-83.

Macivor, J.A. & Navarro Medina, J.D. (1972). Sobre la invernada de Sterna sandvicensis en el Sureste. Ardeola 16: 263-264.

Muselet, D. (1982). Les quartiers d'hivernage des Sternes pierregarins Sterna hirundo européennes. L'Oiseau et R.F.O. 52 (3): 219-235.

Pardo, R. (1985). Censo de lardios del Pais Valenciano. Butll. Agro. 1: 23-29.

Sargatal, J. & Llinas, R. (1978). Els ocells de l'Emporda. Grafiques Montserrat. Figueres.

Vilagrasa, X., Carrera, E. & Pardo, R. (1982). Hivernage de la Sterne caugek Sterna sandvicensis sur les côtes de la Catalogne et du Levant (Espagne) Alauda 50 (2): 108-113.

CHAPTER 5 : IMPACT ASSESSMENT

Chairman: Dr. Joe SULTANA, Chairman I.C.B.P.
Europe.

BARDAWIL LAGOON BASELINE ENVIRONMENTAL STUDY AND VULNERABILITY TO OIL-POLLUTION

By G.M. Dunnet & H.Q.P. Crick,
Culterty Field Station, Zoology Dept.,
Aberdeen University, Scotland, and
Sherif Baha El Din, B.P., Egypt

Introduction

The Bardawil Lagoon in North Sinai is reported as one of the very important wetland sites in Egypt, as far as bird populations are concerned. It has been clearly demonstrated that large numbers of migratory birds pass through the area in both spring and autumn and it has been claimed that the Lagoon is itself an important site for resting and feeding for migratory waterbirds, and also that it is an important site for wintering waterbirds. The populations of breeding birds in Bardawil are reportedly very small. These are the conclusions of a number of studies made in different seasons and in different years, and which have been concerned primarily with counting the numbers of birds passing Zaranik at the eastern end of the Bardawil Lagoon. The main accounts of ornithological studies in the area are the following:

Paran & Paz (1978) for the period 13 August to 29 September 1973
Paran (1980) for the period 15 August to 7 October 1978
Petersen & Sørensen (1980) for the period 11 to 20 October 1979
Petersen & Sørensen (1981) for the period 7 to 30 September 1980
Petersen & Sørensen (1982) for the period 16 August to 24 September 1981
Paz & Paran (1974) for the period 13 to 15 September 1974
Baha el Din & Salama (1984) for the period 2 to 24 September 1982 (and including a summary of earlier work)
Rabinowitz, Gorney & Matthews (1982) summarizing autumn information for the years 1973, 1974, 1978 and 1979
Fry, Atta, Baha el Din & McKay (1985) covering the period 15 April to 5 May 1985

NATO ASI Series, Vol. G 12
Mediterranean Marine Avifauna
Edited by MEDMARAVIS and X. Monbailliu

A number of reports to the Ministry of Agriculture by Dr. Gemil Atta in the autumn and winter of 1983
Zu-Aretz, Sh. (1980) covering the period 8-12 October 1980 on the Yamit coast 80 km east of Zaranik.

The studies of autumn movements of birds reported in the above publications have consisted of regular counts, from dawn to dusk of all the birds passing a point (usually at the eastern end of the lagoon system at Zaranik). None of the counts is complete for the season but it is clear that hundreds of thousands of birds of many species pass along this coast from east to west each autumn. Various claims have been made that migrating waterbirds use the Bardawil Lagoon as a staging post during migration, but this claim has not been well substantiated in any published work. Accordingly, the main objective of the present investigation was to relate the number and species of birds on the lagoon itself to the observed migrations taking place concurrently. To provide a useful assessment of the importance of the lagoon, information acquired would have to be quantitative, and therefore appropriate methods of sampling and recording were developed.

A second objective was to describe the habitats occupied by birds on the lagoon and its shores, and to record what use the birds were making of the lagoon - for example, for feeding or resting or both.

A third objective of this report was to reconsider a number of matters relating to the spring situation which were reported by Fry et al. (1985), and to provide an overview of all the information so far collected on the ornithology of the Bardawil Lagoon.

• Description of Bardawil Lagoon and its habitats

It is not necessary to provide here a detailed description of the geographical, topographical, hydrological and climatological features of the Bardawil Lagoon. These have already been outlined in the report of the Direction des Relations Economiques extérieures (1982), and the manuscript compiled by Dr. Moustafa Fouda (1985). The following quote from Fouda (1985) adequately describes the lagoon for present purposes: *"The Bardawil Lagoon is a shallow hypersaline lagoon, located in the middle of the Mediterranean coast of Sinai. It spreads from 32°40' to 33°30' eastern longitude and from 31°03' to 31°14' northern latitude. The approximate length is 75 km and the maximum width is 19 km. The surface area is about 595 km2. The lagoon is extremely shallow and the water depth never exceeds 3 m. A long sand bar, 1 km wide maximum, separates the lagoon from the adjacent Mediterranean. Three openings (locally named bugaz) connect the lagoon with the sea. Two of them are man-made: the Western Bugaz (I) at El Rawak (El Rouaq) and the Middle Eastern Bugaz (II) at El Tulul (El Teloul), while the third one, the Extreme Eastern Bugaz (III) El Zaranik, is natural."*

With no water draining into the lagoon from the land, and the small accesses direct to the Mediterranean, evaporation greatly exceeds the influx of water into the lagoon. This leads not only to high salinities, but also to long-term fluctuations of the shoreline with a variety of habitats including bare sand, salt pan, patches of mud, some of which are covered by mats of algae, through to some densely vegetated islets and shorelines with well developed stands of halophytes, mainly species of Arthrocnemum and Halocnemum (we have not checked the identity of these plants, but are quoting from the references given above).

Apart from general unquantified statements about variations in water level, and some maps showing the distribution of different salinities, we have been unable to find information on the seasonal variation in water level and therefore on the distribution and amount of exposed shoreline and its suitability for birds. Our expectation would be that the distribution and nature of these habitats would vary dramatically according to the hydrological circumstances, and that these would not be precisely predictable, although there would be a broad seasonal pattern. We attempted to classify the shorelines into a number of defined habitats during our aerial surveys, and this has enabled us to attach some crude quantitative values to the amounts of these habitats in the lagoon during the period of our survey. We have used these values to make various estimates of the size of the populations of birds which we saw using the lagoon shores, but we would emphasize the danger of extrapolating too readily from these in view of the transient nature and the presumed instability of at least some of the habitats throughout the year.

• Definition of Major Habitats

The main habitats in Bardawil described by Fry et al. (1985) include the following, which in our view are the most important for birds using the lagoon and its shores.

(1) Open Water. Lake Bardawil itself (excluding Lake Zaranik) has extensive areas of open water, all of it shallow, 0.5 to 3.0 m deep, with no emergent vegetation but with patches of the sea grass Ruppia spirilis and filmentous green algae growing on the mainly sandy bottom. Lake Zaranik at the eastern end of the system with a maximum depth of 1.5 m and thick patches of Ruppia, consists of a series of narrower waterways threading between islands and mudflats. In October the extent of open water in this part of the system may be at a seasonal minimum. These waterways connect with the Mediterranean via four narrow passages (bugaz) through the sand bar.

(2) Saline and Hypersaline Shallows and Salt Pans. In October, especially in Lake Zaranik and the eastern end of Lake Bardawil, but also scattered elsewhere around the lagoon, were isolated shallow pools of highly saline water merging into dry salt pans.

(3) Bare Sandy Shores. Much of the shoreline of Lake Bardawil and some of the shores of Lake Zaranik consist of pure

sand. In some parts the sand dips steeply into the water at the margin, in others there are areas of relatively flat sandy shores. In most cases the sand was not vegetated though scattered bushes of halophytes may occur.

(4) Mudflat. In some areas there is a substantial accumulation of silt among the sand to create extensive areas of mudflats. These are most common in Lake Zaranik, where they may be very extensive, but are patchily distributed also on other shore lines of Lake Bardawil itself. In some areas the mudflats are completely blanketed in an algal mat, and this mat can become fragmented into pancake-like patches which may peel away from the underlying mud to provide a broken surface which in some situations appeared to be particularly attractive to birds. Such broken surfaces occurred on both wet and dry mud.

(5) Halophytic Vegetation. On some coastlines and indeed on some of the islets within Lake Zaranik there are extensive areas of closed canopy *Arthrocnemum* and *Halocnemum* vegetation. These habitats can be linearly distributed along the shore line, can completely blanket some of the islets, to obscure the soil and emerge from the surrounding water, and may be patchily distributed over gravel and shell deposits, for example on Mat Eblis. These vegetated patches provided important cover and possibly food for a large variety of scattered passerine passage-migrants in October.

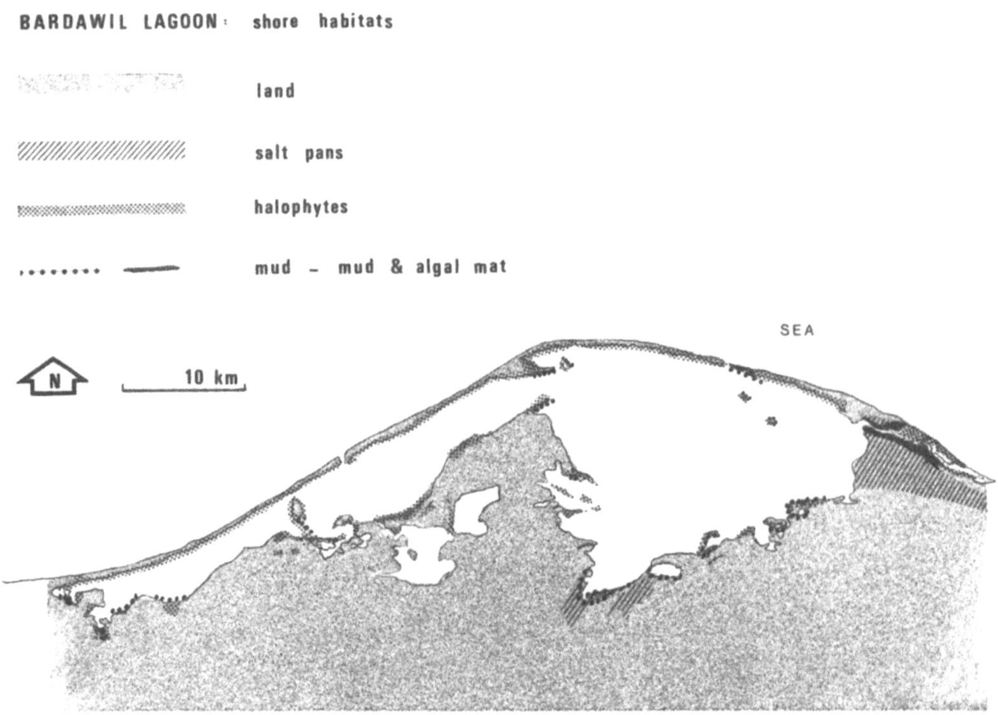

Map 1: Bardawil Lagoon. Shore habitats

Methods

Methods were developed to obtain quantitative and representative data on the distribution and abundance of birds in the various habitats. These are briefly described below, but it should be noted that all were constrained by the official requirement for us to be in our camp on Zaranik beach, from 18.00-06.00 hours each night, and while there to be quiet, unilluminated and inactive.

• Aerial Surveys

These surveys were carried out from a Cesna 172 aircraft which had room for three passengers as well as the pilot. We had prepared in advance a flight plan which would cover a large proportion of the Bardawil Lagoon and was designed to include substantial areas of all the major habitats. It was intended that the aircraft should fly along the flight plan at heights of 500 feet or less and that the actual track covered should be monitored and recorded by one of the team (Waheed Salama or Sherif Baha el Din) who sat in the front alongside the pilot, whom he instructed. The track actually followed was recorded on a map, with fixed points recorded and timed accurately to the second. Also kept were notes and diagrams of the habitats which were covered during the aerial transect. The height and speed were recorded at intervals of every few minutes. The two members of the team sitting in the back each recorded into a tape recorder what they saw on their side of the aircraft, and each observation was accompanied by a time, accurate to the second. Records were made of all birds detected, as well as comments on habitat. After the flight an accurate plan of the track was prepared from the information collected by all three observers.

The first flight took place on 3 October, departing El Arish at 11.38 hours and returning to El Arish at 13.35 hours. The second survey was carried out on 8 October, departing El Arish at 11.30 hours and leaving the western end of the lagoon, heading for Cairo at 12.30 hours.

All observers took a considerable number of photographs to record habitats and other features of the lagoon as detected from the air. These photographs have been assembled to provide information on various habitats and details of the bugaz. They were also used in compiling the distribution of habitats around the lagoon. While theoretically it would have been possible to record the numbers of birds seen in precise time intervals, in the event so few birds were seen that no quantitative data on bird abundance were obtained by this method.

• Walking Surveys

To measure the distribution and abundance of birds along the shoreline, surveys were carried out by observers who walked along the shore and recorded all birds seen and what they were doing. A careful watch was also maintained offshore and any birds seen flying or sitting on the water were also recorded.

In these surveys, data were recorded separately for each 200 m length of shore.

The outer bar of Bardawil Lagoon was out of bounds for such activities owing to military restrictions. Accordingly, sites were selected for these walking surveys with reference to accessibility and the habitats which they contained. General notes were made of habitats but since our habitat classification is fairly broad, detailed descriptions were not taken.

• Boat Surveys

We hired boats from local fishermen to carry out two different kinds of survey:

- Open Water. The main areas of open water are in Lake Bardawil. These are very extensive and are used by fishermen who have a number of settlements around the southern shore of the lagoon. On 7 October we hired a boat from El Teloul and from 09.40 to 16.17 hours we covered extensive areas of open water. All three observers maintained a continuous watch to detect and identify all birds seen. There was no attempt made to limit the width of the strip within which birds would be counted and everything seen was recorded. Records were made in relation to intervals of five minutes so that estimates of abundance and variability could be obtained.

- Boats were also used to study the birds present in the Lake Zaranik area. This area has much less extensive open water, but many channels between small halophyte-covered islets and also channels between more extensive areas of mudflat and saltpan. In Lake Zaranik most of the birds seen from the boats were on the shore and again records were maintained in units of five minutes of time. In this case, the habitats on which the birds occurred were also identified and recorded as bare sand, halophyte vegetation, mudflat, open water and flying birds.

• Migration Observations

To be consistent with the extensive observations made previously on migration passing Zaranik, we watched migrants passing the camp from a position just above high water mark. Observers maintained a watch and recorded all birds seen moving, and the direction in which they moved. Once again, to provide some quantitative aspects to this data all observations were recorded in ten minute periods.

Distribution and Abundance of Water Birds

• Open Water

Open water is extensive in Lake Bardawil (map 1), but is confined to relatively narrow channels in Lake Zaranik. Observations on birds using open water were obtained from:

- aerial surveys covering both Lakes Zaranik and Bardawil;
- boat survey from El Teloul;
- observations offshore during all beach walking surveys; and
- boat and walking surveys in Zaranik.

These surveys included, between them, large areas which are representative, in our view, of the whole lagoon system.

Aerial Surveys, October 3 and 8

Though height and speed varied somewhat, we flew mostly between 200' and 600' above the water at 90-100 knots. At this height and speed it was possible to see all gulls, terns and herons whether flying or on the ground or water, but small waders were difficult to detect unless they flew, but they could not be identified with confidence. We felt sure that we would detect any flocks of duck on the water (garganeys were expected).

In fact, very few birds were seen from the aircraft. Totals seen by two observers, looking from different sides of the aircraft, on each trip are as follows:

	Garganey	Terns	Gulls	Waders	Herons
3 Oct.	0	70	407	53	16
	250	42	55	480	49
8 Oct.	0	0	100	373	9
	3	7	169	565	59

On open water only a few gulls were seen, mainly in Lake Zaranik and near Bugaz No. 1 and 2. Only one flock of 250 garganey was detected. Our conclusion from aerial surveys is that birds are very scarce indeed on open water.

Boat Survey from El Teloul, October 7

This survey covered extensive areas of open water, during the one hour leg from El Teloul to El Ofra; 50 minutes from El Ofra to Mat Eblis and 2h10 min. from Mat Eblis back to El Teloul - a total of 4 hours in open water. All birds seen were recorded in 5-minute intervals (48 intervals in all). In 43 of the 5-minute intervals no birds were seen either flying or on the water. Birds seen in the remaining 5-minute intervals were:

- 6-white-winged black terns and 1 hoopoe;
- 1 tern and 6 "teal";
- 1 tern and 1 duck;
- 1 unidentified passerine;
- 2 little terns and 1 gull-billed tern.

These data confirm the conclusion from the aerial surveys, that birds are few and far between on open water. There is no evidence of large numbers of gulls, terns or ducks feeding or resting on open water.

Beach Surveys

Observations of birds sitting on or flying over open water made on all beach surveys, and only occasional gulls or terns, singly or in small parties were recorded. No large flocks or concentrations of birds were seen on open water.

Beach and Boat Surveys in Lake Zaranik

In the narrower channel of water in Lake Zaranik our records (tab. 1) confirm the general scarcity of birds on open water though a small number of flocks of garganey and scattered groups of slender-billed gulls did occur here.

Date	Length (km)	Habitat	Total Number	Predominant Species
4.10.85	15	Water	24	Sanderling (60)
		Sand	198	Dunlin (51)
				Slender-billed Gull (32)
				Kentish Plover (8)
		Halophytes	85	Grey Heron (26)
				Little Egret (20
		Mudflat	1638	Dunlin (892)
				Waders (573)
				Slender-billed Gull (71)
		Flying	712	Waders (376)
				Curlew (90)
				Dunlin (67)
6.10.85	11	Water	90	Slender-billed Gull (71)
		Sand	55	Kentish Plover (17)
				Dunlin (16)
		Halophytes	37	
		Mudflat	2322	Waders (1442)
				Dunlin (771)
		Flying	98	Garganey (78)
				Slender-billed Gull (27)

Tab. 1: Surveys of Zaranik from Boats

Our conclusion is that during the autumn migration the open-water habitats of the Bardawil Lagoon system do not support large numbers of waterbirds for either feeding or roosting by day.

• Bare Sand

Distribution

Bare sand is the dominant type of shoreline in Bardawil Lagoon. It occurs mainly along the southern shore and is found also at the bugaz, where water can flow between the lagoon and the Mediterranean Sea.

Sample Areas

Walking surveys of sandy shore were made at representative and accessible areas along the southern shore of Lake Bardawil and Lake Zaranik. The outer bar, including the bugaz, was not surveyed because it is an area restricted to military use.

Results

No significant concentrations of birds were seen on bare sandy shores during aerial surveys. Small numbers of birds (less than 10) were seen at the bugaz.

Initial walking surveys made at Lake Zaranik recorded the numbers of birds seen on each habitat (tab. 3) although the quantity of each habitat surveyed was not measured. Dunlin and Kentish Plovers were the commonest birds on bare sand. Further surveys made from boats in Lake Zaranik also recorded sanderlings as occurring frequently (tab. 1). Walking surveys beside Lake Bardawil consisted of counts of birds in 200 m transects along the shore. 67 transects were made overall and the numbers seen of each of the four commonest species are in tab. 4. The four wader species occurred in flocks containing up to half the total number of birds counted. From a total of 13.4 km surveyed, the mean number of the four commonest waders was 3 per 200 m transect.

	Bardawil & Zaranik		Zaraniq	
	%	Length (km)	%	Length (km)
Bare Sand	46	166	40	20
Halophytes	38	137	28	14
Mud Algal mat	13 3 } 16	47 11 } 58	} 32	16
		360 km		50 km

Tab. 2: Relative Abundance of Habitat Types.

Date	Length (km)	Habitat	Total Number	Predominant Species
29.9.86	4.1	Water	144	Garganey (125)
		Sand	60	Dunlin (40)
		Halophyte	14	
		Mudflat	287	Dunlin (203) Grey Heron (62)
		Flying	343	Waders (230)
1.10.85	7.0	Water	184	Garganey (133) Slender-billed Gull (51)
		Sand	261	Dunlin (15) Kentish Plover (100)
		Halophyte	8	
		Mudflat	1337	Dunlin (650) Kentish Plover (494) Grey Heron (77)
		Flying	107	Common Tern (49) Waders (50)
3.10.85	3.0	Water	97	Slender-billed Gull (55) Garganey (40)
		Sand	30	
		Halophyte	4	
		Mudflat	621	Dunlin (340) Little Stint (123) Kentish Plover (123) Grey Heron (59)
		Flying	0	
6.10.85	1.4	Mudflat	1036	Waders (740) Kentish Plover (224) Dunlin (54)
		Flying	400	Waders

Tab. 3: Counts of birds on each habitat at Zaranik on walking surveys

Species	Mean	Number of birds/transect		
		Minimum	Maximum	Total No. of Birds
Little Stint	0.9	0	38	62
Dunlin	0.6	0	17	39
Kentish Plover	1.2	0	34	77
Ringed Plover	0.2	0	4	11

Tab. 4: Counts of waders on bare sandy beach in Lake Bardawil. Counts are summarized from 67 x 200 m transects.

- <u>Halophyte Communities</u>

This habitat takes two forms - islets completely covered with vegetation in Lake Zaranik, and mainly sandy shorelines with a linear distribution of scattered halophyte plants along them.

These habitats are widespread (map 1) and our data were obtained from:

Beach walks at El Teloul		October 4
	Romana	October 2
Boat survey and beach walks at		
	Zaranik	September 29 October 1, 3, 4, 6
Beach walk at Mat Eblis		October 7.

Our best quantitative data for linear habitats were obtained at El Teloul and Romana. At the former we have numbers/200 m over a distance of 1.8 km, and at the latter over 1.4 km. The full data are given in tab. 5.

	El Teloul									Total	Romana							Total
Kentish Plover	(17)	4	-	(15)	-	(120)	(9)	4	1	170	-	-	-	-	-	-	3	3
Dunlin	(45)	-	-	(110)	-	(60)	-	-	-	215	-	-	4	-	-	-	-	4
Little Stint	-	1	-	-	-	-	-	-	2	3								
Redshank	-	-	-	-	-	(23)	-	-	-	23								
Greater sand Plover	-	-	-	-	-	-	-	1	-	1								
Little Tern	-	-	-	-	-	-	3	-	-	3								
Kingfisher	-	-	1	-	-	-	-	-	-	1								

( ) = in a flock. (In addition, 12 passerines - wagtails, wheatears and larks were recorded).

Tab. 5: Numbers of waterbirds seen in 200 m lengths of halophyte dominated shores at El Teloul and Romana

These data indicate a very patchy distribution with small flocks of waders gathered either to roost or to feed. The area at El Teloul was an enclosed shallow marginal lagoon with extensive patches of halophytes, approaching the habitats in Lake Zaranik in character. It seemed to be favoured by waders. By contrast, at Romana the shore was sandy and open to the main lagoon, with a narrow band of halophytes and was typical of the great majority of the halophyte habitats around the edge of Lake Bardawil. This area was not nearly so densely populated and supported fewer species of birds. The mean number of Kentish Plover per km length of such shoreline is approximately 2, and of Dunlin 2.8. The distribution and abundance of patches like that at the lagoon east of El Teloul is not known, but from our aerial survey they are certainly few and scattered along the southern shores of Lake Bardawil.

Mat Eblis

This is an unusual habitat in the Bardawil system, consisting of very low ridges of shell and gravel on which halophytes are well established, and separated by elongated pools and wet areas, some of which have become salt pans. The area was relatively attractive to waterbirds and migrant passerines, which were scattered over the whole area. On October 7, between 13.09 and 14.02 hours we walked about 3 km along part of the peninsula and the following birds were recorded:

Waterbirds		Landbirds	
Kentish Plover	37	Wagtail sp.	1
Turnstone	7	Common Redstart	2
Dunlin	1	Willow Warbler	6
Little Stint	19	Whinchat	1
Grey Plover	1	Sedge Warbler	5
Redshank	3	Red-backed Shrike	4
Temmincks's Stint	1	"Reedy Warbler"	1
Common Snipe	1	Crested Lark	3
Marsh Sandpiper	3	Meadow Pipit	1
Greater Sand Plover	1	Common Wheatear	1
Common Kingfisher	2	"Willow-chiff"	2
Grey Heron	1		

It is difficult to relate these to a precise area, and the figures do not constitute a total count of birds present on the area covered. However, they do indicate the variety and relative abundance of species present.

Lake Zaranik

In Lake Zaranik, halophyte habitats include almost completely vegetated islets which were fringed with shores of mud, sand or emergent halophytes and with halophytic vegetation growing close to the waterline. Our information on birds using these habitats is derived from:

- counts made from a boat moving between the islets and offshore (when all birds seen were recorded, per 5-minute interval, according to the habitat on which they occurred); and

- walks through the area when the total distance covered was measured and the numbers of birds seen on each habitat were recorded.

A summary of the data for waterbirds at Zaranik are presented in tab. 1 and 3. It is not possible to be precisely quantitative using these data. Although the boat progressed at approximately 6 km/hour, so that 5-minute intervals approximated to 500 m, at times two shorelines were available for counting, and at others only one or even none. Further, a record was not kept of the amount of shoreline in each habitat category which we surveyed. However, analysis of aerial photographs have enabled us to calculate the approximate relative abundance of the habitat types in Lake Zaranik (tab. 2), and our boat surveys were extensive enough to be approximately representative of the whole.

Combining data for waterbirds from the two boat counts on October 4 and 6, we get 3,960 on mud, 122 on halophytic habitats, and 253 on bare sand - a ratio of 32:1:2. The relative availability of these habitats is in the ratio of 1:1:1.5, indicating that the halophytic vegetation, like bare sand, supports many fewer waterbirds than mudflats. No species of bird seemed to be especially associated with this habitat.

At the end of September SBED walked 15.5 km of shoreline habitats in the Zaranik area and recorded the birds seen in each habitat. He covered the same or closely similar areas on each transect so that there are statistical problems in lumping the counts. However, using the same approach as in the previous paragraph, we find that he recorded 3,281 waterbirds on mud, 26 on halophytic habitats and 351 on bare sand, a ratio of 126:1:14 compared with a habitat availability of 1:1:1.5. Again, the relative unattractiveness of halophytic vegetation habitats for waterbirds is clearly demonstrated.

- Mudflats

Mudflat habitats take a variety of forms, mainly dependent on the width of the area of exposed mud. At its simplest it is a strip a few metres wide above the waterline, but as the water edge becomes convoluted quite extensive areas of wet mud can occur. In some of these the surface of the mud may become "pancaked" and broken and these are particularly attractive to waders, e.g. Phalorope Bay, west of El Teloul; some areas towards the western end of Lake Zaranik. In addition, the southern and western shores of Lake Zaranik have extensive tracts of dried-out mud which may or may not be blanketed by an algal mat, either continuous or pancaked. These extensive areas of dry mud are used by waders for resting and small flocks were encountered in serveral areas. At the south-west end of Lake Zaranik the dry mud-flats merge into extensive salt pans.

The distribution of mudflats and muddy shores is shown on map 1 but the different categories are not distinguished there.

Muddy Shores

We sampled these habitats in Lake Zaranik from boat surveys and beach walks, but no in such a way as to get good quantitative data on numbers/km of shore. The summarized data have been presented above (tab. 1 and 3) and it is clear that in Zaranik, when the specially favourable muddy areas are included, mud habitats are many times more densely inhabited by waders than any other Bardawil habitat. We have, however, very little quantitative data on abundance along narrow muddy shores because such shores are uncommon outside Lake Zaranik. One muddy patch on the otherwise sandy shore at Romana, was approximately 200 m long and contained a flock of 18 in addition to 4 separate ringed Plovers and a group of 5 Dunlin.

Within Lake Zaranik, the two boat surveys revealed a total of approximately 2,000 birds on mudflats on each occasion. Some flocks of small waders were too distant to be identified to species, but almost certainly were composed largely of Dunlin. Of the 1,000 or so identified to species, by far the most numerous - 700-900 - were Dunlin, the great majority of which occurred in one or two large flocks, while the remainder were scattered singly or in small parties along the shoreline. The next most common was the Kentish Plover with totals of 40-50, and dispersed more evenly with occasionally a group of 20 or so. Of the eight other species of Waders recorded on muddy shores on these surveys, none was abundant, and all were scattered sparsely through the habitat.

Most of the waders occurred on rather broad expenses of "broken" wet mud and at Zaranik were seen in the same sites on both days of the boat surveys and also during earlier beach walks. Such concentrations may occur predictably on these areas and while the habitats there appear distinctive to us, we have no precise measurements of the features which make them attractive to waders.

The following are examples of the size and composition of wader flocks at five sites in Lake Zaranik - see map 2 for favoured areas:

- 475 Dunlin, 2 Common Snipe
- 600-800 Waders - 95% Dunlin plus Grey Plover, Ringed Plover, Kentish Plover, Redshank, Common Snipe
- 1,000 Small Waders, mainly Dunlin
- 500-550 Dunlin plus 5 Kentish Plovers, 1 Grey Plover, 1 Redshank, 1 Ruff
- 25 Dunlin, 3 Redshank, 6 Grey Plover, 1 Greater Sand Plover, 1 Bar-tailed Godwit, 1 Greenshank.

Similar concentrations of several species of waders were found near El Teloul: one area consisted mainly of halophytes and sand (the inner lagoon to the east of the village) and the other near the entrance to the almost enclosed area of Lake Bardawil lying to the south west of El Teloul. The latter, which

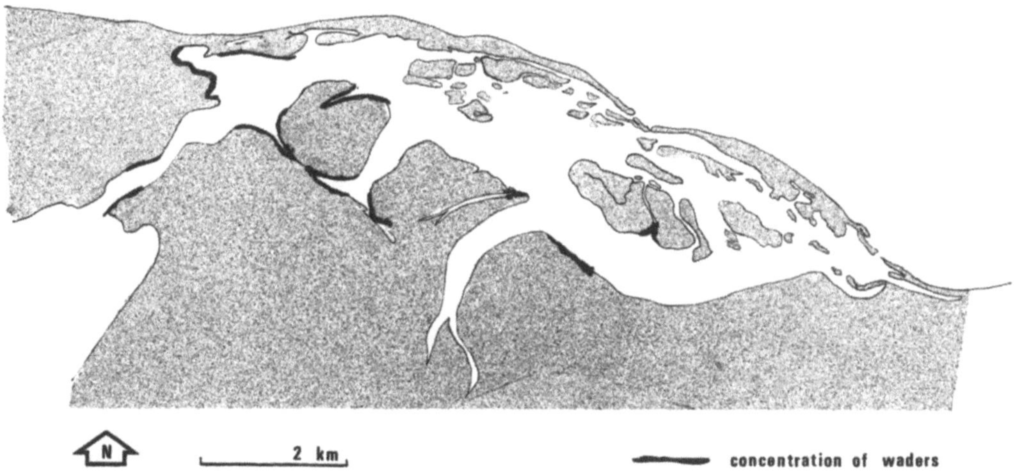

Map 2: Areas of high concentrations of Waders in Lake Zaranik.

we called Phalarope Bay, was a muddy beach, with several shallow wet patches, surrounded by bare sand. In these areas, where we found flocks of waders, accurate counts were difficult to obtain and we made estimates of the total numbers present. Sample counts of individual birds enabled us to determine the species composition of the flocks (tab. 6).

Clearly, waders are not abundantly distributed over all the shoreline of the Bardawil Lagoon, but occur in high density only in restricted areas of the mudflat habitat. Such habitats are commonest in Lake Zaranik and occur in a few isolated patches on the southern shore of Lake Bardawil (e.g. Phalarope Bay). In early October the total number of waders utilizing the shores would be of the order of 10,000 at any one time - equivalent to a small European estuary.

	4 October afternoon	5 October midday
Grey Heron	5	-
Great White Heron	-	1
Avocet	2	-
Little Ringed Plover	-	1
Ringed Plover	6	11
Kentish Plover	98	142
Little Stint	112	19
Dunlin	122	33
Redshank	8	12
Green Sandpiper	2	-
Common Sandpiper	-	1
Red-necked phalarope	5	8
Waders sp. (mainly Dunlin and Kentish Plover)	228	73
White-winged Black Tern	3	2
Little Tern	21	1
Slender-billed Gull	221	20
Approximate Total	833	324

Tab. 6: Waders using Phalarope Bay, West of El Teloul

Migration

All-day counts of migrating birds were made at Zaranik for 6 days from 26 September 1985. The vast majority of birds flew westwards and the low numbers which flew to the east are not included in the analysis below.

A total of 24,355 (tab. 7, Appendix) birds was counted of which 71% were Garganey, 13% Waders, 6% Herons and 5% Terns. The majority of Garganey occurred in the first two days (66%) with peak numbers occurring at 06.00 hours, but generally there was a steady passage of c. 180 birds hr^{-1} throughout the period. Of the Waders, Dunlin and Little Stint were observed to peak in numbers between 06.00 and 09.00 at c. 24 birds hr^{-1} and then decreased gradually during the rest of each day; Sanderling numbers peaked at 06.00, 13.00 and 15.00 hours; Ringed and Kentish Plovers peaked in numbers at 07.00 hours at c. 42 hr^{-1}. The majority of Herons passed Zaranik between 10.00-17.00 at a rate of c. 20 hr^{-1}. The Gulls flew past at a rate c. 16 hr^{-1} during 06.00-09.00 hours and 12.00-17.00 hours. The diurnal variations in numbers of each type of migrating bird is in fig. 1.

Additional observations of the passage of birds were made from 06.16 to 08.30 hours on 4 October and 06.04 to 07.50 on 6 October at Zaranik when 477 and 257 birds were counted respectively. On the first of these counts an unusual passage of 254

	Number	%	Major Species (n)
Garganey	17,346	71.2	
Waders	3,063	12.6	Dunlin (666), Sanderling (606), Little Stint (407), Ringed Plover (334), Kentish Plover (278)
Herons	1,379	5.7	Grey (492), Little Egret (366), Night (279), Purple (104)
Terns	1,262	5.2	Common (940), White-winged Black (194)
Gulls	942	3.9	Slender-billed (303), Lesser Black-backed (261), Yellow-legged (161)
Other Ducks	192	0.8	Teal (122)
Other Birds	108	0.4	Kingfisher (77)
Skuas	63	0.3	
	24,355		

Tab. 7: Numbers of birds seen at Zaranik migrating westwards from 26 September to 1 October 1985

Avocets was observed. The numbers of birds passing per hour was less than had been counted at the end of September because of considerably fewer Garganey (see below). A short period of observations was made also at Romana, at the western end of the lagoon, from 07.45 to 09.45 hours on 2 October. A total of 438 birds flew westwards at Romana, including 373 Garganey.

Overall, the number of birds passing Zaranik during the end of September was c. 650 hr^{-1} from 06.00 to 08.00 hours and the c. 270 hr^{-1} during the rest of the day. If Garganey are excluded then peak numbers occurred at 07.00 hours (170 hr^{-1}) and numbers were constant at c. 100 hr^{-1} in other hours.

During migration watches, a small number of migrants were seen to land in the eastern end of Lake Zaranik. In addition, large flocks (hundreds) of Garganey were seen to land on the sea offshore from Zaranik, where they stayed for a few hours before continuing their migration.

In addition to waterbirds, there was evidence of landbird migration through the area. Migrants, mostly passerines, were observed on all terrestrial habitats, even on bare saltpans, and especially in halophytic vegetation. These birds were usual-

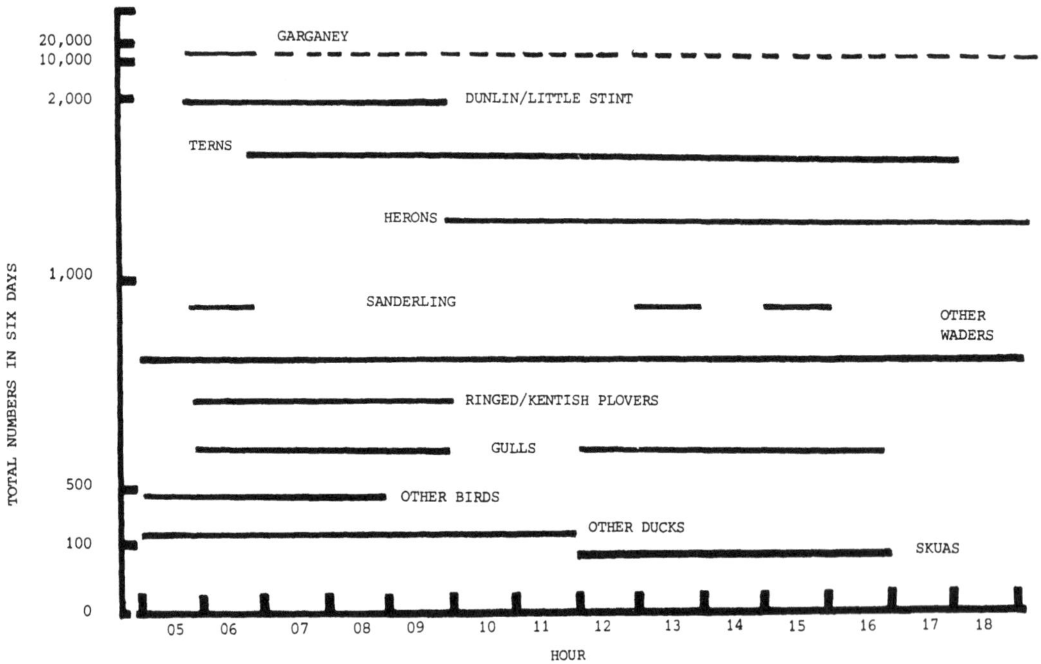

Fig. 1: Daily patterns of migration. Zaranik, September/October 1985.

ly at very low densities compared with Waders and often appeared to be seeking shade under vegetation. Among the birds less commonly seen during migration watches, the number of Kingfishers moving westward was notable: 77 seen over 6 days. Kingfishers were frequently seen fishing inside the lagoon: 6 in 77 walked transects on bare sandy beach (i.e., 1 per 2.5 km); during boat and walked surveys in Zaranik 26 were seen on sand, 21 on halophytic vegetation, 4 on mud, thus showing a marked preference for the sand and halophytic shores.

Several species of raptors were seen on passage; some ospreys fed in the lagoon, and some falcons, harriers and buzzards rested on shores.

Commentary on the major species of birds in Bardawil Lagoon

• Garganey (*Anas querquedula*)

Although 17,000 were seen on autumn migration over six days from Zaranik they were rarely seen to settle inside Bardawil Lagoon, but more often they landed on the Mediterranean Sea, close offshore. This is an important area, therefore, for Garganey, but only during a limited period in autumn.

• Dunlin (*Calidris alpina*)

We estimated that Bardawil Lagoon and Lake Zaranik in particular contain up to 10,000 Dunlin at any one time during autumn. These are distributed mainly on certain areas of mudflats, but also they are thinly distributed on bare sand and halophytic shores. This is not an important population for western Europe, which may number about one million (Cramp, 1983).

• Little Stint (*Calidris minuta*)

These were found usually in association with Dunlin. However, they seemed to favour bare sand over muddy shores. We can calculate from our estimates of population density and habitat distribution that there were c. 800 along sandy shores; this is approximately 5% of the total population said to be wintering in Egypt (Cramp, 1983).

• Kentish Plover (*Charadrius alexandrinus*)

Kentish Plovers are declining in population over much of western Europe, but the north African coast is a breeding stronghold and important wintering ground for 20,000 (Cramp, 1983). We calculated that in October there were c. 1,000 on sandy shores and about half that number over the rest of the habitats around Bardawil Lagoon. Fry et al. (1985) observed breeding Kentish Plovers on dry, salty mud around Bardawil Lagoon in spring. They estimated that the breeding population in the area was 3,750-7,500 birds. The higher estimate included three exceptional concentrations of birds which was a source of bias. Their data can be reworked by considering simply the total number of birds seen and the total distance walked, from which a population estimate of c. 3,800 can be derived. We concluded that Bardawil Lagoon is an internationally important breeding and wintering area for this bird in the western Palearctic.

• Little Tern (*Sterna albifrons*)

We saw very few Little Terns migrating past Zaranik (84 in 6 days) in autumn and less than 100 in Bardawil Lagoon overall. Fry et al. (1985) estimated that there was a population of 4,500 of these birds in spring and surmized that breeding was imminent then around the lagoon. Their calculations can be reworked (as above for Kentish Plovers) to give a lower, and possibly less biased population estimate of c. 2,300 birds. Despite this revised lower estimate of breeding population, we can still concur with Fry et al. (1985) when they consider Bardawil Lagoon to contain an important population of Little Terns for the western Palearctic, in the light of marked population declines elsewhere in this region (Cramp, 1985). However, the breeding status and size of the breeding population of this species need to be confirmed.

- Passerine Migrants

The numbers of these could not be accurately assessed by us, but a wide variety were observed, especially in halophytic vegetation, although densities appeared to be low. Our observations appear to be qualitatively similar to Fry et al. (1985) in spring, although we did not netting. However, we do not feel that we have sufficient information to determine the importance of the area for staging.

The Heiligenhafen Conference in 1980 recommended that sites with wader populations of 20,000 or more are of international importance (Prater, 1981). Bardawil did not meet this criterion in October. On the other hand, both Little Terns in the breeding season and Kentish Plover may constitute the "1% of a geographical population" which denotes the status of "high international conservation significance".

Vulnerability to oil pollution

- The Open Sea

Our observations on migration indicated that flocks of Garganey commonly settle on the Mediterranean within a few hundred metres offshore of the outer bar of Bardawil Lagoon, and possibly further offshore. We were unable to make quantitative observations of this activity but in view of the very large numbers of Garganey passing through the area in autumn, it is possible that large numbers of these ducks could be vulnerable to oil floating on the sea near Bardawil Lagoon. We are not aware of any other concentrations of birds offshore in this area, but several species of Gull and Tern use the sea.

- The Outer Bar

In the event of oil being spilled in the Mediterranean, it could well end up on the outer shores of the Bardawil bar. Because of the military situation we were unable to make observations on the shores of the outer bar, either inside or outside the lagoon, and it would not be appropriate to extrapolate from the sandy shores inside the lagoon in order to predict the birds on the outer shore. However, general observations on the sandy coast to the west of El Arish suggest that the shoreline would be inhabited by Sanderlings in autumn and winter and by Kentish Plovers throughout the year. We would expect both these species to be rather thinly distributed along the shorelines so that any mortality would be likely to be quite small. In that situation, neither species is likely to occur in numbers of national or international significance.

• The Bugaz

The only way that oil spilt in the Mediterranean could enter the Bardawil Lagoon system would be through the four bugaz. These are narrow sandy channels, at least some of which require to be dredged in order to remain open. At the eastern end of the lagoon, the two bugaz opening into Lake Zaranik consist of meandering channels among sandy islets and the connection with the sea is not direct. In the event of spilled oil approaching any of the four bugaz, the setting up of a booming system should be possible which would prevent the entry of oil into the lagoon. This would be justified in terms of fishery in the lagoon if nothing else. There does not seem to be, at least in spring and autumn, any strong flow of water through the bugaz, but the transport of oil would be strongly related to wind direction. In spring, Fry et al. (1985) found concentrations of Terns using the areas of water around the bugaz, and in our aerial surveys we found flocks of Gulls and Terns in their vicinity also. If Gulls and Terns do find the bugaz a rich feeding area they could be particularly vulnerable to any oil being transported through these channels, and Little Terns if they breed in large numbers, could be seriously at risk.

• The Bardawil Lagoon

Once oil got into the lagoon, either through the bugaz or by a spill or leak within the lagoon itself, its transport would be almost entirely wind-driven. It is not within the scope of our study to consider the effect of this on the primary production (Ruppia, and algae) and the components of the food chain leading to the several species of fish which constitute the fishery. Clearly such effects are likely to be serious in terms of economic, social and political considerations.

When we consider the likely effects on the birds in the lagoon it is convenient again to sub-divide the system into its constituent habitats:

Open water

All the evidence that we have obtained suggests that in autumn there are no substantial gatherings of birds on the open water. We found remarkably few Garganey on the water, although they were migrating along the Mediterranean coast line at the time, and in October both Gulls and Terns were few. However, in spring, Fry et al. (1985) suggested that relatively large numbers of Little Terns bred in the area and fed in the lagoon and claimed that this was an internationally important population of that species. There is no doubt that any large mortality of Little Terns in spring would lead to outspoken criticism in the ornithological world. Similarly, though we did not see them, flocks of flamingoes and pelicans are reported to use the lagoon and these large and spectacular birds would also be vulnerable to oil in the lagoon itself.

Sandy shores

We have estimated that approximately 46% of the shoreline of the Bardawil Lagoon is bare sand, and sandy shores are widely distributed throughout the system. The density of shore birds per kilometre of shoreline of bare sand is very small and in both spring and autumn it is clear that neither large populations nor particularly vulnerable species occur in this habitat.

Mudflats

Muddy shores comprise about 19% of the shoreline of the system, and is by no means a uniform habitat. We have identified a number of locations, particularly in Lake Zaranik, but also one or two in the Lake Bardawil where large numbers of Waders occur regularly. These areas have been defined in the report, and are illustrated in map 2 for Lake Zaranik. On these patches of mud there may be hundreds or even a few thousand Waders at any one time, involving a number of species of ornithological interest. It would be important if oil was moving in the lagoon to try to prevent it getting on to such shores, this would be feasible for Lake Zaranik and areas such as Phalarope Bay by the use of oil-booms.

Halophytic communities

About 38% of the Bardawil shore is covered with halophytic vegetation. Again, some of these areas are rich in variety of shore birds, e.g. Mat Eblis, and the lagoon to the east of El Teloul. Patches such as these are of considerable ornithological interest, but they are relatively small and do not assume national or international importance. Speaking more generally, these halophytic communities do not support large populations of shore birds, but they do provide shelter and possibly food for migrating passerine birds which are seen commonly to rest in them. Again, while interesting ornithologically, it is not possible to say that these communities are of national or international importance. It may be worth suggesting that another reason for preventing any spilt oil getting into these communities would be the great difficulty of cleaning it up, and the probability that oil will persist among this vegetation for a long period.

In general, the Bardawil Lagoon does have some important ornithological features, but certainly in autumn, apart from Kentish Plovers, there were no concentrations of birds which were deemed to be nationally or internationally important. The Kentish Plover population, throughout the year, may constitute more than 1% of the population of the western Palaearctic and thus constitutes a feature of international conservation significance. In spring it could be that the breeding population of Little Terns is of international importance but more data would be required to confirm this. Given its size, the lagoon system is not ornithologically rich, but there are some parts of it, and in particular, the Lake Zaranik zone, which do have small areas of real interest which should be protected from oil pollution if at all possible.

Acknowledgements

We wish to thank Mr. David Merrie of BP Petroleum Development, Egypt, for his support and interest in our project, and for making arrangements for transport and accommodation. Mr. Waheed Salama of the North Sinai Governorate, ably assisted in the field and hospitably accommodated us when we were unable to return to our camp. Dr. George Michaelis of the Egyptian Wildlife Service alleviated some of our camping hardship.

References

Baha el Din, S. and Salama, W. (1984). Waterbird migration study at Zaranik, Sinai, 2-24 September 1982. Courser 1: 28-43.

Cramp, S. (1983). Handbook of Birds of Europe, the Middle East and North Africa. Vol. III, Waders to Gulls. Oxford University Press, Oxford.

Cramp, S. (1985). Handbook of Birds of Europe, the Middle East and North Africa. Vol. IV, Terns to Woodpeckers. Oxford University Press, Oxford.

Direction des Relations Economiques Extérieures (1982). Bardawil Lagoon Development Program, Vol. 1, Main Report, France.

Fry, C.H.; Atta, G.A.M.; Baha el Din, S. and McKay, C.R. (1985). Bardawil Lagoon Baseline Studies: 1985 Spring Ornithological Survey. Report to BP Petroleum Development Ltd., Egypt. Centre for Environmental Management and Planning, Aberdeen.

Fouda, M.M. (1985). Ecology of Bardawil Lagoon, Northern Coast of Sinai Peninsular, Unpublished MS, Cairo.

Paran, Y. (1980). Some notes on waterbirds on the north coast of Sinai, 1978/79. Orn. Soc. Mid. East Bulletin 4: 2-5.

Paran, Y. and Paz, U. (1978). Autumn migration of waterbirds on the north coast of Sinai. Poster presentation, XVII Int. Orn. Cong., Berlin.

Paz, U. and Paran, Y. (1974). Summary of migration of waterbirds in the north of Sinai. Report, Nature Reserves Authority in Israel.

Petersen, I. and Sørensen, V.G. (1980). Ornithological observations from the Bardawil Lagoon, North Sinai, Egypt, October 1979. Report, Huble.

Petersen, I. and Sørensen, V.G. (1981). Migration studies from the eastern part of the lagoon "Sabkher el Bardawil" on the north coast of the Sinai Peninsular 7-30 September 1980. Report of the 2nd Danish Ornithological Expedition to Egypt 1981. Report, Copenhagen.

Petersen, I. and Sørensen, V.G. (1982). 3rd Danish Ornithological Expedition to Egypt, Autumn 1981, Prelim. Report, Copenhagen.

Prater, A.J. (1981). Estuary birds of Britain and Ireland. T. & A.D. Poyser, Calton.

Rabinowitz, D.; Gorney, E. and Matthews, G. (1982). Bardawil Lagoon, North Sinai, a summary of environmental characteristics. Sinai Conservation Group Report, UK.

Zu-Aretz, Sh. (1980). Autumn migration survey of waterbirds, October 1980. Report, Nature Reserves Authority in Israel.

POLLUTANT LEVELS AND THEIR EFFECTS ON MEDITERRANEAN SEABIRDS

Marco Lambertini, LIPU Lega Italiana Protezione Uccelli,
Vicolo San Tiburzio 5/A, 43100 Parma, Italy
Centro tecnico scientifico di Idrobiologia e pesca,
Via dell'Ambrogiana 2, 57100 Livorno, Italy

Claudio Leonzio, Dipartimento Biologia ambientale,
Università di Siena, Via delle Cerchia 3, 43100 Siena, Italy

Introduction

The Mediterranean is a semi-enclosed basin with only limited water exchange with the Atlantic ocean through the Straits of Gibraltar.

From the ecological point of view the Mediterranean is an oligotrophic body of water with limited productivity and displays a high species diversity. All these geographic and ecological features make the Mediterranean highly vulnerable to pollutants.

The heavy industrialization of the north western coasts, the high polluted continental waterways feeding the Mediterranean, urban unloading, atmospheric fall-out and rain cleaning are the major sources of pollutant input. So the Mediterranean basin could be considered to be potentially exposed to heavy pollution. For this reason in recent years increasing attention has been focused on contaminant levels. Monitoring surveys of coastal waters have been carried out in an attempt to establish baseline data on the actual pollution level. Water, sediments and several marine organisms (particularly fish and invertebrates) have been investigated (see Fowler,1985).

Data on marine birds have appeared only recently. From these surveys it appears that there are three main categories of contaminants potentially harmful to wild birds: heavy metals (particularly mercury), chlorinated hydrocarbons and crude oil.

NATO ASI Series, Vol. G 12
Mediterranean Marine Avifauna
Edited by MEDMARAVIS and X. Monbailliu

The aim of this paper is to review the presently available data on mediterranean seabirds and to discuss levels and effects looking at the whole basin.

Sampling Procedures

Both methodological and ethic problems occur in studying pollution levels and their effects on wildlife. Firstly, in order to correctly understand and discuss the results, it is necessary to know the ecology of the species very well, particularly the feeding ecology. Almost the total body burden of contaminants is taken with the food. Seabirds show a great variety in diet and foraging habits. In order to avoid intraspecific competition adaptive pressure has produced a lot of feeding strategies and fishing techniques resulting in the zoning of birds with respect to type of habitat and prey (Ashmole, 1971). Several ecological niches appear in apparently homogeneous habitats such as the sea. Species-specific diets and behaviours are evident but the occurrence of variation within the same species is also well documented: sex, age and local learned behaviour are known to strongly influence the composition of diet through biometric features, experience, search images, etc.

Specific phenology is also very important in interpreting the data obtained and planning research. Resident species reflect the local status of pollution better than migratory ones which may retain pollutants from wintering/breeding areas or show low levels when coming from non-polluted regions. The contaminant content of eggs appears to come directly from food and migratory species seem to reflect the contamination status of the foraging areas of the pre-laying period (2-3 weeks) (Clark and Peakall, 1976).

Different intake, regulation, retention and protection mechanisms occur among birds such as the biological effects of specific sensibility to a toxin.

An ethical problem arises in the study of a threatened species where disturbance and stress must be kept to a minimum. For example Audouin's Gull (*Larus audouinii*), a very rare and threatened bird endemic in the Mediterranean (global population about 5,000 pairs), was monitored by taking only deserted or unhatched eggs, nestling's down, dead chicks and adults occasionally found in the colony. Visits to the breeding colony were also reduced to a minimum (three per reproductive season) to avoid disturbance (Lambertini et al., in press).

Pollution in the Mediterraean

- General features

Surface water, sediment and marine biota taken from several mediterranean regions were analyzed for metals and organochlorines (see Fowler, 1985). Mercury concentrations in certain pelagic fishes are higher than in corresponding specimens from the Atlantic (Baldi et al., 1978). On the other hand mercury levels in open waters and deep sediments are similar to those outside the Mediterranean. Particularly high mercury levels are found in marine organisms from the Tuscan archipelago. As for chlorinated hydrocarbons, the content of DDT and its metabolites is low while the levels of PCB's are higher, especially in samples from the northern Tyrrhenian coasts (Focardi et al., 1981).

Oil pollutants are also another serious problem in the Mediterranean. This problem is long-standing but only now quantitative data are being made available. Fowler (1985) indicates that about 0.5 - 1.0 x 10^6 tons per year of crude oil are spilled into the Mediterranean. If this is true the Mediterranean is the most oil polluted sea in the world. Oil pollution is typically in patches but an unbroken oil film is beginning to spread over the whole basin surface. The highest risk areas for oil pollution are not those for other chemicals. The Alboran sea and the north african coast are the most endangered. This is due to the heavy tanker traffic and the many scattered oil-loading ports along the coast.

Even if the data are insufficient to present a complete picture of the situation it appears that the north-western region is more polluted by chemical contaminants. A lot of "hot spots" exist and are related to geological and more often to anthropogenic sources (industrial sites, river outflows, harbours, coastal towns).

Data exist on water, sediments, fish and small marine biota but there is less information on the contamination of mediterranean seabirds.

- Seabirds

Birds are known to be good indicators and early warning systems of ecosystem and habitat modifications, including the effects of pollution, especially in aquatic habitats (Fimreite, 1971; Prestt, 1971). Warm blood, high metabolic rate and high trophic levels (all seabirds are at least secondary consumers) make it possible to obtain integrated information on pollution in the ecosystem from the study of wild birds. Their diet spectrum contains many man-edible items and could often give us direct information on the risk of contamination for man, especially coastal populations which depend more on sea products.

Until some years ago the data available on Mediterranean avifauna was very poor and scattered. Recently a long term study performed by the Department of Environmental Biology in Siena has provided information on the contamination of wild birds. This study revealed the distribution of trace elements and chlorinated hydrocarbons in many bird species, only a few of which however are strictly linked to the marine environment.

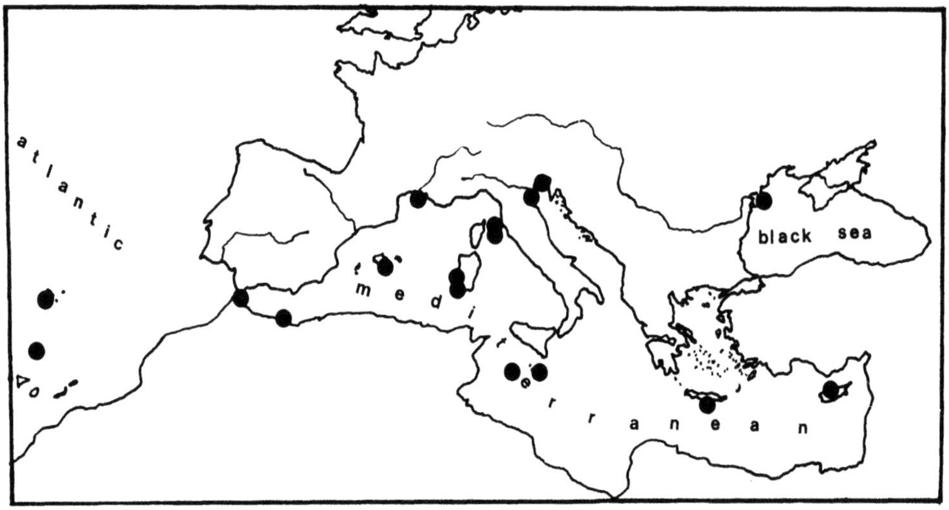

Fig. 1: Sampling sites for studies on seabird pollution in the Mediterranean and neighbouring regions.

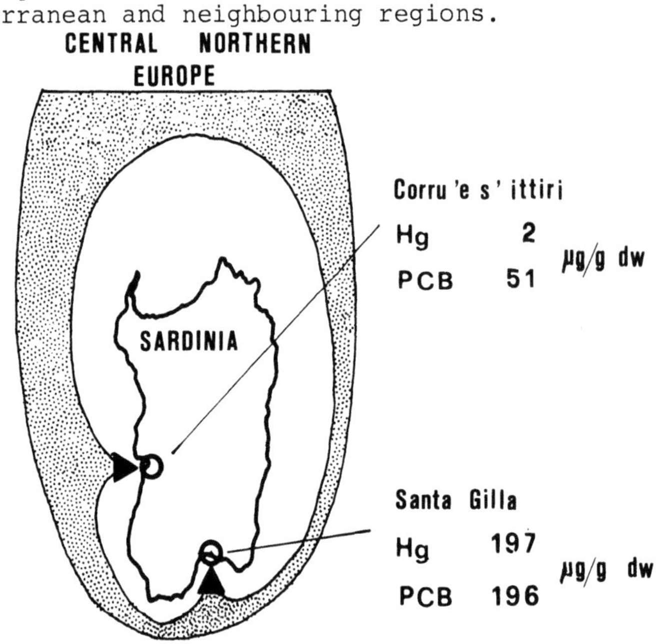

Fig. 2: Average concentrations of mercury in the liver and of polichlorobiphenyls in the fat of cormorants wintering in a polluted and in a not-polluted lagoon (modified from Cottiglia et al., 1982).

Heavy metals

The Mediterranean has been shown to be an area with high mercury levels. The reason lies not only in the anthropogenic input but also in geochemical and geographical features (International Register of Potentially Toxic Chemicals, 1978). High mercury levels have been recorded in several marine organisms (Leonzio et al., 1981). Some predatory fish such as tuna (Thunnus thynnus) and swordfish (Xiphias glaudius) accumulate very high amounts of the metal (Renzoni et al., 1978). Similar or higher levels have been recorded in the eggs and tissues of Cory's shearwater (Calonectris diomedea): about five times those recorded in specimens living in the north-western Atlantic. Slightly higher levels have been found in the eggs of another fish-eating species, the Audouin's gull (Larus audouinii) collected in the Tuscan archipelago. The mercury found in the eggs of piscivorous birds from the Danube delta in the Black sea is very low and of the same order as that found in atlantic Cory's Shearwaters.

Some species which are not strictly predatory such as Yellow-legged Gull (Larus cachinnans) have a mercury content of the eggs and the tissues 3-5 times lower than the fish eating birds. Analysis of down samples from newly hatched Yellow-legged Gull and Audouin's Gull chicks collected at the same time on Capraia island (north Tyrrhenian sea) showed a lower mercury contamination in the first species (Lambertini, 1982) which largely feeds at refuse dump and on non marine material.

The existence of "hot spots" for mercury mainly due to human activity is well documented by the results on Cormorants (Phalacrocorax carbo), Little Terns (Sterna albifrons) and Gull-billed Terns (Sterna nilotica) collected in a sardinian lagoon polluted by a chlor-alkali plant: in the eggs and tissues of the birds mercury reaches the highest levels recorded for the Mediterranean (fig. 2).

In the tissues and organs examined mercury residues show a similar distribution for all species (fig. 3). The highest concentrations are found in the liver with decreasing levels in the kidney, muscle, brain, uropygial gland and fat. In almost all the material examined so far the differences existing in the mercury levels of egg components are confirmed: mercury residues are about five times higher in the albumen than in the yolk. In the albumen most of the mercury is in organic form and associated with protein of the same molecular weight as albumen. The data don't permit conclusive remarks on the trend in contamination during the year nor a forecast for the future. Nevertheless mercury seems to be increasing in the eggs of Audouin's Gull collected in 1981-82-83 on two islands in Tuscan archipelago (Lambertini et al., in press).

Data on cadmium in the Mediterranean basin has revealed very low levels in aquatic invertebrates and fish (see Fowler, 1985). In the bird eggs cadmium never exceeded 0.05 ppm. d.w. In target organs such as the liver and kidney cadmium is high

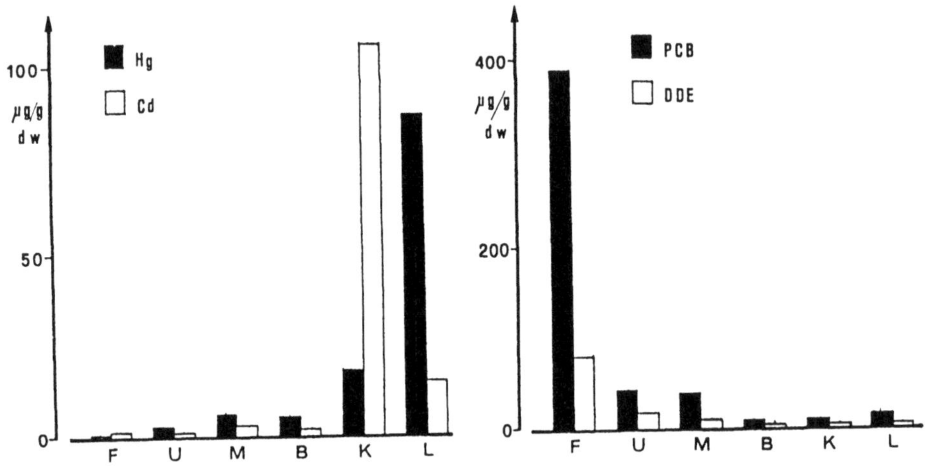

Fig. 3: Distribution pattern of metals and chlorinated hydrocarbons in tissues and organs of Cory's Shearwater. F: subcutaneous fat; U: uropygial gland; M: muscle; B: brain; K: kidney; L: liver. (Modified from Renzoni et al., 1986).

only in the Cory's Shearwater. Birds belonging to a similar trophic level such as cormorants and shags have little cadmium in their tissues. The pertinent differences among these predatory species are their feeding zones and the different species of organisms they prey on. Similar observations have been reported by Anderlini et al. (1972) and Hutton (1981) for other seabirds.

In the numerous results from very different areas with a representative group of species, lead never exceeds 0.25 ppm d.w. in eggs and 1.5 ppm d.w. in the liver. Lead concentrations in some tissues of marine vertebrates are probably regulated (Phillips, 1977) but the liver, kidney and particularly the bone accumulate lead available in the environment (Longcore et al., 1974). Present data clearly reflect very low levels of lead or its non-availability in ecosystems of several areas of the Mediterranean.

Chlorinated hydrocarbons

The most widespread chlorinated hydrocarbon residues in marine organisms of the Mediterranean are PCB's and DDE. They are found from plankton to the highest vertebrates and seem to be increasing along the various steps of the food chain (see Fowler, 1985). Seabirds, particularly the fish-eating species, have considerable pesticide residues in the tissues and eggs (tab. 1 and 2). The general connection between feeding habits and residue levels in eggs and tissues is evident except in the Yellow-legged Gull (Larus cachinnans), a typical scavenger (eggs values are 1.20-12.60 ppm d.w. for PCB's and 1.00-8.90 for DDE). This anomalous situation may be due to the altered feeding habits of this species (Focardi et al., 1980); in fact this gull has shifted its feeding from the sea to urban refuse dumps.

Many authors report that the levels of organichlorines in the eggs reflect the situation of these compounds in the bird's diet and thus in the environment where they live and feed (Clark and Peakall, 1976; Focardi et al., 1980). From our data it can be seen that throughout the Mediterranean basin the most abundant organochlorines are the PCB's (in all samples the PCB's/DDE ratio is greater than 1). This is due to the heavy use of these compounds in many industrial activities. By contrast in the Danube delta (Fossi et al., 1984) the most abundant residue is DDT and its by-products, e.g. pp'DDE and pp'DDD (PCB's/DDE ratio less than 1). This is a most interesting result examplified in the cormorants and white pelicans because it permits us, by the PCB's/DDE (or PCB's/DDT's) ratio to separate areas of prevalently agricultural contamination from areas of great industrial development. In fact the same range of PCB's/DDE ratios as in the Mediterranean (2-4) was found in many atlantic seabirds (Risebrough et al., 1976) and in eggs (1.1-7.9) of seabirds collected in eastern Canada (Pearce et al., 1979).

Our data show two "hot spots" for chlorinated hydrocarbons, the Tuscan archipelago and the Balearic islands. The high levels found in the Tuscan samples are explained by the presence in this area of big harbours like Livorno and La Spezia and large industrial centres (Focardi et al., 1980). It is difficult to understand the Balearic situation, particularly the high PCB's and DDE levels found in eggs and tissues of Cory's Shearwater. In this species the data collected in the mediterranean area when compared to the Atlantic (both for eggs and tissues) confirm the high contamination of the Mediterranean for many pollutants such as mercury and PCB's. In addition a time dependent increase of PCB's has been noted in the long term monitoring of eggs and tissues of marine and terrestrial birds (Boldreghini et al., in press; Lambertini et al., in press; Massi, in prep.).

Crude oil

In contrast to northern Europe and America, very little data have been gathered on the effects of oil pollution on me-

Species	Site	Reference	Hg	Se	Cd	Pb	DDE	PCB's	Tissue	N
Cory's Shearwater (Calonectris diomedea)	Selvagen	Renzoni et al.1986	15.53 2.28	33.64 19.62	26.45 1.30	0.97 0.66	0.35 0.22	0.78 0.48	L	3
"	Majorca	"	49.69 23.08	25.38 5.48	7.52 2.38	0.67 0.20	4.74 2.11	18.75 7.28	L	5
"	Linosa	"	86.16 99.16	132.68 72.27	13.91 15.17	0.81 0.48	1.84 1.50	3.39 2.95	L	5
	Crete	"	74.85 50.94	32.99 21.29	55.92 39.76	0.76 0.47	2.65 0.92	4.23 2.02	L	5
Manx Shearwater (Puffinus puffinus)	Tyrrhe-nian Sea	Lambertini in prep.	16.07	44.00	1.16	1.18	2.33	7.96	L	1
"	Gibraltar	Bourne & Bogan,1976					0.93-1.80*	0.99-1.10*	M	2
Storm Petrel (Hydrobates pelagicus)	Tyrrhe-nian Sea	Lambertini in prep.	3.57	36.00	1.03	2.33	19.32	72.94	L	1
Shag (Phalacrocorax aristotelis)	Gibraltar	Bourne & Bogan,1976					2.00-5.40*	4.10-7.10*	M	2
"	Capraia	Lambertini in prep.	65.58 83.70	47.73 38.10	0.72 0.48	1.05 0.78	1.24 0.98	26.41 23.20	L	3
Cormorant (Phala-crocorax carbo)	Santa Gilla	Focardi et al., 1986	197.04 118.38	54.37 58.74	0.60 0.42	0.83 0.71	1.89 2.06	5.08 6.20	L	7
"	Corru'e s'Ittiri	"	10.94 10.01	2.57 0.84	0.58 0.37	0.16 0.19	0.72 0.31	2.20 1.12	L	3
"	Marano Lagoon	"	42.43 42.88	37.71 47.85	0.19 0.19	0.47 0.13	2.82 1.93	4.06 3.90	L	3
Mediterranean Yellow-legged Gull (Larus cachinnans)	Elba	Leonzio in prep.	13.30 9.25	17.03 6.81	0.73 0.37	1.82 2.01	2.12 1.32	23.59 24.56	L	7
	Po delta	"	13.08 6.62	15.87 1.68	3.54 1.87	3.78 3.40	0.89 0.51	8.56 6.83	L	4
"	Capraia	Lambertini 1981	5.07* 3.96						D	40
"	Capraia	Lambertini 1982	8.76* 4.08						D	23
"	Madeira	Bourne & Bogan,1976					0.40*	0.40*	M	1
"	Gibraltar	"					0.63-1.80*	0.82-3.80*	M	2
Audouins gull (Larus audouinii)	Capraia	Lambertini, 1982	14.96* 3.16						D	20
Lesser black-backed Gull (Larus fuscus)	Gibraltar	Bourne & Bogan,1976					0.57*	0.52*	M	1

Tab. 1: Trace elements and organochlorines concentrations in tissues of mediterranean seabirds (L: liver; M: muscle; D: down of newly hatched nestlings). Levels are expressed as arithmetic mean ± standard deviation or min.-max. measurements (mg/kg dry weight or wet weight when countersigned by the asterisk).

diterranean seabirds. The reason for this is that there have been no ecological disasters so far from incidents on the oil drilling platforms or tankers. The Mediterranean is crossed daily by an enormous number of tankers and a considerable amount of crude oil is pumped into the sea with water taken aboard as ballast.

The results of the Spanish Beached Bird Survey (unpublished data in De Juana et al., 1984) lead to the conclusion that along the spanish mediterranean coasts oil is not yet the major problem for seabirds. On the other side, during field observations every coastal birdwatcher has often seen gulls and terns dirty with black oil spots. All the eleven shags (Phalacrocorax aristotelis) found dead in the fishing nets of Capraia island (northern Tyrrhenian sea) had different degrees of oil on the feathers (Lambertini, pers. comm.). Guyot and Thibault (1985) during 1980-82 found several oily dead specimens of Shag and Manx Shearwater (Puffinus puffinus) along the Bocche de Bonifacio coast, in Corsica. Baha el Din et al. (1986) consider oil pollution to be the first threatening factor for the seabird populations in mediterranean Egypt and Araujo et al. (1977) hypotized oil effects on seabirds in the Calabrian archipelago.

Impact assessment

Following and modifying Nisbet's (1980) classification we divide the effects of environmental contaminants into four categories listed in diminishing order of severity: a) acute (deadly or otherwise) poisoning of adults and chicks; b) delayed lethal effects from accumulation and damage to organs and tissues; c) sub-lethal effects (not reproductive); d) effects on reproductive performance in the absence of overt mortality.

We must remember that all we know about the effects of pollutants on birds comes from laboratory studies. This information is often contradictory and gathered under controlled conditions in a very short period of time. Synergistic and antagonistic effects of contaminants, intermittent intake due to migratory or erratic movements, efficient mechanisms of detoxication and interspecific sensitivity have to be taken into account in extrapolating from experimental data. In addition the similarity of the symptoms produced by a wide range of pollutants poses some difficulties in assessing the cause-effect relation.

The effect of pollution on wild birds in the Mediterranean region has not been sufficiently investigated so that this discussion is necessarily brief and mainly speculative. It may be useful to survey the effects of the main types of mediterranean pollutants.

Species	Site	Reference	Hg	Se	Cd	Pb	DDE	PCB's	N
Cory's Shearwater (Calonectris diomedea)	Selvagen	Renzoni et al., 1986	2.09 1.04	3.43 0.80	< 0.03	< 0.30	1.49 0.68	4.19 1.70	20
"	Majorca	"	7.29 1.94	5.47 0.81	< 0.03	< 0.30	18.77 12.46	95.08 56.72	10
"	Linosa	"	5.90 1.45	5.89 1.43	< 0.03	< 0.30	3.75 2.08	10.59 4.18	11
"	Crete	"	4.78 - 5.76	3.29 - 4.32	< 0.03	< 0.30	4.79 - 9.99	7.96 - 20.81	2
"	Crete	Bourne & Bogan, 1976					2.90 - 3.50*	3.20 - 4.10*	2
Shag (Phalacrocorax aristotelis)	Capraia	Lambertini et al., in press	2.14 - 5.00	2.48 - 2.70	< 0.03	< 0.30	6.36 - 7.05	23.77 - 25.92	2
Cormorant (Phalacrocorax carbo)	Danube delta	Fossi et al., 1985	4.98 2.23	4.48 1.74	< 0.03	< 0.30	57.20 19.35	23.58 12.86	19
White pelican (Pelecanus onocrotalus)	Danube delta	"	3.55 1.94	3.49 0.78	< 0.03	< 0.30	15.38 5.34	5.58 2.23	12
Yellow-legged Gull (Larus cachinnans)	Elba	Focardi et al., 1980	2.47 1.20	1.67 0.46	< 0.02	< 0.20	5.52 2.71	29.40 15.55	13
"	Capraia	"	3.37 2.29	1.32 0.30	< 0.02	< 0.20	7.50 2.52	42.60 14.53	12
"	Capraia	Lambertini 1981	1.09* 0.86						
"	Capraia	Lambertini 1982	0.64* 0.41						
"	Rhone delta	Mendola et al., 1977					1.90 - 10.00*	16.00 - 160.00*	6
"	Crete	Bourne & Bogan, 1976					2.58*	5.32*	1
"	Cyprus	"					8.90 - 29.00*	1.20 - 4.20*	2
"	Balearic	Bijleveld et al, 1979	1.12*				3.55*	16.75*	1
Audouin's Gull (Larus audouinii)	Chafarinas	"	0.70 - 0.95*				1.00 - 2.80*	3.10 - 4.20*	3
"	Balearic	"	0.87 - 1.39*				2.20 - 5.40*	12.80 - 20.60*	2
"	Capraia	Lambertini 1982	1.27* 0.31						5
"	Capraia	Lambertini et al. in press	11.26 3.46	6.63 1.23	< 0.02	< 0.20	6.78 2.63	41.16 15.01	11
"	Elba	"	9.40 3.08	5.70 1.06	< 0.02	< 0.20	3.30 0.81	21.88 5.19	5
Little Tern (Sterna albifrons)	Mistras lagoon	Focardi et al., 1980	7.87 5.13	1.14 0.52	< 0.02	< 0.20	5.05 0.67	8.69 2.18	6
"	Macchiareddu lagoon	"	38.83 23.85	3.44 1.28	< 0.02	< 0.20	5.09 0.57	15.25 2.20	6
"	Po delta	Boldreghini et al., in press	3.83 2.59	2.08 0.68	< 0.02	< 0.20	1.95 0.45	7.92 3.61	6
"	Po delta	Renzoni et al., 1982	2.26 1.15	2.52 1.56	< 0.02	< 0.20	11.29 5.77	12.40 5.14	13
"	Po delta	"	6.76 4.80	2.19 0.63	< 0.02	< 0.20	3.83 2.24	7.79 7.41	16
Common Tern (Sterna hirundo)	Po delta	Boldreghini et al., in press	2.19 1.01	1.95 0.58	< 0.02	< 0.20	6.04 4.81	24.52 15.28	12
"	Marano lagoon	Renzoni et al., 1982	10.16 3.56	3.63 1.28	< 0.02	< 0.20	3.74 0.80	9.67 1.65	22
Gull-billed Tern (Sterna nilotica)	Po delta	"	1.25 0.54	1.49 0.47	< 0.02	< 0.20	11.47 4.69	4.95 5.48	15
"	Macchiareddu lagoon	"	15.23 6.62	7.04 2.65	< 0.02	< 0.20	2.97 2.11	3.19 1.95	7

Tab. 2: Trace elements and organichlorines concentrations in the eggs of mediterranean seabirds. Levels are expressed as arithmetic mean ± standard deviation or min.-max. measurements (mg/kg dry weight or wet weight when countersigned by the asterisk).

- Heavy metals

Many experimental studies point out the wide range of serious biological effects of metals on birds. Acute mercury poisoning damages liver, kidney and central nervous system. Sublethal doses affect physiological, bahavioural and reproductive performance (see Fimreite, 1980). Neuro-pathological dose-dependent effects are the most typical and dangerous (Casarett and Doull, 1980). Experimentally Fimreite (1971) found many reproductive alterations in Red-necked pheasants treated with dietary levels of methylmercury, the most common organic form of mercury in biological organisms and eggs (Fimreite et al., 1974): low hatchability, decreased egg production, reduced egg weight and size, inhibitory effects on fertilization, shell, less eggs and abnormal egg coloration, the latter at the highest doses. Mercury concentrations in egg samples with significantly decreased hatchability were in the range 0.5-1.5 ppm. w.w.

Lambertini (1981) found a negative correlation between volume and mercury levels in 22 eggs of Yellow-legged Gull (Larus cachinnans) from Capraia island in the northern Tyrrhenian sea, a highly mercury polluted area. The NOEL (no-effect-level) appeared to be about 1 ppm. w.w. These results were not confirmed in subsequent years (Lambertini unpublished data) nor in Audouin's Gull (Larus audouinii) (Lambertini et al., in press). The down of unhatched or just-hatched dead chicks of Larus cachinnans showed higher mercury levels than successfully hatched ones (Lambertini, 1981). Confirmation via a larger statistical sample is necessary.

Mercury is also known to cause behavioural changes in wild birds (Peakall, 1985). Although the effects could be dose-dependant, it is more likely that they occur above a certain NOEL. Experiments in which methyl-mercury was injected into eggs at the beginning of incubation demonstrated a NOEL between 0.05-0.5 ppm for nestling behavioural responsiveness (Hughes et al., 1976). As suggested by Murton et al. (1977) for cadmium in the Atlantic ocean, the natural occurrence of mercury in the Mediterranean could have led to the development of special defence mechanisms in certain mediterranean seabird populations. So the high levels of mercury detected in eggs of Audouin's Gull (Larus audouinii), an endemic mediterranean gull (Lambertini et al., in press) do not seem to compromise the hatching success of the species in the Tyrrhenian sea (Arcamone et al., in press).

In general there is no direct evidence that the high mercury background in the Mediterranean influences the bird biology. Nevertheless environments heavily polluted by human activities may be harmful to local wildlife (fig. 3). In these places regular field monitoring is urgently required.

Cadmium is found in significantly high levels only in the tissues of Cory's Shearwater (Calonectris diomedea) (Renzoni et al., 1986). In the eggs of the same species cadmium was always below the limits of sensitivity of the instrument (0.03 g g^{-1}) as a result of an efficient barrier. Less is known about the effects of cadmium on birds. It would also appear from mediterranean data that pelagic seabirds such as procellariformes accumulate exceptional amounts of this element (Fleischer et al., 1974, in Bourne, 1976). Probably the cadmium is of natural origin and these birds may have evolved a protective mechanism by protein banding and subsequent elimination by the kidney (Hutton, 1981).

- Chlorinated hydrocarbons

Organochlorines are well known to be toxic to birds. They can induce a great number of acute and sub-lethal effects. Dieldrin and relative cyclodiene insecticides are the most toxic and they are hardly present in the Mediterranean. DDT and PCB compounds are the most widespread molecules of the group. They are toxic at relatively high concentrations compared to other pollutants but their toxic metabolites are very persistent and they accumulate in the organism. After the uptake organochlorines tend to concentrate in both intestinal and subcutaneous fat. During feeding stress, starvation, high activity or hormonal stimulation the fat is consumed and the pollutants are mobilized and the toxins take effect. So it is difficult to correctly explain the tissue levels of these contaminants because the wide variation in fat content. In addition great caution should be used in interpreting the results since responses vary considerably from one species to another and a wide range of sensitivity is shown. Gulls, for example, seem to have a remarkable tolerance to organichlorines (Keith and Gruchy, 1972) while pelecaniformes are more sensitive (Peakall, 1975).

The toxic effects of DDT include several behavioural alterations in feeding, social and breeding activities (see Peakall, 1985). Acute poisoning in adults manifests in increasing lack of coordination of movements as results of central nervous system damage. The distribution trend in tissues and organs shows an increase in DDE in the kidney, brain, liver, muscle, uropygial gland and fat (Renzoni et al., in press) (fig. 3). Eggshell thinning is a well-documented effect presumably a result of DDE interference with calcium metabolism in several bird species (Ractliffe,1970; King et al., 1978). This is not a dose-dependant effect nor any research allows a NOEL value to be established. No eggshell thinning is revealed in Yellow-legged Gull (Larus cachinnans) eggs from Tyrrhenian sea (Lambertini, unpublished data) with an average value of 6.50 ppm d.w. DDE. Brain levels approaching 10 ppm DDT's are often lethal and don't occur in mediterranean seabirds.

PCB's have a relatively lower toxicity than DDT compounds. The distribution trend is the same as DDT's. Acute biological effects include liver and brain damage at hundreds of ppm. The death of a number of cormorants (Phalacrocorax carbo) in the Netherlands was attributed to high levels of PCB's in liver (390 ppm, mean value) and brain (190 ppm, mean value) (Koeman et al., 1973). There is little direct evidence that PCB's alone have had significant effects on breeding performance in seabirds. Eggshell thinning induced by PCB's requires confirmation but it is certain that there is a delay in the onset of breeding (Peakall, 1975).

The PCB's/DDE ratio demontrates the anthropogenic source of these contaminants. In western Europe agricultural use of DDT has not been permitted for many years but it is still employed in eastern countries. On the other hand PCB's are widespread and more abundant in the western and more industrialized areas. We could therefore expect a low ratio in the eastern and a high ratio in the western Mediterranean (Fossi et al., 1984; Renzoni et al., 1986). In the west a progressive decrease of the ratio is now evident: this is due to the decrease in DDE year by year and the opposite constant increasing input of PCB's. This trend must be attentively monitored.

- Crude oil

Oil pollution is a particular serious hazard to seabirds because oil floats on the surface of the water where it comes into contact with swimming birds. Oil firstly destroys the buoyancy of the plumage and insulation, making the bird flightless and unable to thermoregulate. Birds in contact with fresh oil may inhale volatile toxic components and may swallow oil during preening. Respiratory and gastro-intestinal irritation are the first effects. More systematic absorption can lead to damage of the liver, kidney, pancreas and adrenals (Gorman and Milne, 1971, in Bourne, 1976). Small tar balls of aggregated oil can be ingested as mistaken for food (Diamond, 1974, in Bourne, 1976). Oil pullution has more subtle effects too. Oily nest material, breast/abdomen feathers and feet can transfer oil to the eggs during incubation, killing the developing embryo by hindering respiratory processes through the shell (White et al., 1979). The high levels of oil pollution in the Mediterranean presuppose the occurrence of local acute and deadly poisoning, oiling episodes and sublethal effects due to systematic absorption or eggshell smearing. More data are needed especially from the african coast.

Other Hazards

In addition there are several other pollution hazards for mediterranean seabirds. We can only list the main types, discussing them where possible because very little information and field data are available.

Sewage is a hazard more serious for man than for seabirds. Human derived pathogenic agents can be dangerous to wild birds but more often the contrary is true. No relative data are available for the Mediterranean. Sewage pollution is typically concentrated about the coast near big towns or the estuaries of rivers coming from highly polluted areas. It is general knowledge that the sewage situation in Mediterranean is worsening. The toxin produced by the anaerobic bacterium *Clostridium botulinum* is a particularly well-known cause of mortality among waterbirds occurring more often in the warmer seasons and latitudes (see Bourne, 1976 and Fowler, 1985).

Radioactivity superimposed on the very low natural background can be attributed entirely to atmospheric fall out. Mediterranean water levels are similar to those of other seas and far from being dangerous (Fowler, 1985).

Several artefacts such as floating nylon lines, plastic bags and plastic particles are known to be deadly for some cetaceans and turtles and some authors have also suggested for seabirds (Furness, 1985).

Natural Defences

For a proper estimation of the environmental impact of pollution on seabirds a very important aspect must be taken into account: the natural defence mechanisms.

The efficiency of such mechanisms seems to be very high considering that in wild birds levels of contaminants considered lethal or highly toxic for experimental birds have often been recorded without apparent effects.

The natural reaction against pollution involves behavioural, metabolic and biochemical processes. The latter forms the basis of a detoxification system. In injury derived from heavy metals, metallothioneins play a central role. These low molecular weight proteins synthesized in the liver have emerged as a key compound in regulating the flow of metals through the organism in health and disease. Recently a cadmium binding protein with properties similar to metallothionein was demonstrated in the kidney of pelagic seabirds (Hutton, 1981). The minimal residues of cadmium and lead in eggs suggest that the female is able to regulate metal deposition in egg components, perhaps via an efficient binding mechanism.

Mercury that occurs in the natural environment mainly in the methylated form (Buffoni et al., 1982) cannot be bound to metallothioneins. Nevertheless there is evidence that combination with selenium leads to neutralization of its toxicity. The process could be performed by the demethylation of methylmercury and the complexation of mercury and selenium in a non-toxic compound accumulated in the liver cells (Thibaud in press). The recovery in wild birds of proportional amounts of selenium and mercury in the liver is considered a sign of this detoxification process. In some mediterranean areas heavily polluted by mercury, the Cormorant (*Phalacrocorax carbo*) and the Black-necked grebe (*Podiceps nigricollis*) showed a highly significant selenium/mercury correlation in the liver (Cottiglia et al., 1982).

Many foreign compounds of anthropogenic origin undergo mechanisms for their detoxification and/or elimination. This may be achieved by a number of apparently universally distributed enzyme systems that function to convert apolar chemicals to more water soluble and excretable metabolites. The first phase is a biotransformation catalyzed by the cytocrome P-450 or mixed function oxidase (MFO) enzyme system; in the second phase the resulting product may be conjugated for further elimination.

Recently enzymatic activity of this type correlated to xenobiotic metabolism has been reported in a number of mediterranean marine birds (Fossi et al., in press).

Conclusions

Crude oil and derivates are widespread and quite abundant in the Mediterranean. So local episodes of acute poisoning/oiling are likely to occur. No sufficient data are available to assess reproductive effects of alifatic hydrocarbons but they are probably more than potential. At the present minor forms of pollution like heat, sewage and radioactivity don't seem to be relevant.

Among toxic chemicals, mercury and organochlorines (PCB's and DDE) are the most widespread and potentially dangerous pollutants. The Mediterranean is characterized by high mercury levels from geological sources and it is possible that certain mediterranean seabird populations developed defence mechanisms. PCB's are more present in the north-western share of the basin while DDE in the eastern one. The PCB's/DDT's ratio is generally increasing year by year for constant/increasing input of PCB's and gradual breakdown of DDT's. All these compounds don't seem to reach sufficient levels to rise acute poisoning in seabirds. So we can exclude acute poisoning in adult birds except in sporadic episodes like near industrial sites or heavily polluted river outflows. Instead it is likely that chemical pollutants harm some species, especially the more sensitive, at sublethal and reproductive level.

As Nisbet (1980) said we agree that without detailed study of potential sublethal toxic effects, no mediterranean seabird species can be assumed to be normal in its reproductive performance, physiology and behaviour. More field data are required, considering the high vulnerable status of mediterranean ecosystems.

References

Anderlini, V.C., Connors, P.G., Risebrough, R.W. and Martin, J.H. (1972). Concentrations of heavy metals in some Antarctic and North American seabirds. Proc. Colloq. Conserv. Probl. Antarctica, Blacksburg, Virginia.

Araujo, J., Munoz-Cobo, J. and Purroy, F.S. (1977). Las rapaces y aves marinas del archipelago de Cabrera. Naturalia Hispanica 12, 94 pp., ICONA, Madrid.

Aracmone, E., Lambertini, M. and Meschini, E. (in press). Il Gabbiano corso (Larus audouinii Payr.) nidificante nell'arcipelago toscano (1977-1983). Proc. 2° Conv. It. Ornit., Parma, 1983.

Ashmole, N.P. (1971). Seabird ecology and the marine environment. In Avian Biology vol. I (D.S. Farner and J.R. King, eds.), pp. 223-286, Academic Press, New York.

Baldi, F., Renzoni, A. and Bernhard, M. (1978). Mercury concentrations in pelagic fishes (anchovy, mackerel and sardine) from the Italian coast and strait of Gibraltar. IVes Journées Etud. Pollutions, pp. 251-254, CIESM, Antalya.

Boldreghini, P., Focardi, S., Fossi, C. and Renzoni, A. (in press). Elementi in tracce e idrocarburi clorurati in uova di uccelli raccolte nel delta del Po. Proc. II° Conv. It. Ornit., Parma, 1983.

Bourne, W.R.P. (1976). Seabirds and pollution. In Marine Pollution (R. Johnston ed.), pp. 403-502, Academic Press, London.

Buffoni, G., Bernhard, M. and Renzoni, A. (1982). Mercury in the mediterranean tuna. Why is their level higher than in Atlantic tuna? A model. Thalassia Jugosl. 18: 231-243.

Casarett, L.J. and Doul, J. (1980). Toxicology, Mc Millan Publ., New York.

Clark, A.L. and Peakall, D. (1976). Organochlorine residues in Eleonora's falcon (Falco eleonorae), its eggs and its prey. Ibis 119: 353-358.

Cottiglia, M., Focardi, S., Leonzio, C., Mascia, C., Renzoni, A. and Fossi, C. (1982). Contaminants in tissues of shorebirds from a polluted lagoon of the island of Sardinia. Proc. 6es Journées Etud. Pollut., pp. 293-298, CIESM, Cannes.

Fimreite, N. (1971). Effects of dietary methylmercury on ring necked pheasants. Can. Wildl. Serv. Occas. Pap. 9, pp. 39.

Fimreite, N. (1980). Accumulation and effects of mercury on birds. In The biogeochemistry of mercury in the environment (J.O. Nriagu ed.), pp. 601-627. Elsevier Biomedical Press, Amsterdam.

Fimreite, N., Holswarth, W.N., Keith, J.A., Pearce, P.A. and Gruchy, I.M. (1971). Mercury in fish and fish-eating birds near sites of industrial contamination in Canada. Can. Field Nat. 85: 211-220.

Fimreite, N., Brun, E., Froslie, A., Frederichsen, P. and Gundersen, N. (1974). Mercury in eggs of norvegian seabirds. Astarte 1: 71-75.

Focardi, S., Leonzio, C. and Renzoni, A. (1980). Chlorinated hydrocarbons and trace metals in eggs of Larus argentatus michahellis Naumann and Sterna albifrons Pallas. Proc. 1° Cong. SITE, pp. 411-414, Parma.

Focardi, S., Bacci, E., Leonzio, C. and Crisetig, G. (1984). Chlorinated hydrocarbons in marine animals from the northern Tyrrhenian sea (NW Mediterranean). Thalassia Jugoslavica 20: 37-43.

Fossi, C., Focardi, S.C., Leonzio, C. and Renzoni, A. (1984). Trace metals and chlorinated hydrocarbons in birds' eggs from the delta of the Danube. Environm. Cons. 11: 345-350.

Fossi, C., Leonzio, C. and Focardi, S. (in press). Increase of organochlorines and MFO activity in water birds wintering in an italian lagoon. Bull. Env. Contam. Toxicology.

Fowler, S.W. (1985). Assessing pollution in the mediterranean sea. In Pollutants and their ecotoxicological significance (H.W. Nurnberg ed.), pp. 269-287, Wiley and Sons Ltd., London.

Furness, R.W. (1985). Plastic particle pollution: accumulation by procellariformes seabirds at scottish colonies. Mar. Poll. Bull. 16: 103-106.

Guyot, I. and Thibault, J.-C. (1985). Nouvelles données sur la protection des oiseaux marins nicheurs en Corse. In Oiseaux marins nicheurs du midi et de la Corse, pp. 20-22. Annales du CROP n° 2, Aix-en-Provence.

Hughes, S.A., Rosenthal, E. and Sparber, S.B. (1976). Time dependent effects produced in chicks after prenatal injection of methylmercury. Pharm. Biochem. Behav. 4: 507-513.

Hutton, M. (1981). Accumulation of heavy metals and selenium in three seabird species from the United Kingdom. Environ. Pollut. 26: 129-145.

Keith, J.A. and Gruchy, I.M. (1972). Residue levels of chemical pollutants in North american wildlife. Proc. XV Int. Orn. Cong. pp. 437-454.

King, K.A., Flickinger, E.L. and Hildebrand, H.H. (1978). Shell thinning and pesticides residues in Texas aquatic bird eggs 1970. Pestic. Monitor. J. 12: 16-21.

Koeman, H., Van Velzen-Blad, W.C., De Vries, R. and Vos, J.G. (1973). Effects of PCB and DDE in cormorants and evaluation of PCB residues from an experimental study. J. Reprod. Fert. 19: 353-364.

International Register of Potentially Toxic Chemicals (1978). Data profiles for chemicals for the evaluation of their hazards to the environment of the Mediterranean Sea. UNEP 1, pp. 281-367.

Juana de E., Varela, J. and Witt, H.-H. (1984). The conservation of seabirds at the Chafarinas islands. In Status and conservation of the world's seabirds, ICBP Technical Pub. n° 2, pp. 363-370.

Lambertini, M. (1981). Inquinamento da mercurio in uova e nidiacei di Gabbiano reale (Larus argentatus) nell'isola di Capraia (Livorno). Proc. I Conv. It. Orn., pp. 93-96, Aulla.

Lambertini, M. (1982). Mercury levels in Larus audouinii and Larus argentatus michahellis breeding in Capraia island (Tyrrhenian sea). Riv. It. Orn. 52: 75-79.

Lambertini, M., Focardi, S., Fossi, C. and Leonzio, C. (in press). Il Gabbiano corso (Larus audouinii) nidificante nell'arcipelago toscano: contaminazione da elementi in tracce (Hg, Pb, Se, Cd) e idrocarburi clorurati (HCB, DDE, DDT, PCB's). Proc. II Conv. It. Orn., 1983, Parma.

Leonzio, C., Bacci, E., Focardi, S. and Renzoni, A. (1981). Heavy metals in organisms from the northern Tyrrhenian sea. Sci. Total. Env. 24: 249-254.

Leonzio, C., Focardi, S., Fossi, C. and Renzoni, A. (1984). Seabirds as indicators of mercury pollution in the Mediterranean. Meeting on the biogeochemical cycle of mercury in the Mediterranean, Siena.

Leonzio, C., Fossi, C. and Focardi, S. (in press). Heavy metals and selenium variation in a migratory bird wintering in a mercury polluted area. Bull. Env. Contam. Toxicol.

Livingstone, D.R. (1985). Response of the detoxication/toxication enzyme systems of molluscs to organic pollutant and xenobioties. Mar. Poll. Bull. 16: 158-164.

Longcore, J.R., Locke, L.N., Begley, G.E. and Andrews, R. (1974) Significance of lead residues in mallard tissues. Spec. Scient. Rep. Wildlife n° 182, pp. 1-24.

MED POL Phase II (1983). Assessment of the present state of pollution by mercury in the Mediterranean sea and proposed control measures. In Long-term programme for pollution monitoring and research in the Mediterranean sea, United Nations Environment Programme, Athens.

Murton, R.K., Osborn, D. and Ward, P. (1978). Are heavy pollutants in atlantic seabirds? Ibis 120: 106-107.

Nisbet, I.C.T. (1980). Effects of toxic pollutants on productivity in colonial waterbirds. Trans. Linn. Soc. New York, IX: 103-114.

Peakall, D.B. (1975). Physiological effects of chlorinated hydrocarbons on avian species. In Environmental dynamics of pesticides (R. Jaque and V.H. Freed Eds.), pp. 343-360. Plenum Press, New York.

Peakall, D.B. (1985). Behavioural responses of birds to pesticides and other contaminants. Residue Reviews 96: 45-77.

Pearce, P.A., Peakall, D.B. and Reynolds, L.M. (1979). Shell thinning and residues of organochlorines and mercury in seabirds eggs. Eastern Canada, 1970-76. Pest. Monit. J. 13: 61-68.

Phillips, D.H.J. (1977). The use of biological indicator organisms to monitor trace metal pollution in the marine and estuarine environments. Environ. Pollut. 13: 281-317.

Prestt, I. (1971). Techniques for assessment of pollution effects on seabirds. Proc. Roy. Soc. Lond. B 177: 287-294.

Ratcliffe, D.A. (1970). Changes attributable to pesticides in eggs breakage frequency and eggshell thickness in some british birds. J. Appl. Ecol. 7: 67-115.

Renzoni, A., Bernhard, M., Sara, R. and Stoeppler, M. (1978). Comparison between the Hg body burden of Thunnus thynnus from the Mediterranean and the Atlantic. IV Journées Etud. Pollutions, pp. 255-260, CIESM, Autalya.

Renzoni, A., Focardi, S., Fossi, C., Leonzio, C. and Majol, J. (1986). Comparison between concentrations of mercury and other contaminants in eggs and tissues of Cory's Shearwater Calonectris diomedea collected on atlantic and mediterranean islands. Environ. Pollut. 40: 17-37.

Risebrough, R.W., De Lappe, B.W. and Walker, II W. (1976). Transfert of higher molecular weight chlorinated hydrocarbons to the marine environment. In Marine Pollutant transfer (H.L. Windom and R.A. Duce eds.), pp. 261-321. Lexington Books, Lexington.

Thibaud, Y. (in press). The role of biochemical processes in the accumulation of mercury by marine organisms. Report of the meeting on the Biogeochemical cycle of mercury in the Mediterranean, Siena, 1984.

White, D.H., King, A. and Coon, N.C. (1979). Effects of n° 2 fuel oil on hatchability of marine and estuarine birds eggs. Bull. Environm. Contam. Toxicol. 21: 7-10.

HUMAN IMPACT ON SEABIRDS IN THE BALEARIC ISLANDS

Joan Mayol Serra, Conselleria d'Agricultura i Pesca,
Pje. Guillermo de Torrella, Palma de Mallorca

Introduction

The geographical location of the Balearics, situated in the middle of the western Mediterranean, attracts a rich marine avifauna: the rugged coastline consisting of calcareous cliffs and interspersed with several islets (see tab. 1) as well as the importance of its continental platform determine the composition and distribution of numerous seabirds. The relative abundance of seabirds as well as their peculiarity within the Spanish and West-European context justify that we pay a great attention to this faunistical group especially since the literature contains only a few generalized publications which are out of print.

This lack of data is due to the fact that most ornithologists are concentrating their efforts on localized areas such as the Cabrera archipelago. Most bibliographical references consist of short notes or observations without a great value.

Island	Circum-ference	Suitable breeding habitat for sea-birds	Satellite is-lets (uninha-bited)	Continental platform down to 50m depth
Mallorca	555 km	60% (coast higher than 20 m)	36	2,000 sq.km
Menorca	285	50%	23	360
Ibiza	210	60%	42	480 (Ibiza and Formentera)
Formentera	69	50%	12	
Cabrera	38	50%	17	see Mallorca

Tab. 1: Physical characteristics of the Balearic Islands.

NATO ASI Series, Vol. G 12
Mediterranean Marine Avifauna
Edited by MEDMARAVIS and X. Monbailliu

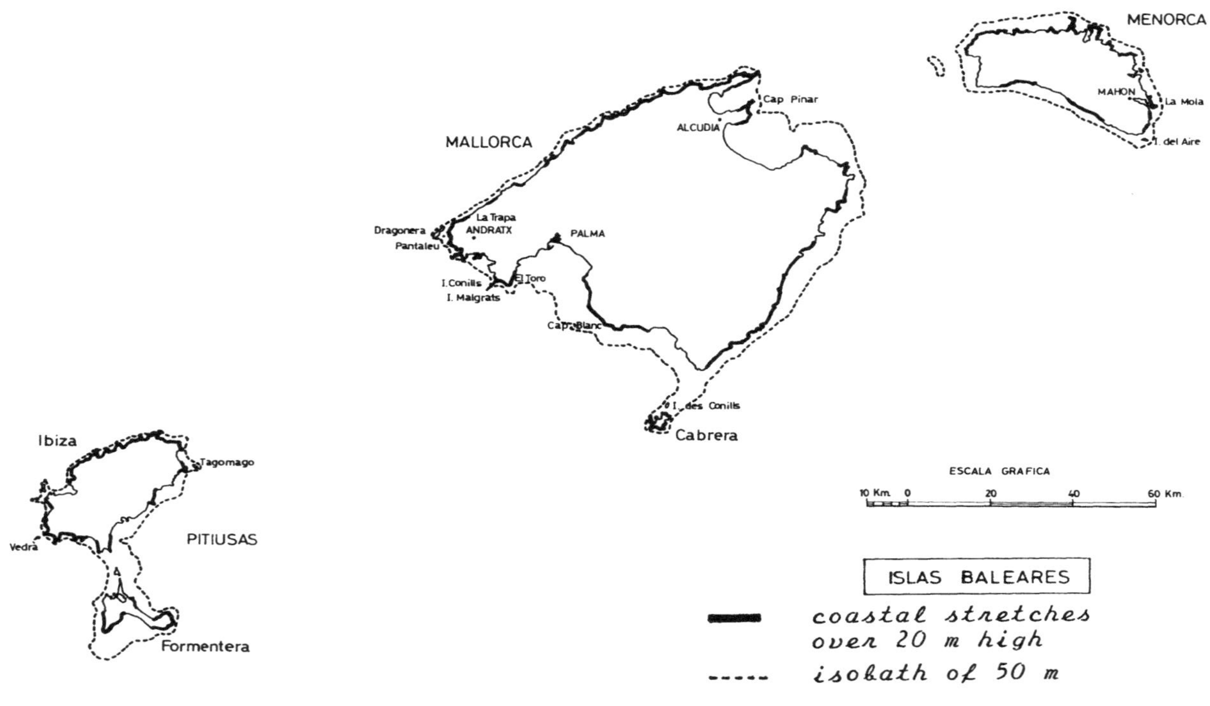

Fig. 1: Geographical situation of the Balearics showing 50 m isobaths and coastal stretches higher than 20 m.

Island	Population	Tourist capacity	Fishing boats
Mallorca	562,193	196,263 beds	435
Menorca	60,006	15,309	144
Ibiza	65,228	50,962 }	232
Formentera	4,513	5,060 }	

4,416,221 tourists visited the Balearics in 1983

Tab. 2: Human parameters which may interfere with seabirds (1983)

Island	Population	Fishing boats
Mallorca	341,450	590
Menorca	42,500	152
Ibizi & Formentera	38,139	385

98,127 travellers visited the Balearics in 1950

Tab. 3: Human parameters of importance to seabirds in 1950.

The objective of this paper is to evaluate the current state of the marine avifauna in relation to the history and to the socio-economical structure of the Balearics (see tab. 2 & 3).

It would be of great value to compare the population fluctuations of each species with the demographical growth of man. Unfortunately, we do not dispose of enough data of the numbers of seabirds during the last century. Accordingly, our analysis reflects a personal point of view especially since we had to interpret some dubious references. Yet, as far as the available data sustain the results, we think that our qualitative conclusions are meaningful. Further, we hope that future researchers will make use of our results which may lead to a more accurate analysis.

The socio-economical development of the Balearics is clearly marked by tourism which has modified a virtual autarctical economy based on the use of local natural resources. At the beginning of this century people emigrated from the Balearics. The development of the services sector has reversed this trend since many immigrants arrive from the continent.

Especially the low coasts of Mallorca and Ibiza have been developed for residential and tourist purposes. This brought some pollution including eutrophization of coastal waters. Very few industrial plants have been installed in the archipelago. If the number of fishing boats is relatively stable during this century, modern fishing equipment and the motorization of vessels have led to overfishing of the commercial fish species. An important fleet of boats from the peninsula fish in Balearic waters which also contributes to overfishing. Sport fishing, either using rods or underwater, is also very popular. No statistics are available but some specialists have estimated that their catch is equivalent to the amount fished by professionals during summer.

Species	Status
Calonectris diomedea	dispersive, abundant
Puffinus gravis	accidental
Puffinus puffinus mauretanicus	dispersive, frequent
Hydrobates pelagicus melitensis	dispersive, frequent
Oceanodroma leucorrhoa	accidental
Phalacrocorax aristotelis desmarestii	dispersive, abundant
Sula bassana	winter, rare
Stercorarius parasiticus	accidental
Stercorarius skua	wintering, rare
Larus melanocephalus	migrating, very rare
Larus minutus	wintering, very rare
Larus ridibundus	abundant in winter
Larus genei	accidental
Larus audouinii	dispersive, rare
Larus canus	accidental
Larus fuscus	wintering, rare
Larus cachinnans michahellis	dispersive, abundant
Larus marinus	accidental
Rissa tridactyla	wintering, very rare
Sterna sandvicensis	wintering, frequent
Uria aalge	accidental
Alca torda	wintering, rare
Fratercula arctica	wintering, rare

Tab. 4: Status of seabirds in the Balearics

In this paper the following species have been excluded since most of them do not depend, during most of their lifetime, on the marine environment: Great Cormorant, Osprey, Gavia sp., Chlidonias sp. and Sterna sp. except for the Sandwich Tern.

Accidental species have not been taken into account as their occurrence is insignificant. We deal mainly with the breeding population of the archipelago as we consider the reproductive status to be the most important parameter with regard to the subject of this paper. Some relevant data of other species will be discussed if found appropriate.

The numbers observed in September 1978 shown on tab. 5 allow us a first evaluation of the population of seabirds in the Balearics.

Larus cachinnans	2,054
Phalacrocorax aristotelis	1,253
Calonectris diomedea	124
Hydrobates pelagicus	18
Larus audouinii	17

Tab. 5: Seabirds observed in September 1978 during the oceanographical survey R.O.M. along the Balearic coast (485 miles).

Material and Methods

Our analysis is based on various bibliographical references, most of them being incomplete surveys, on unpublished personal data and on several censuses carried out by the administration (ICONA, SECONA) or by ornithologists of G.O.B. or S.E.O. We have analyzed all available data for all breeding seabird species in the Balearics. We have evaluated these data in relation to environmental impact caused by man in the Balearics. Finally we project a general outline of the population fluctuations of seabirds which might have been influenced by human activities over the last few years. We have tried to carry out a realistic prognosis which we hope will be useful for a conservation management plan.

Data analysis per species

• Calonectris diomedea (Cory's shearwater)

Nineteenth century writers considered the Cory's shearwater to be a very common species (Barcelo and Combis, 1866; Hansburgo-Lorena, 1869-1891). In the 1920's this shearwater is thought to be common in Alcudia (Munn, 1921) while, in 1957, it is reported as common in Menorca (Moll, 1957). Several other observations confirm its abundance (Munn, 1926; ICONA, 1978). Yet, it is difficult to consider today this species to be the most numerous seabird, not because its population diminished recently but because the Yellow-legged gull has increased its population tremendously. However (Moll, 1957) fewer numbers have been reported in Menorca due to the consumption of its eggs by local people. Apparently eggs of Cory's shearwaters are favoured because they make pastry more "spongy" as compared to chicken eggs. Munn (1924) and Muntaner & Congost (1984) confirm this practice of egg consumption although it has stopped more recently.

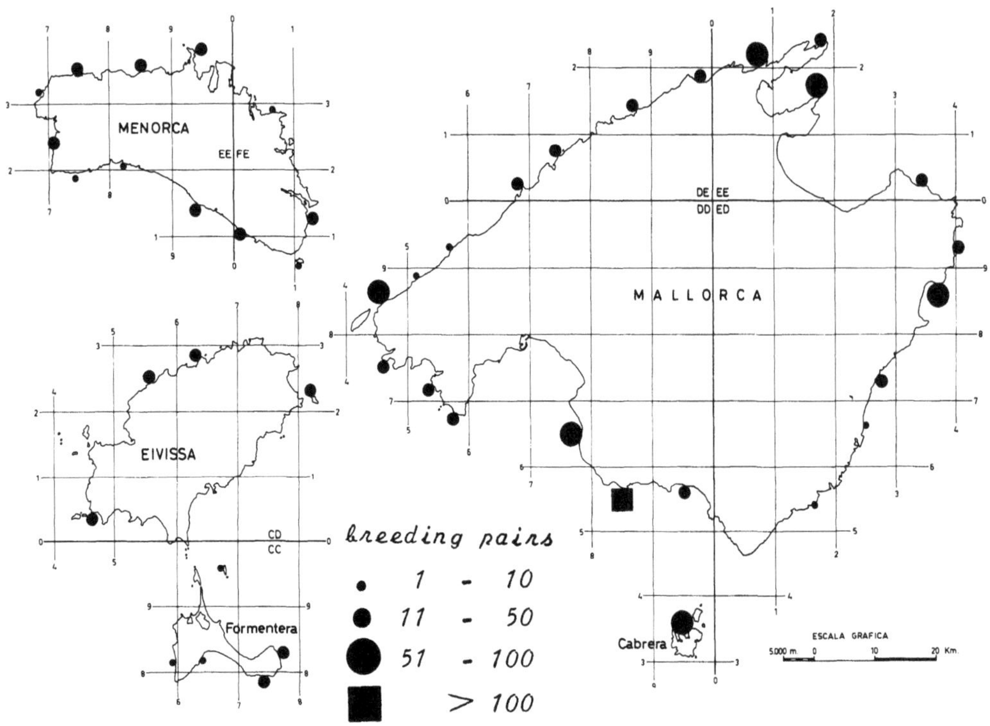

Fig. 2: The Balearic islands mapped out per 100 sq. km quadrates as used for the census of seabirds.

The isle of Aire (Menorca) is the only colony reported in the literature of the beginning of this century (Moll, 1957). This may be doubtful since the geomorphology of the islet does not permit a large colony of shearwaters to settle down: the island habitat is flat and lacks a rugged coastline. The disappearance of Cory's shearwaters at Aire might have been brought about by the intensive use of the islet as a game reserve for introduced rabbits. Besides the disturbance caused by hunting and egg collection, population restriction was also caused by capture of chicks for gastronomic purposes. This culinary habit has disappeared in Mallorca (Duran, 1956; Mayol, 1977; Ribas, 1981) at the beginning of this century while it is still a residual activity in the Pitiusas. Accidentally shearwaters are being caught at sea by fishing tackle, a phenomenon reported as being frequent by Munn (1924) and Ribas (1981) although we have not been able so far to prove this. Yet, some fishermen seem to use chicks as bait. Our inquiry about hunting and sport fishing at the Cabrera archipelago does not confirm this practice. However, occasionally, we have found Cory's shearwaters killed on fishing lines. Others were unvoluntarily captured by other fishing equipment. At Cabrera, some colonies can be damaged by military activities. At the islet of Ses Rates, 37% of the breeding population has been killed by bomb shells.

Vegetation fire can also be destructive as happened on an islet in 1985 during the incubation period where 10 breeding adults of the 100 occupied nests situated under vegetation suffocated or burnt to death.

The total breeding population of the archipelago is not exactly known but is estimated at 1,250 to 2,500 pairs. Recent population fluctuations tend to indicate an increase as measured from some pilot samples in Cabrera during the last three years where we registered a 28% population growth (Araujo, 1977). We also noticed a population expansion of another colony situated near Mallorca but we do not dispose of any exact data.

As far as pollutants are concerned such as heavy metals, the shearwater population of Cabrera is considered to be among the most contaminated in the Mediterranean (Renzoni, 1986). Yet, we have not been able so far to detect a sensible reduction in the fertility of Cory's shearwaters.

• Puffinus puffinus mauretanicus (Balearic shearwater)

Most of the authors who observed the Balearic shearwater in the past seem to consider it to be less common at present. Some of them conclude that this shearwater is rare (Barcelo, 1866; Saunders, 1872-74; Moll, 1957; Jordans, 1928; Munn, 1943).

We agree with their analysis although this shearwater is probably the most abundant seabird species in some parts of the Pitiusas such as at Formentera and along the southern coast of Ibiza. We observed frequently recent bone remains in various caves of Cabrera, Dragonera and Ibiza. This leads us to think

Year	Number of ringed Cory's shearwaters
1977	26 (during the first three years we only ringed part of the birds found)
1978	34
1980	48
1981	55
1982	62
1983	39 (ringing incomplete)
1984	71
1985	64 (after the fire)

Tab. 6: Number of chicks of Cory's shearwaters ringed in Mallorca

of an important decrease of this species over a long lapse of time. Personal observations recently confirmed by Alcover (pers.com.) make us also believe that the Balearic shearwater population has diminished. An oral proof confirms also this population reduction: fishermen of Andratx collected "bags full of chicks" (sic) on Malgrats Islet and were used for bait as happened in the 1940's. Today, as this shearwater is rarer, it is getting more difficult to catch chicks. The main reason for its decline is the presence of predators especially black rats (Rattus rattus) which can be very numerous on some islets, for example at the islet of Conills de Calvià, Mallorca, where we noticed an important reproduction failure.

In the Pitiusas, Balearic shearwaters were caught for human consumption (Ticehurst & Whistler, 1930; Mayol, 1977-78; Ribas, 1981). This kind of predation by man has recently fallen into disuse although it was a common practice up to the sixties. Accidental capture does not seem to have any effects on the population of Balearic shearwaters although in some cases we may have our reservations about this. Oil pollution may cause some problems in the regions where they pass the summer (Cantabrico, French Atlantic coast).

The most exact population census has been carried out in Formentera where some 1,500 pairs bred in 1976 (ICONA, 1978). Some 100 couples breed at La Trapa, Mallorca (G.O.B., 1983). The total Balearic population may be estimated at 1,800 to 2,500 pairs. It is important to consider that this population represents the total world population of P.p. mauretanicus, an endemic subspecies for the archipelago.

Reproduction at Menorca has been reported by several authors at the turn of the century (islets of Aire and La Mola de Mahon). However we do not dispose of any valid proof. The Yelkouan subspecies may frequent the northern and western coasts of Menorca during the breeding period (Mackrill, in lit.). We have no explanation for its absence elsewhere in the archipelago.

• Hydrobates pelagicus (Storm Petrel)

Only Barcelo & Combis in 1866 considered this species common. Other authors give incomplete statements of their observations. The Storm petrel seems to be especially rare at Menorca (Moll, 1957; Muntaner & Congost, 1984; Mackrill, 1984). We localized some small colonies in Mallorca: at Pantaleu, El Toro and Cabrera. Their population size, usually fluctuating over the years, tends to decrease (Araujo, 1977). It may have disappeared at Vedrà where storm petrels and their eggs were seen in the 1920's (Jordans, 1928). No recent observations have been made there since, possibly because of the presence of rats. We are unable to evaluate the balearic population which may be situated between 100 and 1,000 breeding pairs.

• Phalacrocorax aristotelis (Shag)

All authors consider the shag to be common. Its coastal distribution and characteristic behaviour don't allow it to pass unnoticed. We foresee to be able to define its demographical structure in the near future. At present a complete census of the breeding population is being completed (see Capella et al., this volume).

The shag is reported as common in Menorca since the first naturalists published their data. Hansburgo-Lorena (1869-91) quotes a series of nesting sites which coincide with the data of the 1930's and of our more recent observations. Yet, Munn (1924) indicates that all colonies at Menorca are small although he might not have visited the most favourable breeding sites. The estimate of Muntaner and Congost (1984) is probably incomplete (156 breeding pairs) and we judge their conclusion of a population decrease as erroneous. The estimate of 400 breeding pairs reported by Mackrill (1984) seems to be more realistic and corresponds to population numbers cited by the earlier authors. Thus we consider the shag population of Menorca to be steady.

At Mallorca we think that the shag population has increased. Munn (1921) reports two to three couples at Cap Pinar where nowadays a colony has settled down 20 times as large. During the last decade we observed an irregular increment in several colonies at Mallorca leading us to think that the shag population is expanding. Some other colonies however have diminished in numbers. The total population of Mallorca including Cabrera may be approaching 1,000 pairs. The fluctuating population of Cabrera tends to diminish in size (tab. 7). Ringing and population studies carried out within the colonies can disturb shags leading to change of breeding sites. However, the continuous military manoeuvres in the Cabrera archipelago (active raids, gunfire) during one week every spring disrupt more seriously the shag colonies. We observed very large concentrations of shags along the Mallorca coast (2,000 shags in June 1978) while military exercises were carried out at Cabrera. Although military raids may not directly cause victims among shags, they may disrupt the normal life cycle of young birds just about to fledge.

Year	Population of 2 islet colonies censused all over the year	
1974	274	154
1975	227	142
1976		62

Tab. 7: Decrease of shags in two colonies at Cabrera. According to Araujo et al., 1976.

At the Pitiusas, Hansburgo-Lorena (1869-91) considers the shag to be common in 1860. Ticehurst and Whistler (1930) mention that shags, especially adults, are not rare. Human consumption of birds, including shags, has been a constant practice there (Mayol, 1977). Ribas (1981) indicates that dynamite is used for catching shags: this may be an anecdotal but also revealing detail.

We do not think that captures of shags for gastronomic purposes explain their scarcity in the Pitiusas. Fishing gear, mainly composed of gill nets, constitutes the main cause of shag mortality since we estimate that 250 and 500 shags, especially first year birds, die every year because of accidental capture. Consequently the shag population between Formentera and Ibiza does not exceed 100 pairs. It is worth mentioning that, thanks to legal hunting restrictions, shooting shags at sea, practiced till a few years ago, when shooting rock doves (Columba livia) from a boat, has decreased considerably.

- Larus cachinnans (Yellow-legged gull)

Without any doubt this species has enjoyed the most noticeable expansion. All authors consider the Yellow-legged gull as common (Hansburgo-Lorena, 1869-91; Munn, 1924; Moll, 1957 and Bernis et al., 1958), or very common while indicating that this gull can be seen "all year round, mostly in the Bay of Palma and in the harbour" (Barcelo & Combis, 1866). These authors refer to this gull as being common but not abundant. Munn (1921 (1924) and Ticehurst & Whistler (1930) indicate that the Yellow-legged gull breeds on high cliffs in small groups. Population numbers seem to be constant till the sixties coinciding with a progressive abandonment or decrease in collection of eggs for human consumption. At the same time the production of solid domestic waste increased due to urbanization and tourist development.

The population boom of Yellow-legged gulls which started first in the continental colonies is probably a third cause of their increase in the Balearics. In 1983 the official nature conservation authorities organized the first general census of the breeding population in the archipelago: 8,000 pairs were censused then. We consider that this population is still increasing. A surprising case in a small colony in northern Menorca illustrates this population boom: almost 50% more gulls bred there as compared to the previous year despite the fact that we destroyed their nests since the neighbouring colony of Audouin's gulls suffered from the influx of Yellowlegs. The results of this census is summarized in the table below:

Mallorca & Cabrera	3,030 (± 210) pairs
Menorca	1,998
Ibiza & Formentera	2,672 (± 142)
TOTAL:	7,700 (± 352)

Tab. 8: Census of Yellow-legged gulls in the Balearics in 1983 (ICONA)

Various authors (Munn, 1943; Muntaner & Congost, 1984; Ribas, 1981 and Ticehurst & Whistler, 1930) indicate that predation of Yellow-legged gulls has directly decreased at the islet of Conills (Cabrera): pigs are no longer introduced there for fattening by eating eggs and chicks of gulls.

Besides, offal from trawlers and waste produced by tourists contribute to the unrestrained population increase of Yellowlegs. Consequently, today the population of Yellow-legged gulls is considered to have broken the state of ecological balance which is worrying the public opinion: press and television have commented that these gulls constitute a "plague" and they suggest to control further population growth of this species.

As an example, Araujo et al. (1976) illustrate this explosive growth of Yellowlegs in the colony of Conills islet (Cabrera): where 354 pairs bred in 1974, 442 pairs in 1975 and 709 couples in 1976. In 1983 we have censused there 880 (± 30) breeding pairs. Culling of adult Yellow-legged gulls might be applied during the forthcoming season by administering narcotic drugs.

• *Larus audouinii* (Audouin's gull)

In a previous paper several observations and annotated references explain the population growth of the Audouin's gull (Mayol, 1978). Nesting sites of Audouins are in general relatively accessible which make them highly vulnerable to disturbance by man. The most important colonies are known to fishermen who used to pick up their eggs till the beginning of the seventies as was the case at the Pitiusas. Nowadays, this practice has been abandoned.

The balearic population of Audouin's gulls may be estimated at 500 pairs. This breeding population fluctuates every year but shows an overall tendency to increase. The more important colonies are situated in the Pitiusas where they change breeding sites every year although some traditional nesting areas are reoccupied each year. Fishing equipment may cause some deaths among Audouins since they get occasionally caught on fishing lines. The impact by the Yellow-legged gull is more important and has been documented by Bradley (this volume). In the Balearics, reproductive failure and mortality are mainly caused by military manoeuvres during the breeding period.

At Cabrera, breeding success of the 150 pairs there is lower than at other colonies. During some years, as was the case in 1984, there is no reproduction at all since the main colony is being assailed by bomb shells from the sea and from land. During the incubation period, at the end of May, the loss of eggs is not so much to be attributed to military manoeuvres but to predation by other species especially the Yellow-legged gull. The latter frightens the Audouins off from their nests and gets hold of their eggs.

From above observations we may conclude that military raids can interfere with breeding success of Audouins. As a matter of fact, the most important colony in the world is situated in the Chafarinas, an archipelago occupied by army units. Cabrera has a similar military status and another Mallorcan colony of Audouins is also installed in a military camp. There the population increased from 4 breeding pairs in 1978 to 75 in 1985. The military status of some islands or coasts implicate that visits, and thus interference by the public, are limited whereby Audouins and other seabirds are protected. It is useful to maintain these protective measures in the Audouin's colonies. However, the negative impact caused by military raids should be avoided during the breeding season.

The population increase of the Mallorcan colony mentioned above deserves further attention. In 1978 the Audouins occupied a small islet. This colony expanded to 62 breeding pairs. The following year the colony moved to the mainland coast of Mallorca where it occupied a prominent cliff 15 to 20 m high and appearing less accessible than the former islet breeding site. Unfortunately we were not able to evaluate the productivity of this colony which would be interesting to compare with our data of breeding success in islet colonies. Bijleveld (1979) measured the level of contamination by pesticides and heavy metals in eggs and concluded that impact by contaminants is higher among Audouins from the Balearics than from the Chafarinas.

Wintering seabirds

The estimate of the number of wintering seabirds is more difficult than censusing breeding pairs. Over the last years the balearic ornithological group G.O.B. has started to count wintering seabirds. These data cover only a few years and thus cannot be interpreted with certainty. Yet, we think that the number of wintering Sandwich terns is on the increase since Munn considered this species very rare in 1930. Today, Sandwich terns can easily be observed in the harbours or along the coasts of the archipelago.

Mortality because of oil pollution is rare. Nadal (1958) indicates that 100 puffins (Fratercula arctica) were found oiled on the east coast of Mallorca in February 1957. Such a catastrophe has not occurred since. Oil slicks are relatively rare in the western Mediterranean probably because less tankers pass through the Suez Canal. Muntaner and Congost (1984) give further details on pollution by hydrocarbons.

Although shooting takes place in the Pitiusas hunting does not seem to affect population numbers of seabirds. The wintering seabird population is relatively large in the Pitiusas including the Gannet (Sula bassana) and Great Skua (Stercocarius skua). There are less seabirds wintering along the Mallorcan coast while very few frequent Menorca. This can be explained by the proximity of the Strait of Gibraltar (Mayol, 1978) or by the geography of the continental platform and thus by the presence of fishing territories for these seabirds (Muntaner, pers.com.).

Conclusion: Human impact

From the above varied but incomplete analysis we will try to evaluate impact brought by man as a predator, competitor or inductor of this interesting ornithological community. Tab. 9 summarizes the various human incidences for each seabird species. In the first place, urban development is responsible for the modification of the island habitat as referred to "Balearization". However, seabirds have hardly suffered so far from urbanization since few of them breed on beaches or low coastal stretches. Besides, in the Balearics, this habitat frequented by man is proportionally less important than cliffs and islets. Nevertheless, since low coastal stretches have become increasingly rare for residential development, some important human settlements have been built on rocky coast lines. As a more extreme example: the islet of Tagomago (Pitiusas) has been developed recently and planning permission has been granted for construction of a residential estate at Cap Blanc (Mallorca) where important seabird colonies are situated.

If up till now seabirds have not suffered from urbanization as was the case with other birds (waders for instance), the current expansion of the building industry will be decisive for the survival of seabirds in the Balearics. Further urbanization represents major irrevocable threats if no conservation measures are being implemented.

Nature of impact:

1) Intentional capture
2) Human consumption of eggs
3) Human consumption or use of chicks as bait
4) Accidental capture
5) Competition with rivals or predators
6) Pollution with pesticide or heavy metals
7) Pollution with hydrocarbons
8) Availability of food
9) Military manoeuvres interfering with reproduction
10) Urbanization

Species	Factors									
	1	2	3	4	5	6	7	8	9	10
Calonectris diomedea	I>	>>	I>	2	-	3	-	+	2<	-
Puffinus puffinus	I>	>>	2>>	+	3	+	+	+	I<	+
Hydrobates pelagicus	-	-	-	-	3	+	+	-	I<	-
Phalacrocorax aristotelis	I>	-	I>	2	-	+	-	+	I	I<
Larus cachinnans	I>	>>	-	-	-	+	-	<<	I	I
Larus audouinii	-	>>	-	I	2	I	-	-	3<	-
Wintering species species	I>	-	-	I	-	+	I	-	-	-

Every parameter has been evaluated for each species as follows:

+ applicable but not measured

I seldom   2 noticeable impact   3 important impact

< rising   << rising rapidly

> decreasing   >> decreasing rapidly

Tab. 9: Evaluation of human impact on seabirds in the Balearics

Pollutants such as pesticides and heavy metals constitute a second form of impact to most seabirds, followed up by military manoeuvres, accidental capture and shooting even if the Spanish law protects all seabird species except the Yellow-legged gull. The species that suffers the most from various sources of impact is the Balearic shearwater, followed by the shag and

Cory's shearwater. If we consider the net impact all over the archipelago, despite the important effects on the first two species mentioned above, the Cory's shearwater is most affected by human interventions followed up by the shag, the Balearic shearwater and the Audouin's gull. As we have indicated in our introduction, the results are to be considered as transitory and would benefit from a more profound knowledge of the marine avifauna.

References

Alcover, J.A. et al. (1981). Les quimeres del passat. Els vertebrats fossils del Plio-quaternari de les Balears i Pitiuses. Ed. Moll Palma de Mallorca.

Araujo et al. (1976). Populations of seabirds in the Cabrera archipelago (Balearic islands). Ardea, 64: 83-84.

Araujo et al. (1977). Las rapaces y aves marinas del archipielago de Cabrera. Naturalia hispanica, 12, 94 p.

Bannerman, D.A. & Bannerman, W.M. (1983). The Birds of the Balearics. Croom Helm ed. London.

Barcelo y Combis, F. (1866). Catalogo metodico de las aves observadas en las islas Baleares. Rev. Progr. Cienc. Exac. Fis. Nat. 16: 45-62 & 101-123.

Bernis, F.; Diez, P.M. & Tato, J. (1958). Guion de la avifauna balear. Ardeola, 4: 25-97.

Bijleveld, M. et al. (1979). Persistent pollutants in Audouin's Gull (*Larus audouinii*) in the Western Mediterranean: A case-study with wide implications? Environmental Conservation, 6 (2): 139-142.

Blondel, J. & Vielliard, J. (1965). Sobre migracion y avifauna en Mallorca, primavera de 1983, Ardeola, 11 (2): 85-94.

Congreve, W.M. (1957-58). Six springtime weeks at Puerto de Alcudia (Majorca). Oologist records, 31: 17-21, 34-39 & 56-60; 32: 1-8.

Duran, D. (1956). El modo de vida del pescador de Cala Manacor. Trabajos de Geografia, 28. 52 p. Palma.

Garcias Font, L. (1954). Notas zoologicas: *Sula bassana* L. en Mallorca. Bol. Soc. Hist. Nat. Bal., 1954.

G.O.B. (1983). La Trapa, un espacio natural salvado por suscripcion publica. Quercus, 10: 42-44.

Henrici, P. (1926-27). Ornithologische Ergebnisse zweier kurzer Reisen nach Balearen und Pityusen. Beiträge Fortpfl. Vögel, 2: 119-127 & 165-169; 3: 7-13, 48-52 & 98-102.

Hansburgo-Lorena, L.S. (1869-1891). Die Balearen in Wort und Bild. 7 vols. (Consultation of translated version).

Hernandez Ponseti, M. (1911). Catàlogo de las aves observadas en la isla de Menorca. Revista de Menorca. Mahon.

Homeyer, A. (1862-63). Die Balearen. Journal f. ornith., 10: 1-23, 241-285 & 417-434; 11: 81-99.

Hjortnaes, P. & Jensen, J.V. (1974). Algunas observaciones ornitologicas en Mallorca, Abril, 1973. Ardeola, 20: 263-265.

ICONA-Baleares (1978). Estudio del medio fisico de la isla de Formentera, 2 vols. (Inedited).

Jordans, A.V. (1914). Die Vogelfauna Mallorcas. Falco, 162 pps.

Jordans, A.V. (1924-25). Die Ergebnisse meiner zweiten Reise nach Mallorca. Journal f. ornith., 72: 145-170, 381-410 & 518-536; 73: 194-207.

Jordans, A.V. (1928). Die Ergebnisse meiner dritten Reise nach den Balearen. Novitates zool., 34: 262-336.

Jordans, A.V. (1933). Ein weiterer Beitrag zur Kenntnis der Vogelfauna der Balearen und Pityusen. Anz. Ornith. Ges. Bayern, 2: 223-250.

Koening, A. (1929). Nidooloigsches von den Balearen. Beitrag.-Fortpfl. Biol. Vögel, 5: 90-106, 137-141 & 174-180.

Lentz, D. (1934). Aus der Vogelwelt Mallorcas. Ornith. Beobachter, 31: 57-72.

Mackrill, ed. (1984). A checklist of the species of birds recorded on Menorca up to June 1984. (Inedited manuscript).

Mayol, J. (1977). La ornitofagia tradicional en las islas Pityusas. Vida silvestre, 24: 242-247.

Mayol, J. (1978). Els aucells de les Balears. Manuals d'introduccio a la Naturalesa, 2. Ed. Moll. Palma.

Mayol, J. (1978). Observaciones sobre la Gaviota de Audouin, _Larus adouinii_ Payr. en el Mediterraneo occidental (primavera de 1978). Naturalia hispanica, 20: 34 pgs.

Mester, H. (1971). Die Vogelwelt der Pityusen. Bonn. Zool.-Beitr.,22: 28-89.

Moll, J. (1957). Aves de Menorca. Est. Gral. Luliano. Palma.

Moltoni, E. & Brichetti, P. (1979). Osservacioni ornitologiche nell'isola di Mallorca (Isole Baleari) agli inizi dell'aprile 1979 ed elenco degli ucelli inanellati presi alle isole Baleari. Riv. Ital. di Ornitologia, 49 (2): 117-186.

Moreau, R.E. & Moreau, W.M. (1954). Notas otonales sobre aves de Levante e islas Pitiusas. Ardeola, 1: 86-115.

Munn, Ph.W. (1921). Notes on the birds of Alcudia (Majorca). Ibis, 1921: 672-719.

Munn, Ph.W. (1924). Notes on the birds of Minorca. Ibis, 1924: 446-467.

Munn, Ph.W. (1925). Additional notes on the birds of Alcudia. Ibis, 1925: 39-47.

Munn, Ph.W. (1926). Additional notes on the birds of the Balearic Islands. Ibis, 1926: 467-477.

Munn, Ph.W. (1928). Further notes on the birds of the Balearic islands. Ibis, 1928: 17-22.

Munn, Ph.W. (1920). Further notes on the birds of the Balearic islands. Ibis, 1930: 123-128.

Munn, Ph.W. (1931). The birds of the Balearic Islands. Novitates Zool. 37: 53-132.

Munn, Ph.W. (1932). Further notes on the birds of the Balearic islands. Ibis, 1932: 262-266.

Munn, Ph.W. (1934). Notes on the birds of the Balearic islands. Ibis, 1934: 717-723.

Munn, Ph.W. (1936). Further notes on the birds of the Balearic islands. Ibis, 1936: 591-594.

Munn, Ph.W. (1941). Further notes on the birds of the Balearic islands. Ibis, 1941: 174-175.

Munn, Ph.W. (1943). Notes on the birds of the Balearic islands. Ibis, 1943: 341-343.

Munn, Ph.W. (1948). Notes on the birds of the Balearic islands. Ibis, 1948: 593-595.

Muntaner, J. & Congost, J. (1984). Avifauna de Menorca. (2nd ed.) Treballs del Museu de Zoologia, 1 Ajuntament de Barcelona, 205 p.

Muntaner, J.; Escandell, A.; Ramos, E. & Orfila, G. (1984). Adicion y revision faunistica de las especies de la publicacion "Avifauna de Menorca". In Muntaner, J. & Congost, J., 1984.

Murphy, R.R. (1926). A cruise to Mallorca. Amer. Museum Nat. Hist. 26: 552-559.

Nadal, B. (1958). El problema del petroleo en el mar. Balearica, 2: 49.

Renzoni, A. et al. (1986). Comparison between concentrations of mercury and other contaminants in eggs and tissues of Cory's shearwater Calonectris diomedea collected on Atlantic and Mediterranean islands. Environmental Pollution (A) 40: 17-35.

Ribas, V et al. (1981). Avifauna d'Eivissa. Col. Nit de Sant Joan, 5. 37 pgs. Institut d'Estudis Eivissencs.

Saunders, H. (1872-74). Ornithological rambles in Spain. The Field. (15 articles in total).

Tato, J.J. (1957). Aves observadas en Baleares durante el ano ornitologico 1956-57. Balearica, 1: 9-23.

Ticehurst, C.B. (1930). About Puffinus puffinus mauretanicus. Bull. B.O.C., 343: 83-84.

Ticehurst, C.B. & Whistler, H. (1930). A spring tour in Eastern Spain and the Pityusae Islands. Ibis, 1930: 638-677.

Westernhaben,W.v. (1957). Zur Vogelwelt Mallorcas. Booner. Zool.-Beiträg. 8: 178-192.

CHRISTMAS ISLAND (PACIFIC OCEAN) SEABIRDS AND THE EL NINO SOUTHERN OSCILLATION (ENSO) : 1984 PERSPECTIVES.

Ralph W. Schreiber and Elizabeth Anne Schreiber
Los Angeles County Museum of Natural History
900 Exposition Blvd., Los Angeles, CA 90007

Most seabirds breed on oceanic islands, foraging at the uppermost levels of the marine food chain. Thus their evolution is intricately tied to atmospheric and oceanic conditions influencing the fish and squid populations upon which the birds feed (Ashmole, 1971) but we can primarily only study them on nesting islands. Christmas Island (2 N, 157 W), the largest coral atoll in the world, supports 18 species of breeding seabirds (Schreiber and Ashmole, 1980; Schreiber and Schreiber, 1984).

Several species are normally found in only small numbers (tens to hundreds of pairs), many thousands of pairs of other species are found, and the Sooty Tern (scientific names of all species are in Table 1) population was estimated at 14,000,000 individuals in the 1960s (Table 1). The birds nest primarily on lagoon islets, although some do nest on the mainland of the atoll. Each species breeds on extended but fairly regular, yet independent, nesting cycles (Schreiber & Ashmole, 1970). Data available for 11 years between 1950 and 1982 indicate that no massive reproductive failure occurred, although individual species have shown an occasional partial failure. Since the summer of 1982 parameters of seabird breeding on Christmas Island have been significantly different.

In mid-1982 an unusually severe El Nino-Southern Oscillation (ENSO) began which caused a total breeding failure of the birds on Christmas Island (International Conference on the TOGA Scientific Programme, 1985). We have, since then, been studying the gradual return of the birds to the island.

Seabirds have a low reproductive output in any one breeding season since they generally raise only one chick per year, but have a high potential over their total life span. They feed

NATO ASI Series, Vol. G 12
Mediterranean Marine Avifauna
Edited by MEDMARAVIS and X. Monbailliu

Species	"Normal" (a) numbers of pairs, prior to the ENSO, November 1982 - July 1983	Maximum total population estimates in 1984
Phoenix Petrel *Pterodroma alba*	12,000	800 adults
Wedge-tailed Shearwater *Puffinus pacificus*	500,000	? (few)
Christmas Shearwater *Puffinus nativitatis*	6,000	1,000 adults
Audubon Shearwater *Puffinus lherminieri*	1,000	not seen
White-throated Storm Petrel *Nesofregetta albigularis*	500	50 adults
Red-tailed Tropicbird *Phaethon rubridauda*	4,000	500 pairs
Masked Booby *Sula dactylatra*	1,500	600 adults
Brown Booby *Sula leucogaster*	50	20 adults
Red-footed Booby *Sula sula*	6,000	500 pairs
Great Frigatebird *Fregata minor*	6,000	700 pairs
Lesser Frigatebird *Fregata ariel*	4,000	3,500 pairs
Sooty Tern *Sterna fuscata*	3-4 million	40,000 adults
Gray-backed Tern *Sterna lunata*	3,000	400 adults
Crested Tern *Thalasseus bergii*	350	350 pairs
Blue-gray Noddy *Procelsterna cerulea*	2,000	200 pairs
Brown Noddy *Anous stolidus*	3,000	400 pairs
Black Noddy *Anous minutus*	10,000	3,000 adults
White Tern *Gygis alba*	4,000	1,000 adults

(a) See Schreiber and Schreiber (1984) Tables 1 and 2 for discussion of methodology for arriving at these estimates, and estimates for 1983.

Tab. 1: Seabird population estimates for Christmas Island, Central Pacific Ocean

on an erratically available food supply that is controlled by predator fish/marine mammals driving their fish and squid prey to near the ocean surface, and whose ultimate availability and abundance is related to atmospheric and oceanographic factors which affect primary productivity and thermocline depth. Seabirds are opportunistic in feeding, confined to within 500 - 1,000 kms of the nest site during breeding, and more wide ranging throughout the rest of the year probably only confined by the occurence of certain water types where their food is found.

Most tropical seabirds lay one large egg per nesting attempt and breed at most once per year. They have long incubation periods (30-60 days depending on species) and long periods of nestling dependence, requiring the efforts of both parents for successful fledging (2 to 4 months in most species but up to 18 months in frigatebirds). Mortality is high during the first year of inedependent life (up to 90%) as young learn where and how to feed. For individuals who survive the first year or two of independence, a life span of 30-50 years is probably not unusual. Most species do not begin to breed until they are at least 3 years of age and some species do not breed before 8-12 years old. This suite of biological characteristics is generally assumed to reflect breeding biology responses to the sporadic nature of the birds' food supply (Lack 1954, Ashmole 1971). Prior to 1982 no known ENSO effects on seabirds had been documented in equatorial regions away from coastal upwelling (Boersma 1978) and tropical biologists did not consider ENSO to be a significant event for seabirds. Studies of the responses of bird populations to the oceanographic conditions of the 1982-83 ENSO episode lead us to believe that ENSO's are a major evolutionary forcing mechanism shaping the lives of tropical seabirds.

ENSO is a complicated series of interactions between ocean and atmosphere, primarily in equatorial regions, that involves massive changes in atmospheric pressure, wind direction, sea surface level, sea surface temperature, depth of the thermocline, primary productivity, and eventually seabird populations (for summaries see National Research Council 1983, Cane 1983, Rasmusson and Wallace 1983, Barber and Chavez 1983, various articles in Oceanus 1984 and IC TOGA SP 1985). While minor ENSO events occur each decade and more severe events seem to occur once or twice per century, the 1982-83 ENSO is considered to be a major event occurring only every 400 to 600 years.

While the data are not as complete as we desire, important historic studies of seabirds have been done on Christmas Island, and Diamond (1984) considered the seabird community one of the best studied in the tropics in relation to diet. M. D. Gallagher (1960) recorded observations on the island from mid-June 1958 until mid-June 1959. N. Philip Ashmole visited the island periodically between March 1963 and February 1964 and studied in detail the diet of the smaller species present (Ashmole and Ashmole 1967). The Pacific Ocean Biological Survey Program of the Smithsonian Institution visited the island periodically in 1963 through 1967. As part of that study R.W. Schreiber spent from April through September 1967 and late October 1967 through early January 1968 there studying breeding cycles, growth and development, and diet (Schreiber and Ashmole 1970, Schreiber and Hensley 1976). We surveyed the island in March and August 1979, July 1980, and June 1982. All those data present a pattern of consistent timing of nesting and estimates of various species population sizes.

During a visit in November 1982 we discovered a virtual total reproductive failure and disappearance from the atoll of most of the birds (Schreiber and Schreiber 1983). Our visits in June and November 1983, February, June and October 1984 have allowed us to document the bird populations since this disappearance.

In Tab. 1 we summarize the available data on numbers of individuals or pairs estimated prior to ENSO and our best estimate for 1984. Christmas Island is 250 square miles in area and the birds nest in widely scattered, often inaccessible areas. Our research in the past three years has clearly indicated to us the difficulty in attempting to count or estimate total populations present on the atoll. Rather, we are now studying specific colony sites in detail, while surveying other regions of the atoll in order to obtain some indication of population size and reprodictive success. Because of the drastic differences between the populations in previous years and 1982-83 we do make population estimates for the whole atoll to indicate general trends, and we believe these have some value in studying the recovery of the seabird community.

On Christmas Island three orders of birds are present (see Tab. 1, Schreiber and Schreiber 1984 for population estimates): The five species of Procellariformes (shearwaters, petrels, storm-petrels) are small (wing spans 30 to 80 cm, weights 50 to 300 grams) mostly dark birds that nest on the ground, burrow shallowly under vegetation, or dig deep burrows in suitable soil. They are primarily nocturnal and feed on invertebrates. The Pelecaniformes (tropicbirds, boobies, frigatebirds) are represented by six species, the largest birds present (wing spans 150 to 200 cm, weights 900 to 2,000 grams) and are either white or brown/black and white. Two boobies and the tropicbird nest on the ground with little or no nesting material but the frigatebirds and Red-footed Boobies build nests on bushes or shrubs. Their activities are primarily diurnal although they are capable of all flight activities at night. They eat fish and squid. Seven species of Charadriiformes (terns and noddies) are present, all are rather similar in size and morphology (wing spans 50 to 80 cm, weights 50 to 300 grams). They tend to be grey, white and black, white, or black in color and either nest on the ground or in bushes. They eat fish, squid, and other invertebrates.

Seabirds exhibit a wide variety of morphological characteristics, both in flight capabilities and feeding apparatus. Their feeding methods have been well described by Ashmole (1971). Ashmole and Ashmole (1967), Schreiber and Hensley (1976), Diamond (1984), and Harrison et al. (1983) have discussed the diet of various tropical seabird communities. Seabird species are tied to certain water types, as defined by temperature and salinity, and different feeding methods predominate in those various water types (Ainley and Boekelheide, 1984). Species feeding at, near to, or above the waters surface are most abundant in ocean areas whose surface water temperatures are above 23°C. We suspect that the depth of the thermocline, and thus the density of prey items, are critical to the birds effective feeding. Individual species of birds eat a variety of food items and considerable overlap in diet occurs between species in a given area. Flying fish (Exocoetidae) and squid (Ommastrephidae) are the major diet items although a wide diversity of prey is taken (Fefer et al., 1984). Prey size varies directly with the size of the bird. Prey items vary diurnally, seasonally, and by locality. That seabirds require a large volume of food is indicated for example by the Northwestern Hawaiian Islands population of 19 species, where an estimated 14,000,000 individual birds consume an estimated 400,000 metric tons of fish, squid, and crustaceans annually (Fefer et al., 1984).

Our counts on Christmas Island indicated a diversity of recovery responses of the birds from ENSO. Crested Terns and Lesser Frigatebirds show essentially no change in total population after a failed breeding season in 1982. All other species exhibit only slow recovery. Generally the larger species (Masked Booby, Red-footed Booby, Red-tailed Tropicbird, Great Frigatebird, and Lesser Frigatebird) are proportionately more common now than the smaller species (shearwaters, petrels, and terns). The Sooty Tern has suffered the most dramatic population decline. The primarily nocturnal Wedge-tailed Shearwater and the rare Audubon's Shearwater and White-throated Storm Petrel are difficult to census. We do not consider our data on them sufficient to discuss population changes except perhaps to note that all three species are present and breeding in 1984 after virtual absence in 1983.

In Tab. 2 we summarize our data on the presence and absence of eggs and young for our three visits to the atoll in 1984 based on surveys of specific study areas which we visit consistently. Rates of reproductive success in Red-tailed Tropicbird, Masked Booby and Great Frigate colonies are among the highest we have ever measured. We interpret this to mean, that while total populations may be reduced, those individuals which are breeding can find sufficient food to maintain themselves and raise young. Why the smaller species are present in such small numbers and having little reproductive success remains unclear. However, we believe it is related to a lack of fish and squid of the size which these birds eat.

The break in the food chain caused by ENSO has complex ramifications throughout the marine food web and points out a great need for further research on fish resources. We do know that as the warm western Pacific waters moved eastward across the Pacific, flying fish (Exocoetidae) and squid (Hemiramphidae) populations underwent several changes from normal. Individuals became more dispersed as the thermocline deepened and as produc-

Table 2. Seabird population breeding cycle for 1984 on Christmas Island. Numbers are compared to "normal" for that season in specific study areas.

Species	February 1984		July 1984		October/November 1984	
	Adult numbers	Breeding cycle	Adult numbers	Breeding cycle	Adult numbers	Breeding cycle
Phoenix Petrel	60%	eggs and young	5%	eggs and young	50%	eggs and young
Wedge-tailed Shearwater	10%	a few large young	no estimate made	eggs	no estimate made	large young
Christmas Shearwater	50%	a few eggs and young	5%	eggs and small young	50%	eggs
Audubon Shearwater	none seen	---	present	none seen	none seen	----
White-throated Storm-Petrel	none seen	---	present	none seen	present	eggs
Red-tailed Tropicbird	present	courting, a few eggs	100% (1)	eggs and a few young	present	young
Masked Booby	5%	no eggs, no nestlings, a few juveniles	30% (1)	eggs	50%	eggs and young
Red-footed Booby	30%	eggs, a few nestlings, a few juveniles	50%	eggs and young	50%	eggs and young
Brown Booby	5 adults	none	8 adults	none	11 adults	none
Great Frigatebird	10%	eggs, a few juveniles	2-%	eggs and nestlings	30%	nestlings and juveniles
Lesser Frigatebird	normal	juveniles	normal, colony changed location	eggs and small nestlings	normal	a few nestlings, many juveniles
Sooty Tern	new colony locations at Southeast point and North west point. 30,000	a few eggs, oldest nestlings flying	40,000 NW 60,000 SE	very few eggs (100s) a few nestlings (1000s	15,000 total	a few eggs (100s) a few nestlings (10s)
Gray-backed Tern	20%	eggs	present	a few eggs and nestlings	1%	none
Crested Tern	present	none expected	100%	14 eggs and 75 nestlings and juveniles	present	none expected
Blue-gray Noddy	5%	a few eggs	few	a few eggs and nestlings	1%	none
Brown Noddy	5%	none	10%	few eggs and nestlings (10s)	1%	none
Black Noddy (2)	10%	eggs	75%	nestlings and juveniles	10%	none
White Tern	25%	few eggs and young (10s)	5%	few eggs and young (10s)	50%	eggs and nestlings (100s)
Christmas Island Warbler (3)	present	no nesting	present	nesting	present	no nesting
Shorebirds (4)	no change from usual numbers at any time except for Bristle-thighed Curlews which were very common and the most numerous we have ever recorded them in October/November 1984.					

(1) few found in non-study areas.
(2) Motu Tabu colony much smaller than "normal" but Cook Islet and Motu Upua supported large numbers.
(3) no nesting occurred in 1983 but numbers seemed normal in 1984.
(4) Golden Plover (*Pluvialis dominica*), Bristle-thighed Curlew (*Numenius tahitiensis*), Wandering Tattler (*Heteroscelus incanus*), Ruddy Turnstone (*Arenaria interpres*), and Sanderling (*Calidris alba*).

tivity of the phytoplankton ceased large numbers probably died, others ceased growing and did not spawn. When colder waters returned and nutrients became available the process was reversed: larger fish and squid moved back into the region around Christmas Island and their numbers became concentrated again as the thermocline moved upward, they began to grow and started to reproduce. Larger birds had their food supply back and were able to begin nesting. Small birds would have to wait until the reproductive process provided them with larval fish. Many of them may have died due to lack of food.

We have been collecting regurgitation samples from all birds present on the atoll in an effort to compare the current diet with previous data. Thus far several changes have been documented in the food since ENSO: 1) percentages of each species taken have changed, 2) new species are being taken, and 3) size of prey taken is reduced. By continuing to monitor seabird colonies on Christmas Island and the food being eaten we will be able to examine the recovery of the equatorial marine system.

Tab. 2 illustrates the recovery of the Christmas Island population through October 1984 and some specifics are briefly discussed here. In the fall of 1982 a few Phoenix Petrels and Christmas Shearwaters were courting but never laid eggs. Normally during this time they would have had chicks. By November 1984 neither of these species nor Wedge-tailed Shearwaters were back in their historic numbers and reproduction has been very limited. Audubon's Shearwaters have not been found nesting since July 1982. White-throated Storm Petrels were absent from the fall of 1982 through the summer of 1983 but by fall 1984 nested again in small numbers.

Red-tailed Tropicbirds, which nest from late June through October, fledged only a few young early in the 1982 season, but by June 1983 the species appeared to be breeding normally, though timing of nesting was delayed, and the population remains low.

The booby species that breed on Christmas Island, Masked, Brown and Red-footed, lost all their young in fall 1982 and most adults left the island. Red-foots were the first to return and commence breeding in limited numbers in June 1983, and in June 1984 about 25% of the usual population was breeding. About 50% of the Masked Booby population returned in May-June 1984 and began laying eggs. We have only seen 10 to 15 adult Brown Boobies and 2 fledging on the island since ENSO.

In June 1982 approximately 8,000 Great Frigatebirds were nesting. By November fewer than 100 adults were present and all the young in our study areas were dead or dying of starvation. The Great Frigates have had one of the slowest recoveries of any bird on Christmas Island so that by October 1984 only about 10% of the normal population had returned to nest. Young fledged successfully though, in 1984. As with the petrels, and for that matter, Red-footed Boobies, few frigates are present in the surrounding waters (David Ainley, pers. comm.).

Another phenomenon that has occurred on the atoll is that many frigates switched from eating marine fish and squid to taking large numbers of Tilapia (Sarotherodon mossambicus) (Loebel 1980) which live in the land-locked lagoons and ponds of the atoll. We had not seen this feeding behaviour before.

We estimated 3,500 pairs of Lesser Frigates with eggs and small young in June 1982. In November most young were dead, lying in their nests, and the few left were starving. About 300 young fledged before ENSO arrived. Many adults were nesting by May-June 1983 but there was high mortality of young due to starvation. Approximately 2,500 pairs nested in 1984 and many young fledged. During the 1957-58 ENSO a British military officer stationed on the island, M.D. Gallagher, noted that he could find only 500 Lessers nesting, perhaps indicating that they were somewhat affected by the ENSO that year.

Sooty Terns are normally just beginning their nesting season in November, but in 1982 they never arrived. They have two breeding seasons per year on Christmas Island and by the June 1983 season only 1% had returned. Since then, their normal highly synchronized nesting cycle has been greatly extended, and only another 2 to 3% have returned. In October 1983 we found 5,000 large Sooty nestlings starved to death in one colony, while another small colony 30 km away had large, healthy young, which we don't understand unless adults from the two colonies feed in different areas.

Most tern species were severly affected and it wasn't until July 1984 that some successful nesting occurred. These species included Grey-backed Terns, Blue-grey Noddies, Common Noddies and White Terns. Crested Terns appear to have been totally unaffected by ENSO, finishing a nesting season before November 1982 and beginning again normally in April 1983.

In June 1983 40 to 50% of the previous Black Noddy population attempted nesting. They were not successful, however, and in October 1983 large numbers of skeletons of chicks were present in nests. In July 1984 10 to 20% of the previous population nested and fledged chicks successfully.

At this point we are not sure whether the birds who have not returned died or are just remaining at sea for some reason. The food supply may not yet be back to normal and many birds may be remaining in other areas where they have found food. Seabirds are adapted to life at sea and do not necessarily need to come to land.

Breeding biology characteristics of seabirds (small clutch size, long incubation period, extended parental care, deferred maturity, stable pair bonds, and long life span) are generally assumed to indicate dependence on a distant, limited and/or ephemeral food supply of fish and squid. As evolutionary biologists we accept that these characteristics evolved through natural selection. We have been schooled to think that environmental factors affecting natural selection interact with individuals in density dependent ways. That is, when a population reaches a certain density relative to a specific need, such as nesting space or food, then those individuals possessing the "right" characteristics compete better for these resources and survive, leaving behind the most offspring who share those characteristics.

Now, in light of the 1982-83 ENSO, we must rethink our interpretations. What appeared to be normally behaving populations in balance with their resources, suddenly were decimated. We are perhaps wasting our time trying to unravel ecological relationships, based on inter- and intra-specific competition for resources. Is it only during conditions such as ENSO that the real nature of seabird's ecological adaptations are exhibited? Seabird populations may, in fact, be continually increasing until a "crash" occurs, such as that caused by ENSO. The concept of environmental variability has taken on a new dimension to say the least! In addition to rethinking our interpretations of results in the future we have begun to look back at data from previous years in this new light and are finding population declines and high chick mortality in other ENSO years, although since those ENSO's were less severe the effects also were less severe.

The events of 1982-83 have shown us that only long term studies (probably on the order of 20 to 50 years, or one generation for these birds) will allow proper interpretation of how the ocean and atmosphere interaction affects seabirds. The 1982-83 ENSO has given us perspective to appreciate the role variability has on the environment and birds. We believe ENSO effects to be the major evolutionary bottleneck through which individuals and populations periodically pass. ENSO provides us with an opportunity to determine the effect a forcing mechanism has on regulating the population dynamics of higher marine vertebrates, such as birds.

Our data clearly indicate that some species at least (and our analysis continues) predicted the upcoming ENSO event two to three months prior to any measured changes in the marine environment (Schreiber and Schreiber, 1985). Thus, through careful study of the seabirds in the tropics it may be possible to add to the effectiveness of our monitoring systems of global atmospheric-oceanic changes. The seabird community of Christmas Island is recovering from the ENSO perturbation, but many "abnormal" events continue to occur. Because of the lack of seasonality on the atoll and the variability normally present in the reproductive parameters of the species involved it will be several years before we expect to see reestablished the once huge populations.

The responses of seabirds to the 1983 ENSO conditions in the remainder of the world oceans were dramatic and have been reported for the Hawaiian Islands, the Galapagos Islands, Ecuador, Peru, Chile, Panama, the entire west coast of North America to the Gulf of Alaska, in the Caribbean, in Great Britain, and in South Africa (for summary, Ainley et al., in prep.). Time and distance from the onset of ENSO alter the degree of response and the central Pacific data appear at present to be the only "predictive" information. This may, however, merely reflect incomplete analyses to date for those other regions. The 1982-83 ENSO is stimulating further comprehensive analyses.

Unless a total reproductive failure occurs (as on Christmas Island in fall 1982 through spring 1983), environmental variability may cause only subtle changes in what ornithologists can measure such as decreased asymptotic weights or decreased growth rates. Seabird population or island productivity/production in any one year is relatively unimportant to the population in the long term, but such data do serve as a monitor of perturbations in the environment.

We as ornithologists need to use the available data on the ocean and atmosphere, especially as regards changes in thermocline depth, primary productivity and fish-squid abundance. These are all indicative of the state of the birds' food supply and thus are critical to understanding seabird biology.

Long term studies are required to understand even one generation of these species, and especially the recruitment needed to maintain a population. Only through banding and marking large numbers of individuals of both nestlings and adults and then recapturing many of those individuals over several years shall we gain an understanding of how birds carry on their lives on island nesting colonies. Island life is, of course, only a portion of their biology. But it is a portion which is possible to study, and as we have seen, immediately reflects changes in the at-sea environment where the birds feed for their living. Over 80% of a seabird's life is spent on or over the ocean. Thus, it is that ocean which we must understand in order to gain an appreciation for the role of seabirds in the marine environment.

Acknowledgments: We thank Mr. T.L. Cross, National Science Foundation grants OCE-8308756 and OCE-8404152, the National Geographic Society Committee for Research and Exploration Grant 2657-83 and Earthwatch for financial assistance. The Minister for the Line and Phoenix Islands Groups, Republic of Kiribati, kindly gave permission to study on Christmas Island. We have benefited distinctly from discussions with, among others, D. Ainley, R. Barber, and K. Wyrtki.

References:

Ainley, D.G., and J. Boekelheide (1984). An ecological comparison of oceanic seabird communities of the South Pacific Ocean. Pp. 2-23, in R.W. Schreiber, Ed. Tropical Seabird Biology, Studies in Avian Biology No. 8.

Ashmole, N. Philip (1971). Seabird ecology and the marine environment. Pp. 223-286, in D.S. Farner and J.R. King, Eds. Avian Biology, Vol. 1, Academic Press, New York.

Ashmole, N. Philip, and Myrtle J. Ashmole (1967). Comparative feeding ecology of seabirds of a tropical oceanic island. Peabody Museum of Natural History Yale University Bull. 24: 1-131.1 History Yale University Bull. 24: 1-131.

Barber, R.T., and F.P. Chavez (1983). Biological consequences of El Nino. Science, 222: 1203-1210.

Boersma, P. Dee (1978). Breeding patterns of Galapagos penguins as an indicator of oceanographic conditions. Science, 200: 1189-1195.

Cane, M.A. (1983). Oceanographic events during El Nino. Science, 222: 1189-1195.

Clark, L., R.E. Ricklefs, and R.W. Schreiber (1983). Nest-site selection by the Red-tailed Tropicbird. Auk 100: 953-959.

Diamond, A.W. (1984). Feeding overlap in some tropical and temperate seabird communities. Pp. 24-26, in Tropical Seabird Biology, Studies in Avian Biology No. 8 (R.W. Schreiber, Ed.).

Fefer, S.I., C.S. Harrison, M.B. Naughton, and R.J. Shallenberger (1984). Synopsis of results of recent seabird research conducted in the Northwestern Hawaiian Islands. Pp. 9-76, in R.W. Grigg, and K.Y. Tanoue, Eds. Proc. 2nd Symp. on Resource Investigations in the Northwestern Hawaiian Islands, Vol. 1. UHIHI-Sea Grant-MR-84-01.

Gallagher, M.D. (1960). Bird notes from Christmas Island, Pacific Ocean. Ibis, 102: 489-502.

Harrison, C.S., T.S. Hida, and M.P. Seki (1983). Hawaiian seabird feeding ecology. Wildlife Monographs No. 85, The Wildlife Society, Washington D.C.

International Conference on the TOGA Scientific Programme, 1985. Papers presented at the Joint Scientific Committee for the World Climate Research Programme, Paris, September 1984.

Lobel, Philip S. (1980). Invasion by the Mozambique Tilapia (Saratherodon mossambicus); Pisces; Chiclidae) of a Pacific atoll marine ecosystem. Micronesica, 16: 349-355.

National Academy of Sciences (1983). El Nino and the Southern Oscillation, a scientific plan. Climate Research Committee, National Academy Press, pp. 1-72.

Oceanus (1984). El Nino 1984. Vol. 27, No. 2. Woods Hole Oceanographic Institution, Woods Hole, Mass.

Rasmusson, E.M., and J.M. Wallace (1983). Meteorological aspects of the El Nino/Southern Oscillation. Science, 222: 1195-1202.

Schreiber, E.A., D.G. Ainley, and R.W. Schreiber (1986). Seabird ecology: Insights from a global "natural experiment". Scientific American, in press.

Schreiber, R.W., and N.P. Ashmole (1970). Sea-bird breeding seasons on Christmas Island, Pacific Ocean. Ibis, 112: 363-394.

Schreiber, R.W., and E.A. Schreiber (1983). Reproductive failure of marine birds on Christmas Island, Fall 1982. Tropical Ocean Atmosphere Newsletter. February No. 16, pp. 10-12.

Schreiber, R.W., and E.A. Schreiber (1984). Central pacific seabirds and the El Nino Southern Oscillation: 1982 to 1983 Perspectives. Science, 225: 713-716.

Schreiber, R.W., and D.A. Hensley (1976). The diets of Sula dactylatra, Sula sula, and Fregata minor on Christmas Island, Pacific Ocean. Pacific Science, 30: 241-248.

Schreiber, R.W., and E.A. Schreiber (1985). The role of central pacific seabirds in predicting the 1982-83 ENSO. Pp. VI-32-37, in International Conference on the TOGA Scientific Programme, Joint Scientific Committee for the World Climate Research Programme, Paris, September 1984.

WINTER MORTALITY OF SEABIRDS ON THE PORTUGUESE COAST

Antonio M. Teixeira, Serviço Nacional de Parques,
Reservas e Conservaçao de Natureza
Rua da Lapa, 73, 1200 Lisboa, Portugal

In the winter of 1982/83, the Portuguese Secretary of State for the Environment has decided to set up a national scheme to monitor seabird deaths. This was continued in subsequent years, based on examination of corpses found ashore at selected sample areas regularly visited by teams of volunteers in the autumn and winter months (November to March). Techniques have been adapted from the RSPB and NSO schemes in the United Kingdom and the Netherlands (Stowe, 1982; Camphuysen, 1983; Teixeira, 1985b).

Support and co-ordination by a State department proved quite essential. However, this programme really became established thanks to the enthusiastic participation of a few dozen birdwatchers having specialized in the identification of seabirds (tab. 1). Their interest and technical ability have been improving in successive years and this evolution is stimulated by a series of feedback mechanisms based on the very success of the programme they are helping to build up.

	1982/83	1983/84	1984/85	1985/86
Number of participants	17	43	24	24
Sample areas covered	32	36	73	53
Mean km/month walked	85	117	147	104

Tab. 1: The development of the Portuguese coastal surveys in recent winters.

NATO ASI Series, Vol. G 12
Mediterranean Marine Avifauna
Edited by MEDMARAVIS and X. Monbailliu

	1982/83	1983/84	1984/85	1985/86
Gavia immer	-	1	2	-
Gavia stellata	-	-	-	1
Fulmarus glacialis	2	9	3	1
Puffinus puffinus	-	4	3	1
Calonectris diomedea	-	1	-	-
Hydrobates pelagicus	-	4	1	13
Oceanodroma leucorhoa	-	105	6	1
Sula bassana	4	54	28	37
Phalacrocorax carbo	-	-	-	2
Phalacrocorax aristotelis	-	3	-	-
Melanitta nigra	6	27	7	11
Phalaropus fulicarius	-	1	7	1
Larus ridibundus	5	71	28	22
Larus minutus	-	4	8	-
Larus melanocephalus	-	3	-	1
Larus argentatus	8	25	22	16
Larus fuscus	9	50	47	24
Larus argentatus/ L. fuscus (1st winter immatures)	39	94	118	76
Larus marinus	-	-	2	-
Larus glaucoides	-	-	1	-
Larus canus	2	-	2	1
Rissa tridactyla	6	1,011	28	13
Sterna sandvicensis	-	1	4	4
Sterna hirundo	1	-	-	1
Sterna albifrons	-	1	1	-
Alca torda	92	599	634	286
Uria aalge	12	17	42	10
Fratercula arctica	16	10	182	10
Alle alle	-	1	1	1
Other (species not assessed)	4	111	29	7
Number of corpses found	206	2,207	1,206	(540)
Mean km/month walked	-	117	147	104
Number of corpses/km	2.4	3.8	1.6	1.7
% corpses oiled	6.8	9.2	7.4	7.6

Tab. 2: Results of the national censuses in Portugal (November to March). Totals for 1985/86 include only three months of surveys (November to January) as these are still in progress.

Fig. 1 : Seabirds found dead ashore are regularly monitored on the Portuguese coast by teams of volunteers walking selected areas on beaches from November to March. Spray paint is a convenient form of marking the corpses found to avoid duplicating the counts in subsequent visits.

In the first few years (1982 to 1986) these censuses have monitored twenty-eight species of seabirds, with more than forty-one hundred corpses found ashore (tab. 2).

Caution must be taken when comparing the numbers of corpses found in different years, because the distances covered were not always the same and techniques did also change slightly. All sample areas have been visited five times a year since 1983/84, November to March at around the last week-end of the month (Teixeira 1985c). All the corpses found are removed from the beach or permanently marked to avoid duplicating the counts. The number of seabirds found per kilometre may thus be used to compare data between different counts.

One of the most stimulating facts emerging from these censuses relate to the comparatively low incidence of oiling on seabirds found dead on the Portuguese coast, when compared with other areas further north in Western Europe (Monnat 1978, Stowe & Underwood 1984, Camphuysen 1983). Since the censuses were ini-

Fig. 2: Many dead Puffins Fratercula arctica are occasionally found in winter wrecked on beaches after storms. These birds may also be affected by oiling incidents.

tiated,the number of corpses showing externally any traces of oiling has remained well below ten per cent in our sample (tab.2). Furthermore, it is quite obvious that an undetermined part of the corpses analyzed had been in contact with patches of floating oil only after the birds have died from other causes. This would suggest that oiling may be responsible only for a rather small part of all seabird deaths occuring on south-western Iberian shores.

The species apparently most affected by oil pollution include the Gannet Sula bassana, the Kittiwake Rissa tridactyla and two auks, the Guillemot Uria aalge and the Puffin Fratercula arctica. It is worth noting that the Razorbill Alca torda seems comparatively unaffected by this form of pollution (tab. 3). This may be caused by markedly coastal habits of Razorbills in this country where many can be observed very close to the shore during autumn and winter.

Great care is needed however because many of the Razorbill corpses found ashore come from gill-net fisheries, at the mouth of the Tagus estuary and elsewhere on the Portuguese coast.

	1982/83	1983/84	1984/85	1985/86
Fulmarus glacialis	50	-	-	-
Hydrobates pelagicus	-	-	-	-
Oceanodroma leucorhoa	-	1	-	-
Sula bassana	-	37.5	17.9	13.5
Melanitta nigra	33.3	-	-	54.6
Larus ridibundus	-	8	-	-
Larus argentatus	-	15.8	13.6	-
Larus fuscus	-	7.3	8.5	-
Larus argentatus/ L. fuscus (1st winter immatures)	2.6	19.5	0.9	6.6
Rissa tridactyla	16.7	8.6	25.0	23.1
Alca torda	4.4	7	2.5	3.9
Uria aalge	-	40	19.1	40
Fratercula arctica	31.3	60	19.2	60

Tab. 3: The incidence of oiling (%) on seabird species most frequently found ashore. Data for 1985/86 include only the first three months of censuses (November to January).

This probably increases the number of "clean" birds found dead on the beaches, thus reducing the importance in the sample of any birds eventually killed in oiling incidents.

The Portuguese coast occupies a nodal geographical position in Western Europe, with many kilometres facing the strong winds and high energy waves of the NE Atlantic. However, in the southern province of Algarve more than one hundred and fifty kilometres of uninterrupted sandy beaches lie quite sheltered from these influences, thus benefiting from conditions similar to those found on some western Mediterranean shores.

The difference shows up clearly when comparing the results of censuses on western and southern shores (tab. 4). The mean number of dead seabirds per kilometre found in the south has averaged regularly much lower than for western shores. The mean number of corpses oiled in the south also averages much lower (tab. 4), suggesting that seabirds there may be comparatively safe from oil pollution.

	1983/84	1984/85	1985/86
Mean km/month walked			
West Coast	108.7	126	90.4
South Coast	8.3	21	13.6
Total	117	147	104
Number of seabirds found			
West Coast	2,148	1,095	493
South Coast	59	110	47
Total	2,207	1,206	(540)
Mean number of seabirds/km			
West Coast	3.96	1.76	1.82
South Coast	1.37	0.99	1.15
Total	3.77	1.64	1.73
Number of seabirds found oiled			
West Coast	157	83	38
South Coast	-	6	3
Total	157	89	41
Mean number of seabirds oiled/km			
West Coast	0.29	0.13	0.14
South Coast	-	0.05	0.07
Total	0.27	0.12	0.13

Tab. 4: Comparison of the results from beaches on western and southern shores. The latter have been notorius for the reduced seabird mortalities monitored in recent winters.

The Portuguese censuses are contributing also to our understanding of the migration strategies and biology of individual seabird species. For instance, recent work at Ponte da Telha has demonstrated that many Razorbills found in winter on the Portuguese coast are immature and first-winter birds (Teixeira, 1985c) as discussed in tab. 5. This had been anticipated by Mead (1974) and Lloyd (1974), based on ringing results.

Recoveries of ringed birds and biometric criteria also suggest that the vast majority of Razorbills regularly found wintering in Portugal originate from western and northern colonies in the British Isles. The entrance of the Tagus estuary is im-

	1983/84	1984/85	1985/86
First winter	301	74	46
Immature + adult	144	224	134
Age not assessed	154	124	30
Sample	599	422	210
% corpses oiled	1.4	0.3	-

Tab. 5: Composition of the Razorbill (Alca torda) catch sampled at Fonte da Telha in recent years (November to March). The gill-net fisheries there are affecting mostly healthy birds. Ageing based on bill-grooves (Hope Jones et al 1982). Data for 1985 refer only to three months (November to January).

portant for seabirds and vast mortalities of Razorbills wintering there have been documented in recent years (Hope Jones 1984, Bourne 1984, Teixeira 1985a, 1985c). Most of these deaths are caused involuntarily when the auks get casually entangled in the gill-nets underwater when diving.

Exceptionally high numbers of the Puffin Fratercula arctica have occurred in 1984/85, with 182 dead birds found ashore (Teixeira in press.). This mortality first became obvious in late December 1984, with most of the corpses being found on southwestern shores in late January and early February 1985.

Reports sent in by the participants in the national census suggest that up to 64% of all the Puffins found ashore may have been adult birds. These are recognizable from immatures by differences in the shape and colours of the bill (Harris, 1981). This proportion is comparable to that found on the French coast of Brittany by Hope Jones et al. (1982) following the wreck there of the "Amoco Cadiz" tanker on 16 March 1978.

The number of grooves in the bill could only be observed for 24 dead Puffins. Birds with two or more grooves were considered adults and clearly dominated in our sample (tab. 6).

Adult Puffins and the older immatures had their primaries very worn and bleached, thus contrasting sharply with the much darker black-coloured upper wing-coverts and body feathers. Puffins with distinctly smaller dark-coloured bills showing no grooves were considered first-winter birds. These invariably had fresh primaries, indistinguishable in colour from their black upper wing-coverts.

	juvenile	immature	adult
Wing Length			
Number	7	9	19
Mean	157.1	160.9	160.7
Range	152-165	156-168	148-168
S.D.	4.45	3.98	4.81
Upper Bill Length			
Number	7	9	22
Mean	36.7	39.1	39.6
Range	34-38	36-44	36-44
S.D.	1.25	3.14	2.01
Straight Bill Length			
Number	4	7	11
Mean	31	31.7	31.7
Range	29-33	38-35	29-34
S.D.	1.63	2.43	1.62
Beak Depth			
Number	4	7	11
Mean	20.8	22.4	23.3
Range	20-21	21-24	21-25
S.D.	0.5	1.13	1.35

Tab. 6: Measurements (in mm) of Puffins *Fratercula arctica* found dead on Portuguese beaches, January to March 1985. Ageing based on bill criteria (Harris 1981). The two sexes are combined in the sample. S.D. = standard deviation.

It is worth noting that all the Puffins found dead ashore in January and February 1985 were in winter plumage, showing reduced bill ornaments and a distinctive patch of grey feathers around the eye (Harris, 1984).

One of the birds found in late February was in full moult, having dropped all the primary feathers simultaneously. With a winglength of only 110 mm (maximum chord) it would have been unable to fly. Another Puffin observed moulting to summer plumage in early April had fresh primaries growing simultaneously on both wings and is also supposed to have been flightless at this stage (right wing = 98 mm, left wing = 132 mm, both maximum chord). Winglength measurements from such birds have been excluded from analysis in tab. 6.

Three Puffins involved in this wreck wore rings of the British Museum. All had been marked fully grown at northern and western Scottish colonies, two at Sule Skerry in Orkney and another at St. Kilda, North-West of Scotland.

The winglength of birds with two or more bill-grooves found washed on beaches in 1985 (tab. 6) compares well with measurements in Harris (1984) for colonies in northern Britain. The diferences observed in the size of the bill are caused by the loss of the horny plates in winter (Harris, 1984) and may have been exaggerated by our measuring techniques, not strictly coincident with those employed by Harris (1981).

Only 19 per cent of the Puffins found ashore in 1984/85 did show externally any signs of oil contamination (tab. 3). Storms at sea and rough weather further north are supposed to have caused this wreck.

Although regrettable and apparently involving a large proportion of birds quite capable of breeding, these mortalities may not have any noticeable effect on numbers in the colonies (Harris & Wanless 1984). With an estimated world population of the order of five million pairs (Harris 1984), the Puffin is in fact one of the most numerous species of auk breeding in the North Atlantic (Evans 1984, Cramp 1985).

While there are only some 0.7 million pairs now thought to be breeding in the British Isles, changes in the numbers of Puffins monitored at some Scottish colonies may well suggest that population levels remain stable there or have been increasing in recent years (Harris 1984).

All seabird species are protected by the Portuguese law. However, enforcement of the existing regulations may be considered poor even by north-western Mediterranean standards. While seabirds are not considered traditional game species in this country many are still shot for target practice of just 'fun'. This affects mostly the more coastal species regularly present in estuarine areas, like Black-headed Gulls Larus ridibundus and some of the terns (Sterna and Chlidonias spp.). An intensification of efforts currently being invested in environmental education and bird conservation is still needed to help improve this situation.

Fishermen destroy also large numbers of seabirds. Most of these mortalities are not intentional, being caused by accidental swallowing of baited hooks on long lines afloat at sea (mostly affecting the Gannet Sula bassana) or by casual drowning after entanglement in gill-nets set underwater near the shore (where the Razorbill Alca torda is the commonest victim). Some of the fishermen at underdeveloped coastal settlements still retain a tradition of catching seabirds for food using baited hooks or running nooses on the beach. These practices have been decreasing as a consequence of improved standards in economic status and education in recent years.

The magnitude of seabird losses caused by man ashore and in coastal waters is difficult to assess correctly. However, this is estimated to be only of secondary importance when compared to the numbers of seabirds killed by storms at sea. The ways in which these storms affect the birds in Portuguese waters are still a matter for speculation. Prolonged battering and reduced feeding opportunities in rough seas may cause starvation and are now thought to be responsible for many deaths generally attributed to 'bad weather'. However, the storms are likely to affect differently the birds already debilitated by other causes or having become temporarily flightless when replacing their primary feathers simultaneously (as happens with the divers and most auks).

Casual examination of stomach contents in recent years has shown that seabirds may eat small particles of plastic, including spheroidal pellets of translucent industrial polyethylene. Presumably, these materials are ingested by the birds when found floating at sea and they are not thought to have any acute toxic properties (Franeker 1983). However, the accumulation of comparative large masses of non-digestible materials in the stomach of seabirds may have a debilitating effect, leaving the individuals concerned more vulnerable to unfavourable feeding conditions and disease, thus resulting in increased mortality.

Pellets of industrial plastic have been found in the stomachs of Fulmars *Fulmarus glacialis* and Razorbills *A. torda* picked up dead on beaches. More investigation is still needed before the real impact of the ingested plastics can be assessed for seabirds on the Portuguese coast. In the present state of our knowledge, all attempts to relate this form of pollution to any mortalities of seabirds so far monitored are probably only speculative.

Acknowledgement

Our gratitude goes to all the participants in the surveys who have been walking so many kilometres on Portuguese beaches. Generous help was also received from several people involved in seabird research and monitoring schemes in western Europe including Dr. W.R.P. Bourne, P. Hope Jones, Leslie Underwood, Tim Stowe, S. Wanless, C.J. Camphuysen, J.-Y. Monnat and others having contributed at several stages to the development of this project.

References

Bourne, W.R.P. (1984). Wrecked Seabirds Died in Iberian Nets. Marine Pollution Bulletin 15: 277, August 1984.

Camphuysen, C.J. (1983). Review of Seabird Mortality at the Netherlands Coast, Winter 1982/83. Nbr. NSO 4: 97-122.

Cramp, S., ed. (1985). The Birds of the Western Palearctic, vol. IV.

Evans, P.G.H. (1984). Status and Conservation of Seabirds in Northwest Europe (excluding Norway and the USSR). ICBP Technical Publication No. 2: 293-321.

Franeker (van), J.A. (1983). Plastics - een bedreiging voor zeevogels. Nbr. NSO 4: 41-61.

Harris, M.P. (1981). Age determination and first breeding of British Puffins. Brit. Birds 74: 246-256, June 1981.

Harris, M.P. (1984). The Puffin. Calton: T & AD Poyser.

Harris, M.P. & Wanless, S. (1984). The effect of the wreck of seabirds in February 1983 on auk populations on the Isle of May (Fife). Bird Study 31: 103-110, July 1984.

Hope Jones, P. (1984). Skins of Guillemots Uria aalge and Razorbills Alca torda examined at Cascais, Portugal, in May 1982. Mems. Mus. Mar 27: 1-10.

Hope Jones, P., Monnat, J.-Y. & Harris, M.P. (1982). Origins, age and sex of auks (Alcidae) killed in the 'Amoco Cadiz' oiling incident in Britanny, March 1978. Seabird Report 6: 122-130.

Lloyd, C.S. (1974). Movement and Survival of British Razorbills. Bird Study 21: 102-116.

Mead, C.J. (1974). The Results of Ringing Auks in Britain and Ireland. Bird Study 21: 45-86.

Monnat, J.-Y. (1978). Mortalités d'oiseaux à la suite du naufrage du pétrolier Amoco Cadiz. Penn ar Bed 11: 339-360.

Stowe, T.J. (1982). Beached Bird Surveys and Surveillance of cliff-breeding seabirds. RSPB, NCC - contract HF3/03/141, The Lodge, Sandy.

Stowe, T.J. & UNDERWOOD, L.A. (1984). Oil spillages affecting Seabirds in the United Kingdom, 1966-1983. Marine Pollution Bulletin 15: 147-152.

Teixeira, A.M. (1985a). More auk deaths in Iberian nets. BTO News 138, May 1985.

Teixeira, A.M. (1985b). Inspecçoes invernais da avifauna costeira em Portugal, 1983-84. Cyanopica 3: 373-386.

Teixeira, A.M. (1985c). The Portuguese beached bird surveys. A progress report. Cyanopica 3: 479-486.

Teixeira, A.M. (in press.). Inspecçoes costeiras para detecçao de aves marinhas mortas. Cyanopica 3.

CHAPTER 6 : CONSERVATION AND WILDLIFE MANAGEMENT

Chairman: Dr. Paul ISENMANN, Centre National de la Recherche Scientifique. Montpellier.

SEABIRD CONSERVATION PROBLEMS IN THE MALTESE ISLANDS

By Joe Sultana, I.C.B.P. Chairman Europe
Ornithological Society of Malta, Bir Kirkara - Malta

The Maltese Islands are a small archipelago situated in the centre of the Mediterranean about 89 km south of Sicily and about 334 km north of the Libyan coast, at approximately 36°N and 14°E. The principal islands are Malta (249 sq. km.), Gozo (70 sq. km.) and Comino (2.6 sq. km.). There are also two other smaller, but ornithologically important, islets, Cominotto and Filfla.

The surrounding sea is temperate and the coast, which is generally steep and erosive, is typically Mediterranean. The total length of the coastline is 190 km, 38 km of which are made up of sheer cliffs. 64% of the coastline is natural, 18% semi-natural, with another 18% artificial, the latter consisting mainly of harbour shorelines. There is no intertidal zone and only 5% of the coastline is made up of narrow sandy beaches with hardly any sand dunes of any significance, the rest being mainly rocks.

The population of the Maltese Islands is about 330,000, i.e. about 1,000 person per sq. km. It is therefore not surpising that not only the sandy beaches are heavily used for recreational purposes, but also most of the coastline including cliffs. This fact alone poses various conservation problems for seabirds. Add overfishing as well as indiscriminate bird shooting, which is prevalent on the islands, and one can imagine how enormous the conservation problems are.

In spite of this the Maltese Islands can boast of four breeding seabird species namely the Cory's Shearwater (Calonectris diomedea),Manx Shearwater (Puffinus puffinus yelkouan), Storm Petrel (Hydrobates pelagicus) and the Yellow-leg. Gull (Larus cachinnans michahellis). There is no evidence that the islands receive more than a small representative cross-section of migrating seabirds in the Mediterranean, and apart from three Laridae species which congregate in good numbers in the harbours during strong winds in winter, the number of seabird species observed from land is relatively poor.

NATO ASI Series, Vol. G 12
Mediterranean Marine Avifauna
Edited by MEDMARAVIS and X. Monbailliu

Filfla Islet holds a dense colony of *Hydrobates pelagicus*.

Ta' Cenic Cliffs. The largest colony of *Calonectris diomedea* in Malta is found at these cliffs.

Tab. 1 summarizes the status of the seabird species of the Maltese Islands (vagrants recorded less than 20 times are not included).

Calonectris diomedea	: common breeding visitor - very common offshore early March to October.
Puffinus puffinus yelkouan	: common breeding visitor - frequent offshore from February to August.
Hydrobates pelagicus	: large breeding colony of Filfla.
Sula bassana	: scarce winter visitor.
Phalacrocorax carbo	: scarce winter visitor.
Stercorarius pomarinus	: rare but almost annual offshore.
Larus melanocephalus	: common visitor from December to mid-March.
Larus minutus	: scarce winter visitor.
Larus ridibundus	: very common winter visitor.
Larus genei	: very scarce visitor.
Larus fuscus	: fairly common winter visitor.
Larus cachinnans	: common winter visitor and fairly common breeding resident.
Sterna nilotica	: very scarce and irregular passage migrant.
Sterna caspia	: rare and irregular migrant.
Sterna sandvincensis	: scarce autumn and winter visitor.
Sterna hirundo	: rare and irregular.
Chlidonias niger	: scarce autumn migrant - rare and irregular in spring.
Chlidonias leucopterus	: scarce in spring - very scarce and irregular in autumn.

Tab. 1

In view of the hostile environment for birds, the breeding seabirds are found breeding in totally inaccessible places or in places which are very difficult to reach. The sheer high cliffs, some of which reaching 120 metres above sea level and over, bound the south side of the islands and are ideal in providing sanctuary to the shearwaters and the Yellow-legged Gull. Malta's rock is sedimentary and due to erosion of the softer layers various ledges, burrows, crevices and caves are formed naturally and provide excellent nesting sites. Ta' Cenc Cliffs, which are the most important of all the cliffs in Malta, hold the largest colony of Cory's Shearwater, some Yelkouan Shearwaters and a small colony of Yellow-legged Gulls. It is not only the southern coast which provides adequate breeding facilities for shearwaters. Some of the low erosive cliffs in the northern parts of the islands are used by the Yelkouan Shearwaters. In fact the largest colony which is partly accessible is found at L-Ahrax headland, a relatively out of the way area, in the northern part of Malta. Unlike the Cory's Shearwater, the Yelkouan seems to be more widely scattered along the sea cliffs, and a small colony is also found on Cominotto.

However, the finest seabird colony is found on Filfla, which holds an extremely dense colony of Storm Petrels. This 2.5 ha islet lies about 5 km south of Malta. It is a massive rock rising in precipitous cliffs up to 60 metres high, surrounded by numerous rocks and boulders. For many years Filfla was battered by sea and air bombing practices resulting in the creation of huge rubble slopes beneath the cliffs. A number of birds used to be killed by the blasts as well as squashed or trapped by boulders. A large number of unhatched and cracked eggs could also be attributed to the bombing and shelling practices, which were stopped altogether in 1970.

Bombing was definitely harming the whole ecosystem of the islet and this can best be judged by comparing pictures of the islet which were taken in 1968, when shelling was still carried out frequently, with those taken 14 years after the bombing was stopped. The Cory's Shearwater suffered immensely from this disturbance and the colony dwindled to less than 100 pairs. Once bombing practice has stopped it stands a good chance of slowly recovering to a larger colony, if it is not molested. It could be said that, on the other hand, the Storm Petrel benefited from the creation of the huge rubble slopes which provide adequate habitat for safe nesting sites. Bombing also contributed indirectly by keeping man away. Although Filfla, since 1980, has been included in the list of bird sanctuaries, where all forms of bird killing and trapping are prohibited, there are no restrictions to visitors for recreational purposes. Recreational visits to the islet now pose a different threat to its avifauna.

The population of Yellow-legged Gulls on Filfla has also gone up since bombing has stopped. This increase coincided with a decrease of the population found at Ta' Cenc Cliffs in Gozo, where human activity above the cliffs increased during the past decade. It is not yet known whether the increase in the population of Yellow-legged Gulls on the top of the islet of Filfla poses a threat to the Storm Petrel's colony. Yellow-legged Gulls' regurgitated pellets containing the remains of Storm Petrels are frequently noted on the islet

A different threat to the young of shearwaters breeding on Malta and Gozo is posed by the presence of a large number of rats which infest some parts of the islands. With the decline of the Barn Owl _Tyto alba_ and other predators, rats have multiplied unchecked and are present in the few remote areas, some of which hold shearwaters' colonies. In one particular year, the young of Yelkouans in a partly accessible colony, were all found mutilated and partly eaten by rats. Fortunately, rats are not present on the islet of Filfla. However an increase in human activity there could result in an accidental introduction of these rats which would result in irreparable harm to the Storm Petrel colony.

In an area as densely populated as the Maltese Islands, it is to be expected that bird-life will be under constant and severe pressure. For seabirds this takes two forms: direct human interference and modification or development at or near seabirds habitat.

Human interference takes place mainly in the form of bird shooting which is very popular in the Maltese Islands. There are about 14,000 licensed shooters, about 4% of the whole population. Many of these are indiscriminate shooters. Bird shooting is carried out for fun and not for food. Mounted stuffed birds collection is very popular and seabirds figure prominently in such collections, which are one of the main causes of indiscriminate bird shooting. Even fishermen when out with boats fishing carry out at sea with them their gun and do not hesitate to shoot at any seabird that may venture near. Seabird shooting can best be highlighted by the number of ringed seabirds recovered shot in Maltese waters. To date, to mention two common Laridae species, 21 Black-headed Gulls and 27 Mediterranean Gulls bearing rings, have been reported shot during winter in Malta.

In recent years Malta has also seen a dramatic increase in shooting at sea. Equipped with powerful motorized rubber dinghies, shooters are killing birds at sea in alarming numbers. The worst sufferers are seabirds.

Tab. 2 lists estimated numbers of seabirds killed annually. The totals were estimated with the help of 13 Maltese ornithologists.

Shearwaters	: 1,000 to 2,000 (mainly Cory's Shearwaters)
Gannet	: 20 to 50
Cormorant	: 25 to 50
Skuas spp.	: 5 to 20
Gull spp.	: 1,000 (mainly Black-headed Gull, followed by Yellow-legged Gull and Mediterranean Gull).
Tern spp.	: 100 to 300

Tab. 2

In previous years fishermen used to set up nets in front of Cory's Shearwaters burrows to trap the birds and pluck the underwing white feathers which they used as bait for line-fishing. This habit seems to have stopped but instead shearwaters shooting has taken up. Although some of the shooting of shearwaters takes place for the use of the underwing feathers, most of the shooting is carried out for the fun of killing! In the seven-

Cory's Shearwater

Pullus of Storm Petrel

ties the Ornithological Society of Malta conducted a campaign on the local mass media against this senseless killing and educational material was published to create an awareness of what was happening around us at sea.

In 1980 new bird laws were enacted and all breeding birds became legally protected. Furthermore no seabirds figured in the list of species that could be shot. Shooting at sea within 3 km offshore also became illegal. But we have learnt, through bitter experience that enacting laws and their enforcement are two different things altogether. Illegal bird shooting is still the order of the day.

Oil pollution round the Maltese Islands has so far never posed a threat to the breeding seabirds, and shearwaters, unlike auks, seem to be able to detect and avoid oil slicks. A threat to some of our colonies could be the result of development above the cliffs. One case in point is an industrial estate part of which extends above a Cory's Shearwaters colony. The lights at night definitely disturbed the colony. A small colony of Cory's Shearwater was deserted at Xlendi Bay when electricity was introduced in the area.

Due to the fact that the seabird colonies are mostly inaccessible, and considering the prevailing hostile environment, good studies of Malta's breeding seabirds are difficult to conduct. It is not surprising therefore that only very limited studies have been carried out, which could otherwise be beneficial for seabird conservation.

References

Sultana, J. & Gauci, C. (1970). Bird Studies on Filfla. Malta Ornithological Society, Malta.

Sultana, J. & Gauci, C. (1982). A New Guide to the Birds of Malta. The Ornithological Society, Valletta.

DISTRIBUTION, STATUS AND CONSERVATION OF THE MEDITERRANEAN GULL (LARUS MELANOCEPHALUS) IN GREECE

By Vassilis Goutner, Department of Zoology,
University of Thessaloniki, 540 06 Thessaloniki, Greece

Introduction

The Mediterranean gull is a species with relatively limited breeding distribution in Europe. Greece lies within the eastern European region where the species enjoys high breeding concentrations (Cramp & Simmons, 1983). Literature relevant to these birds in Greece is limited.

This paper was written in order to provide original data from recent field research with the purpose to contribute to our knowledge of this species and to promote its conservation.

Methods

Data on the breeding biology of the Mediterranean gulls were collected at two presently known northern Greek colonies: in the Evros delta from 1981 to 1985 and at Alyki in the 1985 breeding season. In the Evros delta the data are related to site selection (especially vegetation preferences), breeding population numbers, phenology and breeding success. Similar data were also taken at Alyki but to a lesser extent since much effort was put in ringing chicks just before flying: a total of 371 young ones were ringed using metal rings.

For diet analysis, regurgitations were collected during 1984 and 1985 from the Evros delta colonies. The availability of fish species in the lagoon where the colonies are located was studied with the help of local fishermen using several sampling techniques and appropriate nets. The fish material was identified at the Laboratory of Zoology in Thessaloniki and the insect material at the Department of Entomology of the National Museum in Prague, Czechoslovakia. Data on the distribution of the species in Greece are based upon various published and unpublished papers.

NATO ASI Series, Vol. G 12
Mediterranean Marine Avifauna
Edited by MEDMARAVIS and X. Monbailliu

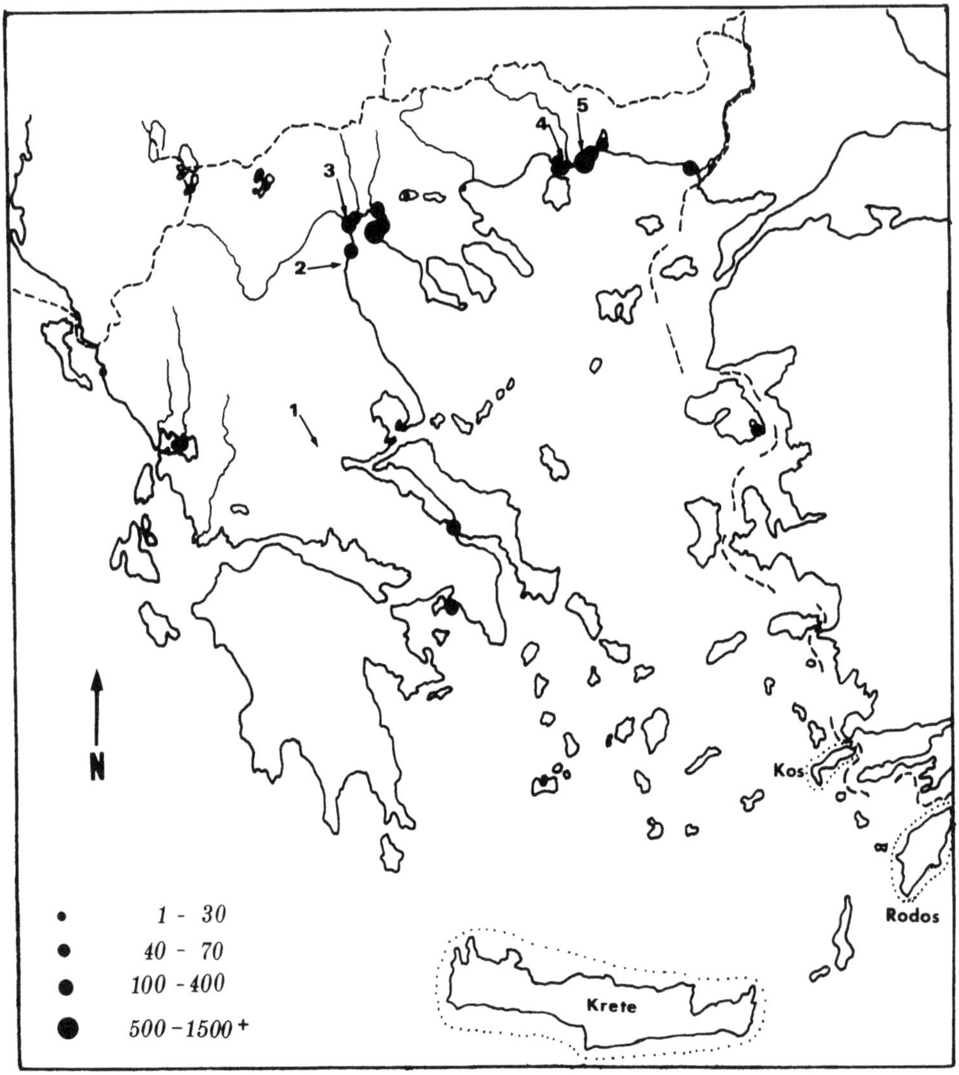

Fig. 1: Non breeding distribution (black dots) and former breeding sites (numbered indicators) of Mediterranean gulls in Greece. 1: Karla region, 2: Korinos, 3: Axios-Loudias delta, 4: Nestos delta region, 5: Porto Lago. Dotted line: southern limit of wintering areas.

Results and discussion

1. Distribution

1.1. Non breeding population

The areas in Greece where the Mediterranean gulls have been recorded during migration and winter are indicated in fig. 1. The early record by Kumerloeve (1957) refers to relatively small numbers of birds observed during early April and mid-September in the Gulf of Thessaloniki during 1953 and 1956 respectively. Nisbet & Smout (1956) state major movements through Bosporus, Dardanelles and Aegean especially in September and according to Bauer et al. (1969), the Greek islands across the Turkish coast (up to Kos and Rhodos) are generally normal migrating and wintering areas whereas relatively small numbers winter in Crete. The zone along the turkish coast coincides with one of the three well known migration routes through Greece. Other islands like Northern Sporades and Kyklades are rather accidentally visited by these gulls (Bauer et al., 1969).

Some unpublished manuscripts and some published literature (Schuster et al., 1959; Makatsch, 1963; Stubbs et al., 1980) as well as several personal observations, indicate that northern Greece and especially the regions of the Gulf of Thessaloniki, the Nestos delta and Porto Lago attract the main bulk of the migrating and wintering population (fig. 1). Within this region, most records in winter (mainly unpublished I.W.R.B. midwinter counts) come from the Axios delta, whereas during the same period of the year limited numbers have been observed in western Greece (Joensen & Madsen, 1985).

Most reports indicate that bird movements are more intensive in April and these congregations in northern Greece may well relate to the arrival at the traditional breeding grounds. After the end of the breeding season (within the first fortnight of August) the largest groups have been observed at Porto Lago (Stubbs et al., 1980). However, their origin and destination are not yet known. Although it appears that the highest numbers of Mediterranean gulls have been observed in northern Greece, we must notice that the availability of records from the Aegean islands is very limited and more research must be done within this region.

1.2. Breeding population

The first breeding record of Mediterranean gulls in Greece is by Reiser 1905 (in Makatsch, 1963), according to which the birds were breeding at lake Karla, close to Lamia (fig. 1). This lake was drained in 1962 and lost forever as a breeding habitat for these gulls which however continued appearing there (Makatsch, 1968). The areas of Axios-Loudias delta, Korinos (Gulf of Thessaloniki), Porto Lago and Nestos delta constituted the only known breeding areas of these birds in Greece for at least the past 20-25 years (tab. 1, fig. 1). All of these colony sites were deserted due to habitat loss by agricultural development

Site Name	Number of pairs	Year	Literature
A. <u>DESERTED</u>			
Karla Lake	no data	1903	Reiser in Makatsch, 1963 Makatsch, 1968
Axios-Loudias delta	104	1966	Conradty & Hohlt, 1967
	no data	1967	Makatsch, 1978
	4,000	1973	Conradty, in lit.
	400	1976	Sterbetz, 1980
Korinos	no data	1961	Makatsch, 1963
Porto Lago	no data	1966	Conradty & Hohlt, 1967
	several hundreds	1982	Centre Ornithologique Rhône, in lit.
Nestos delta	104	1966	Conradty & Hohlt, 1967
	c. 1,000	1968	Bauer et al., in Cramp & Simmons, 1983
Macedonia (exact site not stated)	c. 250	1966	Makatsch, 1968
B. <u>PRESENT</u>			
Alyki	356	1975	Isenmann, 1975
	1,000	1980	Stubbs et al., 1980
	6,000	1981	Müller, in lit.
	2,500	1985	Personal observation
Evros delta	c. 845	1981	Goutner & Kattoulas, 1984
	c. 160	1983	Goutner & Kattoulas, 1984
	c. 220	1984	Goutner in press, b
	c. 1,000	1985	Goutner in press, b

Tab. 1: Recorded breeding sites and numbers of Mediterranean gull pairs in Greece.

including drainage and alterations of the flow of rivers. At Porto Lago, birds formed a large colony in 1983 at a site South West of Vistonis lake. It is possible that this region may still be used as a breeding area.

Nowadays, there are two well known breeding areas of the Mediterranean gulls. The first one, at Alyki, was established in 1974 constituted of 356 pairs (Isenmann, 1975). Within the same year, the breeding site at the closeby Axios area was deserted. It is possible that a part of the population moved to Alyki. Despite the annual fluctuations in numbers of breeding pairs (tab. 1), this colony is today the largest one in the Mediterranean basin. The second colony in the Evros delta was established for the first time in 1981 (Goutner in press, a) and since 1983 its population has been increasing continuously (tab. 1).

2. Breeding Ecology

2.1. Colony site selection

• Alyki is a shallow saline lake of approximately 10 km^2 including saltmarshes and lagoons. A part of the lagoons is used for salt extraction, which has expanded ever since. The area is fully described in Stubbs et al. (1980). The Mediterranean gull colony was made on isolated islets within the lagoon during 1974 (Isenmann, 1975), 1980 (Stubbs et al., 1980) and 1985 (pers. observation). In 1979 the colony was established in the salt pans (Stubbs et al., 1980). The nests were built with the vegetation available on the islets. In 1974 this vegetation constituted of *Salicornia sp.*, *Hordeum maritimum* and *Aeluropus litoralis* (Isenmann, 1975) whilst in 1985 this was *Halocnemum strobilaceum*.

• For the first time, during 1981, the birds in the Evros Delta bred on a part of an isolated coastal islet including low sand dunes covered by ammophilous and halophytic vegetation. The birds used this vegetation as well as dry plant material for nesting. This colony site was deserted in 1982 due to vegetation growth (Goutner, in press, a). Since 1983 the colony was established on an isolated islet in a delta lagoon used by local people as a fishpond (Goutner & Kattoulas, 1984 (photo 1). Here, during 1984 and 1985 the colonies were made with preference for high cover (>31%) halophytic vegetation clearly avoiding uncovered sites (Goutner in press, b). In 1985, a small colony of about 50 pairs started breeding at another site of the coastal islet where breeding took place in 1981. The birds also here preferred breeding in high cover vegetation (Goutner in press, b).

2.2. Heterospecifics

• We only dispose of recent data at the Alyki colony. During the 1985 breeding season we found 30 nests of Gull-billed terns (*Sterna nilotica*) and 23 of Slender-billed gulls (*Larus genei*) at a distance of c. 150 m away of the Mediterranean gull colony. These birds bred on the closest to the Mediterranean gull colony available islet. During summer, when the water level in the lagoon falls due to evaporation, the two islets form one island as they are separated by a very shallow strait of water in winter. However, these two species breed much later than the Mediterranean gull so that all the chicks of the latter are

Phot. 1: Colony of Mediterranean gulls in the Evros delta, 1985.

fledging whereas at the same time hatching of the former just started (end of June). No nests of Gull-billed terns and Slender-billed gulls were found in or around the Mediterranean gull colony.

• In the Evros delta, both islets (coastal and lagoon) where the gulls bred are very important sites for breeding of many hundreds of other larids and waders (Goutner, 1983, Goutner & Kattoulas, 1984). During the course of our study since 1981, we have seen that the change of colony sites by Mediterranean gulls was generally followed by most of other breeding larids (Goutner & Kattoulas, 1984). The establishment of the gull colony on the lagoon islet directly displaced breeding waders like avocet (Recurvirostra avosetta, Goutner, 1985) and rather indirectly other larids (Goutner & Kattoulas, 1984). We have never found nests of other larids within the colonies of Mediterranean gulls and their colonies were distinctly separated (Goutner & Kattoulas, 1984; Goutner in press a, b). During 1985 we found a nest of Mallard (Anas platyrhynchos) and two nests of Redshanks (Tringa totanus) made within the Mediterranean gull colony and very close to gull nests. Makatsch (1968) and Bauer et al. in Cramp & Simmons (1983) also recorded waders and larids breeding with Mediterranean gulls in the same locality.

Food	Dry weight (gr)	% on total food weight (323.3 gr)	Number in the sample
PISCES			
<u>Sea species</u>			
Cepola rubescens	65.4	20.2	23
Diplodus annularis	19.3	6.0	5
Sardinia pilchardus	16.8	5.2	13
Pagellus acarne	15.8	4.9	5
Trachurus mediterraneus	9.2	2.8	3
Trigla sp.	3.4	1.0	1
Spiraca smaris	2.5	0.8	1
Symphodus sp.	1.8	0.5	1
Engraulis engrasiolus	1.3	0.4	2
<u>Brackish water species</u>			
Gobius jozo	51.9	16.0	17
Gobius sp.	24.4	7.5	26
Atherina sp.	2.9	0.9	2
Pomatoschistus sp.	1.5	0.5	1
Mugil sp.	1.4	0.4	1
<u>Fresh water species</u>			
Esox lucius	24.5	7.6	5
Leuciscus cephalus	6.5	2.0	1
Hypophthalmichthys sp.	6.0	1.8	2
Abramis brama	3.2	1.0	1
Cyprinus carpio	1.3	0.4	1
Perca fluviatilis	0.5	0.1	1
Unidentified fish	49.5	15.3	-
MOLLUSCA			
Sepia sp. (shells)	10.4	3.2	23
AMPHIBIA			
Rana ridibunda	3.8	1.2	2

Tab. 2: Diet of Mediterranean gull in the Evros delta colony during the 1984 breeding season.

2.3. <u>Food</u>

Data on the diet of the Mediterranean gull from the Alyki colony have been published by Isenmann (1975) indicating that these birds received a variety of terrestrial and marine food. The food analysis in the Evros delta indicates that:

Food	Frequency in sample	% Frequency	Range of number
Aranea	4	19.0	1 - 2
Isopoda	1	4.8	30
Orthoptera			
Gryllidae			
Gryllidae sp.	1	4.8	1
Tartarogryllus burdigalensis	2	9.5	? - 20
Melanogryllus desertus	1	4.8	25
Acrididae	1	4.8	1
Tetrigidae	1	4.8	1
Dermaptera			
Dermaptera sp.	2	9.5	1 - 2
Labiduna riparia	1	4.8	2
Heteroptera			
Scutelleridae	3	14.3	1
Pentatomidae	3	14.3	1 - 2
Coreidae	2	9.5	1
Leipidoptera			
Noctuidae (larvae)	1	4.8	40
Hymenoptera			
Ichneumonidae	1	4.8	1
Coleoptera			
Carabidae	11	52.4	2 - 38
Dytiscidae	4	19.0	1 - 2
Hydrophilidae	4	19.0	1 - 2
Staphylinidae	2	9.5	1 - 5
Dynastidae			
Pentodon sp.	3	14.3	1 - 3
Oryctes nasicornis	1	4.8	1
Scarabaeidae	2	9.5	3 - 7
Elateridae			
Elateridae sp.	1	4.8	1
Agriotes sp.	1	4.8	1
Tenebrionidae	1	4.8	1
Cerambycidae			
Dorcadion sp.	2	9.5	1
Curculionidae	10	47.6	3 - 50
Unidentified insects	3	14.3	-

Tab. 4: Food items of Mediterranean gulls identified in 21 regurgitations during the breeding season of 1985 in the Evros delta colony.

- during the 1984 breeding season the gulls mainly received fish, constituting 95.6% of the weight of all food consumed. Nine sea water species, five brackish water and six freshwater species were identified (tab. 2). During 1985, the only fish found in the colony was: one specimen of Atherina sp. (1.5 gr dry weight), one of Aphanius fasciatus (l = 49.5 mm, w = 1.3 gr) and a piece of Sardina pilchardus (10 gr).

- The most numerous fish species collected by the gulls were Gobius spp., Cepola rubescens and Sardina pilchardus (tab. 2). Cepola rubescens is a species thrown back to sea from trawlers. We observed that groups of gulls from the colony frequently follow fishing boats. In tab. 3 are indicated the fish species available in the lagoon where the gull colony is located. Of these species, only the first five are of commercial value. A comparison with tab. 2 shows that no fish of commercial value forms part of the gulls' diet. Some of the gulls' food species are found in the lagoon. Especially Gobius spp. may well be taken from the lagoon. Observations indicate that the gulls sometimes feed in the lagoon but these cases are rare.

- Insects constituted the diet of the gulls during 1985 (tab. 4). The birds covered long routes to feed on insects, leaving the colony in the morning in North eastern direction and returning at dusk. The only record of birds (a flock of c. 150 gulls) feeding on insects in the delta area was in mid-May 1985 at a distance of c. 3,800 m from the colony in a cotton field. It was not possible to follow the flocks but a part of the birds may well feed in turkish territory.

Dicentrarchus labrax
Sparus auratus
Mugil cephalus
Mugil ramada
Anguilla anguilla
Engraulis engrasiolus
Atherina boyerii
Sardina pilchardus
Aphanius fasciatus
Gobius spp.

Tab. 3: Fish species available in the Evros delta lagoon where the Mediterranean gull colony is located.

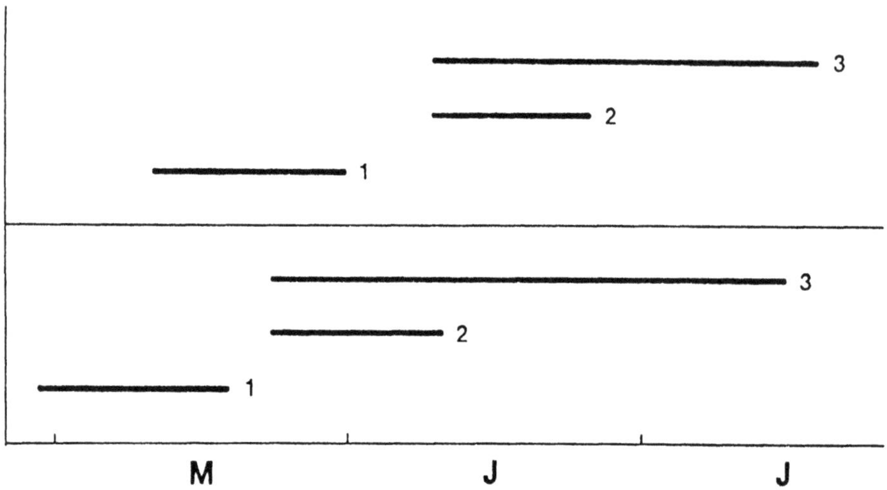

Fig. 2: Timing of breeding at Alyki (lower diagram) and Evros delta (upper diagram) colonies. 1: Laying, 2: Hatching, 3: Fledging period. Data: 1985 for Alyki and 1983-1985 for Evros delta.

Year	Clutch size	S.D.	No. of clutches	P
1981 (*)	2.24	0.73	835	0.05
1983	2.38	0.59	58	0.01
1984	2.62	0.53	217	0.001
1985	2.38	0.72	937	

(*) By Goutner in press, a.

Tab. 6: Clutch size of Mediterranean gulls in the Evros delta

	April	May	June	July
Alyki	14.12 (1.15)	19.93 (1.27)	23.50 (0.67)	25.81 (0.54)
Evros delta	13.40 (1.42)	18.42 (1.20)	22.95 (0.75)	25.95 (0.79)
t	1.89	4.14	2.62	0.70
Significance P	0.05	0.001	0.02	0.30

Tab. 5: Mean air temperatures (C°) during the breeding season at Alyki and Evros delta regions. Data for 24 years by Balafoutis (1977). Figures in parenthesis are standard deviations.

2.4. Timing of breeding

In 1985 the Alyki colony started laying at the end of April, considerably earlier than that in the Evros delta, generally starting at mid-May (fig. 2). These differences may be partly due to different temperature levels between the two areas especially in May (tab. 5).

In 1981 the gulls in the Evros delta started breeding exceptionally earlier: within the first week of May (Goutner in press, a). Makatsch (1968) recorded that during 1966 and 1967 in northern greek colonies these gulls started breeding by mid-May. The fledging period finishes almost simultaneously at both colonies (mid-July, fig. 2). This is mainly due to the fact that the Evros colony suffers many losses by predation late in season.

2.5. Breeding success

Data of clutch size are only available for the Evros delta colonies. The difference in clutch size is considerable from year to year. Tab. 6 indicates similar clutch size in 1983 and 1985. The differences between 1984 and 1985 may be partly due to different diets in these years (tab. 2 and 4).

Of the two colonies, the Alyki one is the most successful. There are no data for previous years but during 1985 we found that c. 2,500 breeding pairs produced c. 4,000 flying chicks: a success of c. 1.6 chicks per pair. Very few destroyed and/or addled eggs and dead chicks were found at the colony.

The Evros delta colony is less successful. Reasons for failure during the breeding season in 1981 have been described (Goutner in press, a) and it appeared that most losses took place in the egg stage during incubation due to predation. In 1981 the success was 0.8 chicks per pair. This success was lower during 1983 and 1984 being between 0.3 - 0.5 chicks per pair in both years. Like 1981, most eggs were predated and/or pecked open during interactions among gulls. In 1983 two very heavy storms took place during incubation resulting in many eggs rolling out of nests. In the lagoon colony predation of adult birds by foxes and jackals is very important. Their droppings and footprints were found among nests. In 1983, 35 adults were found decapitated on or beside their nests, thus preventing hatching of the eggs. One of these adults had been ringed in the Soviet Union (phot. 2). Similarly in 1984 and 1985, 6 and 11 adult gulls were found killed by mammal predators. These predators were able to reach the colony in summer when due to evaporation the water level fell considerably in the lagoon.

In 1985 the two colonies in the delta were absolutely unsuccessful. No young fledged. The small colony at the coastal islet was predated by Yellow-legged gulls (<u>Larus cachinnans</u>) at the beginning of the breeding season. The large lagoon colony was deserted at the end of June; of a total 937 nests made, 20 (2%) hatched, 100 (11%) disappeared and 816 (87%) were found and examined. The results are in tab. 7. Most eggs were again destroyed during interactions among gulls. Predation by other bird predators of the region (gulls, corvids) was unimportant.

Phot. 2: A nest, destroyed eggs, a dead chick and a decapitated adult of Mediterranean gull in the Evros delta colony, 1983. Note the ring on the leg of the bird.

Reasons of failure	No. of eggs	% on failed	% on laid
Pecked open or predated	742	42.6	33.2
Addled (liquid feel when handled)	290	16.6	13.0
Cracked	447	25.7	20.0
Failed to hatch with developed embryo	19	1.1	0.8
Deserted	236	13.6	10.6
Destroyed for unknown reason	7	0.4	0.3
Eggs total	1,741		2,231

Tab. 7: Reasons of failure of Mediterranean gull eggs during the breeding season of 1985 in the Evros delta colony.

Several eggs were "cracked" showing a long crevice between the two poles. It seemed that these eggs broke when the birds sat on them for incubation. 11 freshly hatched chicks were found dead in the nests.

There is no apparent explanation for such a failure. The proportion of eggs, other than predated, which did not hatch (45% of the total) was very great. The hatching failure may be due to pesticides as almost all of the gulls' food in 1985 constituted of insects. The closeby Gull-billed terns suffered from a similar breeding failure. Pesticides are also here the most probable cause as these birds also ate insects, while following the gulls at the feeding grounds. There may be a threshold after incubation: if no chicks are produced or if no massive hatching occurs,the birds desert the colony.

In case chemicals like PCB interferes during egg production through the food chain, deformities of the bill are frequent. Such deformity is the "cross bill" which has been observed in larids (Gilbertson, 1975). No similar deformity was observed till now in the Evros delta colonies but in Alyki a young Mediterranean gull was found having an upper bill deformity (phot. 3). Anyway breeding failure in the Evros delta colony will be better understood after analysis of some of the eggs which were collected after the desertion of the colony.

Phot. 3: Upper bill deformity of a young Mediterranean gull at Alyki colony, 1985.

3. Conservation

The above presented data indicate that at least five well known breeding habitats including the stronghold of the Mediterranean population have been deserted during the last 35 years. Nevertheless, the two colonies at Alyki and in the Evros delta are still constituting the main Mediterranean population with a total of 3,500 pairs in 1985. Agriculture is the main human activity that affects the greek population of Mediterranean gulls. Agriculture is not only directly responsible for habitat destruction but it seems also that pesticides cause breeding failure since an important part of these birds' diet is constituted of insects. Beside pesticides which will be a continuous threat in the next years, the existing colonies will also be threatened by further habitat loss.

The Alyki colony, situated on lagoon islets is threatened by expansion of the salt extraction pans within the lagoon. This expansion was intensified during 1985 and still continues. New dykes are being constructed across the lagoon. If these approach or cross the colony site, the birds will desert their islets. The expansion of the salinas is unauthorized and does not examine the effects on wildlife of the region. In 1980 the surrounding scrub vegetation was destroyed by buldozer and fire by local people since they want to use the area for housing purposes. This led to the destruction of a rare habitat and killed at least 10,000 tortoises (Testudo hermanni) (Stubbs et al., 1980). Thanks to governmental measures this destruction stopped but control must also be enacted to halt the salt work expansion. We submitted a proposal to World Wildlife Fund to study whether the Alyki should be included in the greek list of Ramsar wetlands if no serious protection measures are put into action immediately. The desertion of this site frequented by Mediterranean gulls, Slender-billed gulls, Gull-billed terns, Little terns, Avocets, Pratincoles and other rare wildlife will be rather a matter of time.

The Evros delta colony is now situated on an isolated islet within a very productive area for fish farming. It is quite possible that this use of the lagoon will continue, thus preserving the natural habitat. Additionally, this islet is not grazed by herds during the breeding season in contrast to all the other islets of the same lagoon, an activity resulting in massive destruction of many hundreds of rare wader nests like avocets and pratincoles (Goutner, 1983, 1985). However, there are propositions by local people to cease this use of the lagoon as, due to bad management, sea salt affects the surrounding cultivations. Other propositions by local fishing authorities suggest pouring brackish water in the lagoon and simultaneously closing its entrance off from the sea, an action that should lead to flooding of the islets in the lagoon and loss of the gulls' habitat.

The potential breeding area of the gulls at the coastal islets of the delta, which is also a breeding habitat for many hundreds of other larids and waders, is seriously endangered. This is due to the proposed construction of a huge fishpond by connecting the coastal islets with the mainland. In 1978 biologists from the biological station of Tour de Valat invited by the government to propose conservation measures for the Evros delta, clearly expressed the opinion that such a construction would not only result into an inefficient fishpond but also destroy the few remaining wildlife refuges in the delta. In 1984, during a governmental project for boundary delimitation of national Ramsar wetlands we proposed that a fishpond - if constructed - should be made in such a way that free recycling of sea water will preserve the hydrological conditions within the delta without disrupting the isolation of the coastal islets. However, we are afraid that the project will finally be constructed by the same developer who has already provided many examples of unnecessary destruction of the Evros delta. Despite the fact that the local authorities are increasingly aware of the conservation interests of the wetlands, in a few years the previously famous Evros delta will be transformed into a dry desert without wildlife.

Acknowledgements

I am grateful to Sotiris and Christos Goutner, Yannis Poulopoulos, Dimitris Goutsouras, Alexis Mehtidis and George Mihalakakis for valuable aid in field work. Also to Mr. Byron Antipas and the Hellenic Society for the Protection of Nature for permitting me staying at the Biological Station in the Evros delta. I am especially obliged to Prof. Dr. P.S. Economides for identification of fish material and to Dr. A. Cejchan, Dr. J. Jezek, Dr. S. Bily, Dr. Jelinek, Dr. I. Kovar, Dr. J. Macek and Dr. V. Svihla (National Museum in Prague, Department of Entomology) for identification of insect material. Special thanks to Dr. Jan Jezek for kindly facilitating my stay in Prague and to George Handrinos (Hellenic Ornithological Society) for providing me many published and unpublished articles. Finally many thanks to the local fishermen in the Evros delta for their collaboration.

References

Balafoutis, H.I. (1977). The climate of Macedonia and Western Thrace. Doctoral dissertation, Univ. of Thessaloniki.

Bauer, W.; Helversen, O.v.; Hodge, M. & Martens, J. (1969). Catalogus faunae Graeciae. Paris II, AVES. Kanellis, Thessaloniki.

Conradty, P. & Hohlt, G. (1967). Zur Kenntnis der Vogelwelt Nordgriechenlands. II. Anz. Orn. Ges. Bayern 8: 45-51.

Cramp, I. & Simmons, K.E.L. (eds.) (1983). The birds of the Western Palearctic. Vol. III. Oxford Univ. Press.

Gilbertson, M. (1975). A Great Lakes tragedy. Nature Canada 4 (1): 22-25.

Goutner, V. (1983). The distribution of the waders (Charadrii) in the Evros delta (Greece) during the breeding season. Sci. Annals. Fac. Sciences, Univ. Thessaloniki 23: 37-78.

Goutner, V. & Kattoulas, M. (1984). Breeding distribution of Gulls and Terns (Laridae, Sternidae) in the Evros delta (Greece). Seevögel 5 (3): 40-41.

Goutner, V. (1985). Breeding ecology of the avocet (*Recurvirostra avosetta* L.) in the Evros delta (Greece). Bonn. Zool. Beitr. 36 (1/2): 37-50.

Goutner, V. (in press a). The ecology of the first breeding of the Mediterranean gull (*Larus melanocephalus*, Temminck, 1820) in the Evros delta (Greece). Ökologie der Vögel.

Goutner, V. (in press b). Vegetation preferences by colonies of Mediterranean gulls (*Larus melanocephalus*) and Gull-billed terns (Sterna nilotica) in the Evros delta. Seevögel.

Isenmann, P. (1975). Contribution à l'étude de la biologie de reproduction de la Mouette mélanocephale Larus melanocephalus. Nos Oiseaux 33: 66-73.

Joensen, A.H. & Madsen, J. (1985). Waterfowl and raptors wintering in wetlands of western Greece, 1983-1985. Natura Jutl. 21 (11): 169-200.

Kumerloeve, H. (1957). Séjour et passage de la Mouette mélanocéphale (Larus melanocephalus Temminck) en Mer Egée, dans les Dardanelles et le Bosphore. Alauda 25 (2): 143-145.

Makatsch, W. (1963). Ornithologische Beobachtungen in Griechenland. Zool. Abh. 26 (8): 135-186.

Makatsch, W. (1968). Die Möwen Griechenlands. Natur und Museum 98 (6): 221-230.

Makatsch, W. (1978). Zur augenblicklichen Situation der Vogelwelt Griechenlands. Nature (Hel. Soc. Prot. Nature) 13: 29-33.

Nisbet, I.C.T. & Smout, T.C. (1956). Observations de la migration d'automne de Larus melanocephalus en Grèce et Turquie. Alauda 24: 306-307.

Schuster, S.v.; Knötzsch, G. & Jacoby, H. (1959). Ornithologische Beobachtungen in Macedonien, Thrazien und Mittelgriechenland. Vogelwelt 12: 170-179.

Sterbetz, I. (1980). Notes from Macedonia and Greece. Larus 31/32: 427-439.

Stubbs, D.; Hailey, A.; Tyler, W. & Pulford, E. (1980). Expedition to Greece, 1980. Univ. of London, Nat. Hist. Society.

BREEDING DISTRIBUTION, NUMBERS AND CONSERVATION OF SEABIRDS IN SARDINIA, 1978-85

by Helmar Schenk, Via Castiglione, Cagliari,
and Antonio Torre, 96 Via S. Satta, Alghero

Introduction

With a surface of 24,090 km2, Sardinia is the second largest island in the Mediterranean where it is centrally situated. The Sardinian coastline extends over nearly 1,900 km, whereof 1,385 km (73%) belong to the main island while the remaining 512 km (27%) belong to the various smaller islands (275 km2), all situated on a common continental shelf and prevalently off the four corners of Sardinia. About 75% of the coastline is rocky, consisting of cliffs and rocks falling into the sea. Particularly sheer stretches can be found in the Gulf of Orosei, around Capo Caccia and Capo Teulada and on the Island of Tavolara. They are all of calcareous formation, abounding in caves, cavities, rock ledges and terraces.

Sandy shores are generally found within the major gulfs, those of Asinara, Oristano, Cagliari and Palmas, which often are interrupted by shallow lagoons linked to the sea and fed by fresh-water coming from small rivers. These coastal wetlands cover a surface of about 12,000 ha. Most of them are situated at the extremities of the great tectonic depression of Campidano: around the Gulfs of Cagliari and Oristano. More than 3,000 ha of these wetlands are used as evaporation basins for the State salt-works (Molentargius, S. Caterina) and for the privately owned salt-works of Macchiareddu.

In spite of the increasing human impact on the smaller islands and along the rocky coasts and notwithstanding the accompanying problems of urban, agricultural and industrial pollution, Sardinia still hosts important seabird populations of considerable interest both nationally and throughout the Mediterranean.

NATO ASI Series, Vol. G 12
Mediterranean Marine Avifauna
Edited by MEDMARAVIS and X. Monbailliu

SPECIES	POPULATION (cp. = pairs ind. = individuals)	POPULATION TREND + = increasing - = decreasing ± = stable ? = unknown	NUMBER OF COLONIES
Calonectris diomedea	2,500 - 4,000 cp. 6,000 - 10,000 ind.	± ; - ?	> 25
Puffinus puffinus	7,500 - 13,000 cp. ? 18,000 - 30,000 ind. ?	- ?	≃ 25
Hydrobates pelagicus	150 - 300 cp. ? 400 - 700 ind. ?	- ?	< 10
Phalacrocorax carbo	35 - 45 cp.	± ; + ?	2
Phalacrocorax aristotelis	1,000 - 1,800 cp.	± ? ; + ?	> 40
Larus ridibundus	50 - 100 cp.	± ?	7
Larus genei	530 - 935 cp.	+	2
Larus audouinii	300 - 500 cp.	± ? - ?	> 15
Larus cachinnans	5,000 - 8,750 cp.	+	> 70
Sterna nilotica	50 - 100 cp.	+	1
Sterna hirundo	150 - 240 cp.	+ ?	≃ 20
Sterna albifrons	380 - 500 cp.	-	≃ 25

Tab. 1: Estimated breeding population of 12 species of seabirds in Sardinia, 1978-85.

The main objective of this paper is to summarize the present knowledge on distribution and numbers of seabirds breeding in Sardinia. It is hoped that this will stimulate systematic research and promote active conservation including the establishment of protected areas, and the qualified management and administration of particularly sensitive biotopes.

Materials and Methods

Within this work the term "seabirds" is to include all the species belonging to Procellariiformes, Pelicaniformes and Laridae, which breed on the small islands, along the coasts and in the shallow coastal lakes of Sardinia, within a limit of up to 10 km from the shore, and which feed, even if for brief periods, in the seawater. The data used generally refer to the period 1978-1985, and have been gathered by the authors involved either in the "Progetto Gabbiano Corso" (The Audouin's Gull Project) organized by the WWF/IUCN between 1978-1980 (Schenk, 1978, 1980), or in the "Progetto Laridae in Italia" (The Italian Laridae Project) run between 1982-1984 by the Department of Animal Biology of the University of Pavia and financed by the INBS: Istituto Nazionale di Biologia della Selvaggina (National Institute of Game Biology) (Fasola, 1986; in press) or, but to a lesser extent, in the "Progetto Atlante Italiano" (The Italian Atlas Project) (1983-1985) managed by CISO, the Italian Centre of Ornithological Studies, and subsidized by INBS (Meschini and Roselli, 1985).

Bibliographical information, data as yet unpublished by the authors or by other ornithologists, and reliable accounts by fishermen have also been used. The zoogeographical maps are based on Italian National Grid (I.G.M.) on a scale of 1:25,000 corresponding to 10 km squares. Three levels have been distinguished: certain breeding (big dot), probable breeding (medium-sized dot), and possible breeding (small dot), in accordance with the national standards of the Progetto Atlante Italiano (The Italian Atlas Project) (INBS and CISO, 1985).

In order to define the frequency of each individual species, reference is made only to the number of 10 km squares, 86 in all, which comprise the coastline of Sardinia. Those which were found to cover a minimal land-surface have been integrated in the adjoining square.

The number given for each individual species refers to the census of nests (Phalacrocorax carbo, Larus ridibundus, L. genei, L. audouinii, Sterna nilotica), to the estimate of the number of nests (Phalacrocorax aristotelis, Larus cachinnans) and to the combination of both methods (Sterna hirundo, Sterna albifrons, and partly Larus cachinnans). The population of Calonectris diomedea and Puffinus puffinus has been estimated by counting individuals which, every evening during the breeding season, assemble at sea in the proximity of established colonies or breeding areas assumed from nocturnal vocal activity. The

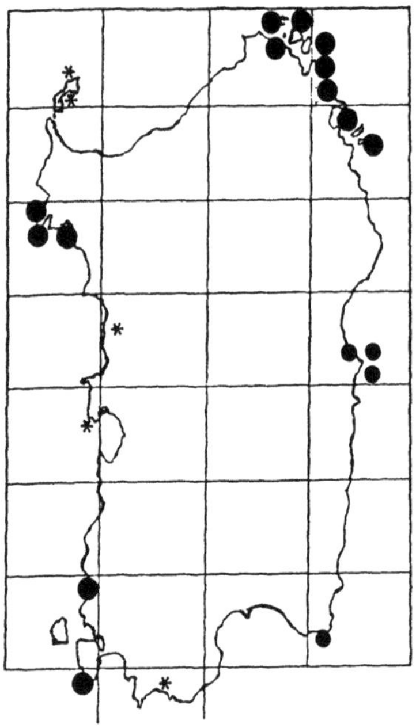

Fig. 1: Calonectris diomedea - Breeding status in coastal squares of 100 km2

Coverage	Certain	Probable	Possible
●●* = 22	● = 13	● = 4	* = 5
= 25.6%	= 59.1%	= 18.2%	= 22.7%

Note: The squares used on the map above represent 16 squares of 100 km2 (40 x 40 km).

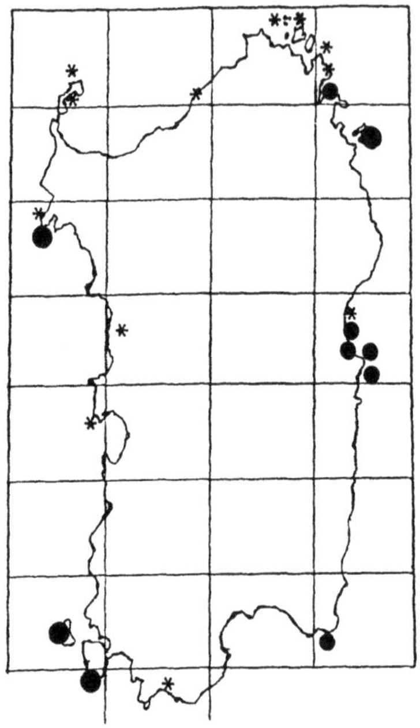

Fig. 2: Puffinus puffinus - Breeding status in coastal squares of 100 km2

Coverage	Certain	Probable	Possible
●●* = 22	● = 4	● = 6	* = 12
= 25.6%	= 18.2%	= 27.3%	= 54.5%

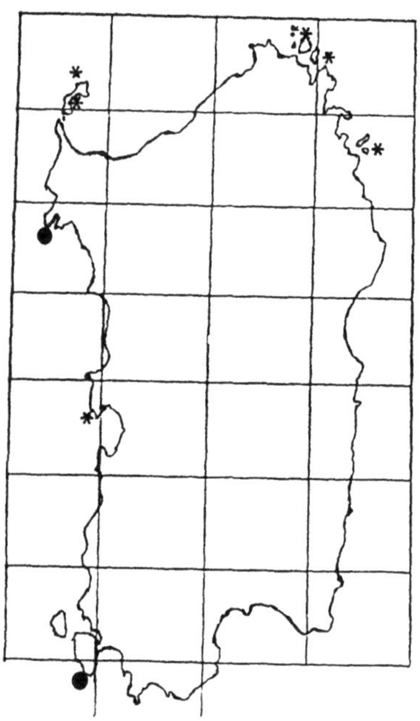

Fig. 3: Hydrobates pelagicus - Breeding status in coastal squares of 100 km2

Coverage	Certain	Probable	Possible
●* = 8	-	● = 2	* = 6
= 9.3%	-	= 25%	= 75%

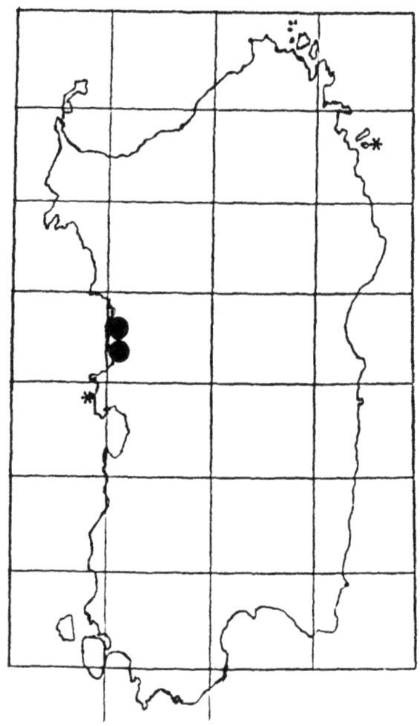

Fig. 4: Phalacrocorax carbo - Breeding status in coastal squares of 100 km2

Coverage	Certain	Probable	Possible
●* = 4	● = 2	-	* = 2
= 4.7%	= 50%	-	= 50%

estimation of Hydrobates pelagicus is based entirely on second hand information and should be interpreted carefully.

The extent of coverage degree hitherto obtained varies considerably from species to species and might be approximately indicated as follows: Calonectris diomedea 60%; Puffinus puffinus 40-50%; Hydrobates pelagicus 30-40%; Phalacrocorax carbo 90%; Phalacrocorax aristotelis 70-80%; Larus ridibundus 85%; Larus genei 100%; Larus audouinii 85%; Larus cachinnans 90%; Sterna nilotica 100%; Sterna hirundo 80-85% and Sterna albifrons 90%. Further details follow here below.

Results

- Distribution and numbers

- Calonectris diomedea (Cory's shearwater) - fig. 1

For the sea-cliffs of Alghero breeding has been recorded by Lubrano and Canu (1977) and by Guillot (1985, pers.com.); for the Archipelago of Maddalena by Cesaraccio (PAI, 1983-1985); for the islands off the Costa Smeralda by Schenk (1979, unpublished); for the Island of Figarolo by Tamponi (1979, pers. com.); for the Archipelago of Tavolara by Bertoleoni (1978, pers.com.); for the Island of Toro by Schenk and Spina (1980, unpublished) and for the islet of Pan di Zucchero by Feola (1980, pers.com.).

The total number has been estimated at approximately 2,500 to 4,000 breeding pairs, of which 1,500-2,000 are in the region of Alghero/Capo Caccia (1980, Schenk and Torre, unpublished) and ca. 300-400 are on the Island of Toro (1980, Schenk and Spina, unpublished).

- Puffinus puffinus (Yelkouan shearwater) - fig. 2

Breeding records come from the Archipelago of Tavolara (Moltoni, 1971; Bertoleoni, 1978, pers.com.); from the Island of Cavoli (Schenk, 1976; unpublished), from the Island of Vacca (Feola, 1980, pers.com.; Giuliani, 1965, pers.com.) and from the Island of San Pietro (Feola, 1980, pers.com.).

The total number has been provisionally estimated at 7,500-13,000 pairs, of which 6,000-9,000 are to be found in the Archipelago of Tavolara (Schenk, 1978/79, unpublished), and 500 on the Island of San Pietro (Feola, 1980, pers.com.). This species merits special attention in the near future.

- Hydrobates pelagicus (Storm petrel) - fig. 3

Local reports confirm the nesting of the Storm Petrel on the islands off the region of Sulcis (Feola, 1980, pers.com.; Aru, 1985, pers.com.); in the zone of Alghero between 1970-1978 (Schenk, 1976, unpublished; Torre, 1983) and in the Archipelago of the Maddalena. Off the coast of Alghero on 23rd May,

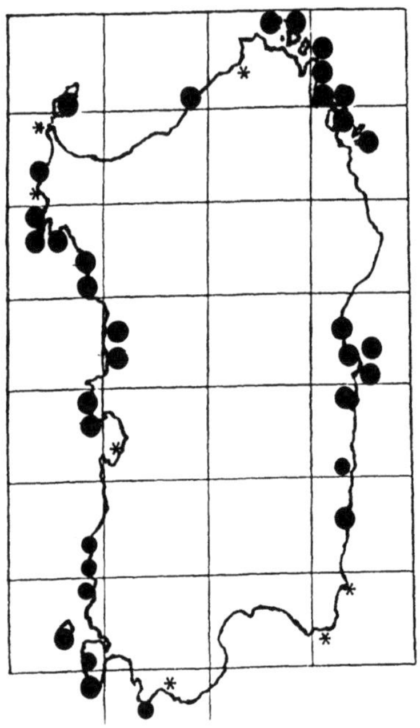

Fig. 5: Phalacrocorax aristotelis - Breeding status in coastal squares of 100 km2

Coverage	Certain	Probable	Possible
● ● * = 41	● = 28	● = 6	* = 7
= 47.7%	= 68.3%	= 14.6%	= 17.1%

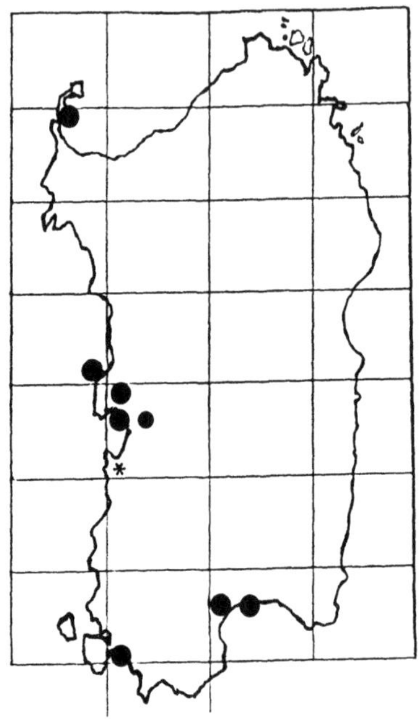

Fig. 6: Larus ridibundus - Breeding status in coastal squares of 100 km2

Coverage	Certain	Probable	Possible
● ● * = 9	● = 7	● = 1	* = 1
= 10.5%	= 77.8%	= 11.1%	= 11.1%

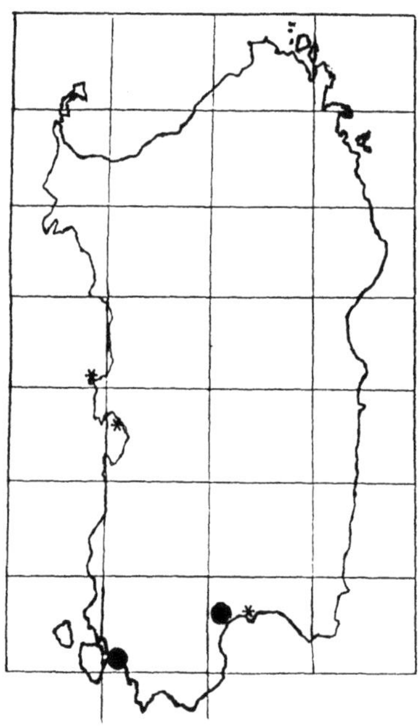

Fig. 7: Larus genei - Breeding status in coastal squares of 100 km2

Coverage	Certain	Probable	Possible
● * = 5	● = 2	-	* = 3
= 5.8%	= 40%	-	= 60%

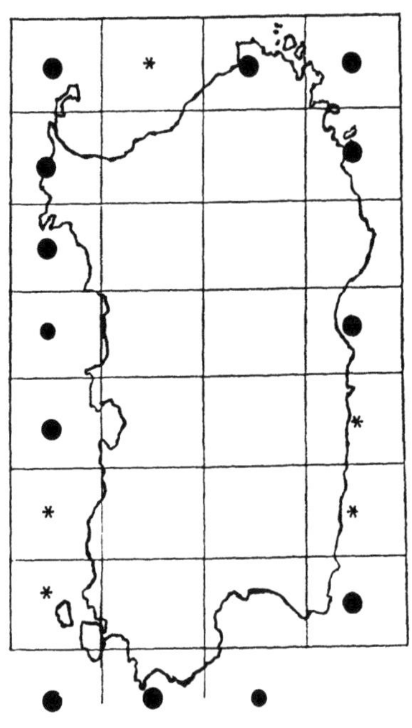

Fig. 8: Larus audouinii - Breeding status in coastal squares of 1,600 km2

Coverage	Certain	Probable	Possible
● ● * = 18	● = 11	● = 2	* = 5
= 90%	= 61.1%	= 11.1%	= 27.8%

For conservation purposes, breeding status is symbolically plotted in the middle of the 1,600 km2 squares.

1981, some 25 Storm petrels together with Cory's shearwaters and Yellow-legged gulls were seen feeding on offal thrown overboard from a fishing boat (Schenk, unpublished). For previous data see Schenk (1976), Brichetti (1979) and Torre (1983).

An estimate of 150-300 pairs should be considered as an attempt to establish the size of the Sardinian population. In the two areas of probable breeding (Alghero, Archipelago off the coast of Sulcis) between 50 and 100 pairs might be breeding according to sightings by local fishermen. The islands of the Archipelago of Maddalena merit further research (Cesaraccio, PAI, 1983-1985).

- Phalacrocorax carbo (Great cormorant) - fig. 4

Breeding has been recorded by Schenk (1976 and unpublished) for the Sinis cliffs (in 1967), for the basaltic rocks of S. Caterina (1971-1985) and for the cliffs of Corona Niedda (1972, 1979). For confirmation with regard to the coast of S. Caterina we refer to Brichetti (1982) and Aresu (PAI, 1983-1985). The population of the two sites presently occupied fluctuates between 35 and 45 pairs per year (1978-1985). Among possible nesting areas, the Archipelago of Tavolara merits special attention (Brichetti, 1982).

- Phalacrocorax aristotelis (Shag) - fig. 5

Breeding records exist for the Island of Asinara (Torre, Monbailliu, 1983-1985, unpublished); for the Archipelago of Maddalena (Schenk, 1978-79, unpublished); Cesaraccio PAI, 1983-85); for the islands off the Costa Smeralda (Schenk, 1978-79, unpublished); for Capo Figari and the Island of Figarolo (Tamponi, 1978/79); for the Archipelago of Tavolara (Schenk, 1978/79; Bertoleoni, 1978, pers.com.); for the Gulf of Orosei (Schenk, 1971, 1973; 1980-85, unpublished); for the Island of Quirra (Schenk, 1979, unpublished); for the Island of Vacca (Schenk, 1979, unpublished); for the Island of San Pietro (Brai, 1982, pers.com.); for the Island of Mal di Ventre (Schenk, 1974, 1984, unpublished); for the coast of S. Caterina and the cliffs of Corona Niedda (Schenk, 1971-1985, unpublished); for the coast between Bosa and Alghero (Schenk, 1978-85, unpublished; and PAI, 1983-85); for the coasts and islands of Alghero (Fasce, 1971, in lit.; Schenk, 1971-80; Guillot and Torre, PAI, 1983-85); and for the Island of Porri (Torre, 1981, unpublished).

The Sardinian population has been estimated at 1,000-1,800 breeding pairs. 200 to 400 pairs have settled in the Gulf of Orosei (including the Island of Ogliastra); 200 to 300 pairs breed in the Archipelago of Tavolara (cf. Moltoni, 1971 and Brichetti, 1982); 200 to 300 pairs nest in the Archipelago of Maddalena while 100 to 130 pairs breed along the Alghero coastline (especially on Isola Piana); 80 to 100 pairs on the Island of Asinara.

There are many difficulties involved in performing a survey of this species, due to both the inaccessibility of several co-

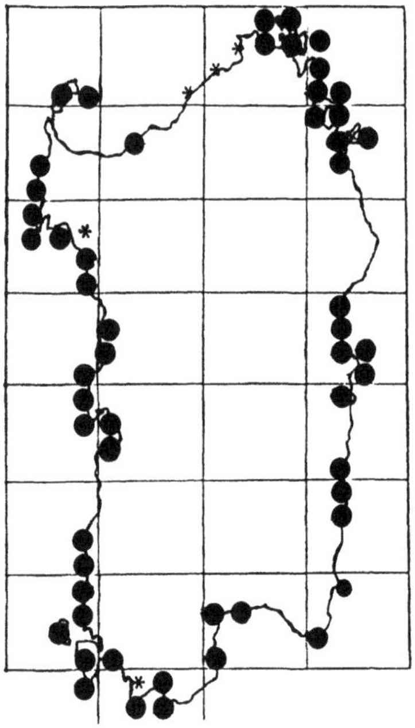

Fig. 9: Larus cachinnans - Breeding status in coastal squares of 100 km2

Coverage	Certain	Probable	Possible
●●* = 62	● = 56	● = 1	* = 5
= 72.1%	= 90.3%	= 1.6%	= 8.1%

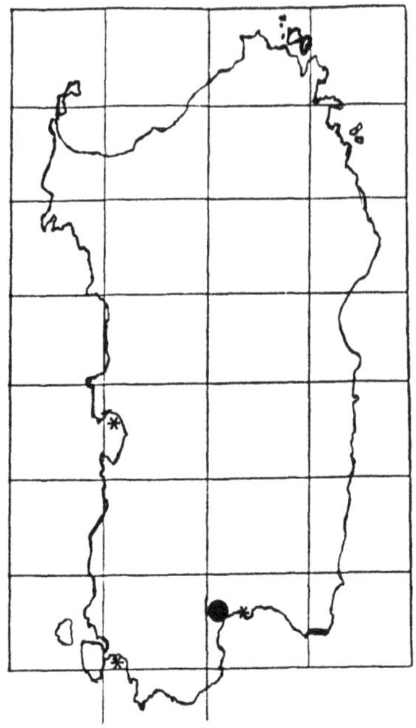

Fig. 10: Sterna nilotica - Breeding status in coastal squares of 100 km2

Coverage	Certain	Probable	Possible
●* = 4	● = 1	-	* = 3
= 4.7%	= 25%	-	= 75%

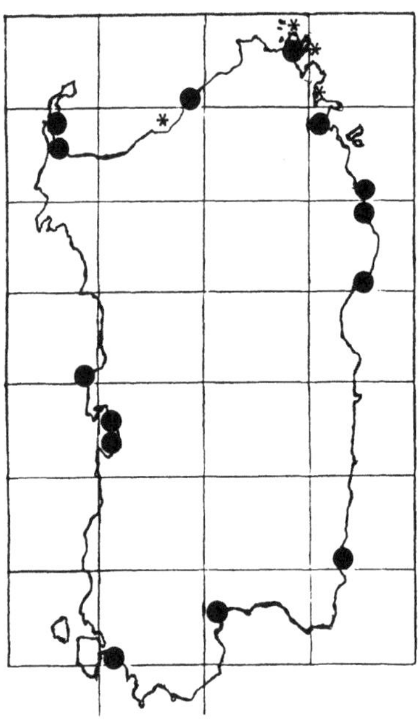

Fig. 11: Sterna hirundo - Breeding status in coastal squares of 100 km2

Coverage	Certain	Probable	Possible
● * = 18	● = 14	-	* = 4
= 20.9%	= 77.8%	-	= 22.2%

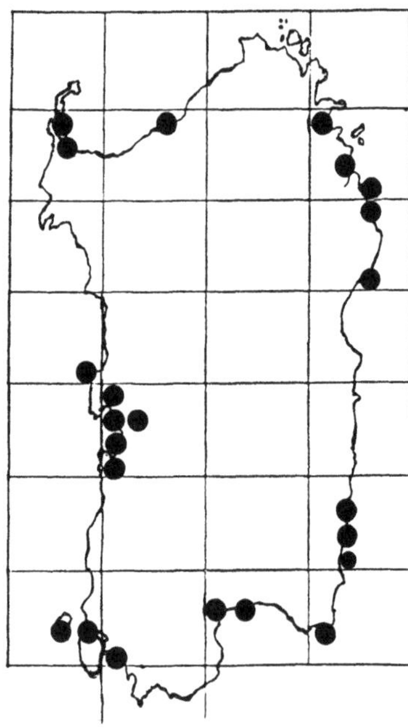

Fig. 12: Sterna albifrons - Breeding status in coastal squares of 100 km2

Coverage	Certain	Probable	Possible
●● = 23	● = 22	● = 1	-
= 26.7%	= 95.7%	= 4.3%	-

lonies situated on cliffs and to the early start of the reproductive cycle in December.

• Larus ridibundus (Black-headed gull) - fig. 6

Breeding of this gull has been confirmed by Schenk for the wetlands of Cagliari (1965-85; cf. also Mossa, Puddu and Schenk, 1983); by Dettori, Pinna and Schenk (PAI, 1983-85, data unpublished) for the coastal lagoons and ponds of Sinis; and by Torre (PAI, 1983-86) for the coast of Stintino.

The total number fluctuates between 50 to 100 pairs. The largest colonies are situated in the salt-works of Macchiareddu (15-30 pairs, 1980-85) and in the region of Sinis near Oristano (30-45 pairs, 1980-85).

• Larus genei (Slender-billed gull) - fig. 7

This species is breeding in Sardinia since 1976 (Schenk, 1976b; Schenk in Fasola, 1986, in press). Breeding was documented from 1978-85 for the coastal lagoons near Cagliari by Schenk, Badas and Baldussi (unpublished; PAI, 1983-85; the Laridae Project, 1983-84) and in the wetlands of Sulcis by Porseo and Aru (PAI, 1984-1985; the Laridae Project, 1984).

The population varied between 530 pairs (1978) and 935 pairs (1984). In 1985 at least 657 pairs were breeding (Schenk, unpublished; Aru, 1986, pers.com.). The largest settlement is to be found in the salt-works of Macchiareddu, where in 1985 there were at least 647 pairs.

• Larus audouinii (Audouin's gull) - fig. 8

Proofs of nesting exist for the Island of Asinara (Monbailliu, Torre, 1983-1985, this volume; PAI... and the Laridae Project, 1983-85); for the Archipelago of Maddalena (Schenk, 1978, 1980; Cesaraccio, PAI, 1983-85; Benussi and Foschi, the Laridae Project, 1984; cf. also Schenk and Meschini in Fasola, 1986, in press); for the islands off the Costa Smeralda (Schenk, 1978, 1980; the Laridae Project, 1983; Benussi and Foschi, the Laridae Project, 1984); for the Gulf of Orosei (Schenk, 1979, 1980; the Laridae Project, 1983); for the islands off south-east Sardinia (the Laridae Project, 1984); for the islands off south-west Sardinia (Mocci Demartis, the Laridae Project, 1983; Perco, 1984, pers.com.); and for the west coast of Sardinia (Mocci Demartis; Benussi and Foschi; Schenk, Spina and Baldussi; Pinna, the Laridae Project, 1984). In 1984, a total of 458 nesting pairs were recorded (cf. Schenk and Meschini in Fasola, 1986, in press).

• Larus cachinnans (Yellow-legged gull) - fig. 9

Breeding has been recorded by the researchers of PAI (1983-85) and of the Laridae Project (1982-84). The population was found to vary between approximately 5,000 and 8,750 pairs. This tendency to increase is difficult to determine due to the

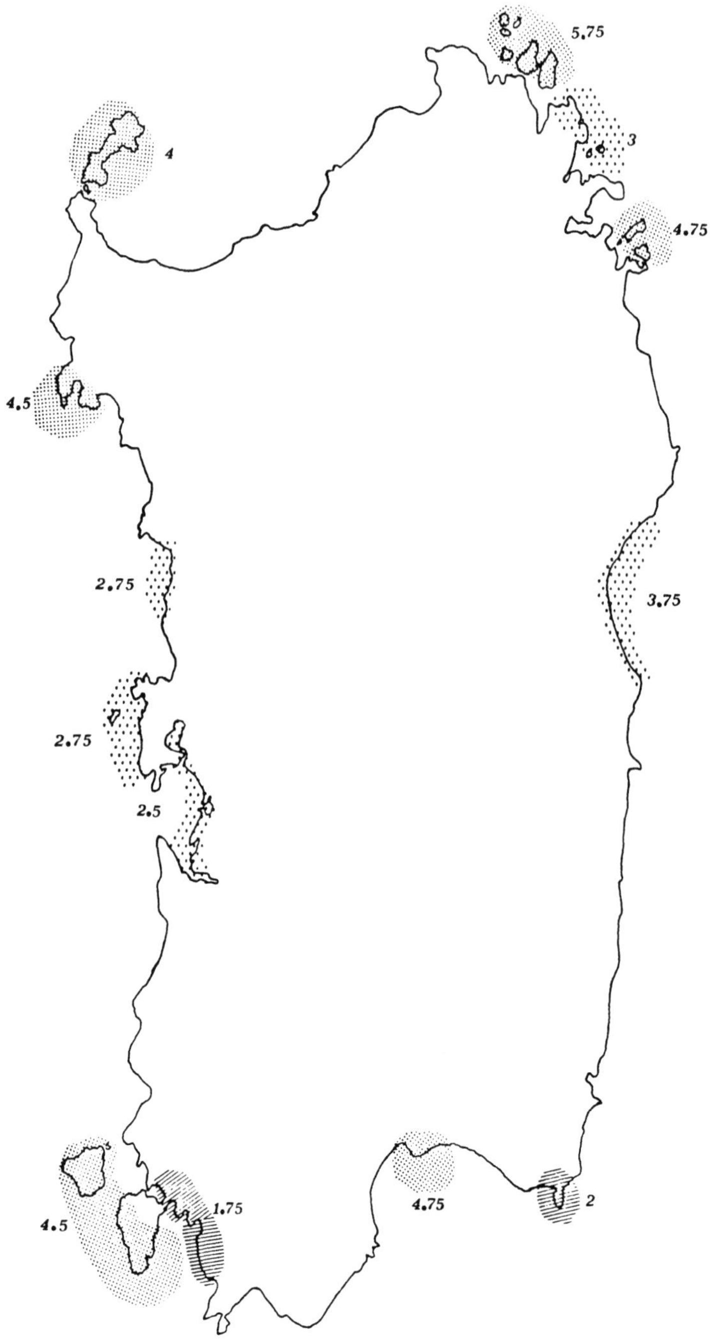

Fig. 13 - Areas of international importance
Areas of national importance
Areas of regional importance

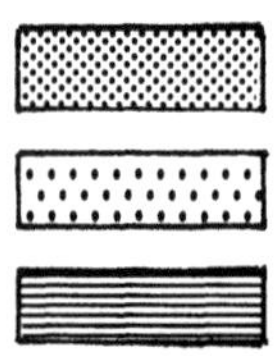

and respective importance values for breeding seabirds in Sardinia, 1978-1985.

lack of complete census in former years. For the period 1971-79, Schenk (1979) estimated between 3,600 and 5,000 pairs, indicating the number of large colonies as being over 30.

The principal colonies are presently situated on the Island of Asinara (1,020 pairs; Torre & Monbailliu, the Laridae Project, 1983-84); on the Island of Tavolara (ca. 1,000 pairs; Schenk, 1979, unpublished); on the coast and islands in the zone of Alghero (over 1,200 pairs; Torre, the Laridae Project, 1983-84); on the Archipelago of Maddalena (over 800 pairs; Schenk, 1979, unpublished); and on the Island of Serpentara (ca. 425 pairs; Schenk, the Laridae Project, 1984). The colonies situated in the coastal wetlands of Sardinia do not exceed a total of 200 pairs.

- Sterna nilotica (Gull-billed tern) - fig. 10

The Gull-billed tern breeds hitherto exclusively in the salt-works of Macchiareddu, where the population has increased from 2 pairs (1971) to 102 pairs (1984; Schenk, the Laridae Project). An incomplete survey revealed 80 pairs in 1985 (Schenk, unpublished).

- Sterna hirundo (Common tern) - fig. 11

Breeding has been recorded by various authors for the coastal lakes of Cagliari (Walter, 1964; Schenk, 1976); for the coast of Stintino (Corti, 1958); and for the coastal lakes of Corru de s'Ittiri and Mistras (Schenk, 1976).

During the research for the Laridae Project further breeding sites were discovered: the Island of Asinara, the coastal lake of Casaraccio and Isola Rosso (Torre, 1983-85); the Archipelago of Maddalena (Benussi and Foschi, 1984; Cesaraccio); Olbia (Schenk, 1984); the Island of Ottiolu (Torre, 1983-1984); the coastal lake of Colostrai (Schenk, 1977); the coastal lake district of S. Caterina (Porseo and Aru, 1983-85); the coastal lake of San Giovanni (Schenk, 1984); the coastal lake of Sale Porcus (Pinna, 1985).

The total population in Sardinia is about 240 pairs (1984), based both on surveys and estimates. The most important settlement is presently in the salt-works of Macchiareddu where, in 1981, there were 96 pairs (Mossa, Puddu and Schenk, 1983) and a minimum of 81 pairs in 1985.

- Sterna albifrons (Little tern) - fig. 12

Nesting has been confirmed in the coastal wetlands of Cagliari by Corti (1958), Kunkel (1963), Walter (1964), and Schenk (1976); for that of S. Caterina, and for the sandy coasts near Pilo and Mistras by Schenk (1976). Since then other colonies have been discovered, especially during the Laridae Project and the PAI (1983-85): in the coastal lake of Casaraccio and the mouth of the River Coghinas (Torre, 1983-84); at the mouth of the Padrongianu/Olbia (Schenk, 1981); on the Island

ZONE	NUMBER OF BREEDING SPECIES (Percentage relative to the total)	NUMBER OF BREEDING SPECIES OF International importance	National importance	Regional importance	IMPORTANCE INDEX WEIGHTS: International = 1 National = 0.5 Regional = 0.25
Asinara	7 (58%)	2	3	2	4
Maddalena - Archipelago	7 (58%)	4	2	1	5.25
Costa Smeralda - Islands	5 (42%)	1	4	0	3
Tavolara - Archipelago	9 (75%)	2	4	3	4.75
Orosei - Gulf	7 (58%)	2	2	3	3.75
S.W. Sardinia - Islands	5 (42%)	0	3	2	2
Cagliari Coastal Lakes	6 (50%)	4	1	1	4.75
S. Antioco Coastal Lakes	5 (42%)	0	2	3	1.75
Sulcis - Islets	6 (50%)	3	3	0	4.5
Oristano Coastal Lakes	6 (50%)	1	1	4	2.5
Sinis - Rocky coasts; Maldiventre and Catalane Islands	7 (58%)	1	1	5	2.75
Cuglieri - Rocky coasts Corona Niedda - Cliffs	6 (50%)	1	2	3	2.75
Alghero Rocky coasts and islets	6 (50%)	3	3	0	4.5

Tab. 2

of Ottiolu (Torre, 1984); on the coast and in the coastal lake district of Orosei (Schenk, 1976 and 1981); in the coastal lake of Colostrai (Schenk, 1978); in the coastal lake of Mulargia (Porseo and Aru, 1983-1985); in the salt lakes of Sale Porcus and Salina Manna (Schenk and Pinna, 1978-85). The total population of the island is approximately 380 pairs (1983-85) as based on census and estimation.

From as early as 1962 (cf. Schenk, 1976; Mossa, Puddu and Schenk, 1983) in the coastal district of Cagliari, a continuous decrease of numbers has been noted, presumably due to pollution by heavy metals (mercury, cadmium, lead).

The most numerous colonies at present (1981-85) are those situated in the coastal lake of Mistras (ca. 80 pairs); in the salt-works of Macchiareddu (ca. 60 pairs; and on the Island of Ottiolu (33 pairs).

- Principal biotopes and their relative importance

After evaluation of the distribution maps (fig. 1-12), 13 major breeding sites have been selected, each of which hosting 5 or more of the 12 breeding species taken into consideration (tab. 2). Of these areas 10 (77%) are made up of archipelagos of small islands and rocky coasts, while 3 (23%) are regions of coastal wetlands (tab. 2).

To select the most important breeding sites of seabirds in Sardinia we applied the criteria of "species density" per area and, provisionally, the "1% criteria" of the total mediterranean (european), italian or sardinian breeding population of each species (Scott, 1980; Evans, this volume).

To stress also quantitative aspects of each of the 13 seabird communities taken into consideration, we attributed the "weight" 1 to all the species reaching the 1% level of their total mediterranean (european) population; the "weight" 0.5 has been attributed to all species reaching the 1% level of their total italian population and all the species reaching the 1% level of their sardinian population have been ponderated with the factor 0.25. The "importance index" thus obtained (tab. 2) varies in the 13 areas evaluated from a minimum of 1.75 for the coastal lagoons of S. Antioco to a maximum value of 5.25 for the Maddalena Archipelago. Areas with an "index" of less than 2.5 have been classified as being of regional importance (n=2); areas with an "index" between 2.5 and 4 are to be considered of national importance (n=5) while the remaining sites with an "index" of 4 or above have been classified as areas of international importance (n=6) for seabirds (tab. 2 and fig. 13).

It should however be noted that for several seabird species their "1% level" will be subject to revision in future years.

Proposals

Recognizing the important gaps in knowledge of the geographical distribution and numbers of several species, a three-year research programme is proposed on seabirds nesting in Sardinia. Special attention should be paid to monitoring the populations of Calonectris diomedea, Larus cachinnans, Larus audouinii, Larus genei and Sterna albifrons which might be considered as biological indicators of environmental quality of the marine and coastal ecosystems of the Mediterranean.

On the basis of the data presently available, the following areas need to be protected urgently as nature reserves: the Maddalena Archipelago, the Archipelago of Tavolara, the Island of Asinara, the islets off the coast of Sulcis, and finally the rocky coasts around Alghero and the sea-cliffs of the Gulf of Orosei.

References

Brichetti, P. (1979). Distribuzione geografica degli uccelli nidificanti in Italia, Corsica e Isole Maltesi. I. Parte introduttiva; Famiglia Podicipedidae, Procellariidae, Hydrobatidae. Ann. Mus. Civ. St. Nat. Brescia 16: 82-158.

Brichetti, P. (1982). 2. Famiglie Phalacrocoracidae, Ciconiidae, Treskiornithidae. Ann. Mus. Civ. St. Nat. Brescia 19: 97-157.

Corti, U.A. (1958). Ergebnisse einer ornithologischen Studienfahrt nach Sardinien. Orn. Beob. 55: 173-187.

INBS & CISO (1985). Istruzioni per i rilevatori. Progetto Atlante Italiano. Livorno.

Kunkel, P. (1963). Beitrag zur Avifauna Sardiniens. Vogelwelt 84: 137-145.

Lubrano, G. & T. Canu (1977). La Berta maggiore in Grotta. Speleologia Sarda 21: 1-4.

Meschini, E. & A. Roselli (1985). Bollettino del Progetto Atlante Italiano. N. 2, suppl. vol. 9 Avocetta.

Moltoni, E. (1971). Gli uccelli ad oggi riscontrati nelle isole di Tavolara, Molara e Molarotto (Sardegna nord-orientale). Riv. It. Orn. 41: 223-372.

Mossa, L.; Puddu, P. & H. Schenk (1983). Studio per la salvaguardia delle Saline di S. Gilla. Regione Autonoma della Sardegna. Assessorato della Difesa dell'Ambiente. Cagliari.

Schenk, H. (1976). Analisi della situazione faunistica in Sardegna. Uccelli e Mammiferi. In: S.O.S. Fauna, Animali in pericolo in Italia: 465-556. Ed. WWF, Camerino.

Schenk, H. (1978). Progress report, WWF/IUCN Project 1.413 - Audouin's Gull, Mediterranean - Sardinia: 1-17 (ciclostilato).

Schenk, H. (1980). Annual report 1979. WWF/IUCN Project 1.413 - Audouin's Gull, Mediterranean - Sardinia: 1-10 (ciclostilato).

Schenk, H. (1986, in press). Gabbiano roseo (Larus genei). In: Fasola, M. (Ed.), Laridae, Italia.

Schenk, H. & E. Meschini (1986, in press). Gabbiano corso (Larus audouinii). In: Fasola, M. (Ed.), Laridae, Italia.

Scott, D.A. (1980). Preliminary Inventory of Wetlands of International Importance for Waterfowl in West Europe and North-West Africa. IWRB Special Publication N. 2.

Torre, A. (1983). Osservazioni di uccelli rari in Sardegna. Boll. Soc. Sarda Scienze Nat., vol. XXII: 169-175.

Walter, H. (1964). Vögel an sardischen Salinen, Bonn. Zool. Beit. 15: 198-210.

CONSERVATION PRIORITIES FOR SEABIRDS IN ITALY

Giuseppe Bogliani, Dipartimento Biologia Animale
University of Pavia - Italy

Introduction

Extinction processes have grown dramatically in recent years; a rough estimate of the number of species involved was recently believed to be about 1,000 per year (Anon. 1974, in Myers, 1979). This figure includes many species not yet described by biologists. The loss of a species is not only of subjective concern (e.g. aesthetic, ethical, philosophical), but is also a threat to the potential future welfare of Mankind (Myers, 1979). In order to avoid the extinction of species, or to reduce the threat of extinction, many conservation programmes have been activated by some internationl organizations (e.g. IUCN, WWF, EEC, etc.) as well as by some national organizations (e.g. national and regional governments, local conservation societies).

Funding, even at a local scale, has proved to be insufficient to assure full coverage of the needed conservation programmes. Furthermore, we are faced with the problem of uneven geographical distribution of funds and of public concern about the conservation and the status of threatened species. The problem is particularly dramatic in the tropics and in many other developing areas of the world. Even in richer countries, as in Europe, there is an unbalanced situation. Therefore it is essential to co-ordinate the efforts of many nations in order to reach the desired goals. The best working contexts would be all countries belonging to a biogeographical region (e.g. the Western Palearctic), but a more realistic aggregation may be the whole of Europe, no matter of politic divisions. Some efforts have been made in recent years by some international agencies, such as the EEC and the Council of Europe, in order to co-ordinate some protection measures by means of, e.g. the Bern Convention. These measures are still insufficient to guarantee an optimal input of funds, of tasks and of "political intervention".

In this paper I suggest a way to optimize the international and national efforts for conservation, as well as the desirable

NATO ASI Series, Vol. G 12
Mediterranean Marine Avifauna
Edited by MEDMARAVIS and X. Monbailliu

role of Italy in this co-ordinated programme.

The continental level

The continent is the best unit to be used in order to co-ordinate the overall conservation efforts at the species level. As a matter of fact, it is highly probable for a species to be confined, or nearly confined, to a continent. Thus, conservation measures undertaken at a continental level will give some acceptable results.

The first goal of the "continental level planners" must be the compilation of lists of species, assigning to each species a "conservational value", depending on some characteristics such as the status, the population trend, the sensitivity to human activities and so on. It is only at a continental level that the "Red Lists" are useful in decision making for conservation.

National level. The example for seabirds in Italy

Many criteria have been proposed for the optimal selection of areas of conservation value (reviewed in Margules and Usher, 1981). Species selection approaches have been rarely attempted. Adamus and Clough (1978) propose some guidelines for rating the "conservation-value" of species during the choice of natural areas to be protected. In spite of many adjustments, some ideas of these authors have been utilized in the following proposals.

The local (national or regional) Red-Lists are often prepared without considering the overall status of a species, and in many cases are more a tool for "twitchers" than a useful source of information for decision makers. The high-rated species are often the locally rarest ones, no matter if they are really in need of protection measures (Fasola and Bogliani, 1986). If the main goal of the national programmes is to contribute to the conservation programmes at the continental level, one must optimize the efforts to reach the maximum results with few available resources. Therefore it is important to:

1. select the species fitting at the best the "continental" objectives (desirable species);
2. analyze the feasibility of conservation measures;
3. compile a list of the species optimizing the desirability and the feasibility at the national level.

Desirability

With the word "desirability" I mean the public concern to protect a species because of the characteristics shown in tab. 1 and discussed below.

	A	B	C	D	E	F
Calonectris diomedea	3	.5	.5	.5	.5	5
Puffinus puffinus	2.5	.5	.5	.5	.5	4.5
Hydrobates pelagicus	2.5	.0	.5	.5	.5	4.5
Phalacrocorax carbo	2	.5	1	1	1	5.5
P. aristotelis	2.5	.5	1	1	1	6
Larus melanocephalus	2	0	1	0	.5	3.5
L. ridibundus	.5	0	1	0	0	1.5
L. genei	1.5	0	1	.5	.5	3.5
L. audouinii	3	.5	1	1	1	6.5
L. cachinnans	3	0	1	0	0	4
Sterna nilotica	1.5	1	1	.5	1	5
S. sandvicensis	1	.5	1	.5	.5	3.5
S. hirundo	1.5	.5	1	.5	1	4.5
S. albifrons	2.5	1	1	.5	1	6

A - population characteristics (see tab. 2)
B - status changes: 0 - increasing; .5 - stable; 1 - decreasing
C - susceptibility to moderate human presence in or around the colony: 0 - neutral; 1 - high.
D - susceptibility to human activities in the surroundings: 0 - beneficial; .5 - neutral; 1 - negative.
E - adaptability to foreseeable changing conditions: 0 - beneficial; .5 - neutral; 1 - negative.
F - total rate.

Tab. 1: Desirability

A. Population characteristics

The first step in choosing a species is to analyze if the conservation measures undertaken at the national level would contribute to the aim of preserving the species at the continental level. Local rarity is not an appreciable feature; for instance, the Sandwich Tern is locally "rare" as a breeder in Italy, but it is common abroad in Europe (Fasola, this volume). As stated by Preston (1948), in an animal community there are more "rare" species than "common" species. The characteristics of being rare do not necessarily mean that a species is in danger of extinction.

	Italy	Population (no. nests)				Rating					
		W Palearct.	%	Mediterr.	%	A	B	C	D	E	F
Calonectris diomedea	7,000**	60,000***	11.7	30,000**	23.3	1	1	1	1	3	3
Puffinus puffinus	15,000**	300,000***	5	50,000***	30	.5	1	.5	1	2	2.5
Hydrobates pelagicus	1,300**	?	?	11,000***	11.8	?	1	.5	1	?	2.5
Phalacrocorax carbo	45	20,000**	.2	250**	18	0	1	1	0	1	2
P. aristotelis	2,600*	50,000***	5.2	5,000**	52	.5	1	1	.5	2	2.5
Larus melanocephalus	260	120,000**	.2	3,800*	6.8	0	.5	.5	1	1.5	2
L. ridibundus	780	946,000**	.1	14,500*	5.4	0	.5	0	0	0	.5
L. genei	950	31,000**	3.1	5,400**	17.6	0	1	0	.5	.5	1.5
L. audouinii	550	5,100*	10.8	5,100*	10.8	1	1	1	1	3	3
L. cachinnans	24,300*	660,000***	3.7	72,000***	33.7	.5	1	1	1	2.5	3
Sterna nilotica	220	4,000*	5	2,100*	9.5	.5	.5	.5	.5	1.5	1.5
Sterna bengalensis	1	?		?							
S. sandvicensis	150	95,000**	0.1	920*	16.3	0	1	0	0	1	1
S. hirundo	4,600	140,000*	3.3	20,500**	22.4	.5	1	.5	0	1	1.5
S. albifrons	6,000	20,000*	30	10,600*	56.6	1	1	1	.5	2.5	2.5

Accuracy of estimates: no asterisk - good estimate of the breeding population; * fairly good estimate; ** probably the size-order only is good; *** the error can be of one or more sized-orders.

A - % Italian Population/Western Palearctic < 2%: 0; 2% - 10%: 0.5; > 10%: 1
B - % Italian Population/Mediterranean rating as A
C - Range position. Central: 1; submarginal: 0.5; marginal: 0.
D - % of Italian range. < 5%: 1; 0.5-5%: 0.5; > 0.5%: 0
E - total rating Western Palearctic
F - total rating Mediterranean

Tab. 2: Population Characteristics

I propose that in rating species to be protected the following features of the local populations should be considered (tab. 2).

• The relative abundance of the Italian population (referred to the Western Palearctic or the Mediterranean populations), data from Brichetti (1979, 1982), Cramp (1985), Cramp/Simmons (1977, 1983), Fasola (this volume), Massa/Catalisano (this volume); species with high percentage of breeding pairs nesting in Italy are rated higher.

• The position of Italian range in the overall range of other species; species with their centre of distribution in Italy are rated higher. For this purpose range categories can be classified as central, sub-marginal and marginal.

The first two points are strongly correlated: as a general rule the highest abundances occur near the centres of the species range and the lowest at the margins (Hengeveld/Haeck, 1982). If we want to preserve a good breeding stock, we must focus on the most abundant populations of the species. Furthermore, at the edges of the range of a given species, the presence of a breeding population may be irregular; as demonstrated for North American seabirds, range increase and contraction may be due to minor short-term climatic changes, while at the centre of the range the population is relatively stable (Myres, 1979). Attention should be paid to the long-term climatic changes, enabling some species to colonize new areas for a long period.

• The importance of the Italian range: the higher the percentage of the relative local range as compared to the overall range, the higher the rating. If existing, an endemic species would be rated very high, while a world-wide distributed species would be rated low.

The total rating is expressed as the sum of the partial rates discussed above, both at the Western Palearctic scale and at the Mediterranean scale. In the discussion and in tab. 1 only the "Mediterranean" rating will be considered.

B. Status changes

Status changes are considered at the overall range scale. A species increasing is not likely to be in threat of extinction, and is rated low. The contrary happens with a decreasing species.

C. Susceptibility to moderate human presence in or around the colony

The three nocturnal and burrow-nesting species, the two Shearwaters and the Storm Petrel, do not seem to be harmed by the presence of humans. This presence is usually restricted to daylight hours. Both the Cory's Shearwater and the Storm Petrel have proved to be tolerant even of heavy disturbance at the nesting sites. Filfla, where the largest Mediterranean colony of the Storm Petrel (10,000 pairs) nests was used for many years as a cannon target by the British Navy; however nesting was not abandoned here (Sultana/Gauci, 1982). The largest known colony of Cory's Shearwater in Italy, on Linosa Island, does not seem to be affected by refuse discharge (Vaughan, 1980). The contrary is true for all the gulls and terns.

D. Susceptibility to human activites in the surroundings

Some species have proved to benefit from human activites. The dramatic increase of the Yellow-legged Gull and the Black-headed Gull all over the Mediterranean is a result of the increased quantity of garbage. Also the Mediterranean Gull is positively affected by some human activites, e.g. commercial fishing and agriculture, as is the case with the Slender-billed Gull, largely depending on small crustaceans captured in saltpans. In the latter species the benefits are counterbalanced by habitat destruction in its main breeding area. Only Cormorants and the Audouin's Gull seem to be negatively affected, mainly by tourism, which is the prevailing human activity in the areas surrounding their colonies. The other species nest in areas not yet "developed" and feed mainly on natural preys.

E. Adaptability to foreseeable changing conditions

With this criterium, I mean the change likely to occur, or already occurring and increasing, in the areas inhabited by the main populations of a species in Italy. The lack of precise information can induce some misleading conclusions. Only our two commonest gulls are likely to benefit from foreseeable changes (e.g. increasing rubbish production, new ways of fish farming, etc.). Cormorants and the Audouin's Gull will suffer from increasing number of motorboats and ships, as well as from the planned new holiday centres along the coasts. The Little Tern is already greatly affected by human disturbance at its colonies in the Po Delta, the main breeding area of this species in Italy (Fasola, this volume). The number of people frequenting here the sandy beaches has increased from a few hundred a year during the 1970's to 176,000 in 1977, 301,600 in 1982, 318,500 in 1983, and 341,200 in 1984. The Regional Government of the Venetian Region, which owns the largest part of the Po Delta, supports increasing touristic activity; no conservation measures have been planned. Common and Gull-billed Terns will be negatively affected by the planned transformation of many natural wetlands into fish-farming and by habitat alteration in their feeding grounds. Other species will be unaffected or the negative effects will be counterbalanced by positive ones.

The total rating of a species is shown in tab. 1. The species having the priority of conservation measures in Italy are, in the following order of priority: 1 - Audouin's Gull; 2 - Little Tern and Shag; 3 - Cormorant; 4 - Cory's Shearwater and Gull-billed Tern, the other seabirds following.

Feasibility

Several characteristics of the species and of the populations must be considered in order to decide which conservation measures must be utilized in the aim of preserving the species. The classic measure is the protection of the breeding grounds by avoiding habitat destruction and human disturbance. Other measures can of course be taken, but colony protection is the easiest one. In tab. 3 some characteristics are shown allowing conservation simply by protecting some defined areas; these points are discussed below.

	Characteristics			Feasibility	
	A	B	C	(A+B+C)	A(B+C)
Calonectris diomedea	1	1	1	3	2
Puffinus puffinus	1	1	.5	2.5	1.5
Hydrobates pelagicus	1	1	1	3	2
Phalacrocorax carbo	1	1	1	3	2
P. aristotelis	1	1	.5	2.5	1.5
Larus melanocephalus	0	1	1	2	0
L. ridibundus	.5	1	1	2.5	1
L. genei	.5	1	1	2.5	1
L. audouinii	0	1	.5-1	2-2.5	0
L. cachinnans	.5	.5	0	1	.25
Sterna nilotica	.5	1	1	2.5	1
Sterna bengalensis	?	1	1	2	?
S. sandvicensis	.5	1	1	2.5	1
S. hirundo	.5	1	.5	2	.75
S. albifrons	0	1	.5-1	1.5-2	0

A - site tenacity: 1 - nests repeatedly in the same site (Turnover Index 0.20); .5 - changes sometimes (T.I. 0.2-0.5); 0 - changes frequently (T.I. 0.5).

B - area size needs: 1 - nests in close aggregations; .5 - nests in loose colonies.

C - dispersion of nesting population: 1 - > 50% of the population in < 5 colonies; .5 - > 50% population in 6-50 colonies; 0 - > 50% population in > 51 colonies.

Tab. 3: Feasibility

A. Site tenacity.

It was defined by McNicholl (1975) as the "tendency to nest repeatedly on or close to the same place". Site tenacity is of course a very important feature to be considered, because the creation of a nature reserve for the protection of a seabird colony may be useless if the population changes frequently its breeding sites. Site tenacity is low in Larid species nesting in stable habitats and high in those species nesting in unstable, or ephemeral, habitats. In tab. 3 the rating has been made using for Laridae the values of the Turnover Index for the Italian populations collected during the "Progetto Laridae" (Fasola, this volume). Site tenacity is high when Turnover Index is low. For the other species, we have no precise data for Italy, but the Turnover Index is presumed to be very low. Unfortunately the two species of great conservation interest cannot be protected simply by establishing a fixed nature reserve. The Little Tern nests in ephemeral habitats and moves every year; the Audouin's Gull, at the margin of its range, is faced probably with sub-optimal conditions and changes often its breeding sites.

B. Area-size needs.

All the Italian seabirds are colonial breeders, so many nests are found within an area. Only the Yellow-legged Gull nests frequently in loose colonies, the other species nesting in dense or fairly dense colonies.

C. Dispersion of the nesting populations.

The more the population is grouped, the higher is the possibility to protect a large part of the nesting areas.

The total rating is expressed in tab. 3 in two ways. The first one is simply the sum of each value; the second is the product of the rating of Site Tenacity and the sum of the other two characteristics. The latter rating seems to be more useful, because it expresses the real possibility to protect a species by establishing a nature reserve.

Conclusions

In Italy no colony of seabirds is protected at present; nevertheless the potential value for conservation of some species here is very high.

Two distinct strategies must be adopted in order to preserve the "desirable" Italian seabirds. First the protection of breeding sites of those species with high or medium site tenacity of conservation value. The choice is actually very easy to do because the most valuable species nest in mixed colonies. The two main areas for this purpose are the wetlands of Southern Sardinia (see Schenk and Torre, this volume) and the Valli di Comacchio (see Brichetti/Foschi, this volume).

The protection of the greatest concentration of nesting Little Terns would be possible by establishing a large nature reserve all over the suitable habitat in the Po Delta. But the concern for nature conservation of the local Regional Government (Venetian Region) is very low and the local people agree with the proposal of opening all the sandy beaches to the tourists. Only a very big effort by the nature conservation societies and by the national government offices may lead to promote law enforcement, enabling the protection of the beaches belonging to the State or some measures of seasonal protection.

References

Adamus, P.R. and Clough, G.C. (1978). Evaluating species for protection in natural areas. Biol. Conserv. 13: 165-178.

Brichetti, P. (1979, 1982). Distribuzione geografica degli Ucelli nidificanti in Italia, Corsica e Isole Maltesi, 1, 2. Natura Bresciana 16: 82-158; 19: 97-157.

Cramp, S. (ed.) (1985). The birds of the Western Palearctic. Vol. IV. Oxford Univ. Press. Oxford.

Cramp, S. and Simmons, K.E.L. (eds.) (1977, 1983). The birds of the Western Palearctic. Vol. I and III. Oxford Univ. Press. Oxford.

Fasola, M. and Bogliani, G. (1986). Proposte sulle priorità delle specie da proteggere. in "Proc. III Italian Ornit. Congr.", in press.

Hengeveld, R. and Haeck, J. (1982). The distribution of abundance. I. Measurements. J. Biogeogr., 9: 303-316.

Margules, C. and Usher, M.B. (1981). Criteria used in assessing wildlife conservation potential: a review. Biol. Conserv. 21: 79-109.

McNicholl, M.K. (1975). Larid site tenacity and group adherence in relation to habitat. Auk 92: 98-104.

Myers, N. (1979). The sinking ark. Pergamon Press. Oxford.

Myres, M.T. (1979). Long-term climatic and oceanographic cycles regulating seabird distributions and numbers. in : Bartonek, J.C./Nettleship, D.N. (eds.) "Conservation of marine birds of Northern North America". USFWS, Wildl. Res. Rep. 11, Washington.

Preston, F.W. (1948). The commonness and rarity of species. Ecology 29: 254-283.

Sultana, J. and Gauci, C. (1982). A new guide to the birds of Malta. The Ornit. Soc., Malta.

Vaughan, R. (1980). Notes on Cory's Shearwater (Calonectris diomedea) and some other birds on Linosa, Pelagic Isles. Riv. ital. Ornit. 50: 143-154.

CHAPTER 7 : PAPERS OF POSTERS PRESENTED AT ALGHERO

SEASONAL FLUCTUATIONS OF GULLS IN LIGURIA

A. Andreotti, A. Borgo, E. Borgo, T. Fior, S. Spano and G. Truffi
Istituto di Zoologia, Via Balbi, 5, Genova

Introduction

Ornithological reports of Ligurian authors of the last century (Calvi, 1828; Durazzo, 1840, 1846; Carazzi, 1887) are either cursory, difficult to interpret or sometimes erroneous as was the case with the breeding status of the Common Gull and Kittiwake.

Also Giglioli's survey of 1890 contains some uncertain data as he reports the Black-headed Gull as being sedentary in the area of Savona. Yet, his conclusions are not very different from the results of the more recent censuses. The observations in Liguria of Salvadori (1872), Martorelli (1906) and Arrigoni (1929) correspond also with our data. From Masnata (1931) we only learn that the Yellow-legged Gull breeds in Liguria.

The status of gulls, after having verified its accuracy, follows here below:

- Larus melanocephalus: present in winter but more numerous in spring. Migratory.
- Larus minutus: present in winter with two main passages. Irregular.
- Larus ridibundus: most common gull; present from the second fortnight of July till the end of April.
- Larus genei: mentioned in literature but without any details. Accidental.
- Larus audouinii: accidental along the coast.
- Larus canus: present in winter with two main passages.
- Larus cachinnans: common and sedentary. We do not exclude the eventual presence in winter of Larus argentatus.

NATO ASI Series, Vol. G 12
Mediterranean Marine Avifauna
Edited by MEDMARAVIS and X. Monbailliu

- *Larus fuscus*: present in every season in small numbers.
- *Larus hyperboreus*: accidental.
- *Larus glaucoides*: accidental.
- *Larus marinus*: accidental.
- *Rissa tridactyla*: irregular and rare. This gull species is more frequent in Liguria than elsewhere in Italy.

The Yellow-legged Gull is the only breeding larid in Liguria. Following the above mentioned authors the Lesser black-backed Gull (*L. fuscus*) has, doubtfully, bred in Liguria. This overview is incomplete and lacks detail. Further research is required related to species distribution and the demographical outburst of some species.

Results of recent observations

The following results are based upon land-based observations between 1972 and 1986 in several places of the central western Ligurian Riviera, and more precisely: Genova (1978-82), Genova Sestri (airport, 1972-73 and 1982-83), Genova Pra (1978, 1979, 1981, 1984, 1986), Varazze (S.V.) (1974-82). See fig. 1.

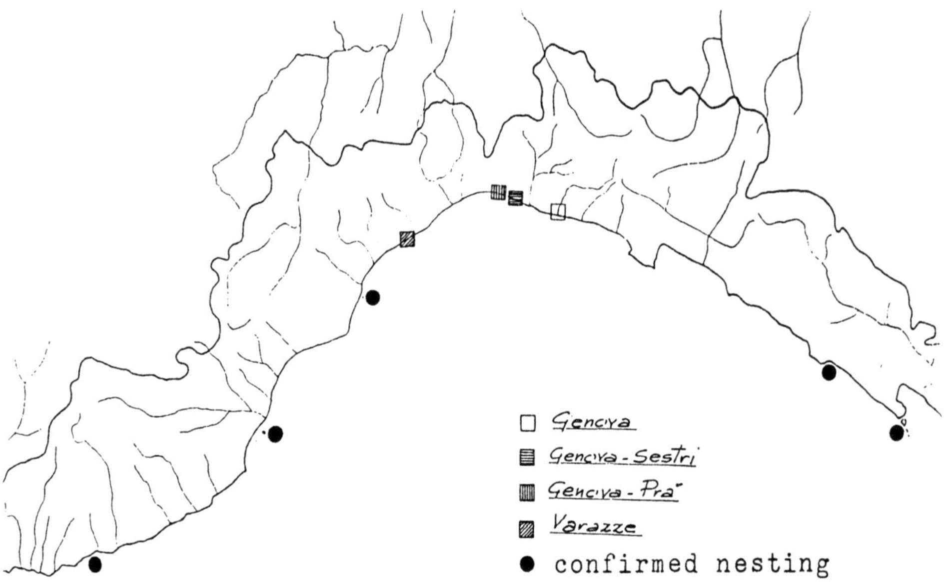

Fig. 1: Study areas and confirmed nesting places of Yellow-legged Gulls (from the left: San Remo, Gallinara, Bergeggi, Monterosso, Porto Venere Islands).

Data come from several observers and are spread over various years. In Genova, this survey was carried out especially along the Biscagno River. The inland basin of this river is frequented by the Black-headed Gull. This area is cultivated and a small winter population (Spano and Toschi, 1969), which has been estimated at 1,500 Black-headed Gulls (Fior, 1982), existed there already in 1968.

Fig. 9 gives an example of the daily movements (Fior, 1982; Fior et al., 1984). Quantitative surveys have been carried out several times in the airport area where the "gull" problem, caused by L. ridibundus and L. cachinnans, interferes with air traffic (Spano and Toschi, 1969; Spano, 1974; Dessessa and Fior, 1984).

The histograms based on fortnightly observations and collected during ten years (1972-73 and 1982-83) illustrate the increase of the wintering contingent of the Black-headed Gull. However, we notice that their maximum numbers have levelled off during the last years. March corresponds to the peak period whereafter the Black-headed Gull virtually disappears. The Yellow-legged Gull is more constant during the whole year and is more numerous in spring and summer. This species shows a maximum after the breeding period as well as in November (fig. 2 and 3).

In Genova Pra our research is linked to the environmental effects caused by the construction of Voltri harbour (modification of inondated areas due to dumping of construction material). A first series of irregular observations was carried out in 1978, 1979 and 1981 (Fior, 1982). We continued more regularly during 1984 and 1985.

Histograms of fortnightly means of numbers of Larus ridibundus and L. cachinnans are given in fig. 4 and 5. They retrace the broad lines of the phenology as observed in the airport area and confirm the fluctuating trend of gulls apart from some predictable differences. Black-headed Gulls, in fact, have been counted in larger numbers from 1978 onward. This species disappears in mid-June.

Fig. 8 summarizes the counts of the other gull species. Larus melanocephalus can be observed all the year long with minimal numbers from April to September and with peaks in March and October. Larus minutus appears only in spring showing a peak during the first fortnight of May. Larus fuscus attains its maximum in March and April where it is almost absent during the other time of the year. Larus canus is a typical winter guest with a peak during the second fortnight of January.

At Varazze (SV) we dispose of processed observations made from 1974 till 1984. Here also the Black-headed Gull disappears during the second fortnight of June and is very numerous in winter. The Yellow-legged Gull follows the same presence pattern as that one of the Black-headed Gull. At Varazze its frequence is thus different from Genova city (see fig. 7).

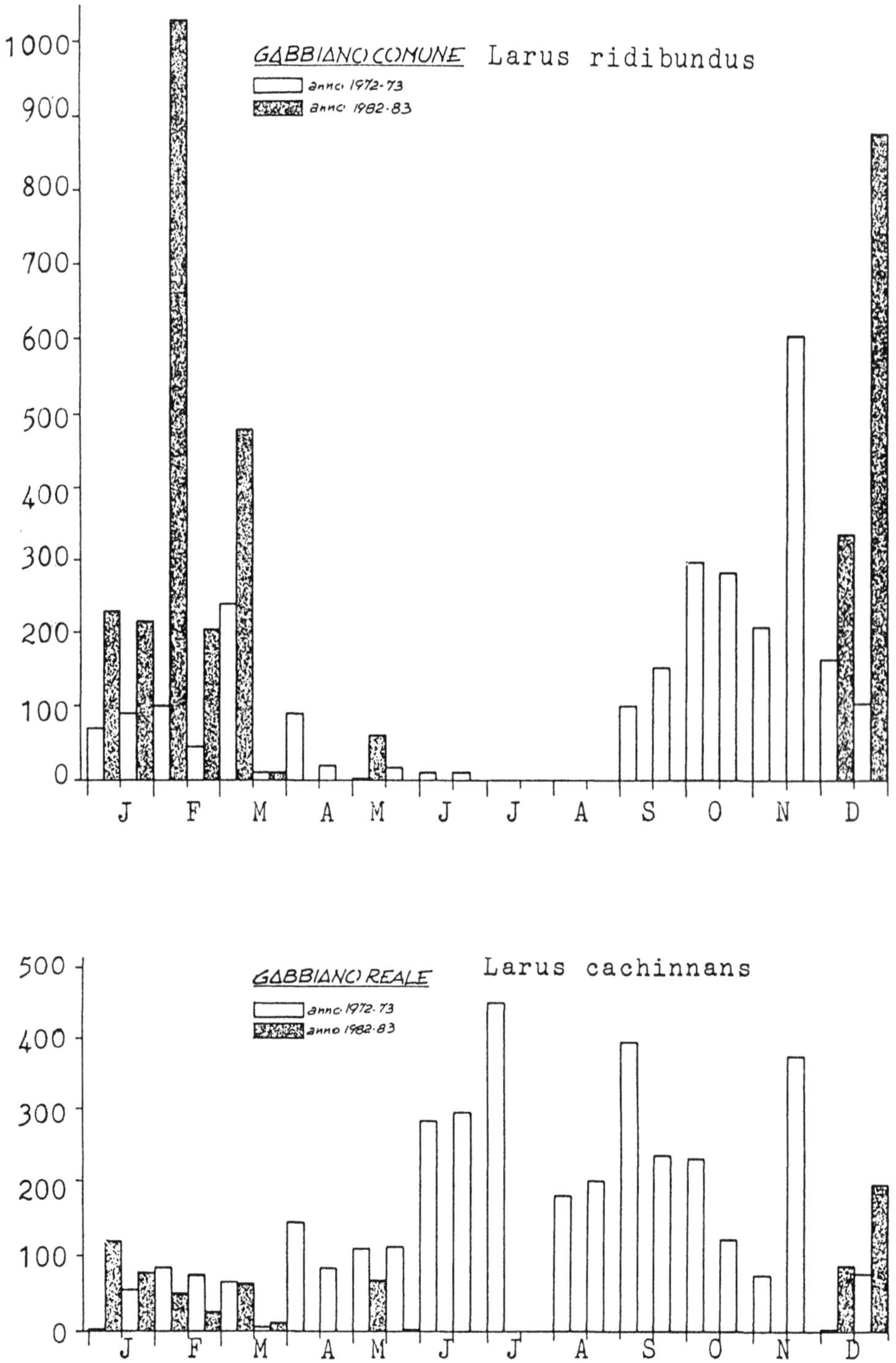

Fig. 2-3: Fluctuations of Larus ridibundus and Larus cachinnans at Genova airport: mean numbers per fortnight for given years.

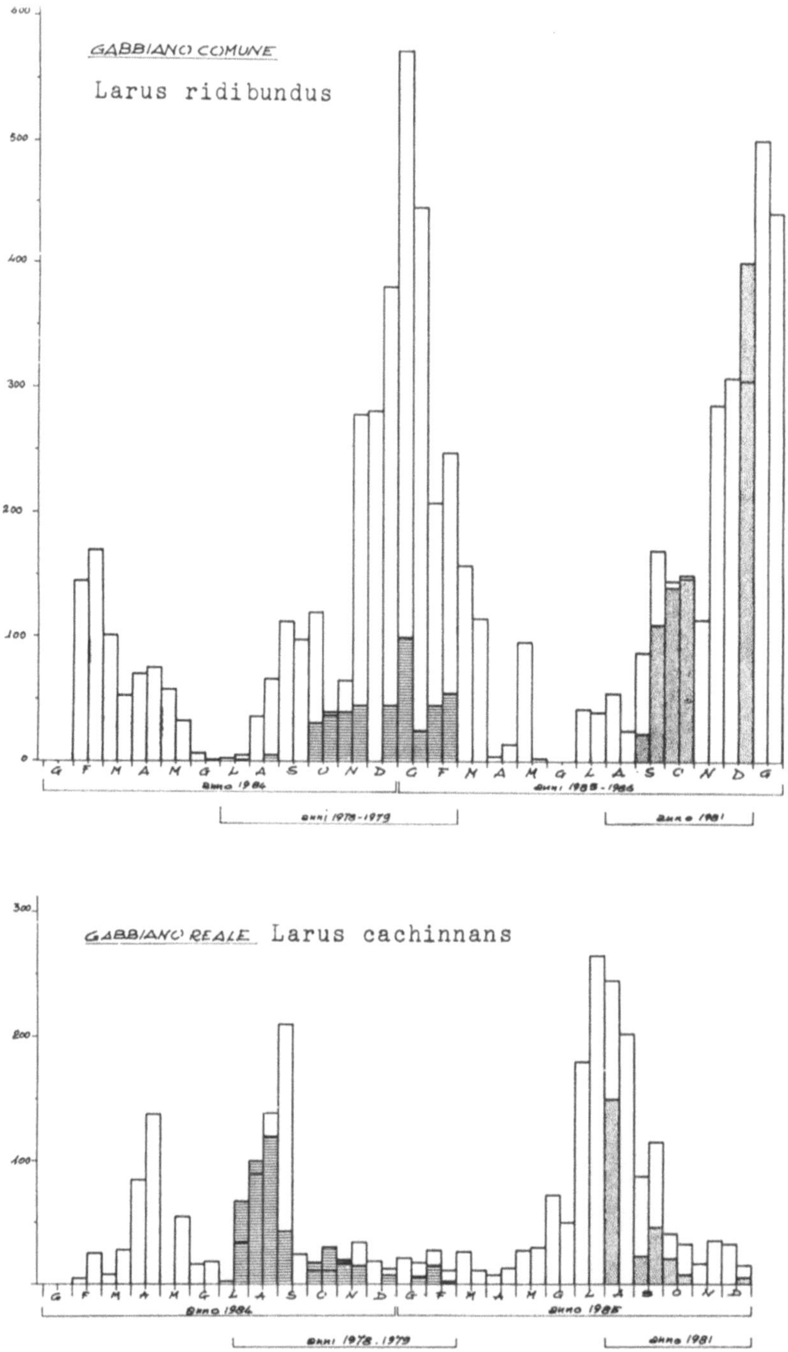

Fig. 4-5: Fluctuations of Larus ridibundus and Larus cachinnans at Genova Pra: mean numbers per fortnight for given years.

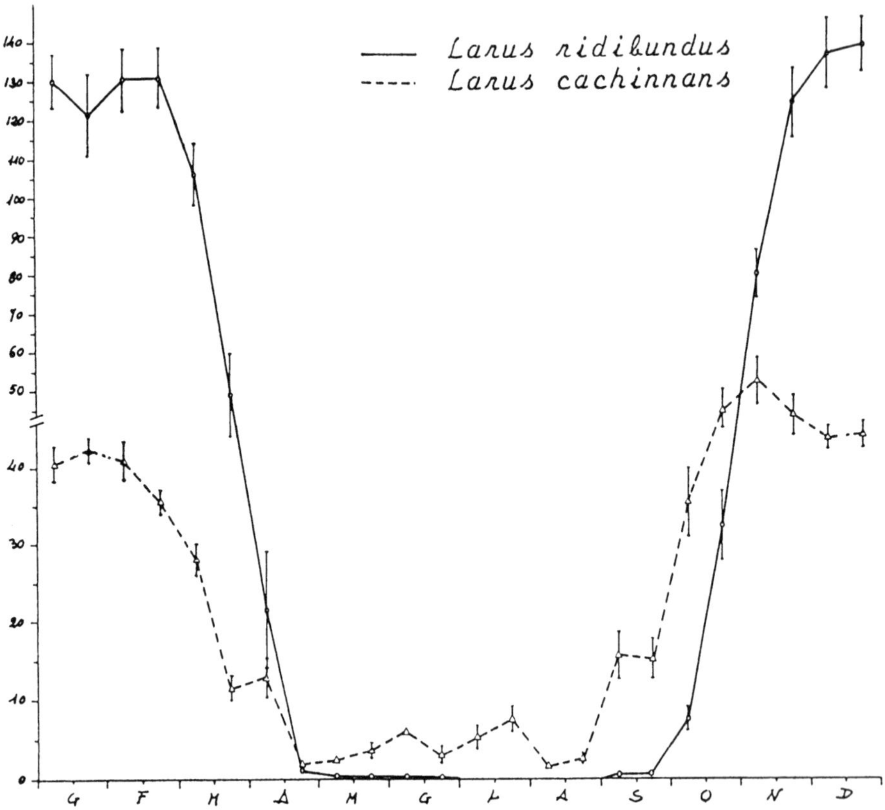

Fig. 6: Fluctuations of Larus ridibundus and Larus cachinnans at Varazze (SV).

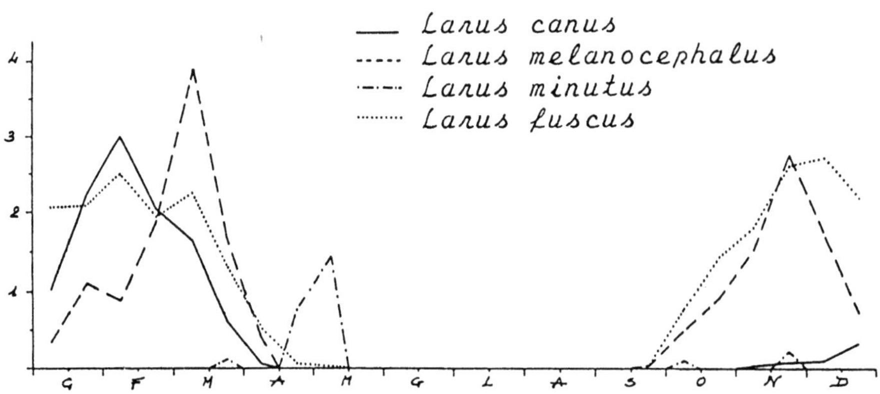

Fig. 7: Fortnightly average of continuous observations carried out from 1974 to 1982 of Larus minutus, melanocephalus, canus and fuscus at Varazze.

The Mediterranean Gull is most numerous in the second fortnight of November and during the first weeks of March. It is present during the whole winter. The Little Gull appears irregularly in autumn. Maximal numbers can be seen during the second halves of November and May. The Lesser Black-backed Gull is a winter guest with maximum numbers during the second fortnight of December. Another relatively rare wintering species is the Common Gull (Larus canus): its maximum is reached in the first fortnight of February. The latter three species are all very rare in Liguria. However, a group of over 100 Mediterranean Gulls has been observed.

Discussion

As far as the occurence of the Black-headed Gull is concerned, we are in agreement with Spano (1974). Larus ridibundus is completely absent during the second fortnight of June and is very rare in summer. Adult birds prevail in autumn-winter while immatures arrive after the second half of March.

In Genova the Yellow-legged Gull is permanently present. Both immatures and adults are most numerous in summer (Gorlier, 1975), but at Varazze its occurence is similar to that one of the Black-headed Gull, since this species leaves this area in summer, as is also the case at Bergeggi and Gallinara. Yellowlegs might leave Varazze because they nest in the neighbouring breeding sites. Later, Yellowlegs are absent at Varazze because the busy tourist season makes it impossible for gulls to frequent the beach.

The data below summarize the most important observations of the most frequent gull species in Liguria.

- Larus melanocephalus: Airport (GE) 452 (15.3.83)
  Genova Pra 200 (10 immature) (2.3.85)
  Varazze 100 (4.3.82)
- Larus minutus: Airport (GE) 31 (24.5.72)
  Genova Pra 55 (54 immature) (12.5.85)
  Varazze 3 (30.4.74)
- Larus ridibundus: Torrente Bisagno (GE) 1,400 (4.1.80)
  Loc. Foce (GE) 800 (8 immature) (16.1.82)
  Airport (GE) 1,030 (15.2.83)
  Genova Pra 660 (30 immature) (2.1.85)
  Varazze 400 (27.2.81)
- Larus cachinnans: Loc. Foce (GE) 50 (25.9.81)
  Airport (GE) 600 (17.9.68)
  Genova Pra 310 (200 immature) (28.7.85)
  Varazze 350 (28.10.79)

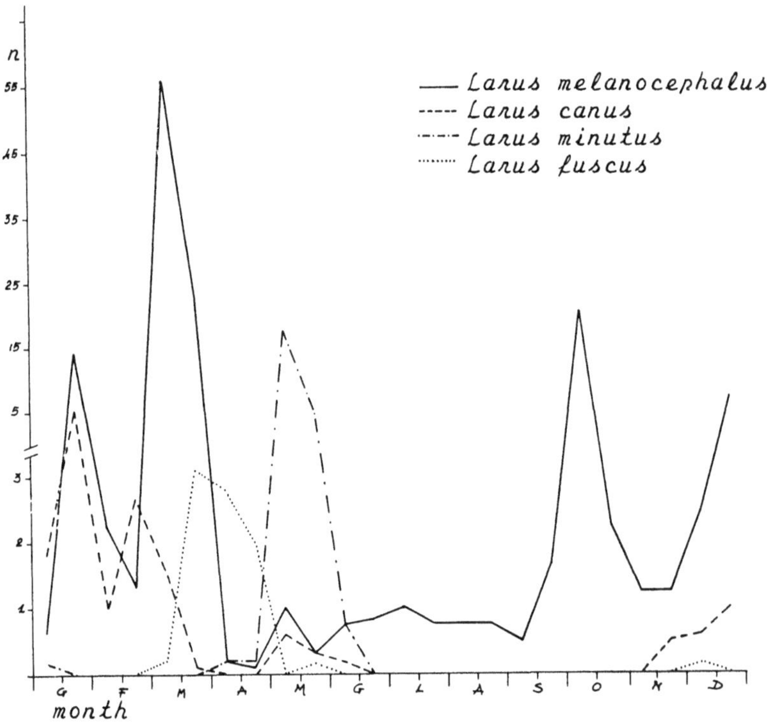

Fig. 8: Fortnightly average 1984-1985 of Larus minutus, Larus melanocephalus, Larus canus and Larus fuscus in Genova Pra.

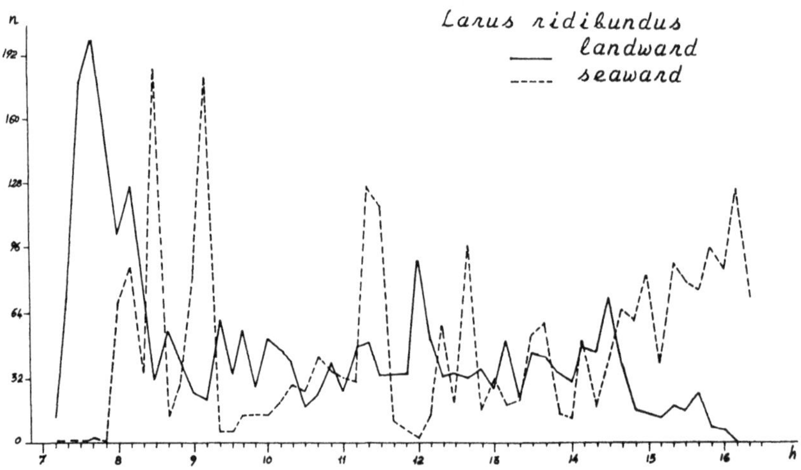

Fig. 9: Daily movements of Larus ridibundus along the Bisagno river (Genova) (2nd December 1977).

- Larus canus: Airport (GE) 12 (15.3.83)
  Genova Pra 8 (3 immature) (23.3.85)
  Varazze 32 (15.2.80)

- Larus fuscus: Airport (GE) 150 (31.3.83)
  Genova Pra 4 (4.5.85)
  Varazze 17 (28.3.81)
  } only adults

These data seem at first sight comparable to those collected by Isenmann and Czajkowski (1978). Yet, as they are restricted to small parts of the Ligurian coast, our data reflect an important underestimate. This aspect has also been underlined by Fasola (1984).

It is to be noted that the Kittiwake has been irregular in Genova Pra: end November (2 adults) and from May to July (never more than 2 immatures). At Varazze a single Kittiwake was observed in January (1982), February (1975) and March (1975, 79). Gorlier (1975) reports also some immatures at Noli (SV) in August and indicates having observed a Pagophila eburnea at Capo Noli (18.8.72). Two Greater black-backed Gulls (Larus marinus) were also seen at the mouth of the river Bisagno on 11.5.81 (Fior).

Daily movements

Gulls regularly occupy stretches of inland water courses during the day. They have been observed at Busalla (GE), along the Scrivia River, at Campo Ligure (GE) along the Stura as well as in the mountain valleys of Giovi and Turchino. We do not know whether there are common roosts. Black-headed Gulls can be observed at night in the harbour of Genova, on the airport runways, on the architraves of public slaughterhouses as well as on the sea.

In winter, at Varazze, all gulls disappear at dusk. Yellow-legged Gulls always leave in the evening for the South-west, probably towards their breeding places that may be used as dormitories.

Origin of gulls observed in Liguria

We only dispose of incomplete data of Italian ringing recoveries. From the 60 recoveries of Black-headed Gulls (Moltoni, 1952, 58, 66, 73; Bendini, 1983; Borgo, unpublished) we can draw the following dispersion map (fig. 10). The number of recoveries is in direct relation to the periods of maximum presence (September 3%, October 2%, November 15%, December 33%, January 18%, February 10%, March 12%, April 7%).

71% of recoveries concern birds found less than one year after ringing; 17% after 1 to 2 years; 5% after 2 to 3 years; 3.5% after 3 to 5 years and the remaining (3.5%) after 5 years.

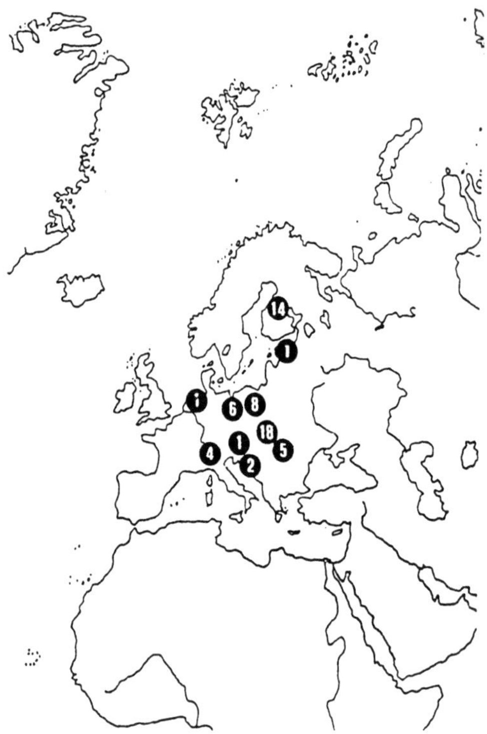

Fig. 10: Origin and relative numbers of Larus ridibundus recovered in Liguria and ringed abroad.

The record of longevity is a Black-headed Gull ringed in 1949 in Switzerland and recaptured in 1957.

Other recoveries concern 11 Mediterranean Gulls all ringed in USSR between 1948 and 1957 and found in March (4), April (3), November (2), May (1) and September (1) (Moltoni, 1949-1958). Only three ringed Yellow-legged Gulls have been recovered so far. One ringed in Tunisia (Moltoni, 1973) and two others ringed on the islands south-east of France (Andreotti & Atturo, unpublished).

Nesting

The Yellow-legged Gull is the only seabird breeding in Liguria. Nesting has recently been recorded at S. Remo, Gallinara and Bergeggi (Balletto and Spano, 1982; Libonati, 1983 in "Progetto Atlante Liguria, unpublished) and in eastern Liguria (Brichetti, 1985). There are probably other breeding sites than shown in fig. 1, especially as the neighbouring French population is expanding (Guyot et al., 1985). We hope to be able to study them in further surveys.

References

Arrigoni degli Oddi, E. (1929). Ornitologia italiana. Hoepli, Milano.

Balletto, E. & Spano, S. (1982). Nidificazione ripetuta di Gabbiano reale (Larus argentatus michahellis) nel porto di Sanremo (Imperia). Riv. Ital. Orn. 52: 191-194.

Bendini, L. (1983). Specie ornitiche inanellate in altri Paesi e riprese in Italia nel 1978. Boll. attività inanellamento, INBS, 2: 77-112.

Borgo, R. (1975). Osservazioni ornitologiche sulla spiaggia di Varazze (Savona). Riv. Ital. Orn. 45: 342-348.

Brichetti, P. (1985). Guida degli uccelli nidificanti in Italia. Scalvi, Bresia.

Calvi, G. (1828). Catalogo di ornitologia di Genova. Pagano, Genova.

Carazzi, D. (1887). Materiali per un'avifauna del golfo di Spezia e della Val di Magra. Tip. Artistica, Spezia.

De Sessa, L. & Fior, T. (1983). Avifauna della zona aeroportuale di Genova. 2° Conv. ital. Orn., Parma (in publ.).

Durazzo, C. (1840).Degli uccelli liguri. Tip. Ponthenier, Genova.

Durazzo, C. (1846). Uccelli. In AA.VV.: Descrizione di Genova e del Genovesato. Vol. I Tip. Ferrando, Genova.

Fasola, M. (1984). Censimento preliminare dei Laridae svernanti in Italia. Avocetta 8 (1): 57-63.

Fior, T. (1982). Studio demo-ecologico sul Gabbiano comune (Larus ridibundus)et sul Gabbiano reale (L. argentatus) nell'area urbana di Genova. Tesi di Laurea in Sc. Biologiche, Univ. Genova.

Fior, T.; Spano, S.; Carretta, L. (1983). Larus ridibundus nell' area urbana di Genova. 2° Conv. ital. Orn., Parma (in publ.).

Fior, T.; Spano, S. & Baino, R. (1984). Wintering gulls (Larus ridibundus, Larus argentatus) in Genoa area. First Conf. Birds wintering Medit. Region, Aulla (in press).

Giglioli, E.H. (1890). Primo resoconto dei risultati dell'inchiesta ornitologica in Italia. Parte II. Succ. Le Monnier, Firenze.

Gorlier, G. (1975). Osservazioni ornitologiche del litorale e della zona di mare compresa tra Vado Ligure (SV) e

Finale Ligure (SV). Riv. ital. Orn. 45: 61-67.

Grosso,G. (1981). Gabbiani d'inverno. Avifauna 4: 129-130.

Guyot, I.; Launay, G. & Vidal, P. (1985). Oiseaux de mer nicheurs du Midi de la France et de Corse: évolution et importance des effectifs. In "Oiseaux marins nicheurs du Midi et de la Corse". Ann. CROP 2, Aix-en-Provence: 31-47.

Isenmann, P. & Czajkowski, M.A. (1978). Note sur un recensement de Laridés entre Nice et Naples en décembre 1977. Riv. ital. Orn. 48: 143-148.

Launay, G. (1985). Nouvelles données sur la biologie du Goéland leucophée Larus cachinnans michaellis dans le Midi de la France. In "Oiseaux marins nicheurs du Midi et de la Corse". Ann. CROP 2, Aix-en-Provence: 31-47.

Libonati, V. (1983). I Gabbiani reali Larus argentatus (Pontopp.) dell'isola Gallinara (Albenga-Savona). Migratori acquatici 6 (4): 7-8.

Martorelli, G. (1906). Gli uccelli d'Italia. Cogliati, Milano.

Masnata, G.B. (1931). Cacce di Liguria. Arti Grafiche Tigullio, Rapallo.

Moltoni, E. (1949). Altre riprese in territorio italiano di uccelli inanellati all'estero. Riv. ital. Orn. 19: 78-84.

Moltoni, E. (1951). Ulteriori notizie su uccelli inanellati all' estero e ripresi in territorio italiano. Riv. ital. Orn. 21: 15-23.

Moltoni, E. (1952). Altre notizie su uccelli inanellati all'estero e ripresi in territorio italiano. Riv. ital. Orn. 22: 1-18.

Moltoni, E. (1953). Uccelli inanellati all'estero e ripresi in territorio italiano e nella Libia. Riv. ital. Orn. 23: 1-12.

Moltoni, E. (1958). Note su alcune riprese in Italia e nella Libia di uccelli inanellati all'estero. Riv. ital. Orn. 28: 1-74.

Moltoni, E. (1966). Altre notizie su uccelli inanellati all'estero e ripresi in Italia e in Libia. Riv. ital. Orn. 36: 109-314.

Moltoni, E. (1973). Elenco di parecchie centinaia di uccelli inanellati all'estero e ripresi in Italia e in Libia. Riv. ital. Orn. 43: suppl.

Moltoni, E. (1976). Nuovi dati su uccelli inanellati all'estero e ripresi in Italia e in Libia. Riv. ital. Orn., 46: suppl.

Salvadori, T. (1872). Fauna d'Italia. Uccelli. Vallardi, Milano.

Spano, S. (1974). II contributo alla conoscenza dei ritmi di occupazione ornitica dell'aeroporto di Genova. Riv. ital. Orn. 44: 81-114.

Spano, S. & Toschi, A. (1969). Ritmi di occupazione ornitica dell'aeroporto di Genova in un ciclo annuale. Riv. ital. Orn. 39: 305-383.

WINTERING POPULATION OF THE COMMON CORMORANT PHALACROCORAX CARBO SINENSIS IN TUSCANY, ITALY

Nicola Baccetti, Istituto di Zoologia, via Mattioli 4,
I - 53100 Siena

Introduction

Common Cormorants of the Continental race Phalacrocorax carbo sinensis have recently increased their breeding populations, since some European countries have stopped control measures against them (Hansen, 1984; Hald-Mortensen, 1985; Rooth, 1985). This fact, not completely analyzed in the 1970's (Cramp & Simmons, 1977) is now also affecting numbers observed on wintering grounds (Hansen, 1984; Suter, 1985).

As regards Italy, little is known about the past and present numerical situations, though evidence suggests a large increase in the number of Cormorants at least in coastal wetlands (Po delta, Southern Sardinia, etc.). Data concerning a single region, Tuscany, are given and briefly commented here.

Methods

In January 1986, Cormorants were counted at least once at all known feeding areas of Tuscany. Several surveys, from November 1985 to March 1986, were made at one site only (Orbetello), where highest numbers occurred in previous years. The birds were counted in the early morning at Orbetello, while they were flying from their roosting places to the lagoons. Elsewhere, less numerous, counts were made during the day, on feeding birds.

Results

High numbers of Cormorants were found at two sites only, Lago di Massaciuccoli and Orbetello, near the northern and southern ends of the Tuscan coastline respectively. At the first place, birds roosted on a wooded river bank, 11 km from the main

NATO ASI Series, Vol. G 12
Mediterranean Marine Avifauna
Edited by MEDMARAVIS and X. Monbailliu

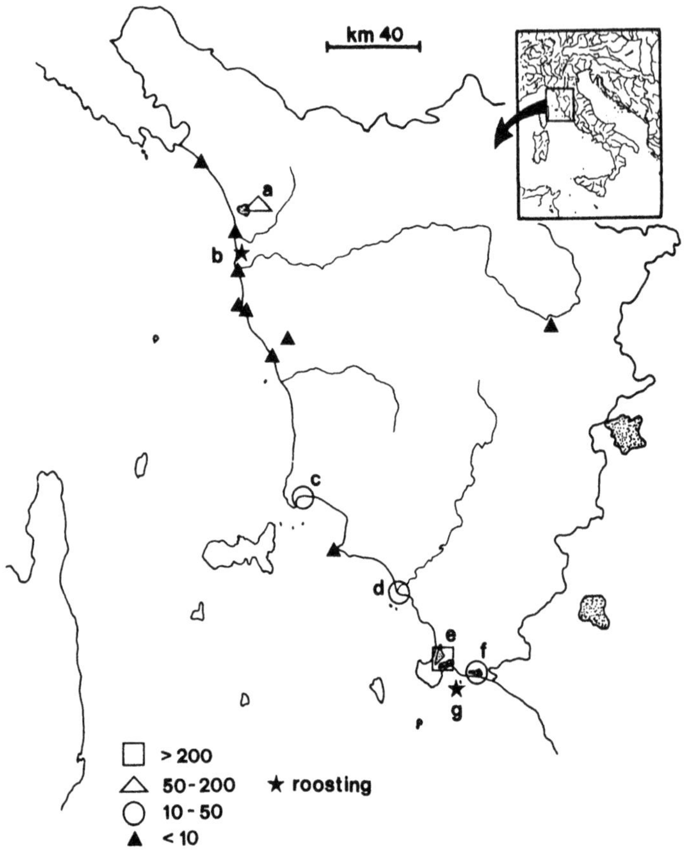

Fig. 1: Situation of Cormorants wintering in Tuscany. Stars indicate the position of roosts, other symbols refer to numbers at feeding sites. Localities: a, Lago di Massaciuccoli; b, San Rossore; c, Piombino; d, Bocca d'Ombrone; e, Lagune di Orbetello; f, Lago di Burano; g, Formica di Burano.

feeding area; at Orbetello, the roost was on a rocky islet in the sea, 6 km off the lagoon. The maximum number of Cormorants at each locality was about 150 and 800, but more than half of the birds of Massaciuccoli disappeared in early January, possibly joining the group of Orbetello or migrating in advance. Other wetlands not far from these two localities had smaller groups of birds, all coming from the same roost; elsewhere, along the coast few Cormorants were found, mainly near Piombino. Only at one inland site a single Cormorant was seen.

The total number of birds counted is around 1,000-1,100. The numbers of birds and the localities are mapped on fig. 1, while fig. 2 shows numerical changes at Orbetello, within the same season. A comparison with the situation in the last five years was made for the two main sites only (tab. 1).

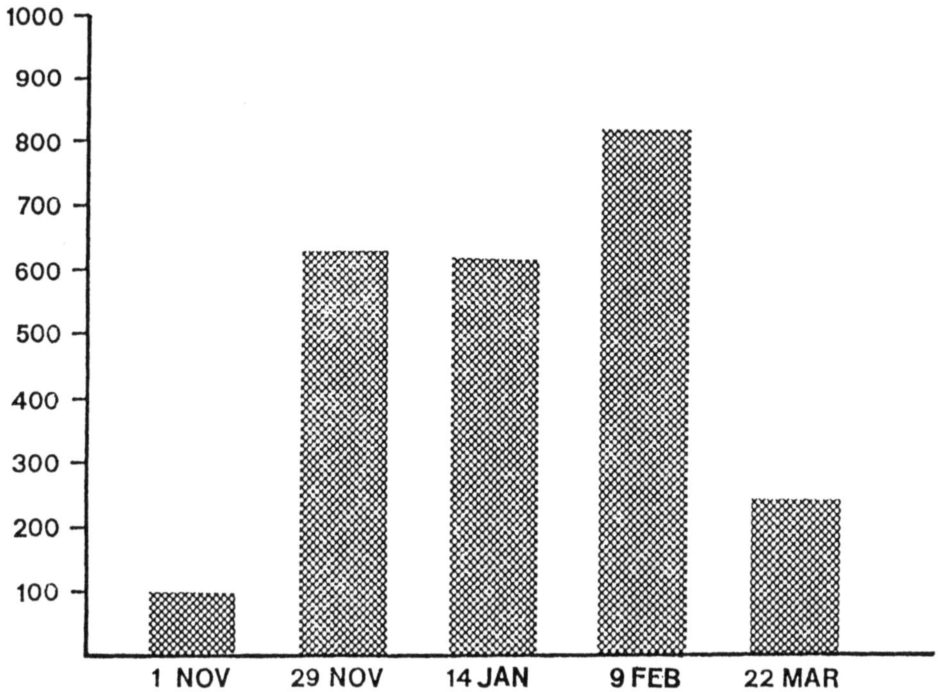

Fig. 2: Numbers of Cormorants at Orbetello, winter 1985-86.

Winter	1980-81	1981-82	1982-83	1983-84	1984-85	1985-86
Site:						
Lago di Massaciuccoli	15	6	(1)	4	12	150
Lagune di Orbetello	200	(350)	(300)	(450)	700	850

Tab. 1: Number of wintering Cormorants at the two main sites, in the last years. Figures in brackets are used for incomplete surveys. Information on the past situation at Orbetello was kindly supplied by Mr. C. Giannella.

Discussion

All coastal lakes, lagoons, as well as some harbours and river-mouths, turned out to be visited by wintering Cormorants, with highest concentration occurring on the lagoons of Orbetello. No regular feeding areas were found at sea, possibly because of depth (Van Eerden, 1986 and pers.comm.). Numbers have much increased in recent years at the main sites, and the presence of singles of small groups in less suitable localities may indicate the same. The presence of birds seems however to depend on the area of protected territory at each site: some inland wetlands were apparently not inhabited because of hunting.

At Orbetello, the wintering population was numerically rather constant between the end of November and mid-February. Birds started arriving each morning just before sunrise, and kept on coming in different sized groups for about half an hour. The daily activity was deeply influenced by fishermen, who tried to chase swimming birds by boat and to drive them away. Such disturbance resulted sometimes in sending the birds back to the roost earlier than usual, and possibly, stimulated them to fish socially in a shorter time. Illegal shooting was moderate. The fishing-net locally used is harmless for diving birds.

The situation at Lago di Massaciuccoli was sensibly different; the reasons of the disappearance of 50% of the flock at mid-winter has not been explained, but is possibly connected to the high number of hunters just beside the lake; lack of fish and few perching places may be two more reasons.

A minimum number of 1,000 birds wintering in Tuscany is by far a small part of the whole Italian population. Thus, the total number is probably now much higher than it was in early 1980's (2,500 ind., Hansen, 1984).

Conclusion

Cormorants were counted in Tuscany during the winter 1985-86. About 1,000-1,100 birds were found, with a minimal percentage of 70% concentrated in a single locality. The present numerical situation is compared with that of the last five years.

References

Cramp, S. & Simmons, K.E.L. (Eds.) (1977). Birds of the Western Palearctic, vol. 1, Oxford Univ. Press.

Hald-Mortensen, P. (1985). Report presented at the Intern. Meeting on Baltic population of Cormorants, Falsterbo, Sweden.

Hansen, K. (1984). The distribution and numbers of the Southern Cormorant Phalacrocorax carbo sinensis in Europe. Dansk. Orn. Foren. Tidsskr. 78: 29-40.

Rooth, J. (1985). Recente ontwkikkelingen van de broedpopulatie van de Aalscholver Phalacrocorax carbo. Limosa 58: 162-163.

Suther, W. (1985). Cormorants in Switzerland. Report presented at the International Meeting on Baltic population of Cormorants, Falsterbo, Sweden.

Van Eerden, M.R. (1986). Aspects of wintering Cormorants in the Mediterranean: first results on movements of colour marked birds. Medmaravis Proc. 1st Medit. Seabird Symposium (Alghero), Springer Verlag, Heidelberg (this volume).

SOME INDICATIONS OF LARIDAE IN ALBANIA

by Francesco Barbieri
Department of Animal Biology, University of Pavia

From 15th till 29th May 1985 during a short sojourn in Albania several ornithological observations were undertaken covering some 106 species (Barbieri et al., 1986).

Although it was not possible to make a complete survey of the coastal zones, three areas are believed to be important for breeding larids:

- wetlands of Shengjin, Lezhe, Kunes;
- lagoon of Karavastase where minimum 11 nests were found of Dalmatian pelicans (Pelicanus crispus);
- lagoon of Nartess, Sazanite island, peninsula of Karaburunit. This lagoon was not visited but observed from a distance. This area covers several salinas and rocky coastlines, all potential breeding habitat for several bird species.

The coasts and lakes of Shkodres, Ohrit, Butrintit as well as the wetlands close to Durres, once important for Larids, are of a minor ornithological importance because of reclamation and cultivation.

In tab. 1 and 2, we computed the rare bibliographical data available of gulls and terns and compared them to our observations.

The series of stamps printed in 1976 as depicted overleaf should witness of the presence of six breeding Laridae species. The behaviour and the environment shown on the 15 q. stamp depicting two White-winged Black Terns but looking more like a pair of rock doves than Chlidonias leucopterus, as well as the nidicolous chicks of the Black-headed gull (stamp of 25 q.) and the erroneous caption of Larus argentatus ressembling a stylish Mediterranean gull lead to confusion and indicate a lack of credibility.

NATO ASI Series, Vol. G 12
Mediterranean Marine Avifauna
Edited by MEDMARAVIS and X. Monbailliu

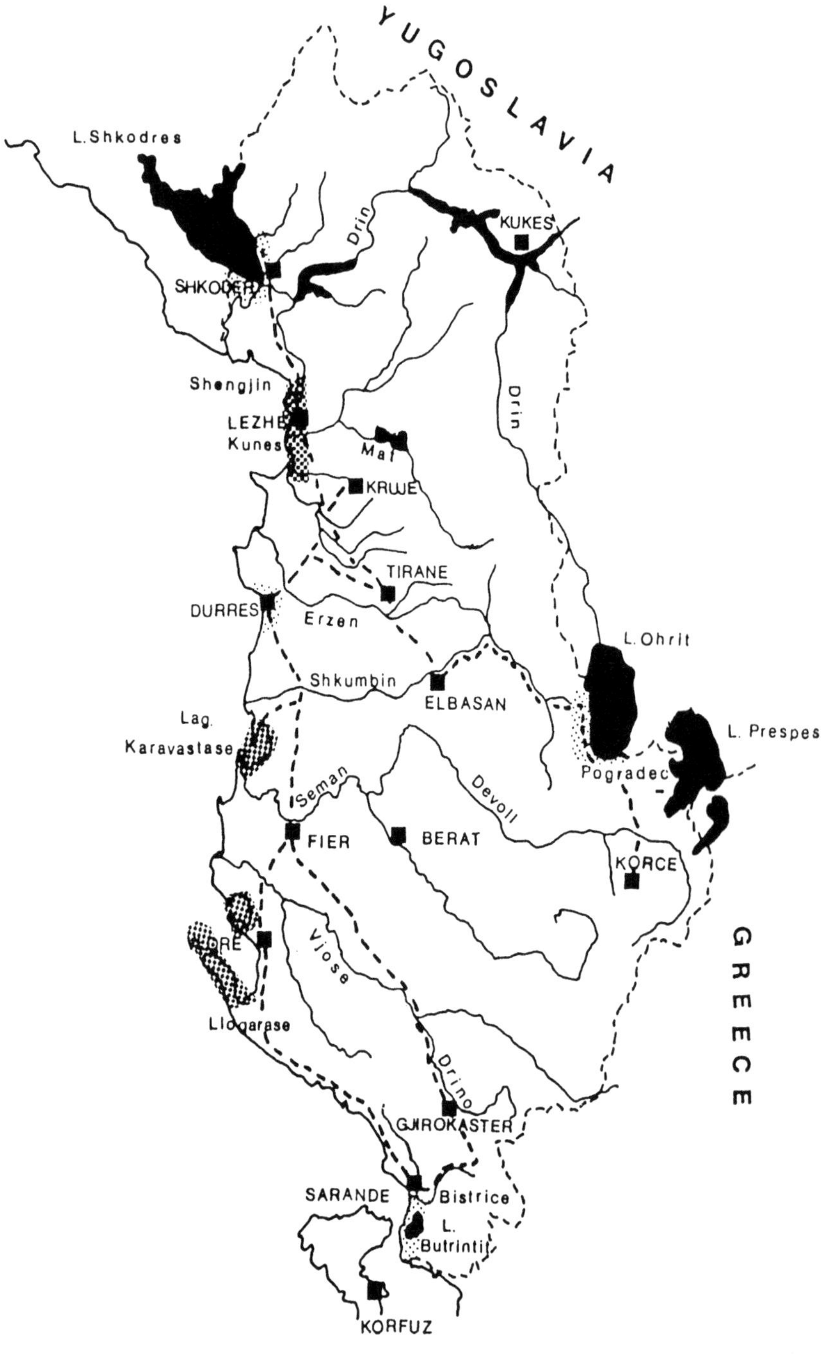

Fig. 1: Zones of ornithological interest of Albania

- Zone of major ornithological importance
- Other areas of ornithological importance
- --- Itinerary

LARIDAE	Ticehurst and Whistler 1932 (8 April - 6 March)	Lamani and Puzanov 1962-1963	Cramp and Simmons 1983	Barbieri and al. 15/29.5.1985
Larus melanocephalus	P	M	M	
Larus minutus	B	M	M	
Larus ridibundus	M	M	M	P Drin, Shengjin
Larus genei		M	M	
Larus canus		M		
Larus fuscus		B	M	
Larus cachinnans	N	N	N	P Drin, Shkoder, Sarande, L. Butrintit
Rissa tridactyla		B		

(N) = Breeding (M) = Migratory (P) = Observed (B) = Bibliography

Tab. 1

STERNIDAE	Ticehurst and Whister 1932 (8 April - 6 March)	Lamani and Puzanov 1962	Cramp and Simmons 1985	Barbieri and al. 15/29.5.1985
Sterna nilotica	N	N	N	
Sterna caspia	P	M	M	
Sterna sandvicensis	M	M	M	
Sterna hirundo	P	N	N	N Kunes P Drin, L. Shkodres
Sterna albifrons	N	N	N	N Karavastase P Drin, L. Butrintit
Chlidonias hybridus	B	B	M	P L. Shkodres
Chlidonias niger	P	N	N	P Shengjin
Chlidonias leucopterus	B	M	M	

(N) = Breeding (M) = Migratory (P) = Observed (B) = Bibliography

Tab. 2

References

Barbieri, F.; Bogliani, G.; Prigioni, C. (1986). Note sull'ornitofauna d'Albania. Riv. ital. Orn. (in stampa).

Cramp, S. (ed.) (1985). The Birds of the Western Palearctic. Vol. IV, Oxford University Press, Oxford.

Cramp, S. and Simmons, K.E.L. (eds.) (1983). The birds of the Western Paleartic. Vol. III, Oxford University Press, Oxford.

Lamani, F.; Puzanov, V. (1962). Inventari i shpendeve te Shqiperise. Bul. Univ. Sht. te Tiranes, Ser. Shk. Nat. 3: 87-103.

Lamani, F.; Puzanov, V. (1963). Shtojce. Liste e shpendeve qe duhet te jene. Bul. Univ. Sht. te Tiranes, Ser. Shk. Nat. 2: 122-124.

STATUS OF THE POPULATIONS OF LARUS CACHINNANS AND STERNA HIRUNDO NESTING IN ISTRIA AND NORTHERN DALMATIA

Enrico Benussi, Civico Museo di Storia Naturale
di Trieste (Italy)

All data registered in the given paper refer to research carried out over a period of four years (1982-85) within the Larus cachinnans and Sterna hirundo colonies of the Istrian peninsula and northern Dalmatian islands, down south to Male Orjule and Rab (Croatia, Yugoslavia).

During this period, single colonies were registered and located, although in some cases it was not possible to carry out yearly regular censuses. There are no relative data for the island of Brioni (south-western Istria) and nearby islands, as the area is inaccessible because of political and military reasons. For a few colonies, Yugoslavian references were consulted.

In the study area, the colonies of the two species are to be found on small- or medium-sized islands or islets, the majority of which are uninhabited and covered with Mediterranean-type vegetation. Only in a few cases the Common Tern bred on small islets, or rather on rocks, lacking vegetation completely.

The most numerous colony of Yellow-legged Gulls is to be found on the island of Zeca and counts about 600-650 couples, while the smallest one appears to be at Male Orjule with 10-15 couples. The population of Sterna hirundo is less numerous. The largest colony, at Terstenik, attracts 30 to 40 couples.

Generally, the two species do not enter into territorial competition and lay their eggs on different, though adjacent, islets. However, as the Common Tern reproduces at a later stage, it is often an easy prey for Larus cachinnans. Both species bred only in seven cases on the same islands, but in different areas.

The population of Yellow-legged Gulls in Istria and Northern Dalmatia is estimated at 3,200 - 3,600 nesting couples, whereas the total number of breeding pairs of Common Terns is 185 - 223. These figures, however, may be an underestimate.

NATO ASI Series, Vol. G 12
Mediterranean Marine Avifauna
Edited by MEDMARAVIS and X. Monbailliu

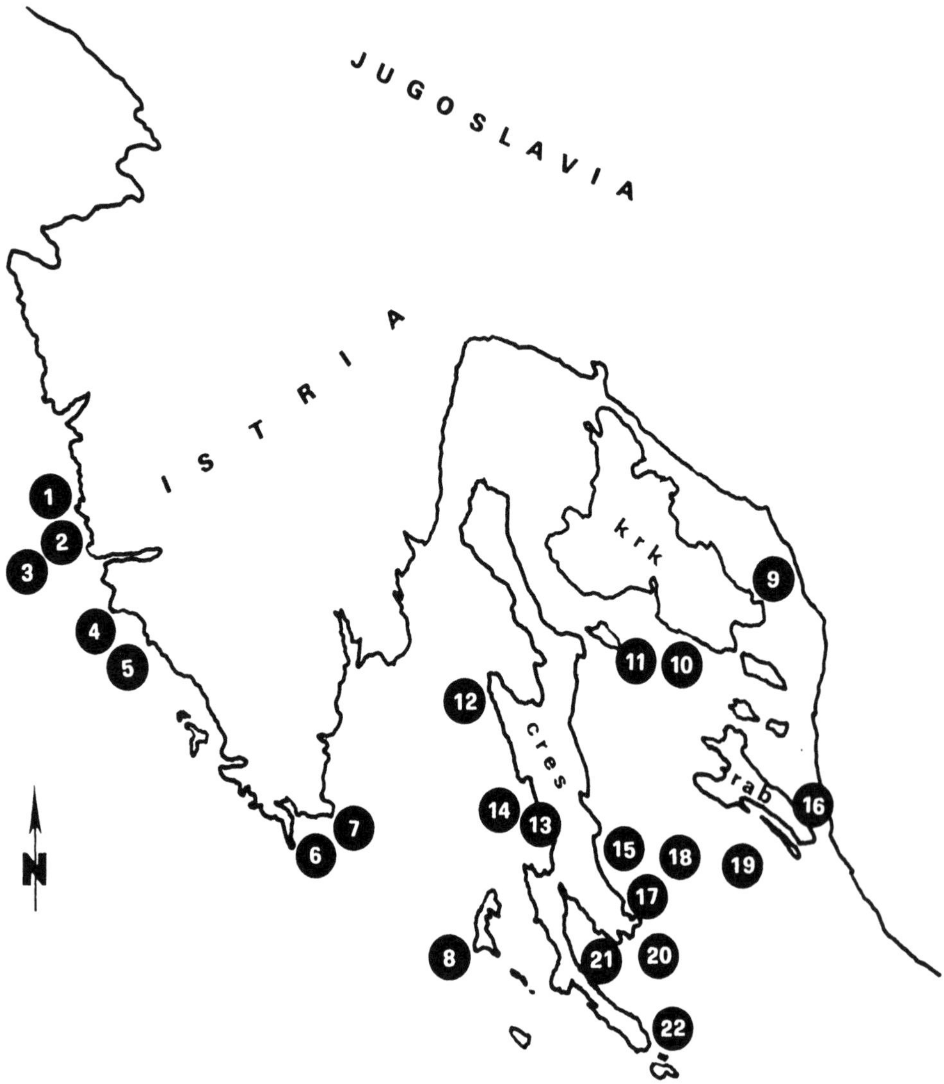

Distribution of colonies of Yellow-legged Gull and Common Terns along the Istria and Northern Dalmatian coasts.

	Place of Colony	Number of breeding pairs	
		Larus cachinnans	*Sterna hirundo*
1	Frzital		15
2	Veliki Skolj	50-70	
3	Lunga	320-350	
4	Sv. Ivan	25	10
5	Mala Velika Sestrice	50-80	15-20
6	Fenera		5
7	Veliki Levan	30-40	
8	Skoljic	15-20	
9	Zecevo	450-480	
10	Galun	260-300	
11	Kormati	300-340	10-15
12	Zaglav	50-60	
13	Visoki		15-20
14	Zeca	600-650	
15	Cutin	400-430	30-35
16	Mag		15-18
17	Ul		5
18	Terstenik	300-350	30-40
19	Dolfin	40-70	
20	Oruda-Palacol	300-320	20
21	Mali Osir		10
22	Male Orjule	10-15	5-10
TOTAL POPULATION		3,200-3,600 pairs	185-223 pairs

Breeding population of *Larus cachinnans* and *Sterna hirundo* in 22 Istrian and Dalmatian colonies (1982-85).

THE 1986 CENSUS OF THE BREEDING POPULATION OF SHAGS IN THE BALEARIC ISLANDS

by Llorenç Capella, Juan Luis Jara,
Joan Mayol, Jorge Muntaner, Miquel Pons,
Conselleria d'Agricultura i Pesca del Govern Balear,
U.V.S. del Secona

Introduction

Before 1986, no complete census had been carried out of the breeding population of shags in the Balearic Islands. Fortunately, data of the Cabrera Archipelago (Araujo et al., 1977) were available, which meant that an analysis could be made of the evolution of the colony there during a period ranging from 10 to 12 years. As for the other islands, a correct estimate existed only for Menorca (Muntaner et al., 1984), while data were lacking for Mallorca, Ibiza and Formentera.

In 1986 the "Wild Life Unit of SECONA" (Agricultural and Fisheries Council of "Govern Balear") decided to carry out a census of the shag population in order to monitor the fluctuation of this species and, if necessary, to apply conservation measures.

Materials and methods

The present census was carried out almost entirely from the sea, using a pneumatic "Zodiac" type boat with an outboard motor. Cabrera, including Conejara, was visited with two boats. Also, two comparative studies were done in Menorca. Two or more observers with long-distance binoculars followed the length of the Islands' coast at a short distance from the shore and at a slow speed, recording the following data:

- Occupied nests
- Nests that were being or had been used but which were empty at the time of the census. The occupancy of some other

NATO ASI Series, Vol. G 12
Mediterranean Marine Avifauna
Edited by MEDMARAVIS and X. Monbailliu

nests was impossible to check from the sea because hidden by rock outcrop.
- Number of adults
- Number of immatures.

For the location of colonies, cartographical maps of scale of 1:50,000 have been used (U.T.M. projection, all nests having been included in a U.T.M. network of 10 x 10 km). In the majority of cases, urbanized and rural sections of sandy coasts have not been surveyed.

To cover the whole Balearic coast the census took several weeks since it was not possible to embark regularly owing to bad weather during the first four months of 1986. The total working days involved amounted to 17, divided up as follows:

Mallorca	8 days	(between 10.01 and 18.03)
Cabrera	1 day	(two teams on 26.02)
Ibiza	2 days	(1 and 2.04)
Formentera	1 day	(3.04)
Menorca	5 days	(between 7 and 12.04 and 1 day from the inland)

In spite of the 3 month duration of the census, since the reproduction of *Phalacrocorax aristotelis* is asynchronous, it was always possible to find a great number of shags in full reproduction.

The censuses were carried out by the staff of the "Wild Life Unit of SECONA" (Agricultural and Fisheries Council). In Menorca we worked in collaboration with the forest agents Carmen Orellana and Salvador Albero. In Cabrera, forest agent Joao Vicenç Lillo assisted us.

Results

In general, the coverage and quality of the census may be considered satisfactory. It was not possible to visit one section of the northern coast of Menorca (from Cavalleria Cape to "Mola de Fornells") nor was it possible to land on certain small barren islands (like "La Imperial de Cabrera" and "Bleda de Menorca"). Since one can not carry out a census from the sea very well, we have to allow for underestimates.

Regarding "possible breeding" cases, one can assume that a nest was not necessarily occupied this year. This may provoke some errors. Yet, in view of the fact that a census always involves an underestimate and that in no way it is possible to calculate exactly the total number of breeding couples, we consider that such errors are largely compensated for.

Tab. 1 indicates graphically the results of the census using U.T.M. units of 10 x 10 km, giving the total of counted and possibly occupied nests.

A minimum of 20 to 40 couples could additionally be estimated for Menorca (15 to 20 for "Isla Bleda", 1 to 10 for Cavalleria Cape, 3 to 10 in "Mola de Fornells", and 1 for the small island of Addaia) and about 20 couples for "Imperial de Cabrera". If one were to take into account both the census and estimates, results would be as follows:

Island	Number of nests counted	Possibly occupied nests	Total counted	Additional estimates	Total
Mallorca	762	233	995	-	995
Cabrera	66	9	75	20	95
Menorca	105	50	155	20-40	175-195
Ibiza	83	22	105	?	105
Formentera	67	4	71	?	71
Baleares	1,083	318	1,401	40-63	1,441-1,464

Tab. 1: Results of census.

Accordingly and considering the scanty bibliographical material, the Phalacrocorax aristotelis desmarestii breeding population in the Balearic Islands would appear to be one of the largest of the Western Mediterranean Sea, if not the second in the Mediterranean. In the Western Mediterranean, the Corsican population comes in third place with 800 to 1,000 couples (Guyot et al., 1985). There is no census available for Sardinia but Schenk and Torre (this volume) estimate there max. 1,800 pairs.

No breeding shags frequent the continental shores of France (Guyot et al., op.cit.) and Italy (Brichetti, 1978). Only a few pairs breed in continental Spain (Ferrer et al., 1986) and in Northern Africa. However, recent and complete data are only available for Algeria (Jacob & Courbet, 1980), where some 20 to 40 couples breed.

In the Balearic Islands, we must highlight the importance of the big colony of "Cabo Blanco" (Mallorca where some 200 breeding couples are concentrated.

References

Araujo, J.; Munoz-Cobo, J. & Purroy, F.J. (1977). Las rapaces y aves marinas del archipiélago de Cabrera. Naturalia Hispanica, 12. 1-94 ICONA Madrid.

Brichetti, P. (1978). Guida degli uccelli nidificanti in Italia. 100 p. F. Lli Scalvi Editori. Brescia.

Ferrer, X.; Martinez, A. & Muntaner, J. (1986). Historia Natural dels Paîsos Catalans. Ocells. Vol. 12, Barcelona.

Guyot, K.; Launay, G. & Vidal, P. (1985). Oiseaux de mer nicheurs du Midi de la France et de Corse: évolution et importance des effectifs. En: Oiseaux marins nicheurs du Midi et de la Corse: 31-47. Annales du C.R.O.P., 2. Aix-en-Provence.

Jacob, J.-P. & Courbet, B. (1980). Oiseaux de mer nicheurs sur la côte algérienne. Le Gerfaut, 70: 385-401.

Muntaner, J.; Escandell, A.; Ramos, E. & Orfila, G. (1984). Adicion y revision faunistica de las especies de la publicacion "Avifauna de Menorca". En: Avifauna de Menorca 2° Ed. (J. Muntaner & J. Congost). Treb. Mus. Zool. 1. Barcelona.

URBAN NESTING OF YELLOW-LEGGED GULLS IN BARCELONA (SPAIN)

Jordi Garcia Petit, M. Elena Marti Gabernet,
Fco. Javier Thomàs Gimeno, Enric Carrera Gallisà,
Seccio d'Ornitologia, Museu de Zoologia,
Apartat de Correus 593, 08080 Barcelona (Spain)

The city of Barcelona, with a population of 4.5 million inhabitants spread over an area of over 10,000 hectares is situated between the mouth of two rivers: Besos and Llobregat. Barcelona has a large fishing port and has always been an important stop over and wintering place for gulls. During summer only two gull species are present: Larus ridibundus and Larus cachinnans.

During the 70's, the Yellow-legged Gull colony at the Medes Islands, close to Barcelona, increased substantially (E.Carrera, 1984). Consequently more adult gulls were observed in the city of Barcelona during the breeding season. In 1975 the first nest was found in the city: the first couple settled down at the zoo, possibly because of the presence of some captured specimen, or because they were fed with plenty of fish. Gradually this reproductive core expanded, spreading over the city. Today the number of urban nesting Yellow-legged Gulls in Barcelona totals 25 pairs.

The main factors which have contributed to the expansion and colonization in the city are:

- large quantity of prominent roofs and terraces of buildings non visited by man;
- proximity of several food sources;
- large winter population, part of which remains as a non breeding core during summer. We think that this core attracts potential breeders to settle down.

The nesting population of Barcelona is not considered as a colony, due to the low number of breeding pairs and since nests are built far apart. There is only one building (Seminari Conciliar) where you can observe pseudo-colonial behaviour amidst nearly 80 pairs en 1983.

NATO ASI Series, Vol. G 12
Mediterranean Marine Avifauna
Edited by MEDMARAVIS and X. Monbailliu

Egg laying occurs simultaneously as is the case at Islas Medes: between the end of March and the middle of April. However, we have discovered two nests where laying began in the middle of May. Out of 8 controlled nests, clutches consisted of 2.8 eggs. The breeding success was checked until fledging: of 16 nests 1.68 chicks flew out per nest. This rate is similar to the one observed in urban nests of the United Kingdom (Sunderland: 1.3 chicks/pair, South Shields 1.6 chicks/pair; P. Monaghan, 1979), and in Adriatic colonies (1.5 chicks/pair; Spitzer, 1978), but is smaller than the rate observed in Camargue (2 chicks/pair; Isenmann, 1976). Mortality was caused first by human intervention which ruined four complete clutches. Secondly, four chicks fell out of the nest and died.

Nests are situated in three different zones:

- Zoological Park and Ciutadella Park: few low buildings interspersed with gardens;
- Cathedral and Ciutat Vella: several monumental and high buildings, the roofs of which are seldom visited by man;
- Seminari Conciliar building and its neighbourhood: buildings of an uniform height consisting of offices and residences, some of them being monumental.

The characteristics of nest site selection on buildings are:

- non visited by man during breeding, if not a certain distance to the nest is kept. However, at the Zoo some nests are situated only 5 meters from visitors;
- a building much higher than the surrounding dwellings;
- near a food source.

Nest material consists of plant debris: Platanus sp seeds, little branches, grass, pine needles, etc. We have also found some other materials such as ropes, feathers, plastic, etc.

The diet of the urban Yellow-legged Gulls of Barcelona is not very different from gulls in breeding colonies nearby (Carrera, 1984; Araujo, Munoz Cobo, Purroy, 1977). However there are some cases of different food sources. An important part of the diet is fish (Cepola rubescens, Trachurus trachurus, Scomber scombrus, Ciprinus carpio). We also found Sepia officinalis, Loligo vulgaris and rests of various crustaceans. From rubbish dumps the Yellow-legged Gull collects parts of chicken, rabbit and beef.

The urban gulls of Barcelona prove to be efficient predators on birds which live in the city: feral pigeons (Columba livia domestica) and to a lesser extent swifts (Apus apus) and sparrows (Passer domesticus). At the Zoo they predate on mallards (Anas platyrhynchos) and other ducks, occasionally on blackbirds (Turdus merula), peacock, etc.

The hunting system to capture pigeons consists of hitting it with the wing to make it fall to the ground where it is pecked in the head and other vulnerable parts until it is being killed. Normally only a little part of the biomass is being consumed.

A detailed feeding study revealed few vegetable rests. Some olive seeds (Olea europaea) were found, as well as cherry seeds (Prunus avium), lemon seeds (Citrus limon), apple rests and gramineae. Finally a small part is composed of some diverse material: hard plastic, glass, silver paper, elastic rubber, rope, stones, etc.

The available food sources are:

- Fish taken at sea or from fishermen throwing back uncommercial fish and offal,
- Remains of the urban municipal market,
- Food given by citizens to cats or (less frequently) directly to gulls,
- The population of pigeons, swifts, passerines and other urban vertebrates.

In order to monitor the gulls born in Barcelona we have ringed a total of 14 chicks. It has been reported that some chicks born at the Zoo can be seen regularly in that area 7 months after leaving the nest.

This work is the first part of on-going research on the L. cachinnans population in Barcelona. We are currently following all wintering populations of Larids in the city estimated in 1985 at 15,388 specimen of 8 different species.

References

Araujo, J.; Munoz-Cobo, J. and Purroy, F. (1977). "Las rapaces y aves marinas del Archipiélago de Cabrera". Naturalia Hispanica n° 12.

Carrera, E. and Vilagrasa, J. (1984). "La colonia de gavià argentat de les illes Medes" in Els Sistemes Naturals de les Illes Medes.

Garcia Petit, J.; Thomas Gimeno, F.J. and Marti Gabernet, E. (1985). (Manuscrit) "Inventari biologic del Gavià argentat a la ciutat de Barcelona".

Martinez, A.; Ferrer, X. and Muntaner (1984). "Atlas dels ocells nidificants de Catalunya i Andorra".

Monaghan, P. (1979). "The breeding ecology of urban nesting gulls", Urban Ecology.

BALEARIC SHEARWATERS SUMMERING IN WESTERN FRANCE

by Pierre Yésou

Office National de la Chasse, F-85340 L'Ile d'Olonne

Balearic Shearwaters P. puffinus mauretanicus (= Puffinus yelkouan mauretanicus, a better name according to Bourne 1982 and unpublished report) have long been known to be regularly present along the French Atlantic coast during the summer months (Mayaud 1932). A preliminary estimate of their numbers was recently obtained by studying the density of birds at sea ("Programme C.R.B.P.O.-M.E.R.": Hémery et al. 1986): 10,000 to 15,000 individuals are thought to be simultaneously present in mid-summer from southern Biscay to southern Brittany (Yésou 1985). Their distribution is far from homogeneous, and a major concentration occurs in the area we studied near Les Sables d'Olonne (46°30 N, 1°50 W: fig. 1). Les Sables d'Olonne is the fifth most important fishing port in France according to total tonnage discharged, the first for inshore fishery. The latter activity attracts foraging shearwaters, and could explain their important numbers here.

In calm sea conditions, shearwaters congregate in rafts near the coast at the end of the day: they can then be censused (methods in Yésou 1984). Although of variable efficiency, the censuses organized during the last four years allow a tentative description of the local abundance cycle of Balearic Shearwaters (fig. 2). Maximum numbers occur in July-August, and vary from year to year between 6,000 to a little over 7,000 birds, i.e. 40% to 50% of the estimated total for the entire French Atlantic waters. Numbers seem almost stable in July-August, as no significant change in density was detected during regular survey at sea in 1984.

From a trawler, shearwaters were carefully observed at sea on 21 dates between 19 June and 12 September 1984. Information was collected on moult process and age composition of the population (fig. 3). The proportion of juveniles increases throughout July and remains fairly stable during August and September, around 25%: this value is perhaps overestimated, however, as juveniles show a greater attraction to trawlers than do older

NATO ASI Series, Vol. G 12
Mediterranean Marine Avifauna
Edited by MEDMARAVIS and X. Monbailliu

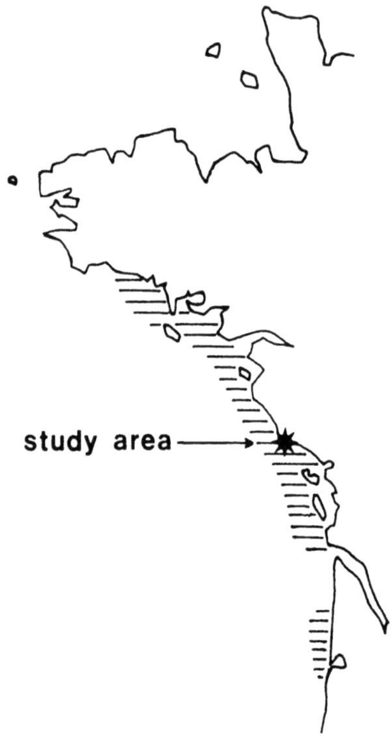

Fig. 1: Summer distribution of Balearic Shearwaters along the French Atlantic coast (shaded. From Hémery et al., 1986, simplified) and location of the study area near Les Sables d'Olonne.

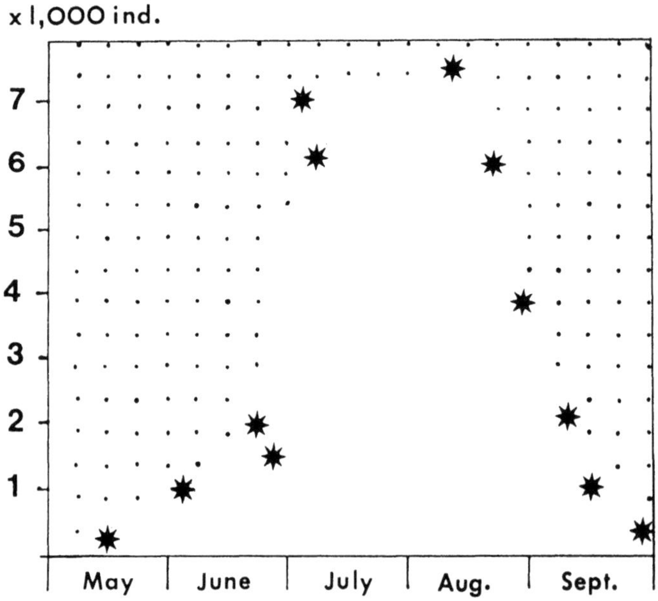

Fig. 2: Numerical importance of Balearic Shearwaters in the study area, from census results (✱) obtained from 1982 to 1985.

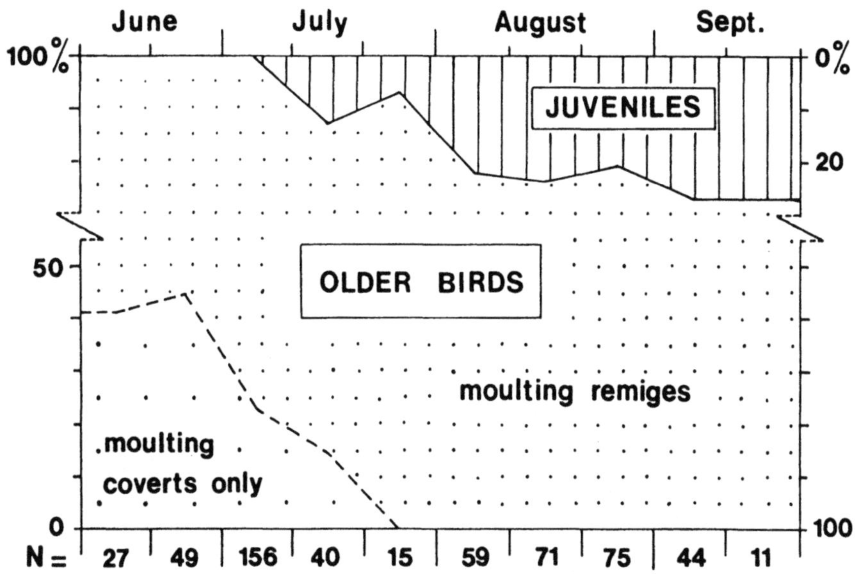

Fig. 3: Variation in the structure of the Balearic Shearwater population, as appearing from direct observation at sea (methods in Yésou, 1985). N = sample size (nbr.ind.) for each ten-days period.

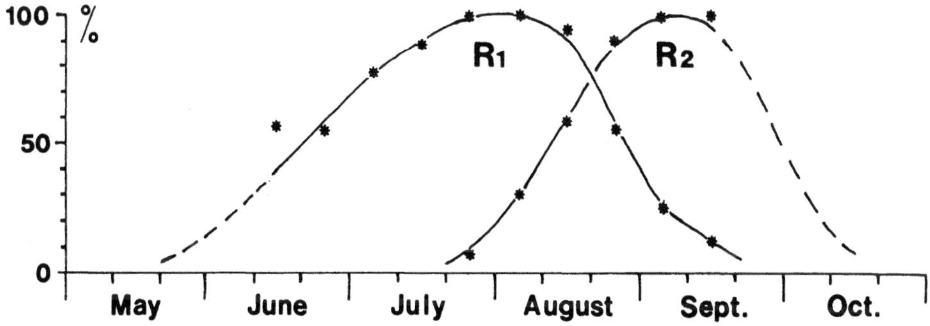

Fig. 4: Proportion of Balearic Shearwaters (second calendar year and older) moulting primaries (R_1) and secondaries (R_2), according to time of the year.
The stars (✱) indicate the observed values, sampled as in fig. 3. Curves are eye-adjusted. Information from other sources indicate that remige moult is over by the end of October.

birds (Yésou, 1985). All these older birds are in moult. Near 60% of them are already moulting remiges in mid-June, and all are doing so by the end of July. Precise observation of individual moult score through the season (n = 435 birds, methods of Yésou, 1985) allows the schematic description of moult progress (fig. 4): secondary moult starts and finishes later than primary moult: flight feather moult is estimated to last three months in an individual; because of individual variations in moult timing, birds in flight feather moult occur from May to October, the whole population being in such a moult from mid-July to at least mid-September.

Some young Balearic Shearwaters are perhaps able to fly by the end of May (Bannerman & Bannerman, 1983), but most leave the colonies at the end of June and during the first half of July (Araujo et al., 1977), i.e. about three weeks before their arrival here. Adults leave their young a week or so before full fledging (Cramp & Simmons, 1977; Vidal, 1985), so they can take part in the massive arrival observed here at the beginning of July. But the reproductive calendar strongly suggests that most of the birds observed earlier are immature. The individual variations in moult timing could then be explained by an earlier moult in immatures than in adults: a situation that would not be surprising, having already been suggested or even demonstrated for other Procellariiformes (Cramp & Simmons, 1977).

References

Ajauro, J.; Munoz-Cobo, J. and Purroy, F.J. (1977). Las rapaces y aves marinas del archipiélago de Cabrera, Madrid.

Bannerman, D.A. and Bannermann, W.M. (1983). The Birds of the Balearics, London.

Bourne, W.R.P. (1982). The relationship of the Manx and Fluttering Shearwaters, Sea Swallow 31: 44-47.

Cramp, S. and Simmons, K.E.L. (eds.) (1977). The Birds of the Western Palearctic, vol. 1, Oxford.

Hémery, G.; Pasquet, E. and Yésou, P. (1986). Data Banks and Population Monitoring in France, in Medmaravis, Proceedings First Mediterranean Seabird Symposium (this volume). Springer Verlag, Heidelberg.

Mayaud, N. (1932). Considérations sur la morphologie et la systémat. de quelques puffins. Alauda 4: 41-78, errata 495.

Vidal, P. (1985). Premières observations sur la biologie de la reproduction du Puffin des Anglais yelkouan Puffinus puffinus yelkouan sur les îles d'Hyères (France). Annales C.R.O.P. 2: 58-62.

Yésou, P. (1984). Premier dénombrement estival des Puffins des Baléares sur le littoral des Sables d'Olonne à Saint-Gilles-Croix-de-Vie, Vendée (France). Gorgebleue, 6: 5-8.

Yésou, P. (1985). Nouvelles données sur la mue de Puffinus p. mauretanicus. L'Oiseau et R.F.O. 55: 177-182.

THE ALGHERO DECLARATION

MEDITERRANEAN SEABIRD DECLARATION OF ALGHERO

The first Mediterranean Seabird Symposium, organized by Medmaravis, Mediterranean Marine Bird Association and held at Alghero (Sardinia) from 26th to 30th March 1986:

- Realizing the urgent need to further scientific research on the marine avifauna in the Mediterranean,

- and being aware that many seabird colonies and roosts face critical levels for survival.

1. URGES

national and regional authorities as well as the international scientific community:

To give a special protection status to the following ten most important seabird sites not fully protected so far in the Mediterranean:

- Cabrera Archipelago, Balearics, Spain
- Chafarinas Islands, Alboran Sea, Spain
- Columbretes Archipelago, Spain
- Dionysiades Island, Greece (Crete)
- Filfla Island, Malta
- Gremdi Islets, Kerkenah, Tunisia
- La Maddalena Archipelago, Sardinia, Italy
- Linosa Island, Italy
- Marettimo Island, Sicily, Italy
- North Sporades Archipelago, Greece.

This list containing the ten key areas for reproduction and wintering of seabirds in the Mediterranean needing protection can be extended to other islands and mainland cliff sea-

NATO ASI Series, Vol. G 12
Mediterranean Marine Avifauna
Edited by MEDMARAVIS and X. Monbailliu

bird colonies as soon as the International Scientific Community can provide biological evidence of the importance of other major seabird sites needing protection.

The above list excludes existing nature reserves or national parks as well as deltas and lagoons already dealt with by the Ramsar Convention of 1971.

The respective governments as well as the international government environmental organizations will be approached to make appropriate measures for their protection.

A complete list of all major seabird areas will complement the above ten key seabird sites in the Mediterranean needing protection.

2. PROPOSES

to the national and regional authorities as well as to the International Scientific Community and to the Secretariat of Medmaravis:

- to further scientific research in the Eastern Mediterranean particularly in regions where census data are virtually non existing;

- to endeavour international seabird studies needing coordination among researchers in various Mediterranean countries.

3. RECOMMENDS

to the Italian Government and to the relevant regional authorities (Veneto, Emilia Romagna) to take urgent protection measures in the "Scanni" of the outer Po Delta holding the largest European population of Little Terns (Sterna albifrons) and representing over 30% of the whole Mediterranean population of this endangered species threatened by increasing touristic development.

4. RECOMMENDS

also to the Italian Government and to the autonomous regional government of Sardinia to attribute a legal protection status to the most important coastal and island ecosystems of Sardinia, being:

- the Maddalena Archipelago
- the island of Asinara
- the cliffs of Capo Caccia, Alghero
- the archipelago of Tavolara.

These marine ecosystems are of international importance for seabirds as well as for other species such as birds of prey, vultures and other animals and plants.

5. INVITES

the Spanish Government as well as the appropriate regional authorities to protect the Ebro Delta and the Charafinas Islands, world famous for holding the largest colonies of the threatened Audouin's Gull (Larus audouinii, 70% of the world population), as well as several other species such as:

- hundreds of Gull-billd Tern (Sterna nilotica), Sandwich Tern (Sterna sandvicensis), Little Tern (Sterna albifrons) and Slender-billed Gull (Larus genei), all breeding in the Ebro Delta,

- thousands of Cory's Shearwater (Calonectris diomedea) and several individuals of Osprey (Pandion haliaetus) and Monk Seal (Monachus monachus), all present at the Islas Chafarinas.

6. Being aware of the unique patrimonium of colour ring codes on birds,

   SUGGESTS to researchers before using colour rings on Mediterranean seabirds to co-ordinate with fellow researchers and to contact the Medmaravis Secretariat as well as Euring (Arnhem, The Netherlands), in order to avoid applying already used colour ring combinations and to optimize the available colour codes on such species as Larus cachinnans and Phalocrocorax aristotelis.

7. PROPOSES TO MEDMARAVIS

- to continue the co-ordination of future activities related to the Mediterranean marine avifauna till the next Mediterranean Seabird Symposium,

- to assist in the organization of the second Mediterranean Seabird Symposium scheduled for September 1989 at Palma di Mallorca, Baleares,

- to organize a technical committee meeting at the seat of a regional office of an international environmental organization,

- to seek regional, national and international co-ordination in order to stimulate further studies and conservation projects of Mediterranean seabirds.

Alghero, 30th March 1986

DECLARATION D'ALGHERO CONCERNANT L'AVIFAUNE MARINE EN MEDITERRANEE

Le premier Symposium sur l'Avifaune Marine Méditerranéenne, organisé par Medmaravis (Association Méditerranéenne pour l'Avifaune Marine) à Alghero (Sardaigne) du 26 au 30 mars 1986:

- se rendant compte de la nécessité urgente de développer des programmes de recherche sur l'avifaune marine en Méditerranée,

- et sachant que plusieurs colonies et remises d'oiseaux marins sont menacées.

1. CONSEILLE VIVEMENT

aux gouvernements nationaux et régionaux ainsi qu'à la communauté scientifique internationale:

De sauvegarder légalement les 10 sites les plus importants en Méditerranée pour l'avifaune marine et non encore protégés, notamment:

- Archipel de Cabrera, Baléares, Espagne
- Iles de Chafarinas, Mer d'Alboran, Espagne
- Archipel des Columbretes, Espagne
- Ile de Dionysiades, Crète, Grèce
- Ile de Filfla, Malta
- Ilots Gremdi, Kerkenah, Tunisie
- Archipel de la Maddalena, Sardaigne, Italie
- Ile de Linosa, Italie
- Ile de Marettimo, Sicile, Italie
- Archipel des Sporades du Nord, Grèce.

Cette liste contient les dix sites clefs pour la reproduction ou l'hivernage de l'avifaune marine en Méditerranée nécessitant des mesures de protection. Elle peut être élargie à d'autres îles et colonies situées sur des falaises continentales dès que la communauté scientifique internationale pourra fournir les données

NATO ASI Series, Vol. G 12
Mediterranean Marine Avifauna
Edited by MEDMARAVIS and X. Monbailliu

biologiques relatives à leur importance majeure pour la survie de l'avifaune marine en Méditerranée.

Cette liste exclut les réserves naturelles ou les parcs nationaux existants ainsi que les deltas et lagunes déjà traités par la Convention de Ramsar (1971).

Les gouvernements respectifs ainsi que les organisations internationales non-gouvernementales sont invités à sauvegarder ces sites d'importance internationale pour l'avifaune marine.

Une liste complète de toutes les aires majeures fréquentées par les oiseaux marins en Méditerranée sera préparée pour être le complément des 10 sites cités ci-dessus.

2. PROPOSE

aux gouvernements nationaux et régionaux ainsi qu'à la communauté scientifique internationale et au secrétariat de Medmaravis:

- de promouvoir les programmes de recherche scientifique sur l'avifaune marine en Méditerranée orientale et plus particulièrement dans les régions dépourvues de données de recensement;

- de mettre en oeuvre des études trans-nationales nécessitant une coordination parmi des chercheurs de plusieurs pays méditerranéens.

3. RECOMMANDE

au Gouvernement italien et aux autorités régionales compétentes (Veneto, Emilia Romagna) de prendre des mesures de protection immédiates pour sauvegarder les "Scanni" du Delta du Pô qui abritent la population la plus importante en Europe de la Sterne naine (Sterna albifrons), soit 30% de la population totale en Méditerranée. La Sterne naine est une espèce en danger en raison du rapide développement touristique des côtes.

4. CONSEILLE

également au Gouvernement italien ainsi qu'au Gouvernement régional de la Sardaigne d'attribuer un satut légal de protection aux écosystèmes littoraux ou insulaires les plus importants de la Sardaigne, comme:

- l'Archipel de la Maddalena
- l'Ile de l'Asinara
- les Falaises de Capo Caccia, Alghero
- l'Archipel de Tavolara.

Ces écosystèmes marins sont d'importance internationale pour l'avifaune marine ainsi que pour d'autres espèces comme des rapaces, des vautours fauves et d'autres animaux et plantes.

5. INVITE

le Gouvernement espagnol et les autorités régionales compétentes à protéger le Delta de l'Ebro et les Iles Chafarinas qui abritent les colonies les plus grandes du Goéland d'Audouin (Larus audouinii, 70% de la population mondiale), ainsi que des colonies de plusieurs autres espèces d'oiseaux marins comme:

- des centaines de couples de Sterne hansel (Sterna nilotica), Sterne caugek (Sterna sandvicensis), Sterne naine (Sterna albifrons) et de Goéland railleur (Larus genei), tous nicheurs dans le Delta de l'Ebre;

- des milliers de Puffin cendré (Calonectris diomedea) et plusieurs individus de Balbuzard (Pandion haliaetus) et de Phoque moine (Monachus monachus), tous présents dans l'archipel des Chafarinas.

6. Se rendant compte de l'enjeu et de l'importance de la codification des bagues en couleurs pour identifier les oiseaux,

SUGGERE aux chercheurs en avifaune marine de se concerter avec des collègues ornithologues avant de s'approprier un code de bagues en couleur, et d'informer les secrétariats de Medmaravis et de Euring (Arnhem, Pays-Bas) afin d'éviter l'emploi de combinaisons de bagues colorées déjà utilisées, et d'optimaliser les codes en couleur disponibles pour telles espèces comme le Goéland leucophée (Larus cachinnans), et le Cormoran huppé (Phalacrocorax aristotelis).

7. PROPOSE AU SECRETARIAT DE MEDMARAVIS DE

- continuer la coordination des activités futures concernant l'avifaune marine en Méditerranée et ceci jusqu'au Symposium prochain (1989),

- d'apporter son aide à l'organisation du second Symposium Méditerranéen sur l'Avifaune Marine, programmé pour Septembre 1989 à Palma di Mallorca, Baléares, Espagne,

- d'organiser une réunion technique au siège d'un bureau régional d'une organisation internationale pour l'environnement,

- d'établir une coordination régionale, nationale et internationale afin de stimuler de nouvelles études et projets de conservation pour l'avifaune marine.

Alghero, 30 mars 1986

DICHIARAZIONE DI ALGHERO SULL'AVIFAUNA MARINA MEDITERRANEA

I partecipanti al I° Simposio sugli Uccelli Marini del Mediterraneo, organizzato dal Medmaravis, Mediterranean Marine Bird Association, tenutosi ad Alghero (Sardegna) dal 26 al 30 marzo 1986:

- considerano urgente la necessità di ulteriori ricerche scientifiche sull'avifauna marina nel Mediterraneo;

- consci del fatto che molte delle colonie di uccelli marini sono minacciate.

1. CHIEDONO

alle autorità nazionali e regionali ed alla comunità scientifica internazionale di accordare una speciale protezione alle seguenti 10 zone, le più importanti nel Mediterraneo le quali non ancora protette:

- Arcipelago di Cabrera, Isole Baleari, Spagna
- Isole di Chafarinas, Mare di Alboran, Spagna
- Arcipelago delle Columbretes, Spagna
- Isola Dionysiades, Creta, Grecia
- Isola di Filfla, Malta
- Isolotti di Gremdi, Kerkenah, Tunisia
- Arcipelago della Maddalena, Sardegna, Italia
- Isola di Linosa, Italia
- Isola di Marettimo, Sicilia, Italia
- Parte Settentrionale dell'Arcipelago delle Sporadi, Grecia.

La lista potrà essere estesa ad altre colonie su isole o coste rocciose qualora la communità scientifica internazionale fornisca un'adeguata documentazione della importanza biologica delle aree stesse.

NATO ASI Series, Vol. G 12
Mediterranean Marine Avifauna
Edited by MEDMARAVIS and X. Monbailliu

La precedente lista esclude le Riserve naturali o i Parchi Nazionali esistenti, ed inoltre esclude i delta e le lagune, che formano l'oggetto della convenzione di Ramsal del 1971.

I rispettivi governi e le organizzazioni internazionali per l'ambiente saranno informati e verranno invitati ad adottare le misure di protezione più adeguate.

Alle dieci zone sopra elencate sarà aggiunta una lista completa delle colonie di uccelli marini più importanti del Mediterraneo che deveno essere protette.

2. PROPONGONO

ai governi nazionali e regionali ed alla comunità scientifica internazionale ed alla segretaria del Medmaravis di:

- approfondire le ricerche scientifiche nel Mediterraneo orientale, particolarmente in quelle regioni per le quali mancano dati;
- di incoraggiare gli studi nei Paesi Mediterranei sugli uccelli marini che richiedono il coordinamento fra i vari ricercatori impegnati.

3. RACCOMANDANO

al Governo italiono ed alle amministrazioni regionali interessate (Veneto, Emilia Romagna), l'adozione di urgenti misure di protezione negli "Scanni" della parte più esterna del Delta del Po, che ospitano la popolazione nidificante di Fraticello (Sterna albifrons) più numerosa d'Europa; questa rappresenta oltre il 30% dell'intera popolazione di una specie minacciata dal crescente afflusso di turisti e dallo sviluppo delle attività balneari.

4. RACCOMANDANO

inoltre al Governo italiono ed al Governo della Regione Autonoma della Sardegna di accordare la protezione legale agli ecosistemi costieri ed insulari più importanti della Sardegna:

- Arcipelago della Maddalena
- Isola dell'Asinara
- Falesie di Capo Caccia, Alghero
- Arcipelago di Tavolara.

Questi ecosistemi rivestono un'importanza internazionale per gli uccelli marini e per rapaci ed altre specie animali e vegetali.

5. INVITANO

il Governo spagnolo e le autorità regionali competenti a proteggere il Delta dello Ebro e le Isole Chafarinas, noti in tutto il mondo in quanto ospitano le maggiori colonie di Gabbiano corso (Larus audouinii, 70% della popolazione mondiale), specie la cui esistenza è minacciata. Inoltre nel Delta dell'Ebro nidificano centinaia di:

- Sterne zampenere (Sterna nilotica)
- Beccapesci (Sterna sandvicensis)
- Fraticelli (Sterna albifrons)
- Gabbiani rosei (Larus genei)

e nelle Isole Chafarinas migliaia di Berte maggiori (Calonectris diomedea) e diversi individui di Falco pescatore (Pandion haliaetus) ed è presente la Foca monaca (Monachus monachus).

6. Consapevoli dell'importanza di marcare gli uccelli con anelli colorati per poterli riconoscere individualmente durante le ricerche scientifiche

INVITANO i ricercatori interessati ad inanellare con anelli colorati a concordare preventivamente i codici con gli altri colleghi impegnati in attività analoghe e a contrattare la Segreteria del Medmaravis e l'Euring (Arnhem, Olanda) per evitare l'uso ripetuto delle stesse combinazioni di colori e per ottimmizzare l'impiego dei colori disponibili in specie quali Larus cachinnans e Pholacrocorax aristotelis.

7. PROPONGONO AL MEDMARAVIS

- di continuare a coordinare le attività future collegate all'avifauna marina del Mediterraneo fino al prossimo "Simposio sugli Uccelli Marini del Mediterraneo",

- di collaborare all'organizzazione del secondo Simposio, programmato per il Settembre 1989 a Palma di Maiorca, Baleari,

- di organizzare una riunione del comitato tecnico presso la sede regionale di una delle organizzazioni internazionali per l'ambiente,

- di provvedere al coordinamento regionale, nazionale ed internazionale al fine di stimolare ulteriori studi e progetti di conservazione sull'avifauna marina del Mediterraneo.

Alghero, li 30 marzo 1986

DECLARACION DE ALGHERO SOBRE LAS AVES MARINAS

El Primer Simposio "Aves Marinas del Mediterraneo", organizado por Medmaravis (Asociación Mediterránea para las Aves Marinas) y celebrado en Alghero (Cerdena) entre el 26 y el 30 de marzo de 1986:

- Consciente de la urgente necesidad de potenciar la investigación cientifica sobre las aves marinas del Mediterráneo,

- Advertido de la crítica situación a la que se enfrentan muchas de las áreas fundamentales para la cría e invernada de dichas aves.

1. <u>URGE</u>

a las autoridades nacionales y regionales, asi como a la comunidad científica internacional,

a proporcionar los medios que garanticen la proteccion legal y efectiva de las siguientes 10 areas de importancia para las aves marinas del Mediterráneo:

- Archipiélago de Cabrera, Baleares, España
- Islas Chafarinas, Mar de Alborán, España
- Islas Columbretes, Castellón, España
- Islas Dionysiades, Creta, Grecia
- Isla de Filfla, Malta
- Islotes de Gremdi, Islas Kerkenah, Tunez
- Archipiélago de la Maddalena, Cerdeña, Italia
- Isla de Linosa, Sicilia, Italia
- Islas Marettimo, Sicilia, Italia
- Archipiélago de las Esporadas del Norte, Grecia.

Esta lista contiene las areas clave para la cría e invernada de las aves marinas del Mediterráneo, y podrá ser aumentada con la inclusion de otras islas o parajes acantilados, siempre que la

NATO ASI Series, Vol. G 12
Mediterranean Marine Avifauna
Edited by MEDMARAVIS and X. Monbailliu

comunidad científica internacional documente adecuadamente su importancia y la necesidad de su protección.

De la mencionada lista se han excluido los parques nacionales y reservas naturales ya existentes, asi como los deltas fluviales y otras zonas húmedas ya incluidas en el convenio de Ramsar de 1971.

Los distintos gobiernos nacionales asi como los organismos internacionales medioambientales serán objeto de contactos en el sentido de gestionar la adopción de medidas apropiadas de protección.

Una lista completa de áreas de interés para las aves marinas completará la ya citada de las 10 áreas claves del Mediterraneo.

2. PROPONE

a las autoridades nacionales y regionales asi como a la comunidad científica internacional y al secretariado de Medmaravis,

- potenciar la investigación científica en el Mediterráneo oriental, especialmente en aquellas regiones donde los censos de población escasean;

- apoyar los estudios coordinados entre investigadores de los distintos países Mediterráneos.

3. RECOMIENDA

al Gobierno italiano y a las principales autoridades regionales (Veneto, Emilia Romagna) adoptar urgentes medidas de protección en el "Scanni" del Delta del Po que mantiene la mayor población europea de Charrancito (Sterna albifrons), cerca de un 30% de la población mediterranea de esta especie amenazada por el creciente desarrollo turistico.

4. RECOMIENDA

al Gobierno italiano y al Gobierno regional autónomo de Cerdeña, la designación como área legalmente protegida de los principales ecosistemas costeros e insulares:

- Archipiélago de Maddalena
- Isla de Asinara
- Acantilados de Capo Caccia (Alghero)
- Archipiélago de Tavolara.

Estas áreas son de importancia internacional para las aves marinas, asi como para otras especies animales y vegetales.

5. INVITA

al Gobierno español y a los gobiernos autónomos correspondientes, a proteger de manera efectiva el Delta del Ebro y las Islas Chafarinas, mundialmente reconocidas por mantener las mayores colonias de la amenazada Gaviota de Audouin (Larus audouinii), reuniendo entre los dos enclaves alrededor del 70% de la poblacion mundial de dicha especie.

Otras especies que allí crían son: la Pagaza Piconegra (Sterna nilotica), el Charrán Patinegro (Sterna sandvicensis), el Charrancito (Sterna albifrons) y la Gaviota Picofina (Larus genei) en el Delta del Ebro y la Pardela Cenicienta (Calonectris diomedea), el Aguila Pescadora (Pandion haliaetus), y la Foca Monje (Monachus monachus) en las Islas Chafarinas.

6. Conscientes del escaso patrimonio constituído por los codigos de colores para el anillamiento,

SUGIERE a los investigadores el uso coordinado del marcaje con colores y la consulta previa con el secretariado de Medmaravis (se crea una comision coordinadora a tal efecto) y Euring (Arnhem, Holanda), con el fin de evitar el empleo de combinaciones de colores ya utilizadas y aprovechar al máximo los códigos disponibles sobre todo en especies como Larus cachinnans y Phalacrocorax aristotelis.

7. PROPONE A MEDMARAVIS

- continuar la coordinación de las actividades futuras relacionadas con la avifauna marina mediterránea, hasta el proximo Simposio sobre Aves Marinas del Mediterráneo,

- colaborar en la organización del segundo Simposio "Aves Marinas del Mediterráneo" a celebrar en septiembre de 1989 en Palma de Mallorca, Baleares,

- organizar el encuentro de un comité técnico en la sede de la oficina regional de alguna organización ambiental internacional,

- promover la coordinacion nacional e internacional con el fin de estimular el desarrollo de futuros estudios y proyectos de conservación sobre las aves marinas del Mediterraneó.

Alghero, a 30 de marzo de 1986

NATO ASI Series G

Vol. 1: **Numerical Taxonomy.** Edited by J. Felsenstein. 644 pages. 1983.

Vol. 2: **Immunotoxicology.** Edited by P.W. Mullen. 161 pages. 1984.

Vol. 3: **In Vitro Effects of Mineral Dusts.** Edited by E.G. Beck and J. Bignon. 548 pages. 1985.

Vol. 4: **Environmental Impact Assessment, Technology Assessment, and Risk Analysis.** Edited by V.T. Covello, J.L. Mumpower, P.J.M. Stallen, and V.R.R. Uppuluri. 1068 pages. 1985.

Vol. 5: **Genetic Differentiation and Dispersal in Plants.** Edited by P. Jacquard, G. Heim, and J. Antonovics. 452 pages. 1985.

Vol. 6: **Chemistry of Multiphase Atmospheric Systems.** Edited by W. Jaeschke. 773 pages. 1986.

Vol. 7: **The Role of Freshwater Outflow in Coastal Marine Ecosystems.** Edited by S. Skreslet. 453 pages. 1986.

Vol. 8: **Stratospheric Ozone Reduction, Solar Ultraviolet Radiation and Plant Life.** Edited by R.C. Worrest and M.M. Caldwell. 374 pages. 1986.

Vol. 9: **Strategies and Advanced Techniques for Marine Pollution Studies: Mediterranean Sea.** Edited by C.S. Giam and H.J.-M. Dou. 475 pages. 1986.

Vol. 10: **Urban Runoff Pollution.** Edited by H.C. Torno, J. Marsalek, and M. Desbordes. 893 pages. 1986.

Vol. 11: **Pest Control: Operations and Systems Analysis in Fruit Fly Management.** Edited by M. Mangel, J.R. Carey, and R.E. Plant. 465 pages. 1986.

Vol. 12: **Mediterranean Marine Avifauna: Population Studies and Conservation.** Edited by MEDMARAVIS and X. Monbailliu. 535 pages. 1986.

MIX
Papier aus verantwortungsvollen Quellen
Paper from responsible sources
FSC® C105338

If you have any concerns about our products,
you can contact us on
ProductSafety@springernature.com

In case Publisher is established outside the EU,
the EU authorized representative is:
Springer Nature Customer Service Center GmbH
Europaplatz 3, 69115 Heidelberg, Germany

Printed by Libri Plureos GmbH
in Hamburg, Germany